Phy 17

D·I·G·I·T·A·L TELEVISION

D·I·G·I·T·A·L TELEVISION

Edited by

C. P. Sandbank
Deputy Director of Engineering
British Broadcasting Corporation

JOHN WILEY & SONS
Chichester · New York · Brisbane · Toronto · Singapore

Copyright © 1990 by John Wiley & Sons Ltd.
Baffins Lane, Chichester
West Sussex PO19 1UD. England

Reprinted March 1992

All rights reserved

No part of this book may be reproduced by any means,
or transmitted, or translated into a machine language
without the written permission of the publisher.

Other Wiley Editorial Offices

John Wiley & Sons, Inc., 605 Third Avenue,
New York, NY 10158-0012, USA

Jacaranda Wiley Ltd, G.P.O. Box 859, Brisbane,
Queensland 4001, Australia

John Wiley & Sons (Canada) Ltd, 22 Worcester Road,
Rexdale, Ontario M9W 1L1, Canada

John Wiley & Sons (SEA) Pte Ltd, 37 Jalan Pemimpin 05-04,
Block B, Union Industrial Building, Singapore 2057

Library of Congress Cataloging-in-Publication Data:

Digital television / edited by C. P. Sandbank.
   p.　cm.
  Includes bibliographical references (p.　).
  ISBN 0 471 92360 5
  1. Digital television.　　I. Sandbank, C. P.
TK6678.D53　1990
621.388—dc20　　　　　　　　　　　　89-28632

British Library Cataloguing in Publication Data:

Digital television.
  1. Digital television equipment
  I. Sandbank, C. P.
  621.388

ISBN 0 471 92360 5

Typeset by APS Ltd., Salisbury, Wiltshire
Printed in Great Britain by Butler & Tanner Ltd., Frome and London

Contents

Preface xiii

1 The Emergence of Digital Television 1
C. P. Sandbank

Pulse Code Modulation (PCM) 2
The Stimulus to Use Digital Techniques 3
Standards 7
   Digital components 8
   Video interface 11
   Recorder 13
   Digital audio 13
Further Developments 14
   High definition television 14
   Digital techniques in domestic receivers 16
   Digitally assisted television (DATV) 18

2 Analogue-to-digital and Digital-to-analogue Conversion and its Effect on Television Signals 21
V. G. Devereux

Fundamental Principles 21
   Analogue-to-digital (A–D) conversion 21
   Digital-to-analogue (D–A) conversion 23
   Sampling and frequency characteristics 23
   Input/output transfer characteristics 26
   Methods of specifying quantising errors 26
Methods Used for A–D and D–A conversion 30
Picture Impairment Caused by Quantisation 37
   Subjective tests on a single codec 40
Test on up to Eight 8-bit Codecs in Tandem with Sampling Close to Nyquist Limit 42
Discussion of Coding Parameters for use in Television Broadcasting 50
   Bits per sample 50
   Sampling frequency 51
Measurement of Performance of A–D and D–A Converters 52
Tolerances for Coding Inaccuracies 59
Specification of Filter Characteristics 61
Circuit Layout Techniques 61
Summary of Factors Affecting Performance of Video A–D and D–A Units 63

3 Semiconductor Storage of TV Signals 65
J. L. Riley

> The need to store television signals 65
> The developing semiconductor memory technology 65
> Developing trends of the technology 74
> Memory device choices 78

Features of Dynamic Semiconductor Memory 79
Design Philosophy 89
> Store size 90
> Multiplex factor 91
> Store configuration 92

Early Designs 94
The 16K Generation 98
> An experimental picture store using 18MHz sampling frequency 98
> Microcomputer-controlled stores 103
> Animation store buffers 106

The 64K Generation 109
> A general purpose digital storage card 109
> A compact random access picture store 117

Delay Applications 122
The 256K Generation 126

4 Digital Encoding and Decoding of Composite Television Signals 131
C. K. P. Clarke

> Digital signal coding for use in studios 131
> PAL decoding with subcarrier-locked sampling 133
> PAL decoding with line-locked sampling 134

Subcarrier Generation From a Line-locked Sampling Frequency 135
> Subcarrier phase generation 135
> Quadrature subcarrier generation 139

Colour Encoding 141
> Features of the PAL signal 141
> Digital PAL encoding 144
> NTSC encoding 153

Chrominance Demodulation 157
> Decoder configurations 157
> Line-locked sampling of composite signals 158
> PAL chrominance demodulation 162
> Signal relationships in decoders 170
> Modulated chrominance processing techniques 174
> NTSC demodulation 178

Chrominance–luminance Separation 181
> Conventional decoding 181
> The three-dimensional spectrum of composite signals 183
> Multi-dimensional luminance and chrominance filters 187
>> Line-delay comb filters 188
>> Field- and picture-delay comb filters 193
>> Improved comb filters 195
>> Complex comb filter design 197

    Performance comparison   204
    Adaptive techniques   208
  Decoders in Digital Studios   209
    Multiple decoding and recoding   209
    Clean coding system   210
    The changeover to digital studios   213

5   Digital Filtering of Television Signals   215
J. O. Drewery

The Two-dimensional Fourier Transform   216
The Three-dimensional Fourier Transform   221
Sampling and the Discrete Fourier Transform   225
General Considerations of Digital Filters   230
Transversal Filters   232
Design of Transversal Filters   234
Hardware Considerations of Transversal Filters   240
Some Applications   244
  Real-time aperture correction   245
  Vertical chrominance filtering   251
  Two-dimensional luminance filtering   254
Noise Reduction   257
  Movement protection   263
  Colour operation   276
  The practical realisation   280
  The prototype equipment   285

6   Interpolation   287
C. K. P. Clarke

Interpolation Theory   288
  Signal sampling and reconstitution   288
  Sample rate changing   293
    Resampling   293
    The interpolation process   295
    Synthesis of aperture functions   298
    Aperture quantisation   301
  Fixed ratio interpolators   304
  Time expansion and compression   311
Television Scan Conversions   317
  Television scanning   317
  Conversion methods   322
    Principles of conversion   322
    Performance of previous converters   335
    Aperture synthesis   341
    Optimised conversion apertures   350
  HDTV conversions   362
  Display improvement   372

7 Multipicture Storage 375
M. G. Croll

Requirements for Multipicture Storage Systems 375
Review of Suitable Storage Media and Technologies 376
Decoding to Provide a PAL Input to a Multiple Picture Store 381
Application of Electronic Multpicture Storage to a News Central Stills Library 382
Studio Stills Store 385
   Semiconductor picture stores 387
   Disc 388
   Streaming tape drive 390
   System computer hardware 391
   System computer software 392
   Picture processing 395
Television Animation Store 396
   Operation of store 400
   Recording pictures in real time on a parallel transfer disc drive 402
     Record channel code 402
     Available redundancy and bit-rate reduction 406
     Error protection 407
     Data format 409
     Performance 410
Animation of Logos 412
   LaserVision optical disc with data instead of f.m. modulated video 412
   Solid state implementation of animated logo 418

8 Digital Video Tape Recording 423
A. Todorovic

The Role of the VTR 423
Limitations of Analogue Recording 423
Potentialities of Digital Recording 425
Specific Problems of Digital Video Recording 429
Digital Video Tape Recording Format 433
The Cassette 435
Track Pattern 437
Recording of Digital Video Signals 442
Recording of Digital Audio Signals 449
Recording of Synchronisation Signals 453
Channel Coding 455
Recording of Cue Audio Signals 455
Recording of Control Signals 456
Recording of Time-code Signals 456
Commercial Products 457

9 Electronic Graphics for Television 465
N. E. Tanton

The Effects of Scanning 467
   Sampling theory 467
   Scanning as sampling 467
   Scanning artwork 469

Combining scanned text with pictures 471
Two-level electronic images 472
Multilevel digital images 472
The Spectral Contributions of Picture Detail 473
Stationary detail 473
Moving detail 473
The Causes of Aliasing 474
Picture frequency flicker (interlace twitter) 475
Stepping or 'jaggies' 475
Movement portrayal 476
Filtered Text Images 480
Source of data 480
Choice of filter function 481
Practical Investigation 482
Filters with monotonic step responses 483
Filters with non-monotonic step responses 486
Combined filtering 490
Use of Filtered Text for Teletext and VDUs 494
Applications of Electronic Graphics 496

10 Telecine and Cameras 499
I. Childs

Types of camera 500
Types of telecine 501
Areas where digital processing may bring benefits 505
Digital Techniques in Telecine Machines 508
A digital telecine processing channel 508
Sequential-to-interlace conversion 518
  Movement interpolation 522
Telecine control systems 524
  Scanning control and variable speed operation 524
  Automatic colour correction 526
Digital Techniques in Cameras 529
A digital camera processing channel 529
Camera control systems 530
Use of solid state area arrays in television cameras 531

11 Digital Chroma-key and Mixer Units 539
V. G. Devereux

Principles of Operation 540
Key generator 540
Foreground suppressor 542
Key processor 544
Mixer 546
Filtering and sample rate changing 546
Other Key Facilities 548
Keying from 12:4:4 YC_BC_R signals 549
Luminance keying 550
Garbage matte 550

x CONTENTS

    Stored key signals 551
    Foreground–background cross-fade 551
  Problems Associated with Digital Processing 552
    Aliasing distortion 552
    Quantising distortion and use of error feedback 553
  Adjustment of Controls 556

12 Digital Video Interfaces 559
D. J. Bradshaw

Interface signal format 560
  Synchronisation 561
The Bit-parallel Interface 564
  Electrical format 564
  Mechanical implementation 566
  Experience with the parallel interface 568
The Bit-serial Interface 568
  Coding strategy 568
  The 8-bit–9-bit block code 569
  Control of the d.c. component 570
  The 8-bit–9-bit map 570
  Link synchronisation 570
  Scrambled 10-bit interface 575
  Electrical format 576
  Mechanical implementation 576
  Experience with the bit-serial interface 576
Ancillary Data Channel 577
Application of Digital Interfaces in Studios 577

13 The Bit-rate Reduction of Television Signals for Long Distance Transmission 583
N. D. Wells

Sample Rate Reduction 585
  Chrominance components 589
  Luminance component 592
  Subsampling of PAL signals 593
  Other forms of comb filter 597
DPCM Coding 598
  Linear prediction 599
  Adaptive prediction 601
  Non-linear quantisation 601
    Minimum mean square error (MMSE) design method 602
    Graphical design method 603
    Adaptive quantisation 604
  Stability of (non-adaptive) DPCM decoders and coders 605
Transform Coding for Bit-rate Reduction 610
  The discrete cosine transform 611
  Two-dimensional transform coding 615
  Bit-rate reduction in the transform domain 617
  Hybrid transform/DPCM coding 620

Variable-Length or Entropy Coding 621
   Construction of typical variable-length codes 622
   Entropy 623
   Practical considerations 625
Vector Quantisation 628
   Codebook generation 630
   Vector quantiser performance 631

References 633

Index 643

Preface

There are several ways of treating a subject like digital television. One way is to find an author who can personally cover the whole subject. It is hard to find someone who has both access to the detail of the wide range of the essential technologies and who also has the time to write a book. Another way is to invite experts on each of the major topics from various organisations (and countries) to contribute chapters on their own subject. This provides a broad coverage but may lack cohesion.

I have chosen an intermediate approach similar to that which I used on a previous occasion [6]. The material has been covered by a group of specialists who have been working closely together in this field for many years and have themselves pioneered some of the applications of digital techniques to broadcasting. We have therefore been able to compare notes as we went along and have tried to make sure that most areas are covered in a way that hangs together, at least as seen in terms of the requirements of one major broadcasting organisation.

The chapter on digital recording has been treated differently. This has been written by a distinguished broadcasting engineer from Yugoslavia who pulled together the activity in which representatives from industry and broadcasting collaborated on a worldwide basis for several years to produce a common approach to digital recording.

My thanks go to all who have contributed to the book. To the BBC for its enlightened attitude to research and development which enabled much of the work described to be applied successfully. To my co-authors who patiently responded to numerous requests for additional material as the contributions came together. To colleagues in other broadcasting organisations and in manufacturing industry from all over the world who provided material for inclusion in the book, and particularly those in the EBU, SMPTE, and CCIR.

I would like to gratefully acknowledge the helpful discussions with many colleagues in the BBC and to thank Miss Eileen Tasker, Mrs Ann Bennet and Mr Ted Hartwell for help with the manuscripts.

Finally my fond thanks go to my wife Audrey for allowing me to cover large amounts of surface area of our home with paper, particularly at a time when she needed some of the space for the manuscript of her own book!

C. P. SANDBANK,
Broadcasting House,
London.

1 The Emergence of Digital Television

C. P. Sandbank

INTRODUCTION

In these days when electronics is having such an impact on our lives, mostly by means of devices based on digital circuits using signals in binary form, it may seem surprising that broadcasting, one of the 'founder members' of the electronics industry, is still based essentially on analogue equipment. The relatively late arrival of digital techniques to television can be ascribed to three main factors:

(i) Semiconductor technology has only recently become available in a form compatible with the demanding requirements of television signal processing.

(ii) Analogue electronics are doing a good job in most areas of the broadcasting chain, namely the studios; the distribution networks; the transmitters and the receivers.

(iii) In broadcasting, change is inhibited by the need to be compatible with what has gone before and what is likely to come next.

Indeed, under these circumstances it is remarkable that the use of digital techniques in broadcasting is already quite extensive and increasing rapidly. This is because digital means are often the only practical and cost-effective way of providing some of the new electronic processes which enhance the production of programmes and have become an essential part of the art of television. In the near future, digital techniques will also become essential for bandwidth reduction in distribution and for domestic receiver enhancements. Eventually we will see digital signals in broadcast and recorded form replacing analogue as a means of delivering TV to the public.

The purpose of this chapter is to give the reader some background to the emergence of digital television. It should also help those who prefer to dip into the chapters for items of specific interest to find their way around the book. The subjects of the chapters are not dealt with in chronological order of the developments or in relation to their position in the TV production chain. The order has been chosen with the aim of treating as early as possible subjects which may provide useful background for the subsequent chapters.

PULSE CODE MODULATION (PCM)

Digital television has its roots in PCM which is now accepted throughout the world as the standard means of encoding signals for most modern communications systems. Although this rapid transition from analogue to digital communication has taken place in the relatively recent history of such techniques as microwave transmission or even satellites, it is interesting to note that its invention predates both by a long time. PCM was invented by Alec Reeves in 1937 but it was not until 30 years later that semiconductor devices became available which made PCM a viable proposition for encoding of speech signals. (It was during this period that the author had the good fortune to be able to count Alec Reeves among his close colleagues.) By the time technology made PCM attractive for extensive application to TV a total of 50 years had elapsed since its invention.

It is all the more remarkable that the very detailed description in Reeves' patent [1], published in the UK in 1939, was so far-sighted that it could serve as an accurate description of many of the concepts treated in this book. The aspects addressed in the patent range from quantisation and encoding to consideration of the noise and bandwidth characteristics of the transmission media right through to the relative merits of parallel and serial interfaces — a subject still giving rise to hot international debate in 1989!

The work of Reeves, though directly relevant to current digital TV systems, since it only required modern semiconductor devices to implement PCM as he described it, was motivated by the desire to solve the problems of voice transmission. It is therefore, in the context of this book, worth drawing attention to another much earlier patent granted to Rainey [2] in 1926 which proposed a means of scanning a picture for transmitting facsimiles or 'telephotography' by means of galvanometers, photocells and relays to produce 'a code combination of electrical impulses'. Although this patent is hardly relevant to the systems described in this book Rainey clearly envisaged the concept of encoding an image in digital form when he first filed his patent in 1921.

PCM is well dealt with in the literature [3]. Suffice it to say, at this stage, it provides a means of describing, or encoding, the continuous analogue waveform as a stream of binary numbers. In this form it is more suited for processing by modern electronic circuitry and more amenable to the application of techniques for efficient retention of essential information and minimising corruption by interfering signals (e.g. noise). To produce the PCM signal, the analogue waveform is first sampled periodically to determine its height at each sample time. The higher the sampling frequency the more accurate is this process and, in the limit, an infinite sampling frequency would describe the original waveform exactly. In practice, the waveform of the signal from, say, a TV camera contains a finite amount of information and there is no advantage in choosing a higher sampling frequency than required to convey this limited information to the accuracy required.

The second step in producing the PCM signal is to assign a numerical magnitude to the amplitude measured at each sample period. Again, if this were done with infinite accuracy, the original analogue waveform could eventually be recovered exactly. However, for the same reasons that an excessive sampling frequency is pointless, the precision with which the amplitude of each sample is described (or, in the language of PCM, the number of discrete levels chosen for the quantisation process) should not be higher than needed to convey the essential information about the image which the TV camera is capable of producing.

The processes of sampling and quantisation are described in Chapter 2 dealing with analogue-to-digital conversion. Having converted the original analogue TV signal into PCM, it is then in a form suitable for signal processing by semiconductor circuits which can now be fabricated as large-scale integrated (LSI) devices capable of carrying out complex functions at speeds commensurate with the high information rate of a TV signal. Of equal importance is the fact that the PCM signal facilitates the presentation of the signal in a form similar to that in which the data is processed and stored in a computer. This has enabled the sophisticated hardware developed for the Information Technology industry to be readily adapted for use in television systems.

THE STIMULUS TO USE DIGITAL TECHNIQUES

The need to provide some memory, which is essential for any basic signal processing, provided the strongest initial stimulus to the development of digital techniques. The requirement which first drew attention to this was the need to convert from one TV scanning standard to another. Before the days of video recording or satellites the normal means of international programme exchange was by the use of 16 mm or 35 mm film shot at 24 frames per second. There was thus no need for standards conversion as the film is speeded up slightly to 25 frames per second and scanned at the required transmission standard using the telecine techniques described in Chapter 10.

To a large extent film has remained the basic medium for worldwide programme distribution. However, for a long time there has been a need for scan conversion. It started with the original cross-channel microwave links connecting French and British TV in 1950 when the problem of exchanging programmes made with different scanning standards first emerged [4]. Here standards conversion was simply carried out by pointing a camera at a screen and relying on the inherent memory and integration times of the display and camera devices to perform the basic functions of interpolation described in Chapter 6. When exchange of programmes produced at 50 Hz/625 lines and 60 Hz/525 lines became possible in electronic form, the desirability to achieve electronic standards conversion became evident [5]. Figure 1.1 illustrates the technological difficulty of providing the required memory for a standards conversion before the availability of semiconductor storage in digital form. The device is a precision cut block of quartz with reflecting facets in which the TV signal travels back and forth as an ultrasonic beam until, after passing through 12 such blocks, it has traversed a path length equivalent in time to two fields of television!

The rapid progress made in the technology of semiconductor digital storage described in Chapter 3 is the main reason for the transition from analogue to digital television. It is interesting to note that ten years after the first use of the device in Figure 1.1, it was attractive to use digital semiconductor field stores even at a cost of £3K per field and requiring a power of over 200 W. Today, field stores are feasible using a few chips costing less than £10 and requiring a few milliwatts of power. Soon it may be possible to use three-dimensional solid state memory where the capacity will be measured not in fields, but in minutes of television.

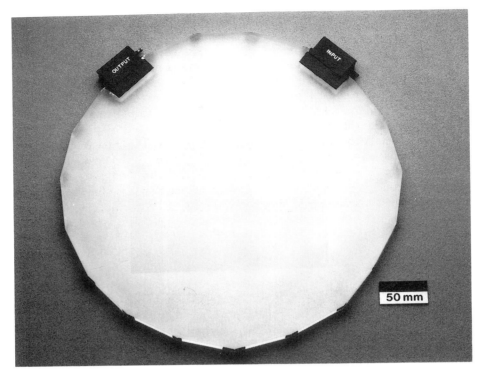

Figure 1.1 Quartz ultrasonic 3.3 ms delay line. It weighed 2.3 kg and 12 of these were used for the picture store in early standards converters

Initially the main application for delay was for timing correction. The two main requirements for this were firstly to correct for the timing imperfections introduced by slight variations in the speed of the tape past the play-back heads in the recorder, and secondly to enable signals arriving from different sources to be locked together in synchronism.

Quite soon after the simple applications of digital delays, more sophisticated digital signal processing techniques began to be introduced to television. A notable example was the use of the digital filtering methods discussed in Chapter 5 for noise reduction at the network output. Also at this time, digital communications were becoming established at bit rates sufficient for the transmission of TV by wide-band media such as optical fibres [6] and satellites [7]. This stimulated the development of digital encoding of TV for transmission, including bandwidth reduction methods, many of which are discussed in Chapter 13 and are only feasible by digital means.

By the end of the 1970s digital TV had become an essential part of production for operational reasons, some of which were mentioned above. The impact on programmes would not have been obvious to the public or even to producers. However, at this time the techniques were beginning to be applied to the art of programme making.

Initially, fairly basic use was made of memory. Producers with access to the stores in the synchronisers started 'grabbing' an occasional frame from the camera and holding it still to heighten dramatic effect. Then there was a combination of storage and image detection used, for instance, by 'teletrack' device [8] used to display the trajectory of objects, such that shown in Figure 1.2. This system detects the position of a moving object in any field by comparing the signal with a stored reference field (e.g. the empty snooker table). The difference represents one 'snapshot' of the moving object. If this is done several times as the object moves about on a stationary background these snapshots of the moving object can all be written into a store and finally superimposed on the original background for display.

Perhaps the digital equipment which first made a great visual impact was that which used a combination of the storage and interpolation processes described in Chapters 3 and 4 to achieve 'special video effects'. The TV industry had long been able to superimpose one image on another either by the traditional means of back projection or by analogue colour separation overlay (see Chapter 11 for an explanation of this process

Figure 1.2 'Teletrack', one of the first special effects units to make use of solid state picture stores

in relation to the more accurate digital method). What it had not been able to do was to reproduce the special optical effects used in the film industry whereby the size and shape of the image could be changed during the optical printing process. Hitherto any change in size and shape could only be made at the TV camera and would appear always as a full screen image. The special effects equipment would carry out the digital interpolation process similar to that used for standards conversion, so that the original full screen image could, for example, be shrunk to occupy a small fraction of the screen, moved around and even rotated about a variety of axes.

When first presented with this new degree of freedom producers used these new techniques to such an extent that the frequent presentation of shrinking rotating images caused the viewers discomfort rather than pleasure. However, after the initial flush of enthusiasm, digital special effects have become a respected and essential part of the art of television and its continuing development.

Since the start of this trend the UK company of Quantel had been at the forefront of this technology to the extent that the name of their company has become a verb used by TV programme directors throughout the world. The capabilities of digital special effects equipment soon went well beyond what was possible with the optical printing methods used with film. Figure 1.3 (see colour plates between pp 282 and 283) is an example of this and was produced by Quantel for the author as the cover illustration for this book. It shows the test card signal being processed so that it appears to be folded back like the leaves of a book—a sequence of images which can only be obtained by digital electronics. (The same image has been used on 'both sides' of the leaves in Figure 1.3 to demonstrate the process, but in an actual production a different 'grabbed' frame would be used for each side of the leaves.)

Many of the digital processes discussed above, such as synchronising, noise reduction and transmission, could be carried out with the TV colour signal encoded in its composite form like PAL, SECAM or NTSC where the *RGB* (red, green, blue) signals have been combined in a bandwidth-efficient way to convey a subjectively acceptable colour picture in a bandwidth similar to that used for the black-and-white transmissions. For the reasons explained in Chapter 6, standards conversion and the interpolation processes used in digital special effects equipment can only be carried out satisfactorily on signals in component coded form. This means that the signal is used in the original *RGB* form using about three times the bandwidth of a composite coded signal. Alternatively, it can be used in a more bandwidth economical form of component coding which relies on the fact that the eye is more tolerant of lower resolution in colour detail than in the detail of the luminance ('brightness'). Here the signal is made up of a luminance signal Y and two 'colour difference' signals C_R, C_B, where C_R is obtained by subtracting the luminance signal from the red signal (i.e. $C_R = R - Y$) and $C_B = B - Y$.

The relationships between Y, C_R and C_B are discussed later. Suffice it to say at this stage that, when C_R and C_B are zero, the picture will go from black to white as Y goes from 0 to 1 and by defining the values of Y, C_R and C_B for the three primary colours at saturation level, R, G and B can always be derived. It is normally sufficient for a process requiring component coded signals to use Y, C_R and C_B where the bandwidth of C_R and C_B is each about half that of Y. Thus, the total bandwidth for a YC_RC_B signal is about twice that for a composite signal.

Since the studios using special effects requiring YC_RC_B signals would in general be using the normal composite coded signals, e.g. PAL, high quality digital decoding and encoding of composite signals became increasingly important. Clearly, since some of the original *RGB* information has had to be discarded in order to squeeze the signal into a composite form occupying about the same bandwidth as the *Y* component only, a full recovery of the original *RGB* or even YC_RC_B is not possible. However, by using digital techniques the various compromises discussed in Chapter 4 can be optimised to produce much better results than is possible with analogue decoders.

The next important step in the field of digital television was the development of random access magnetic storage of the type used in computers (e.g. Winchester or optical discs) in a form suitable for storing large numbers of TV pictures. This elevated the technique of grabbing frames from providing an occasional picture to giving access to a whole library of electronically stored pictures. This also allowed the traditional slide-scanner to be replaced by the much more flexible digital device described in Chapter 7. As soon as magnetic disc storage could be configured to operate at about 200 Mbit/s, the bit-rate required for on-line retrieval of TV pictures, the electronic equivalent of the film rostrum camera became feasible for the production of animated sequences as also discussed in Chapter 7.

The impact of a new technology like digital, which is gradually being implemented throughout the whole television environment, would only be very apparent if one were instantly to turn the clock back to a time before it was used extensively. The field of production enabled by digital techniques whose absence would probably be perceived by the public as the greatest loss is in the general area of electronic graphics covered in Chapters 7 and 9. The quality improvement and additional artistic freedom provided by the direct creation of pictures without the intermediate steps of paint, paper and camera have made an enormous difference to the appearance of images on screen, particularly those required for news and current affairs programmes, often produced at short notice. The ability to draw directly on to the 'screen' with an electronic palette, perform the functions of 'cut and paste' with parts of images selected from the digital stores, combine drawings with live actors, and call up, at a touch of the keyboard, high quality text in a variety of elegant founts, have all given a new dimension to programme production.

Although more and more digital equipment was being used in TV production, the Digital Video Tape Recorder (DVTR), the key element enabling a fully digital studio to be envisaged, did not become commercially available until the late 1980s. The DVTR is described in Chapter 8 with special emphasis on the recording of component signals.

With all the essential items for production available, attention could be turned to the possibility of entirely digital studio operations. Chapter 12 discusses the infrastructure required for a digital studio and gives some early examples of facilities having many of the basic features of a studio.

STANDARDS

For the reasons discussed in the previous section, TV operations use much digital equipment operating in a mainly analogue environment. By the late 1970s it was quite usual for a typical BBC programme to have been through about 15 separate digital processes, some composite and some component. Thus, the signal would encounter many analogue-to-digital and digital-to-analogue conversions and several PAL to *RGB* or YC_RC_B decoding and encoding processes. In some cases where two pieces of digital equipment were adjacent in the sequence it was still necessary to go through an analogue process because there were no industry standards and the only way to interconnect them was by using analogue signals in the established composite or *RGB* format.

This situation stimulated broadcasters on both sides of the Atlantic to start work on defining standards for digital TV so that pieces of equipment could eventually be connected together without the A–D/D–A stages. In Europe the work was led by the European Broadcasting Union and in North America by the Society of Motion Picture and Television Engineers. Initially attention was concentrated on composite standards for the 625-line PAL and 525-line NTSC standards by the EBU and SMPTE respectively.

Digital components

The increasing use of component signals in digital equipment, coupled with the possibility of establishing a substantial area of common ground between component coded signals for the 625 and 525 line encouraged a change in direction. A joint committee of the EBU and the SMPTE chaired by Howard Jones from the BBC Research Department set out, in liaison with industry and broadcasting unions in other parts of the world, to establish a worldwide digital TV standard for component signals. It is worthwhile quoting from the introduction to the document [9] which has now become the basic international digital TV standard: Recommendation 601 of the CCIR (the International Radio Consultative Committee of the International Telecommunication Union). Although the standard was not adopted until 1986, the introduction expresses clearly what prompted the initiative in 1979 which eventually produced the standard.

> The CCIR,
>
> CONSIDERING
> (a) that there are clear advantages for television broadcasters and programme producers in digital studio standards which have the greatest number of significant parameter values common to 525-line and 625-line systems;
> (b) that a world-wide compatible digital approach will permit the development of equipment with many common features, permit operating economies and facilitate the international exchange of programmes;
> (c) that an extensible family of compatible digital coding standards is desirable. Members of such a family could correspond to different quality levels, facilitate additional processing required by present production techniques, and cater for future needs;
> (d) that a system based on the coding of components is able to meet some, and perhaps all, of these desirable objectives;
> (e) that the co-siting of samples representing luminance and colour-difference signals (or, if used, the red, green and blue signals) facilitates the processing of digital component signals, required by present production techniques,
>
> UNANIMOUSLY RECOMMENDS
>
> that the following be used as a basis for digital coding standards for television studios in countries using the 525-line system as well as in those using the 625-line system:

The key to commonality is that, because the products 625 × 50 and 525 × 60 are almost equal, the time taken to scan a TV line in the two systems is very nearly the same. Thus, in the terms discussed earlier, if one samples the analogue waveform described by Y at the same sampling frequency for both scanning systems, it is possible to ensure that there are the same number of samples appearing in the same positions along the line in both the 625-line and 525-line rasters. The circumstances would, of course, be similar for C_R and C_B but these would be sampled at a lower sampling frequency and would, therefore, be spaced further apart than the Y samples.

Most of the points in the 'Considering' quoted above are self-explanatory. In (e) reference is made to the fact that by choosing values for the luminance and chrominance (C_R, C_B) frequencies which are simply related to each other and are integer multiples of 2.25 MHz (the lowest common multiple of the line frequencies in the 625/50 and 525/60 systems) it is possible to co-site the luminance and chrominance samples in the same static orthogonal pattern in both systems.

After much discussion, many experiments and international demonstrations comparing the relative merits of different sampling frequencies as well as different ratios between the sampling frequencies for luminance and chrominance, 13.5 MHz and 6.75 MHz were finally chosen. An early contender, 12 MHz for Y and 4 MHz for C_R and C_B used in one of the systems described in Chapter 7, was rejected on two main grounds. Firstly, 12 MHz, although it stretched the semiconductor device technology at the time, was considered not to provide enough 'headroom' for subsequent processing of a 5.5 MHz base band signal (see Chapter 2). The second reason was that 3:1 for the ratio of luminance to chrominance resolution was considered to be too low particularly because the all-important process of chroma-key described in Chapter 11 could not be carried out with sufficient precision without higher chrominance resolution.

Thus the 13.5:6.75:6.75 sampling structure was chosen. The ratio is normally expressed as 4:2:2, an acknowledgment of point (c) which makes reference to an extensible family of standards. It was always envisaged that the standard could accommodate different quality requirements centred around 4:2:2 as the basic studio standard. For example, in news gathering equipment where a compromise between quality and portability may have to be made, the 2:1:1 member of the family with sampling frequencies for Y, C_R and C_B of 6.75; 3.75 and 3.75 MHz might be used. For RGB the 4:4:4 level would be used and for high definition television a multiple of 4:2:2 might be used. The concept of the extensible family is that it should be possible to interface readily between the different levels.

The following tables give the essence of the Recommendation 601 CCIR digital studio standard. Readers new to the subject may prefer to come back to the detail of these tables after further discussion of the standard in the later chapters.

Table 1.1 gives the main encoding parameters for the 4:2:2 member.

The most important features to note in addition to the sampling frequencies are the choice of 8 bits per sample PCM for each of the three signals and the 720:320:320 samples for Y:C_R:C_B respectively for the active part of the line, i.e. the part carrying the picture rather than the part allocated to the analogue synchronising waveform which does not form part of the digital line specification. The orthogonal sampling structure is shown in Figure 1.4.

Table 1.2 shows the relationship between the 720 luminance sample and the sync. periods. It also illustrates how the different sync-times for the 625/50 and 525/60

Table 1.1
Encoding parameter values for the 4:2:2 member of the digital studio standard CCIR Recommendation 601

Parameters	525-line, 60 field per second systems	625-line, 50 field per second systems
1. Coded signals: Y, C_R, C_B	These signals are obtained from gamma pre-corrected signals, namely: $E'_\gamma, E'_R - E'_\gamma, E'_B - E'_\gamma$	
2. Number of samples per total line: —luminance signal (Y) —each colour-difference signal (C_R, C_B)	858 429	864 432
3. Sampling structure	Orthogonal, line, field and frame repetitive, C_R and C_B samples co-sited with odd (1st, 3rd 5th, etc.) Y samples in each line	
4. Sampling frequency: —luminance signal —each colour-difference signal	13.5 MHz 6.75 MHz The tolerance for the sampling frequencies should coincide with the tolerance for the line frequency of the relevant colour television standard	
5. Form of coding	Uniformly quantised PCM, 8 bits per sample, for the luminance signal and each colour-difference signal	
6. Number of samples per digital active line: —luminance signal —each colour-difference signal	720 360	
7. Analogue-to-digital horizontal timing relationship: —from end of digital active line to O_H	16 luminance clock periods	12 luminance clock periods
8. Correspondence between video signal levels and quantisation levels: —scale —luminance signal	0 to 255 220 quantisation levels with the black level corresponding to level 16 and the peak white level corresponding to the level 235. The signal level may occasionally excurse beyond level 235	
—each colour-difference signal	225 quantisation levels in the centre part of the quantisation scale with zero signal corresponding to level 128	
9. Code-word usage	Code-words corresponding to quantisation levels 0 and 255 are used exclusively for synchronisation. Levels 1 to 254 are available for video	

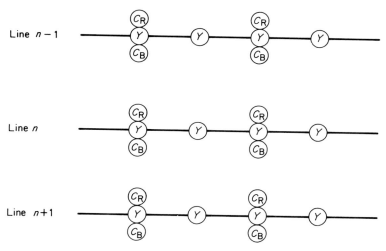

Figure 1.4 The orthogonal sampling structure

systems have been adjusted to facilitate the common sampling structure for the active line. A similar table could of course be constructed for the chrominance signals by replacing the 720 by 360 in each case.

Table 1.2
Relationship of digital active line to analogue sync. reference

525-line 60 field per second systems	122T	720T	16T	
	O_H (leading edge of line syncs, half-amplitude reference)	Digital active-line period	O_H	Next line
625-line, 50 field per second systems	132T	720T	12T	

T: one luminance sampling clock period (74 ns nominal).

The Recommendation 601 document derives and defines in detail the relationships between the parameters in Table 1.1. The normalised values for the boundary conditions are given in Table 1.3.

The other matters defined in Recommendation 601 are aspects of quantisation and the filter characteristics for the signals sampled at 13.5 and 6.75 MHz. These are discussed in detail in Chapter 2 where some of the filter specifications are reproduced.

Video interface

Having established the fundamental encoding parameters for a digital TV standard, two more important standards were established. The first of these is Recommendation 656

Table 1.3

Condition	E'_R	E'_G	E'_B	E'_Y	$E'_R - E'_Y$	$E'_B - E'_Y$
White	1.0	1.0	1.0	1.0	0	0
Black	0	0	0	0	0	0
Red	1.0	0	0	0.299	0.701	−0.299
Green	0	1.0	0	0.587	−0.587	−0.587
Blue	0	0	1.0	0.114	−0.114	0.886
Yellow	1.0	1.0	0	0.886	0.114	−0.886
Cyan	0	1.0	1.0	0.701	−0.701	0.299
Magenta	1.0	0	1.0	0.413	0.587	0.587

[10], the introduction of which starts with some points similar to those quoted earlier from Recommendation 601 and continues:

The CCIR

CONSIDERING

(d) that the practical implementation of Recommendation 601 requires definition of the details of the interfaces and the data streams traversing them;
(e) that such interfaces should have a maximum of commonality between 525-line and 625-line versions;
(f) that in the practical implementation of Recommendation 601 it is desirable that interfaces be defined in both serial and parallel forms;
(g) that digital television signals produced by these interfaces may be a potential source of interference to other services, and due notice must be taken of No. 964 of the Radio Regulations,

UNANIMOUSLY RECOMMENDS

that where interfaces are required for component-coded digital video signals in television studios, the interfaces and the data streams that will traverse them should be in accordance with the following description, defining both bit-parallel and bit-serial implementations.

The definition of the interface is to achieve the objective of enabling digital equipments to be connected directly to each other without the need for an intermediate analogue conversion. For short distances a parallel interface is adequate; for longer distances a serial interface is most suitable although this requires transmission of signals at higher data rates.

It will be clear from Table 1.1 that the bit rate required for the basic $Y + C_R + C_B$ information is

$$(13.5 + 6.75 + 6.75) \times 8 = 216 \text{ Mbit/s}$$

unless the information is transmitted in eight parallel streams in which case each conductor would of course carry data at a rate of 27 Mbit/s. (Both the serial and parallel interfaces make provision for some 'housekeeping' data in the multiplex in addition to the Y, C_R, C_B bit streams.)

The reference to No. 964 of the Ratio Regulations in (g) above draws attention to the need for careful screening of the conductors conveying the digital signals since the ninth and eighteenth harmonies of the 13.5 MHz sampling frequency fall at 121.5 and 243 MHz which are aeronautical emergency channels.

The essentials of Recommendation 656 and its application to the infrastructure of a TV studio are discussed in Chapter 12.

Recorder

The other important digital TV standard is Recommendation 657 [11] which deals with the digital video tape recorder. This is described in detail in Chapter 8 by the engineer who chaired the meetings of the EBU Committee which produced the specifications with the SMPTE and worldwide participation by the industry which manufactures professional VTRs.

Whilst Recommendations 601, 656 and 657 do not solve the problem of unifying the 625/50 and 525/60 scanning standards into a single worldwide standard for direct programme interchange, very important objectives are achieved. It will be clear from the discussion of the hardware considerations such as those described in Chapters 3 and 7–10, that the equipment for both systems is essentially the same. The recorder can be configured to either standard simply by 'throwing a switch', and stores can readily be organised to use common subsystems for operation at 50 and 60 Hz. The requirements for the basic functions like A–D conversion and filtering are identical. This enables the economies of scale to be achieved for digital studio equipment, but above all, creates an environment which encourages the direct interconnection of digital equipments operating with one of the two systems—the normal circumstances in a typical studio complex.

Digital audio

This book does not cover the subject of digital sound. It is, however, appropriate to draw the reader's attention to two other international standards. Broadcasters, for similar reasons to those which led to Recommendation 601, have stimulated the establishment of a digital audio studio standard. This can be truly a worldwide standard since it is not hampered by the need to accommodate two field frequencies. A similar hierarchical approach has been used to recognise different quality levels approriate to different applications.

Recommendation 646 [12] entitled 'Source encoding for digital sound signals in Broadcasting Studios' says:

The CCIR,

CONSIDERING

(a) that the introduction of digital techniques in the studio for broadcasting applications should improve quality as well as operational facilities;

(b) that there is a need to define a common sampling frequency for sound-programmes and for sound accompanying television programmes in studio applications;

(c) that this sampling frequency should be simply related to the 32 kHz sampling frequency recommended for transmission links, and for satelite broadcasting by the CCIR in order to reduce the cost of transcoding equipment;

(d) that the dynamic range has to provide adequate headroom for processing, taking account that at least a 14 bits dynamic range is recommended for transmission links and satellite broadcasting,

UNANIMOUSLY RECOMMENDS

1. that the sampling frequency for the digital encoding of sound signals in broadcasting studio applications including recording should have a nominal value of 48 kHz;
2. that the sampling frequency for the digital encoding of sound signals in television applications should have the same value;
3. that when an item of digital audio equipment is operating in a free-running mode, the maximum tolerance for the internal sampling frequency should be $= 1 \times 10^{-5}$. When items of digital audio equipment are interconnected, in sound broadcasting or television applications, provision must exist for locking the internal sampling frequency clocks to an external sampling frequency (e.g. television synchronizing signals, broadcasting house master clock, high-accuracy clock from a telecommunication network);
4. that the coding used should have a minimum resolution equivalent to 16 bits per sample uniform coding;
5. that no pre-emphasis should be used.

Again, in the same way that Recommendation 601 was of limited use without the definition of the interface standard Recommendation 656, the European Broadcasting Union, this time in conjunction with the Audio Engineering Society, produced the audio interface specification which led to CCIR Recommendation 647 [13]. 'A Digital Audio Interface for Broadcasting Studios'. This is a serial interface having the hierarchical characteristics implied in Recommendation 646 where, for example, whilst paragraph 1 states that 48 kHz is the studio standard sampling frequency and (c) is a reminder of the need to relate to the 32 kHz sampling frequency chosen for transmission links and satellite broadcasting. The need for headroom in the dynamic range is mentioned in (d) and no specific number of bits per sample are specified; paragraph 4 refers only to a minimum of 16.

Thus, the interface described in Recommendation 647 is primarily designed to carry monophonic or stereophonic programmes in a studio environment at a sampling frequency of 48 kHz with a resolution of up to 24 bits per sample. This interface can also be used to carry one or two channels at 32 kHz. In common with the video interface standard Recommendation 656, the audio interface also makes provision for clock and other auxiliary data multiplexed with the audio programme signals. In the spirit of paragraph 2 of Recommendation 646, the specification for the digital video recorder Recommendation 657 incorporates Recommendation 647, as discussed in Chapter 8.

FURTHER DEVELOPMENTS

High definition television

This book deals mainly with the application of the 625-line (and 525-line) digital television signal in the studio and transmission link environment. Two further developments should be drawn to the reader's attention: firstly the extension of the digital standard to HDTV (high definition television) and secondly the use of digital television techniques in domestic receivers. Somewhat surprisingly these two new developments are quite closely linked.

It is generally accepted that HDTV should have twice the vertical and twice the horizontal resolution of the conventional TV systems and with an aspect ratio of about 5:3 (e.g. 16:9) rather than the 4:3 used at the moment. This would give a spatial resolution close to that of 35 mm film and an aspect ratio recognising the subjective improvement achieved by the wider screens now used in the cinema. The field frequencies used for TV would of course also give a much better temporal resolution than achieved with the 24 frames per second used for film, particularly if sequential scan were used for HDTV production.

Figure 1.5 shows a possible extension of Recommendation 601 to HDTV. The HDTV production standard would have twice the number of lines per field compared with the conventional standard or even effectively four times the number, if sequentially scanned (i.e. non-interlaced) pictures were sent at a picture frequency equivalent to the present field frequency. In order to double the spatial resolution along the line, there would have to be an additional luminance sample between every one of the 720 samples in the orthogonal pattern of Recommendation 601 and additional samples because of the wider aspect ratio. For an aspect ratio of 16:9, an additional 480 luminance samples would

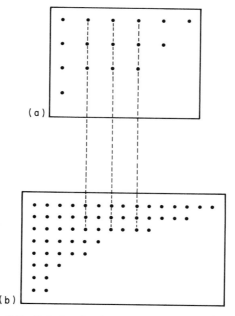

Figure 1.5 Relationship between Recommendation 601 and HDTV digital sampling structure (a) CCIR Recommendation 601, 720 samples per active line for luminance and 4 × 3 aspect ratio. (b) HDTV has twice the vertical and horizontal resolution of Recommendation 601 and 16 × 9 aspect ratio; therefore there are 1920 samples per line

need to be added to the 1440 required to double the resolution of a 4:3 aspect ratio picture. If the criterion of Recommendation 601 for the relationship between colour luminance resolution is retained, then 960 samples of C_R and of C_B would of course be required for each line. It is important to point out that, even in an interlaced mode, the HDTV studio source requires sampling frequencies above 70 MHz and a serial interface for a sequentially scanned HDTV digital studio standard would require a data rate well in excess of 2 Gbit/s.

The discussion so far has avoided the question of field or picture frequency. Broadcasters worldwide would like to seize the opportunity of choosing a new standard for HDTV to establish a truly single production standard rather than the 50/60 duality buried within Recommendation 601. Based on work by NHK in Japan [14] a digital standard has been proposed using the 1920/960/960 $Y/C_R/C_B$ line structure derived from Recommendation 601. It uses 1125 lines interlaced as a sort of compromise between twice 525 and twice 625 but uses a field frequency of 60 Hz, thus making it more difficult to integrate into the 50 Hz environment than the 60 Hz. Because of the need to continue to use programmes produced in HDTV for existing conventional services [15, 16] for many years to come, it is likely that the duality of Recommendation 601 will continue to be imposed on, at least, the initial phases of HDTV production. It will take until well into the 1990s before the establishment of a production standard for HDTV will reach a stage of specification and acceptance comparable to that achieved for conventional standards with Recommendation 601. Complete unification of world-wide standards may not be achieved until it is possible to send HDTV entirely in digital form from studio to the home.

However, most of the techniques described in this book have reached the stage where they can be used for digital HDTV production [17]. Figure 1.6 (see colour plates between pp 282 and 283) shows the improvement in picture quality obtained with the increased vertical and horizontal resolution compared with a conventional picture. The source for the HDTV picture was a store using the techniques described in Chapter 3 but operating at the higher sampling frequency and with the larger capacity appropriate for the increased resolution. The 1250-line and 625-line images were displayed on high resolution and conventional CRTs respectively and it can be seen from the close-up that the shadow mask does not limit the resolution which can be displayed.

The technical challenge is to broadcast this improved resolution without making excessive demands on the radio spectrum. Whether the transmission medium used is terrestrial, satellite, cable, tape or disc, there will always be a premium on the bandwidth occupied. In the case of terrestrial or satellite broadcasting the limit is imposed by the finite allocations. In the case of cable, tape or disc it becomes a matter of installation or running cost.

Digital techniques in domestic receivers

Ultimately, digital techniques will be used to deliver TV to the public using the bandwidth reduction methods discussed in Chapter 13. At this stage it is not possible to obtain sufficient bit-rate reduction to enable a TV signal to be sent digitally in the channels allocated for terrestrial or direct broadcast satellite transmission. A hybrid approach using a combination of analogue and digital transmission is possible, provided that many of the digital processes described in this book could viably be carried out in a domestic receiver.

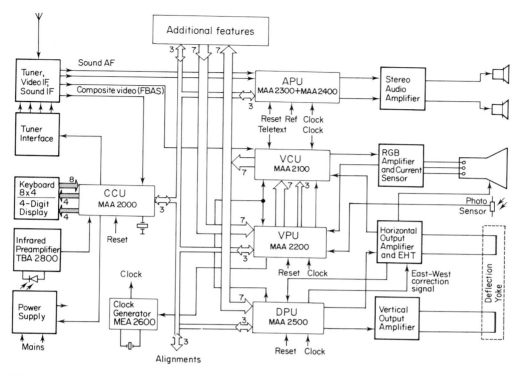

Figure 1.7 The Digivision block diagram for PAL receivers

The feasibility of this was demonstrated in 1983 by ITT–Intermetall in Germany who produced a TV receiver for PAL, SECAM and NTSC, using seven VLSI (very large scale integration) silicon chips to perform most of the receiver functions in digital form. The block diagram of this is shown in Figure 1.7. The key was the development of A–D and D–A converters for vision and sound in the video codec unit and audio processor unit respectively. The video A–D and D–A converters operate at a sampling frequency of 17.73 MHz (see Chapter 4 for the significance of this). Whilst, of course, not up to the standard of the studio quality products described in Chapter 2, their performance was surprisingly good for a mass-produced consumer product. The other significant chips were the video processor unit and the deflection processor unit using digital filtering techniques such as those described in Chapter 5.

The main objective of the 'Digivision' approach introduced by ITT was to use digital television techniques to increase the versatility of the receiver under software control and to reduce the cost by reducing the assembly times and production variables normally associated with analogue circuit boards. However, they recognised that once the main hurdle of producing a domestic receiver with the signals in digital form had been overcome future enhancements were feasible in principle. These include noise reduction, improved decoding, standards conversion (to remove flicker or line structure) and 'enlarging' parts of the picture, all based on the assumption that the cost of field storage will continue to fall to a level where it is no longer a significant fraction of the total cost of the receiver.

The next significant development was the establishment of a standard for DBS transmission which assumed that the studio output would be a Recommendation 601 digital signal rather than a composite PAL or SECAM signal. Thus, rather than encode this into a composite form one would capitalise on the somewhat wider bandwidth of the satellite channel to send a time compressed analogue component signal where the luminance and chrominance were transmitted one after the other during the line period [18, 19]. This strategy relied on two factors. Firstly, that since the public would have to buy satellite dishes and microwave down-converters, they would accept the need to buy new receivers for DBS rather than expect to be able to use their existing PAL receivers. (Alternatively, they would be prepared to buy 'set-top' adaptors for their existing sets which would allow the establishment of a new transmission standard which would eventually lead to the elimination of composite coding at the receiver for DBS reception.) The second factor was the assumption that receivers capable of Recommendation 601 type processing could be produced at reasonable cost to demultiplex luminance and chrominance as well as the sound and data sent as digital packets after the vision signal during the line period.

The first MAC/packet decoder chips which were produced in 1988 by ITT–Intermetall using their 'Digivision' technology but operating in component form at the sampling frequencies of 13.5 and 6.75 MHz specified in Recommendation 601. Figure 1.8 (see colour plates between pp 282 and 283) shows the layout of one of these VLSI chips. Whilst this is still some way from the complexity required to enable sophisticated bandwidth reduction and standards conversion circuitry to be incorporated in domestic TV receivers, it does indicate the feasibility of doing this well before the end of the 20th century. Unless the use of VLSI in TV receivers becomes viable for the decoding of bandwidth compressed signals it is very unlikely that pictures with the definition of Figure 1.6 could be delivered to the public at a level of availability which they have come to expect for TV.

Digitally assisted television (DATV)

The process of DATV [20–22] (digitally assisted TV) has been proposed by the BBC as a means of bridging the gap between the digital signals in the studio and in the receiver for transmission media using channels not able to support all-digital transmission. In this case the same sort of bandwidth reduction concepts as used in Chapter 13 are applied but the signal is split into those parts which need to be sent in analogue form, such as the basic image detail, and those which are best sent in digital form, such as the bandwidth reduction control processes.

One of the most demanding applications of DATV is illustrated in Figure 1.9. In this case, the signal processing is arranged in a way which, in addition to achieving the bandwidth reduction, it also makes the transmitted signal derived from the HDTV source sufficiently like a standard low definition signal, that a conventional receiver can derive a low definition picture. The sophisticated receiver with digital VLSI circuits would display the enhanced definition picture. This approach to higher definition is analogous to the introduction of colour TV in a way which enabled monochrome pictures to continue to be derived from the colour signal. The extent to which the picture produced by the advanced receiver becomes subjectively close to the HDTV original [23] would of course

FURTHER DEVELOPMENTS

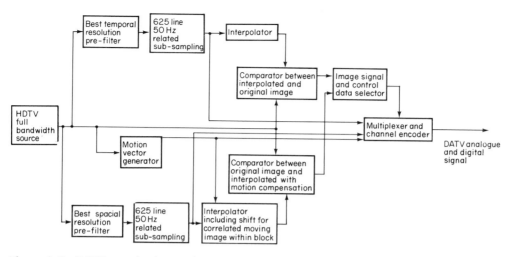

Figure 1.9 DATV encoder for 625-line compatible enhancement

depend on how effective this 'compatible' technology can become at an acceptable cost to the consumer.

The basic concepts of bandwidth reduction are firstly to exploit redundancy in the signal and secondly to exploit the psychophysical properties of the eye–brain combination to produce the most subjectively acceptable compromises. Without going into great detail, the tasks which a future digital TV receiver may have to perform can readily be described by reference to Figure 1.9 which shows how the signal for a 625-line compatible transmission might be produced. The reader may also find this description a useful introduction to some of the image processing concepts discussed in later chapters.

Starting with, say a 1250-line sequentially scanned picture, this would be analysed to determine the characteristics of the image. For instance, if the picture has areas of fine detail which are not moving then, assuming the presence of storage in the receiver, the detail can be built up slowly, thus concentrating the use of the transmission spectrum on conveying fine static (or spatial) detail rather than accurate moving (temporal) detail. On the other hand, for a randomly moving object across the screen (e.g. a bird in flight) the eye requires rapidly up-dated temporal information, but is tolerant of a lower spatial resolution in the moving object. Thus in Figure 1.9, two basic types of processing path are shown, one suited to 'moving' images and the other to 'stationary' ones. In addition, the HDTV source is subsampled to make it similar to a 625-line signal. For example, of the 1250-sequential lines, from the source, every fourth line is transmitted every 50th second so that the original is built up over four fields but would be also recognised as a conventional 625-line interlaced signal with the normal 312-lines per field.

The next part of the encoder simulates both interpolation processes which would be carried out in the receiver to recover something as close as possible to the original from each of the two types of bandwidth-reduced signal, i.e. using the algorithms designed to suit static or moving detail. The result is compared with the original and the signal giving the best result is chosen for transmission. These decisions are made for each part of the picture in small groups of pixels, and the details of the choice transmitted to the receiver by the digital assistance signal.

In addition to the two broad categories of image, static and moving, there is a third category which has to be dealt with for subjectively acceptable results. Having said that the eye is tolerant of loss of detail in moving objects, this does not apply to 'moving static detail' — an apparent contradiction in terms! If the camera is, for example, showing a hand holding an *object d'art* with fine detail it would not be acceptable if the resolution of fine detail were to diminish as the hand moved the object slowly across the field of view. This is because, unlike the detail in the wings of the bird flapping across the screen, the eye can track the 'moving static detail' of the object if it does not move too quickly. Fortunately, it is possible to deal with this situation without the need for additional bandwidth in the analogue channel provided there is enough capacity in the DATV channel to send the receiver the motion vectors giving the speed and direction of the 'moving static detail'. This enables the object to be treated as if it were stationary with the detail built up over several fields and the high resolution images in the intermediate fields interpolated from a knowledge of the speed and direction of the moving detail.

Thus the receiver would take the analogue signal and convert it to a digital signal suitable for processing. Then, using the DATV signal it would process the signal in one of the three ways judged by the encoder to be likely to give the closest approximation to the HDTV original. The concept also illustrates two important principles which will be reflected in many chapters of the book. Firstly, it is often quite acceptable to introduce a complex state-of-the-art process at the studio since it has only to be done a few times rather than be replicated in millions of homes. The DATV encoder with its motion vector measuring circuits, etc, is vastly more complex than the decoder in the receiver. Secondly, developments in semiconductor technology are such that a state-of-the-art digital TV technique which is initially only viable for a few central production installations soon becomes sufficiently cost-effective to be used much more extensively in studios and in many cases can be considered for use in domestic receivers about 10 years later.

2 Analogue-to-digital and Digital-to-analogue Conversion and its Effect on Television Signals

V. G. Devereux

GENERAL

The introduction of digital equipment into basically analogue television studios has resulted in the tandem connection of relatively large numbers of video PCM codecs. (The term 'codec' refers to a combination of an analogue-to-digital (A–D) converter and a digital-to-analogue (D–A) converter.) Furthermore, the number of codecs in tandem is likely to increase for some years before digital paths start to replace analogue connections between digital equipment. Thus a very high performance is required from individual converter units if the use of digital processing is not to cause a degradation in the picture quality obtained from a broadcasting network.

This chapter discusses all aspects of the performance of video PCM codecs, including requirements regarding number of bits per sample and sampling frequency, the meaning and relevance of data given by manufacturers, methods of measuring coding accuracy and acceptable limits for any inaccuracies. By describing tests of several codecs operated in tandem, it is shown that performance is by no means solely dependent on the design of the basic A–D and D–A circuitry. Other factors, such as the quality of the associated video low-pass filters and the use of good construction techniques in surrounding circuits, are also important.

FUNDAMENTAL PRINCIPLES
Analogue-to-digital (A–D) conversion

Analogue-to-digital conversion of video signals involves three main steps, namely quantisation, sampling and coding, as illustrated in Figure 2.1.

The quantiser measures the magnitude of the analogue input signal and divides it into 2^n parts where n is the number of digits in the binary codewords conveyed by the digital output signal. The required $2^n - 1$ decision levels are designed to be equally spaced over the conversion range, assuming that the video signals being processed have been gamma corrected. For A–D conversion of linear video signals prior to gamma correction, a non-linear quantisation characteristic providing greater accuracy near black level than near white level may be used (see Chapter 10 which deals with cameras and telecines). However, the vast majority of applications where A–D conversion of video signals is currently employed occur in the gamma-corrected domain and it will be assumed that this is the case in the remainder of this chapter.

The outputs of the level comparators are sampled at regular intervals by means of digital latches. The outputs of these latches are then transcoded to obtain the n-bit digital signal indicating the highest decision level which has been exceeded at each sampling instant.

It is obviously necessary that the times at which the outputs of the different level comparators are latched must correspond to identical instants in the analogue input signal. The resulting timing problems can be eased by sampling the analogue signal prior to the A–D converters and holding these sample values constant for a significant proportion of the interval between samples. The desirability of employing a sample-and-hold circuit depends on the method of A–D conversion as discussed later.

The sampling frequency f_s which is employed in an A–D converter must normally be at least twice the highest frequency component f_v in the input video signal. This is known as the Nyquist sampling criterion and the reason behind it is to avoid the generation of alias components having frequencies lying within the video frequency baseband of 0 to f_v. The purpose of the low-pass filter shown in Figure 2.1 is to ensure that the sampled video signal does not contain redundant components with frequencies above the specified video bandwidth. This enables the sampling frequency and hence the digital bit-rate to be kept to a minimum without aliasing distortion.

The application of the Nyquist criterion to video signals is complicated by the fact that their spectra can be resolved into three components corresponding to variations in the horizontal and vertical directions on a displayed picture and variations in time due to changing picture content. If suitable two- or three-dimensional filtering is applied, this

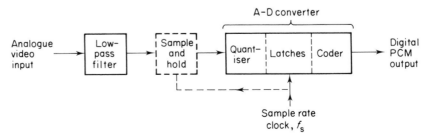

Figure 2.1 Basic elements of A–D converter and associated processes

FUNDAMENTAL PRINCIPLES 23

complex nature of video spectra allows the use of so-called sub-Nyquist sampling frequencies which do not introduce alias components despite being less than twice the highest frequency in the video signal, i.e. the maximum horizontal spatial frequency, f_v. Sub-Nyquist sampling is discussed extensively in Chapter 5 in connection with digital filtering.

Digital-to-analogue (D–A) conversion

In video digital-to-analogue (D–A) converters, each incoming binary codeword is converted to a quantised analogue sample by means of a group of switches operated by the digital input and associated weighting and summing networks, as indicated in Figure 2.2. If the most significant bit (MSB) contributes voltages 0 or $V/2$ to the output analogue sample, the lower significant bits contribute voltages 0 or $V/4$, $V/8$, $V/16$, etc. The resulting analogue sample voltages obtained from video D–A converters are normally held substantially constant for an entire sampling period thus forming a stepwise waveform with each step of width $1/f_s$. The required continuous video waveform is obtained from this stepwise waveform by passing the output from the D–A converter through a low-pass filter and a 'sin $(x)/x$' equaliser.

In practice, the leading edges of the steps corresponding to different code changes can have different overshoots or undershoots, known as glitches. These glitches can be removed by a sample and hold circuit labelled deglitcher in Figure 2.2. With careful design, the energy in the glitches can be made insignificant, thus avoiding the need for a deglitcher. Further details are given later.

Sampling and frequency characteristics

The theoretical sampling frequency and filtering requirements of A–D and D–A conversion will be explained by considering the spectra of the video signal at various stages in these processes. For convenience, it is assumed that the spectrum of the video signal applied to the A–D converter is defined entirely by the preceding low-pass filter whose

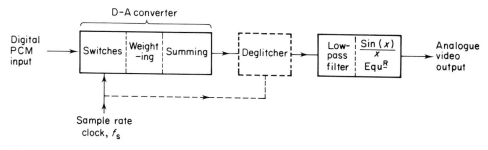

Figure 2.2 Basic elements of D–A converter and associated processes

amplitude/frequency characteristic has a flat pass-band and sharp cut-off at frequency f_v, as illustrated in Figure 2.3b. Sampling of the corresponding video signal $E_i(t)$ as shown in Figure 2.3a is equivalent to multiplying $E_i(t)$ by a train of constant amplitude pulses $E_p(t)$ as shown in Figure 2.3c, the resulting sampled signal being shown in Figure 2.3e. In the frequency domain, the multiplication of $E_i(t)$ by $E_p(t)$ results in a spectrum which contains components at frequencies equal to the sum and difference of all the components in the spectra of $E_i(t)$ and $E_p(t)$. By Fourier analysis, the spectrum of $E_p(t)$ contains components of constant amplitude at all frequencies given by nf_s where n is any integer. Thus the spectrum of video samples $E_s(t)$ is as shown in Figure 2.3f.

As mentioned previously, the output from a D–A converter $E_0(t)$ is a stepwise waveform as shown in Figure 2.3g. This waveform consists of a train of pulses of the same amplitude as the video samples $E_s(t)$ but having a constant width $1/f_s$. The effect of the change in pulse shape on the spectrum of the signal can be explained as follows.

Each impulse in $E_s(t)$ has a flat spectrum at any frequency as shown in Figure 2.4a whereas each rectangular pulse in $E_0(t)$ has a spectrum $A(f)$ of the form shown in Figure 2.4b given by

$$A(f) = \frac{\sin(\pi f/f_s)}{\pi f/f_s} \qquad (2.1)$$

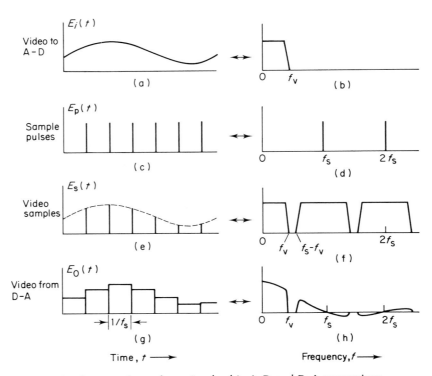

Figure 2.3 Spectra of waveforms involved in A–D and D–A conversions

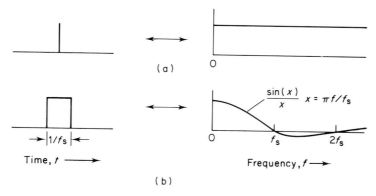

Figure 2.4 Spectra of (a) unit impluse and (b) rectangular pulse

It follows that the process of deriving the D–A converter output $E_o(t)$ from the video samples $E_s(t)$ is equivalent to low-pass filtering $E_s(t)$ using the 'sin $(x)/x$' characteristic of equation 2.1. Thus the spectrum of the video output from the D–A converter is as shown in Figure 2.3h.

By examining the spectrum of the D–A converter output, it can be seen that a signal with the same spectrum as that of the A–D converter input can be obtained by means of a low-pass filter which removes all components above frequency f_v followed by an equaliser whose frequency response over the range 0 to f_v is the inverse of the sin $(x)/x$ characteristic given by equation 2.1. It follows that, if the spectra of the A–D converter input and D–A converter output are the same, the video waveform of the D–A converter output will be the same as the waveform at the input to the A–D converter provided that no group-delay (phase) distortion has been introduced.

Note that the sin $(x)/x$ characteristic of the rectangular pulses has nulls at the sampling frequency f_s and integer multiples thereof. Thus the low-pass filter following the D–A converter does not need to provide particularly high attenuation at those frequencies.

The components in the spectrum of a sampled signal which are centred on frequencies f_s, $2f_s$, etc, i.e. all components except the original baseband components, are often referred to as alias components. The term 'alias components' has no strict definition, however, as far as bandwidth is concerned and is sometimes used to refer to only those unwanted components which lie within the baseband frequency range.

Since the lowest frequency alias component is at $f_s - f_v$, it can be seen that the presence of alias components within the baseband can only be avoided if $f_s - f_v$ is greater than f_v, i.e. if

$$f_s > 2f_v \tag{2.2}$$

In other words, as required by the Nyquist sampling criterion, there must be at least two samples per cycle of the highest video frequency if the original waveform is to be reconstructed without aliasing distortion.

In practice, f_s should be somewhat greater than $2f_v$, to cope with the finite rate of cut-off of realisable filters.

Input/output transfer characteristics

The input/output voltage transfer function of a PCM codec (combination of A–D and D–A converters) is illustrated in Figure 2.5. For an ideal A–D converter, all the steps in the transfer function will have a constant width, L, and for an ideal D–A converter they will have a constant height Q. Thus, instrumental inaccuracies in the A–D and D–A converters cause variations in width and height of the steps respectively. For video signal processing, any constant d.c. offset in all the A–D decision levels or all the D–A quantum levels are not important because it can be compensated for by a corresponding d.c. shift in the applied or decoded video signals.

Methods of specifying quantising errors

It can be seen from Figure 2.5 that the quantising errors caused by an ideal codec, i.e. the differences between its input/output transfer characteristic and a straight line drawn to minimise these differences, have peak values of $\pm Q/2$ and are uniformly distributed between these peak values. The r.m.s magnitude of any set of quantising errors with these peak values and distribution is equal to $Q/\sqrt{12}$.

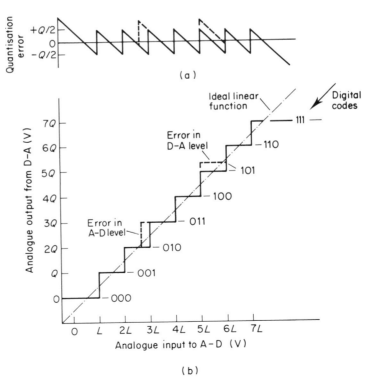

Figure 2.5 Input/output characteristic of 3-bit A–D and D–A converters

In expressions for signal-to-quantising-noise ratios, the peak-to-peak output of a D–A converter fed with n-bit codewords is normally taken to be equal to $2^n \times Q$ although its precise value is $(2^n - 1)Q$. Using this approximation, the peak-to-peak signal S_{pp} to r.m.s. quantising noise Q_{rms} ratio of an ideal codec is given by

$$\left(\frac{S_{pp}}{Q_{rms}}\right)dB = 20 \log_{10}(2^n Q \times \sqrt{12}/Q)$$
$$= 6.02n + 10.8 \text{ dB} \qquad (2.3)$$

This calculation does not take into account the bandwidth of the quantising noise and the effects of sampling. However, if the magnitudes of quantising errors on different samples are uniformly distributed between $\pm Q/2$ in a random manner, it can be shown that the spectrum of the quantising noise is flat and its r.m.s. magnitude in the range 0 to $f_s/2$ remains equal to $Q/\sqrt{12}$ [3]. The above conditions for quantising error distribution are approached more closely as the number of bits per codeword increases and they are substantially satisfied by the number of bits normally used for coding video signals. Thus, when the quantising noise bandwidth is reduced to a video bandwidth f_v less than $f_s/2$, the signal-to-quantising-noise ratio is increased by $10 \log_{10}(f_s/2f_v)$ with respect to equation 2.3.

A further factor to be considered is that the magnitude of the video signal is conventionally taken to be equal to the difference in voltage between black level V_B and white level V_W. Thus, assuming that V_B, V_W and S_{pp} all refer to the same point in the signal chain, the video signal r.m.s. quantising noise ratio of an ideal n-bit codec is given by

Video signal-to-quantising-noise ratio
$$= 6.02n + 10.8 + 10 \log_{10}(f_s/2f_v) - 20 \log_{10}[(V_W - V_B)/S_{pp}] \qquad (2.4)$$

A comparison of ideal and practical video signal-to-quantising-noise ratio provides a very useful guide to the performance of video A–D and D–A converters. A particularly useful feature of signal-to-quantising-noise measurements is that they can be obtained in the presence of rapidly changing video signals. As a result, they take into account incorrect coding of high-frequency signals in addition to static errors in the quantisation characteristic.

Other methods of specifying the accuracy of converters are:

Static errors (or low-frequency) quantising errors are usually specified in terms of 'linearity' errors which, for A–D converters, are based on the deviation of the mid-points of the horizontal parts of the steps shown in Figure 2.5 from their theoretical positions assuming perfect D–A conversion. The deviation of these mid-points from their ideal positions are known as 'integral linearity errors'. Other terms used for defining linearity errors are given below [24].

Terminal-based integral linearity error, E_T, is the maximum linearity error measured with respect to a straight line drawn through the end-points of the input/output characteristic as illustrated in Figure 2.6a. This is the most easily measured and calibrated integral linearity error measurement.

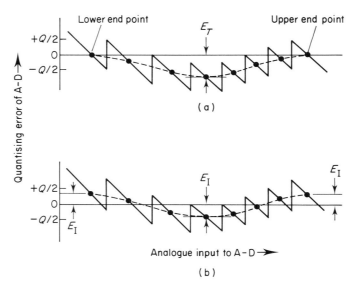

Figure 2.6 Integral based linearity errors of A–D converter (drawn as errors of a codec with ideal D–A converter). (a) Terminal-based integral linearity errors, E_T. (b) Independent integral linearity errors, E_I

Independent integral linearity error, E_I, is the maximum linearity error measured with respect to a straight line drawn to minimise the peak error as illustrated in Figure 2.6b. This is the most useful specification for video A–D converters where absolute gain and offset are not critical.

Differential linearity error is the difference in the distance between two adjacent step midpoints and the step width of an ideal characteristic.

Code size error is the error in the distance between adjacent decision levels i.e. between adjacent vertical transitions in Figure 2.5b.

Quoted specifications for the linearity error of video A–D converters without further qualification, e.g. linearity error is 0.2% of full range, normally refer to the independent integral linearity error.

Another specification of linearity often quoted for video A–D and D–A converters is their peak-to-peak differential gain and phase errors. Differential gain specifications indicate the variation in slope of the input/output characteristic averaged over the range of levels given by the peak-to-peak amplitude of the colour subcarrier used in the video test signal. As a result of the averaging process, differential gain specifications give a better guide to overall linearity than is given by integral linearity error specifications which indicate the maximum error occurring at any single quantum level.

In addition to having a satisfactory static performance, video A–D and D–A converters must be able to accurately convert rapidly changing video signals [25]. Factors affecting this dynamic performance are discussed below.

FUNDAMENTAL PRINCIPLES

If the analogue input of an A–D converter is changing rapidly, any timing error ΔT in the sampling process causes a magnitude error ΔV as indicated in Figure 2.7. For a sinusoidal input of frequency f and peak-to-peak magnitude xV_{FS} where V_{FS} is the full scale magnitude of the conversion range and x lies in the range 0 to 1, the maximum value of ΔV is given by the expression

$$\Delta V = \pi x V_{FS} f \Delta T \tag{2.5}$$

As an example of the significance of timing errors, equation 2.5 shows that a value of ΔT equal to 250 ps will cause a magnitude error ΔV equal to about $V_{FS}/256$ when a full-scale 5 MHz video signal is being digitally encoded. Tests concerned with the effect of timing errors are described later in this chapter.

There are a number of terms containing the word 'aperture' which are used to refer to inaccuracies in A–D conversion obtained only in the presence of rapidly changing analogue signals. These coding errors may occur either because the signal is sampled at an incorrect time or because of instrumental deficiencies, such as too slow a response time or slew-rate limitations in A–D circuit elements. The most commonly used terms are aperture time, aperture jitter, aperture uncertainty and aperture error. There does not seem to be any generally agreed precise meaning for any of these terms and different manufacturers of A–D converters use different terms for, apparently, the same type of error. Difficulties arise because different measurement techniques can yield different results [25–27]. Thus, the application of equation 2.5 to a quoted aperture error ΔT, or similar term, normally gives only a general guide to performance with high-frequency input signals and it will not indicate, for example, that some codes are never generated under these circumstances although they may be generated with slowly changing input signals. In practice, the author has found that the most useful specification of high-frequency performance is given by signal-to-quantising-noise measurements made in the presence of high frequency video signals as discussed above and later in this chapter.

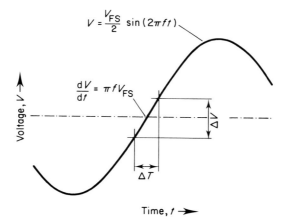

Figure 2.7 Magnitude error ΔV in sampled voltage caused by timing error ΔT in sampling instant for sinusoidal video signal

METHODS USED FOR A–D AND D–A CONVERSION

A–D conversion techniques

The earliest equipment capable of performing reasonably accurate PCM encoding of television signals was almost certainly that constructed in the Bell Research Laboratories in about 1948–9 [28]. The quantisation process in this equipment was performed by means of a specially constructed cathode ray tube in which the anode consisted of a plate containing a pattern of holes of the form shown in Figure 2.8. This plate was scanned by a flat ribbon-shaped beam of electrons whose deflection was proportional to the magnitude of the analogue video signal. When the electron beam passed through any opening in the plate, it hit one of the electrodes situated behind the plate causing a digital 1 to appear at the output. Separate electrodes for each output digit were placed behind each vertical column of holes.

A 'Gray code' rather than binary code pattern of holes was selected because of its property that only one digit changes in any transition between adjacent code levels. Thus quantising errors caused by coding uncertainty when the beam was mid-way between two code levels were restricted to one quantum level. With a binary pattern, large errors could easily be obtained under these conditions; for example an output of 0000 or 1111 could be obtained for a beam position mid-way between 0111 and 1000. The Gray code output from the electrodes was converted to a binary code after it had passed through digital latches triggered by clock pulses. This early equipment was capable of providing 5-bit codewords at a sampling rate of 10 MHz.

Little further work was carried out for the next ten years until the arrival of high-frequency transistors to replace thermionic valves provided the possibility of compact equipment capable of making effective use of digital video signals. An improved version of

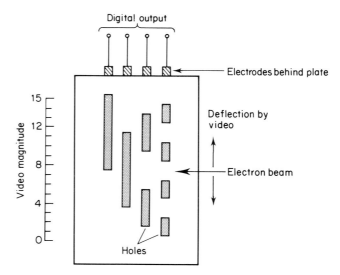

Figure 2.8 4-bit version of aperture plate in a beam coding tube used in earliest television A–D converter

the beam coding tube equipment, with transistors used at least partly in the associated circuitry, was completed in about 1959. This equipment gave 7-bit codewords at a sampling rate of 10 MHz and was used successfully in experimental transmissions of digital video signals over telephone cable pairs [29]. Further improvements led to the construction in about 1963 of high-accuracy 9-bit equipment capable of operating at sampling rates up to at least 12 MHz [30]. In this last equipment employing the beam coding tube, the associated circuitry was entirely solid state, discrete transistors being employed since integrated circuits were not yet available.

At about the same time that the final version of the beam coding tube was being developed, Bell Laboratories started work on the design of a solid state quantiser, a block diagram of which is shown in Figure 2.9 [30]. Each 'folding' amplifier stage in this quantiser performed two functions.

(a) It gave out a digital one or zero depending on whether the analogue input was positive or negative.

(b) It delivered an analogue residue signal to the next stage which was a full-wave rectified version of the input, the output being non-inverted for negative inputs and inverted for positive inputs as shown in Figure 2.9. This rectification process was achieved by means of diodes in the feedback paths of operational amplifiers. The residue was suitably biased and amplified so that it ranged between the same peak values as the input.

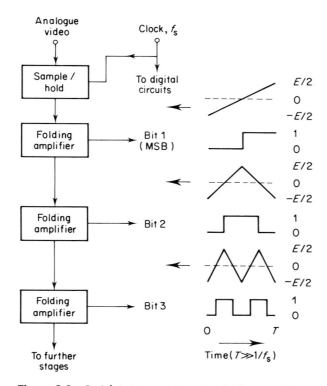

Figure 2.9 Serial A–D converter using folding amplifiers

As in the beam coding tube equipment, the initial digital signal obtained from the folding amplifiers was in the form of a Gray code. In addition to the benefits mentioned above provided by the use of a Gray code, the folding process producing this code also had the useful feature that slowly changing video signals did not provoke large rapid transitions in the residue signal as in the binary code subtraction technique discussed below. Such transitions are difficult to transmit undistorted through a large number of cascaded amplifiers. Nevertheless, the bandwidth of the amplifiers has to be much greater than that of the original analogue video signal in order to handle satisfactorily the rapid transitions obtained with rectified high frequency video signals. Difficulties in encoding high frequency video signals were overcome with the assistance of an initial sample-and-hold circuit. This folding A–D converter provided 9-bit codewords and operated successfully at sampling frequencies up to at least 12 MHz.

While the folding coder was being developed in the Bell Laboratory, work commenced in the BBC Research Department on the construction of a different type of solid state coder [31], a block diagram of which is shown in Figure 2.10. In this parallel–serial type

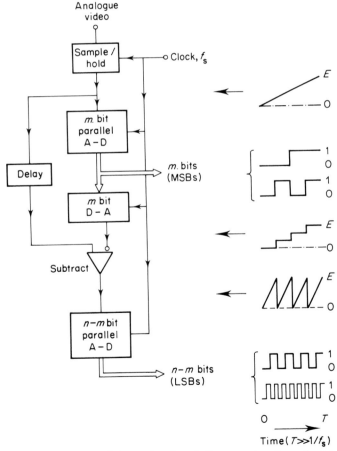

Figure 2.10 Parallel–serial n-bit A–D converter (waveforms given for $n = 4$, $m = 2$)

of coder, the first parallel A–D section coarsely digitises the analogue input to give the m most significant bits of the required n-bit codewords. After the first m bits have been determined, they are converted back to an analogue signal having 2^m quantum levels and this quantised signal is then subtracted from the unquantised input. Finally the resulting difference signal is fed to a second parallel A–D section giving the $n-m$ least significant bits. In a two-stage converter of this type, it is instrumentally convenient to make m equal to $n/2$.

As illustrated in Figure 2.11 each parallel A–D section contains $2^x - 1$ level comparators where x is the number of bits to be generated. These comparators are fed in parallel with the video signal on one of their inputs and reference voltages, uniformly spread over the conversion range, on their other inputs.

In the first parallel–serial converter, 3 bits were generated in each parallel stage giving a total of 6 bits per sample. In theory, a single stage parallel converter could have been used on its own to give all 6 bits, but the number of level comparators required, i.e. 63, was inconveniently large in the days when all the circuitry had to be constructed from discrete transistors. Even the use of seven level comparators per stage led to somewhat unwieldly equipment as illustrated in Figure 2.12. The card being held by the author is a single comparator and tunnel diode latch.

Another alternative would have been to employ only one level comparator in each of six stages. This approach was not adopted because of the difficulty of accurately transmitting the rapid transitions in the difference signal through five or more analogue subtractors connected in cascade.

The 3 + 3-bit equipment which was constructed operated satisfactorily at sampling rates up to at least 13 MHz, good results being obtained with composite PAL colour signals including high amplitudes of the 4.43 MHz colour subcarrier.

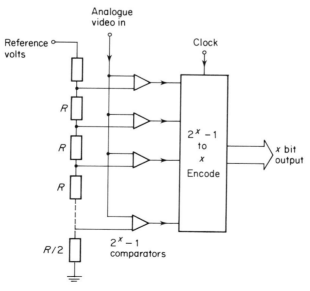

Figure 2.11 Parallel or 'flash' A–D converter

Figure 2.12 Prototype BBC A–D converter constructed in 1965

The BBC parallel–serial equipment was re-engineered about four years later in 1969 [32]. This new equipment made use of small-scale integrated circuits which had recently become available and derived 4 bits per stage to give 8-bit codewords at sampling frequencies up to about 18 MHz. Further improvements enabled correct 8-bit coding to be obtained with decision levels in the first stage significantly less accurate than 1 LSB (1/256 of full range) as required without this modification [33]. The D–A section, however, still needed to be accurate to significantly better than 1 LSB. In this improved circuitry, data obtained from extra level comparators in the second stage was used to modify the digital output of the first stage where necessary. This technique simplified initial adjustment procedures and provided automatic compensation for any long term drifts in the first stage decision levels.

A number of other types of solid state quantisers [34–37] were investigated at about the same time as those mentioned above but the folding and parallel–serial techniques were probably the most successful before large-scale integrated circuit techniques were established.

With the introduction of large-scale integration, the use of a single stage parallel arrangement as shown in Figure 2.11 has become the most commonly used method of A–D conversion for video signals. This type of converter is suited to integrated circuit techniques because the vast majority of the circuitry is digital and the previous disadvantage of the physical size of the large number of level comparators required is no longer a problem. Furthermore, the different comparators can be made sufficiently well matched so that they all examine the video signal at virtually the same instant, thus avoiding the need for an initial sample and hold circuit. For the previous techniques, the use of an initial sample and hold circuit was essential to obtain accurate coding of rapidly changing video signals.

Single chip A–D converters for encoding television signals first became available to broadcasting organisations in 1978. These chips were manufactured in the United States by TRW and employed 255 level comparators in providing 8-bit codewords at sampling frequencies up to 30 MHz. This was the only chip available for some years but they are now available from a number of different manufacturers. The performance of current chips of interest to broadcasters range from 10 bits at 20 MHz sampling to 8 bits at 100 MHz sampling [39–44]. Although this improved performance is not required for encoding normal definition gamma-corrected video signals, there is a requirement for at least 10-bit quantising accuracy for encoding video signals prior to gamma-correction. Also very high sampling frequencies, e.g. 72 MHz, are being used in work on high definition television (HDTV) signals.

Apart from the advantage of size reduction, single chip A–D converters, such as those used in the consumer products described in Chapter 1, also provide the considerable advantages of lower cost, less power consumption and absence of any controls requiring adjustment. The vast reduction in size of digital television coding equipment since the early years can be seen by comparing the 1965 single A–D converter shown in Figure 2.12 with the 1980 equipment shown in Figure 2.13. This latter equipment contains eight A–D and D–A converters plus associated analogue circuitry such as filters, buffer

Figure 2.13 Collection of eight A–D and eight D–A units constructed in 1980, used for tests described later in this chapter

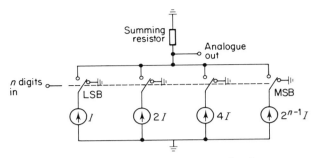

Figure 2.14 D–A converter using weighted current sources

amplifiers and video clamp circuitry. Later versions of the units shown in Figure 2.13 occupied half the rack space. These units were used in the tests described later in this chapter.

D–A conversion techniques

The practical difficulties involved in the D–A conversion of television signals are less severe than those involved in A–D conversion. All the well-known techniques include a number of electronic switches operated by the digital input together with analogue networks for weighting and summing the outputs of these switches.

Generally, the weighting and summing networks are of two types, namely those where weighted currents are added in a single resistor as shown in Figure 2.14 and those where the same magnitude currents are applied to different nodes of a resistor ladder network which is normally arranged as an R–$2R$ ladder as shown in Figure 2.15. Variations of the R–$2R$ ladder technique include the use of switched voltage rather than current sources. The voltage sources are applied via $2R$ resistors to a series chain of resistors with end values of $2R$ and intermediate values of R.

Bell Laboratories employed an R–$2R$ ladder network in the early 1960s to provide accurate 9-bit D–A conversion at sampling rates up to at least 12 MHz [38]. The first BBC D–A converter constructed a few years later employed the weighted current and single resistor technique [31]. Modern single-chip converters still use one of these two

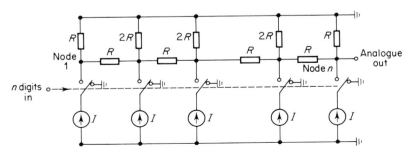

Figure 2.15 D–A converter using R–$2R$ ladder weighting network

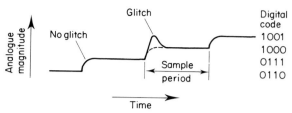

Figure 2.16 Output of D-A converter showing typical glitch at mid-range

techniques. The switched ladder network is generally preferred because matching and tracking problems are eased by the fact that all resistors and currents in the weighting network have similar values.

A desirable feature of the $R-2R$ ladder network is that the input resistance at each node is constant ($=2R/3$). Thus, since each node is driven from an identical design of switch, the delay and rise-time of the signals generated at each node should in theory be constant. Inequality in these delays and/or rise-times is undesirable as it causes overshoots or undershoots, known as glitches, in the output from a D-A converter on steps obtained when a number of digits change simultaneously. These glitches are particularly noticeable when all digits change simultaneously at the middle of the conversion range, e.g., when a 4-bit digital input changes from 0111 to 1000. The resulting effect is illustrated in Figure 2.16.

If the duration of the glitches is short compared to a sampling interval, they can be removed by a following sample and hold circuit which samples the analogue output during the flat part of each step and holds the signal level constant during the transitions. However, considerable care must be taken in the design of this deglitching circuitry and associated buffer amplifiers in order to ensure that no slew-rate distortion is introduced. Any such distortion causes intermodulation between the baseband frequency video components and components centred on the sampling frequency and its harmonics (see Figure 2.3f). The resulting intermodulation products can readily have frequencies lying within the wanted baseband and are, therefore, not removable by the following video low-pass filters. The best way to avoid these problems is to ensure that no significant glitches are generated so that no deglitching circuit is required and to apply video low-pass filtering to the D-A converter output before it passes through any analogue amplifier.

PICTURE IMPAIRMENT CAUSED BY QUANTISATION

General

The picture impairment resulting from the use of an insufficient number of bits per sample can conveniently be divided into the following categories.

(a) *Contouring effects:* Areas of the picture in which brightness varies slowly with position are represented by patches of uniform brightness separated by sharp transitions. These transitions are similar in appearance to contour lines drawn on maps.

(b) *Quantising noise:* In areas of a picture in which quantum levels are crossed at frequent intervals, quantisation errors appear in general as random noise on the picture. In general these errors have a flat spectrum over the entire video bandwidth and their visibility is similar to that given by random, Gaussian white noise with the same r.m.s. magnitude.

(c) *Beat patterns in coloured areas:* When composite colour video signals are digitally encoded, the non-linearity of the quantisation process causes intermodulation between colour subcarrier and sampling frequency components. The resulting distortion products can appear as beat patterns which are mainly noticeable in large areas of uniform hue and saturation. The visibility of this form of distortion depends on the sampling frequency and is minimised if the sampling frequency is an exact multiple of colour subcarrier frequency.

Use of dither signals to reduce visibility of quantising errors

Contouring and beat patterning effects are more visible than random errors of the same r.m.s. magnitude and, therefore, the picture impairment produced by quantisation is reduced if the former types of error can be converted into random form. One method of achieving this effect is to add a random or high frequency 'dither' signal to the video signal at the input of the A–D converter [32, 45, 46].

In tests using a random Gaussian dither signal, it was found that contouring and beat patterning effect were virtually eliminated if the r.m.s. magnitude of the added dither was greater than about $Q/3$ where Q is the spacing between adjacent quantum levels. With 8 bits per sample as normally used in broadcast television, most picture sources produce more than this amount of noise, and there is then no requirement for any additional dither noise. However, exceptions to this rule can occur with electronically generated signals or during a fade to black. Thus while random dither can be beneficial with some critical pictures, its only affect on the majority of pictures is to slightly increase the random noise level. In order to minimise this harmful effect on most pictures, improved forms of dither have been devised.

A very useful form of dither signal is a square-wave with a peak-to-peak magnitude of $Q/2$ and a frequency equal to one half the sampling frequency. It has the effect of apparently doubling the number of quantum levels in plain areas of a picture while causing negligible degradation in detailed or noisy areas. The reason for the apparent doubling of the number of quantum levels is illustrated in Figure 2.17. It can be seen from Figure 2.17a that when the original video signal applied to the A–D converter is within $\pm Q/4$ of a decision level, alternate quantised samples lie on adjacent quantum levels as shown in Figure 2.17b. In these circumstances, the quantised signal has a mean level lying halfway between quantum levels; moreover, the frequency of alternation between quantum levels is at half sampling frequency which is above the cut-off frequency of the video low-pass filter following the D–A converter. As a result, in plain areas of a picture, the output of the filter has twice the number of levels obtained without this form of dither, as shown in Figure 2.17c.

A similar effect to half-sampling frequency dither is given by adding an offset of $Q/4$ and $-Q/4$ to alternate scan lines. In this case, it is left to the human eye to apply low-pass filtering.

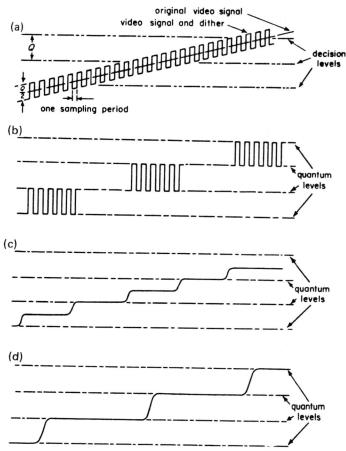

Figure 2.17 Effect of adding half-sampling frequency dither to slowly changing video signal. (a) Relative positions of video signal and decision levels in A–D converter. (b) Output of D–A converter before filtering. (c) Output of D–A converter after filtering. (d) Output of D–A converter if no dither were used

Combinations of dither signals may also be employed. For example, contouring and subcarrier beat patterns are virtually eliminated by a dither signal consisting of a half-sampling frequency component as above plus a random, or more conveniently a pseudo-random, signal of r.m.s. magnitude equal to about $Q/6$. Although the added noise component is beneficial if contouring patterns are visible without it, the resulting increase in noise (about 1.5 dB above that caused by quantisation alone) can slightly degrade picture quality when no contouring is visible without dither. Alternatively, an offset of $\pm Q/8$ applied to alternate scan lines could be used in conjunction with the half-sampling frequency dither of $\pm Q/4$ to effectively quadruple the number of quantum levels in plain areas.

In principle, coherent patterns such as contouring can be randomised with no signal-to-noise penalty by adding a known pseudo-random dither signal before A–D conversion and subtracting an identical signal after D–A conversion [45]. However, this subtraction technique has not been found to be worthwhile in a broadcasting network in view of the instrumental complications involved.

Subjective tests on a single codec

The results of subjective tests on the impairment caused by a single A–D and D–A conversion of 625-line, 5.5 MHz bandwidth (System I) PAL colour and monochrome video signals without and with an added dither signal are shown in Figures 2.18 and 2.19, respectively [47]. For Figure 2.19, the dither signal contained a half-sampling frequency component plus random noise as discussed in the preceding section.

The impairment grading scale used on both figures is given below. This grading was assessed relative to the original analogue signal which was displayed between each test picture.

Grade	Degree of impairment
1	Imperceptible
2	Just perceptible
3	Definitely perceptible but not disturbing
4	Somewhat objectionable
5	Definitely objectionable
6	Unusable

In these tests, the magnitude of the video signal was adjusted so that a 100 % saturated PAL colour bar signal just filled the conversion range. With this adjustment, black-and-white levels occur at quantum levels 62 and 208 respectively in an 8-bit code i.e. full-

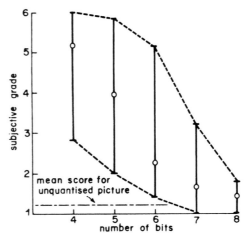

Figure 2.18 Variation in subjective impairment at different numbers of bits per sample. I Variation in grade for different picture sources. o Mean grade for all picture sources

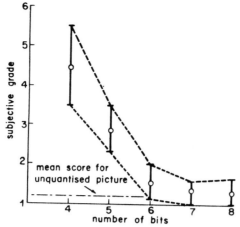

Figure 2.19 Variation in subjective impairment at different numbers of bits per sample with dither in use. I Variation in grade for different picture sources. o Mean grade for all picture sources

range = 256 quantum levels. Three separate sampling frequencies, all in the range 13.3 ± 0.1 MHz, were used in the tests. One frequency was precisely three times the colour subcarrier frequency and thus gave minimum visibility subcarrier beat patterns, the second was locked to 852 times the line-scan frequency and the third was selected to give high visibility of subcarrier beat patterns. The attenuation provided by the low-pass filters at half-sampling frequency (i.e. 30 dB at 6.65 MHz) and at higher frequencies was sufficient to eliminate any visible aliasing distortion. A wide variety of pictures were examined ranging from critical still pictures from a high-quality slide scanner to less-critical moving scenes from an off-air receiver. Unquantised synchronising and blanking information was applied to the quantised signals in order to avoid black-level clamping errors in the monitor.

Comparison of Figures 2.18 and 2.19 shows that the addition of dither signals considerably reduces the variation in the results for different test conditions and, in general, lowers the subjective impairment for a given number of bits, being most effective with the most critical types of picture. However, with 4 and 5 bits, the least critical pictures were rated slightly worse with dither than without. This was because the noise added by the dither produced greater impairment than the slight contouring and/or beat patterning on these pictures without dither.

One of the main conclusions to be drawn from these tests is that, for one codec, there is negligible advantage to be gained by employing more than 7 bits per sample for PCM encoding of composite video signals with dither in use. This result is in good agreement with the results of tests on the impairment caused by analogue random white noise [48]. These tests on analogue white noise showed that an unweighted signal-to-noise ratio of 47.4 dB, as given by equation 2.4 for 7 bits (with an allowance of 1.5 dB for added dither noise) is just beyond the threshold of visibility while a signal-to-noise ratio of 41.4 dB as given by 6 bits causes a just visible impairment.

Although 7 bits per sample may be sufficient for one codec, it is not sufficient for use in a broadcasting network because digital circuitry is at present being introduced in a somewhat piecemeal fashion with the result that it is possible for a broadcast video signal to pass through a comparatively large number of codecs connected in tandem. Assuming that the quantising errors generated in different codecs are uncorrelated, the r.m.s. magnitude of the combined quantising errors obtained from four codecs in tandem, each using 8 bits, will be the same as that obtained from one codec using 7 bits. Thus it can be concluded that 8 bits per sample should be sufficient to give negligible quantising errors after a PAL composite video signal has passed through at least four codecs in tandem. This matter is discussed further in the next section.

With regard to the above conclusions, it should be noted that the specification for encoding the luminance, Y, component of the *YUV* component colour video signal given in CCIR Recommendation 601 [9] states that black-and-white levels should correspond to quantum levels 16 and 235 respectively of an 8-bit code. The resulting spacing of 219 quantum levels between black-and-white levels is 3.5 dB greater than the spacing of 146 quantum levels used for composite video signals in the tests described above and thus the signal-to-noise ratio will be improved by the same factor.

TESTS ON UP TO EIGHT 8-BIT CODECS IN TANDEM WITH SAMPLING CLOSE TO NYQUIST LIMIT

General

The introduction of digital equipment into basically analogue television studios has already resulted in the tandem connection of relatively large numbers of video PCM codecs. The number of codecs in tandem is likely to increase for some years before digital paths start to replace analogue interconnections between digital equipments. Thus it is important to know what picture-quality degradation is caused by multiple codecs. Possible picture impairments include not only the effects of quantisation as discussed in the last section but also the effects of sampling and low-pass filtering. The impairments caused by these latter processes increase as the sampling frequency is decreased towards the Nyquist limit.

In order to investigate the impairments caused by codecs in tandem and by sampling close to the Nyquist limit, work has been carried out in which 625-line, 5.5 MHz bandwidth (System I) PAL colour and monochrome video signals were passed through up to eight PCM codecs connected in tandem using coding parameters of 8 bits per sample and a line-locked sampling frequency of 12 MHz [49]. The equipment used in these tests is shown in Figure 2.13.

The sampling frequency of precisely 12 MHz, was chosen because, at the time, it was the EBU (European Broadcasting Union) proposed frequency for sampling the luminance (Y) component in a *YUV* component video signal. Although a higher frequency of 13.5 MHz for the Y component of *YUV* signals has now been adopted by the CCIR and also not withstanding the fact that most existing digital equipment handling PAL signals uses a subcarrier-locked sampling frequency of 13.3 MHz or 17.7 MHz, the use of the lower line-locked 12 MHz frequency has the advantage for tests aimed at establishing limiting conditions that it is about the most critical sampling frequency which could be used in a broadcasting network handling 5.5 MHz bandwidth video signals

Low-pass filter requirements

For a video bandwidth of 0 to 5.5 MHz, a sampling frequency of 12 MHz is only 9% above the Nyquist limit and, therefore, analogue filters with a fast rate of cut-off are required before and after each codec in order to avoid picture impairments caused by alias components. However, the practical problems of achieving accurate group-delay equalisation of filters escalate rapidly as the rate of cut-off increases. Thus a compromise is required between the problems caused by alias components, group-delay distortion, loss of wanted high-frequency signals and circuit complexity.

Initial experiments with only one codec indicated that the optimum compromise between these various filter requirements would be achieved with filters having a loss at 6 MHz (i.e. half sampling frequency) of about 9 dB. In order to investigate different compromises, two types of filter were constructed having losses at 6 MHz of 6 dB and 12 dB. These filters will be referred to as '6 dB' filters and '12 dB' filters. Both filters were based on seventh-order Cauer attenuating networks. Group-delay equalisation was provided by four all-pass sections in the 6 dB filters and by five sections in the 12 dB filters.

In the eight codecs used in the subjective tests, four codecs contained 6 dB filters (one in the A–D converter and one in the D–A converter) and four codecs contained 12 dB filters.

For both types of filter, the peak-to-peak variation in the amplitude and group-delay characteristics of individual filters were less than 0.05 dB and 6 ns respectively over the pass-band frequency range 0 to 5 MHz. At 5.5 MHz, the attenuation of both types of filter with respect to low frequencies was about 0.5 dB, while the group delay errors were about 50 ns for the 6 dB filters and 100 ns for the 12 dB filters. During tests on a large number of filters in tandem, it was found to be important to achieve very flat pass-band characteristics particularly at low frequencies. This was illustrated by the fact that an error corresponding to a boost of only 0.01 dB in each filter at frequencies below about 0.25 MHz caused a significant slope on the top of the bar in a pulse and bar waveform which had passed through eight codecs. The resulting K-rating for the bar of about 2% was greater than that caused by any other form of frequency response distortion as illustrated in Figure 2.20.

A critical comparison of the 6 dB and 12 dB filters with regard to alias distortion was obtained using a 0 to 7 MHz frequency sweep excursing between black and white levels and repeated on every scan line, thus giving vertical bars on a picture. The effect of the aliasing components with sampling at 12 MHz was most noticeable as a low-frequency beat pattern in picture areas corresponding to input frequencies in the range 5.5 to 6.5 MHz. Beat patterns were obvious with both filters but they were significantly less noticeable with the 12 dB filters as indicated by the waveforms shown in Figure 2.21a and 2.21b. These tests on the effects of aliasing were unduly critical, however, as indicated by the following discussion.

Firstly, it should be noted that the low-frequency alias beat patterns appearing on picture monitors do not exist as low-frequency components in the video signal; they are intermodulation products of high-frequency wanted and alias components, these products being generated by the non-linear gamma characteristic of the display tube. Since they are formed by a multiplicative process, if the magnitude of the wanted component, and hence that of the alias component, is decreased by a factor $1/K$, the magnitude of the beat patterns is reduced by a factor $1/K^2$. As a result, the visibility of the beat patterns

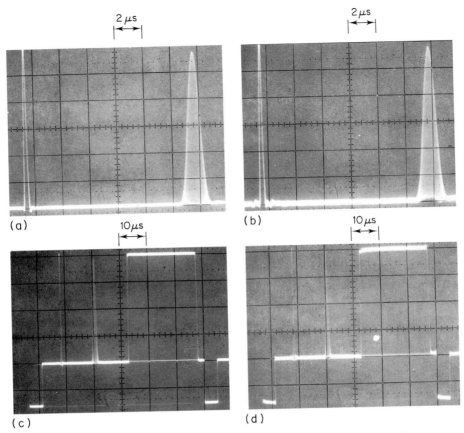

Figure 2.20 Pulse and bar response of eight codecs in tandem. (a) Input and (b) output with 2 T luminance and 10 T chrominance pulses. (c) Input and (d) output with pulses as (a) and (b) and 25 µs bar

decreases rapidly as the wanted signal is reduced below its maximum level. Furthermore, high level high-frequency sinusoidal waveforms are very rarely present in normal picture material including broadcast electronically generated signals.

Secondly, alias components resulting from insufficient attenuation above 5.5 MHz in a D–A low-pass filter will be removed at a later point in the video chain by the normal low-pass filtering present in System I broadcast transmitters and they will be substantially reduced by other filtering such as the input filter of a following codec. This effect is illustrated in Figure 2.21c. The filtering in following codecs also removes some of the alias components resulting from insufficient filtering at the input of an earlier codec. Thus, a single codec is the worst case for aliasing distortion as indicated by comparing Figure 2.21a and 2.21d. Note also that the filtering required in special effects, e.g. zoom and squeeze, equipment will considerably reduce the effects of alias components in the input signals.

In tests using video signals containing very rapid changes in signal level, it was found, as expected, that the 6 dB filters produced less ringing effects near these transitions than

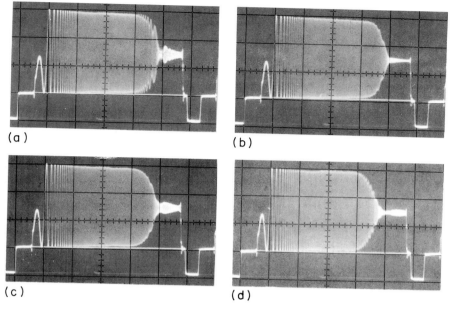

Figure 2.21 Response of codec to 0 to 7 MHz line-sweep signal. (a) One codec, 6 dB filters (b) one codec, 12 dB filters. (c) One codec, 6 dB filter, and extra 6 dB filter at output. (d) Four codecs, 6 dB filters in each codec

the 12 dB filter. Lowering the rate of cut-off filters not only reduces the duration of ringing but it also allows better group-delay correction to be achieved thus making the ringing more symmetrical about the transitions. These effects are illustrated in Figure 2.22. In subjective tests, one of the most obvious effects of ringing occurred near the changes in colour in a PAL encoded colour bar signal (see Figure 2.23). With this signal, the ringing occurred at a frequency equal to the difference between the cut-off frequency of the filters and the colour subcarrier frequency of 4.4 MHz, i.e. at about 1 MHz. Ringing at this frequency is far more visible than the ringing at 5.5 MHz obtained near narrow pulses or sharp steps in the luminance component.

Effects of black-level clamp circuitry

In initial tests on eight codecs in tandem, it was found that an appreciable amount of low-frequency noise, known as clamp streaking, was introduced by the black-level circuits employed in each codec; these clamping errors were caused by the accumulation of 8-bit quantising errors in the backporch of the video signal. To alleviate this problem, the response of the clamp circuits was adjusted so that changes in black level were restored more slowly, i.e. with a time constant of about 30 line-periods instead of the initial 5 line-periods. Secondly, it was found to be desirable to add a colour burst to monochrome as well as composite colour signals and to add a half-sampling frequency dither signal. The purpose of these added signals was to randomise the quantising errors in the backporch

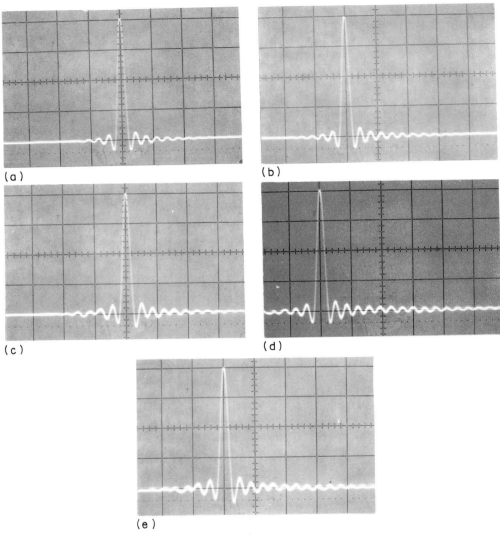

Figure 2.22 Luminance ringing caused by filters in response to a 1 T pulse (1 T = 100 ns). (a) One 6 dB filter. (b) One 12 dB filter. (c) Four codecs using 6 dB filters. (d) Four codecs using 12 dB filters. (e) Eight codecs, i.e. (c) plus (d)

of the video signal with the result that the mean black-level voltage measured during clamp pulses was determined to a greater accuracy than $\pm 1/2$ LSB.

After these clamp modifications had been completed, clamp-streaking effects were very difficult to detect on low-noise video signals used in the subjective tests described later. However, with video signals having significantly higher initial noise levels, the clamp streaking effects to be expected from the tandem connection of the clamp circuits without the intervening PCM coding operations could be clearly seen.

The responses of the clamp circuits in one codec and in eight codecs in tandem with a

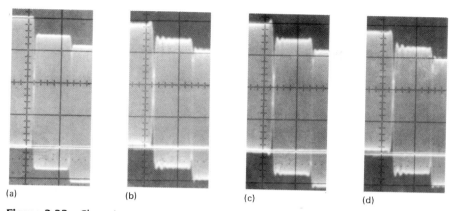

Figure 2.23 Chrominance ringing caused by filters at transitions of the magenta bar on a 100% colour bar signal. (a) Input signal. (b) Output of eight codecs. (c) Output of four codecs using 6 dB filters. (d) Output of four codecs using 12 dB filters

clamp time constant of 30 line-periods are shown in Figure 2.24. For this figure, the video signal supplied to the first codec consisted of a black-level signal (including synchronising pulses) to which a square-wave signal with transitions at 8 ms intervals had been added. It can be seen that the response time of the eight clamp circuits in tandem is much faster than that of the individual clamp circuits. This effect, together with the accumulation of quantising noise in the video signal, causes far greater unwanted variations in black level after eight codecs than after a single codec.

A better solution to eliminating clamp problems would have been to insert new noise-free line-blanking information in the D–A converter of each codec. This would prevent the accumulation of clamping errors assuming no noise was added between codecs. This solution was not adopted in the tests under discussion because of the instrumental complexities involved but fortunately most equipment used in broadcasting studios inserts new line-blanking information for other reasons.

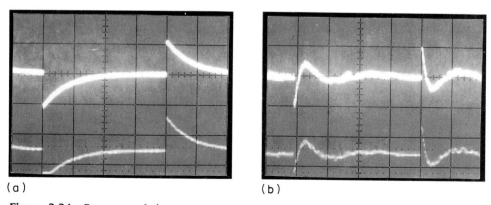

Figure 2.24 Response of clamp circuits to step changes in the d.c. level of a 'black-level' composite video signal. Horizontal scale: 2 ms cm^{-1}. Vertical scale: 20 LSB cm^{-1}. (a) One codec. (b) Eight codecs

Quantising accuracy of codecs

Measurements using the technique described in a later section indicated that the quantising noise introduced by each codec for slowly changing video signals was about 2 dB greater than for an ideal 8-bit codec. For high amplitude, high frequency signals the errors increased to about 7 dB greater than the ideal theoretical value.

Subjective tests on eight codecs

The performance of the codecs discussed above employing coding parameters of 8 bits per sample and a sampling frequency of 12 MHz was examined by means of subjective tests on 625-line, 5.5 MHz bandwidth monochrome and PAL colour video signals. In these tests, the input signal to the codecs was compared with the output signal obtained either from one codec containing 6 dB filters or from eight codecs, four of which contained 6 dB filters, the other four having 12 dB filters. The reason for performing tests on one codec using the 6 dB filters was to examine the worst possible condition for alias components. The signal from eight codecs provided the maximum possible degradation for all other forms of signal distortion.

For any one test condition, the input and output signals were compared using a picture sequence provided by 'Signal A for 10 seconds, Signal B for 10 seconds, Signal A for 10 seconds, Signal B for 10 seconds', with about 2 seconds of a mid-grey video signal between each change.

One of the signals A and B was the input signal, the other being an output signal, the input/output order being varied in a random manner. Picture quality was judged using the CCIR comparison scale:

	Grade
A much better than B	+3
A better than B	+2
A slightly better than B	+1
A the same as B	0
A slightly worse than B	−1
A worse than B	−2
A much worse than B	−3

In addition, for grades other than 0, the observers were asked to indicate the types of picture impairment that they thought were worse in one picture compared to the other using the following list:

(a) aliasing
(b) ringing on chrominance edges
(c) ringing on luminance edges
(d) loss of definition
(e) noise

(f) beat patterns in plain coloured areas
(g) other impairments.

The tests were carried out under CCIR Recommendation 500 viewing conditions using experienced technical observers seated at four times picture height.

The video input signals were obtained either from a 35 mm colour slide-scanner or from a Philips electronic test card generator, type PM 5544. Four slides selected from those recommended by the EBU were used in the subjective tests.

Moving pictures were obtained by means of a device fitted to the slide-scanner which moved the scanning raster from side to side in a sinusoidal manner with a period of about 6 s and a peak-to-peak amplitude of about 10% of the line-scan width.

In the A–D converters, the black level of monochrome and composite PAL video signals was clamped to quantum level 64 and white level was set to quantum level 204.

A summary of the results obtained for various test conditions is shown in Table 2.1. A positive mean grade indicates that the test picture was graded worse than the reference picture.

Additional tests on one codec with 6 dB filters using a monochrome caption slide gave very similar results, the mean grade obtained being 0.00 and 0.09 for the moving and stationary conditions, respectively.

Table 2.1
Summary of overall grades given by subjective tests using 7-point impairment scale

	Colour			Monochrome		
	Stationary	Moving		Stationary	Moving	
Number of codecs	1	8	1	1	8	1
Mean grade	0.10	0.25	−0.14	0.00	0.50	0.05
Standard error, σ_n	0.06	0.10	0.09	0.06	0.11	0.06
Number of observers	20	12	11	20	12	11

σ_n = standard eror of mean grade obtained from n votes where n = (number of observers) × (number of slides).

The results shown in Table 2.1 indicate that the picture impairment introduced by one codec was negligible for all the test conditions. After eight codecs, there was a very small but statistically significant worsening of the picture quality.

An analysis of the causes of impairment which observers were asked to note is given in Table 2.2. This table gives the mean grade of individual impairments (e.g. aliasing or noise) calculated using the scale given below

	Grade
Individual impairment on output voted worse than on input	+1
No observed difference between input and output	0
Individual impairment on input voted worse than on output	−1

Table 2.2
Mean grades of individual impairments using three-point comparison scale

Impairment	Colour			Monochrome		
	Stationary		Moving	Stationary		Moving
Number of codecs	1	8	1	1	8	1
Aliasing	0.04	0.01	−0.02	−0.02	−0.01	0
Chrominance ringing	−0.01	0.02	0	0	0	0
Luminance ringing	−0.01	−0.013	−0.02	−0.01	−0.01	0
Loss of definition	0.03	0.09	−0.05	0	0.16	0.05
Noise	0	0.05	0	0.02	0.25	0
Chrominance beat patterns	0.01	0.01	0	0	0	0
Others or not stated	0	0.02	0	0.02	0.03	0

The only forms of impairment of any significance in Table 2.2 are noise and loss of definition after eight codecs.

It should be noted that factors other than a loss of high-frequency video components may have been mainly responsible for apparent loss of definition after eight codecs. Informal discussions with observers after the tests had been completed, indicated that in some cases the loss of definition was still visible at considerably greater viewing distances than would be expected had the loss of definition resulted solely from the attenuation of frequency components above 5 MHz.

Alternative reasons for the apparent loss of definitions were as follows. Firstly, it may have been confused with a loss in contrast caused by a slight overall non-linearity which increased the brightness of grey levels with respect to black and white levels. Secondly, it may have been caused by the slight attenuation of frequency components over most of the wanted pass-band relative to very low frequencies as indicated by the slope of the bar in the pulse and bar waveform shown in Figure 2.20. This would reduce high-frequency detail because the low-frequency components were used to set up the brightness of the display at black and white levels to known values.

DISCUSSION OF CODING PARAMETERS FOR USE IN TELEVISION BROADCASTING

Bits per sample

The subjective tests previously described have shown that, if 8 bits per sample are used to encode composite video signals, the visibility of quantising noise on normal pictures is close to the threshold of perceptibility after eight codecs in tandem, being virtually undetected on colour pictures but noted by about 25% of observers on monochrome pictures.

The effects of 8-bit quantising were more obvious, however, on critical test signals such as a low-slope line–sawtooth waveform. With these critical signals, in addition to 'contouring' effects which are just visible after only one codec, an even more disturbing impairment could be seen after several codecs in tandem. This additional impairment was similar in appearance to clamp-streaking noise, and resulted from random variations in the contouring on successive pictures. More importantly, similar effects could be seen during the fading to black level of less critical signals. The visibility of these low-frequency changes in contouring was significantly reduced but not made imperceptible by the addition of a half-sampling-frequency dither signal in each codec.

Considering that the use of relatively large numbers of codecs in tandem is not uncommon in current broadcast television chains and that an allowance should be made for other noise sources in these chains, the results given above indicate that 9 bits per sample are desirable for encoding composite colour video signals if quantising errors are to have no perceptible effect on the broadcast picture quality during the period in which television networks contain a mixture of analogue and digital equipment. If codecs are employed with only 8 bits per sample, it is very desirable that a half-sampling-frequency dither signal should be added to the video signal prior to digital encoding; even for 9 bits per sample, dither of this form would provide a beneficial effect on very critical pictures.

For the luminance component of YUV components video signals, additional factors to be considered are as follows.

The absence of a colour subcarrier and synchronising pulses allows an increase in the number of quantum levels between black and white levels. The CCIR Recommendation 601 for YUV coding states that these levels in the luminance signal should be at 8-bit quantum levels 16 and 235 respectively, as opposed to levels 64 and 204 used in the tests on composite video signals. As a result, the signal-to-quantising-noise ratio for luminance signals should be about 4 dB greater than that obtained for composite video signals.

In addition, YUV component codecs are likely to be introduced in a more organised manner than composite codecs and it is, therefore, unlikely that the number of component codecs connected in tandem will ever be very large.

Taking into account all the factors given above, it is reasonable to conclude that the use of 8 bits per sample for coding the luminance component of YUV video signals should be sufficient to avoid any perceptible quantising impairment on normal pictures; any residual impairment on critical test signals should be substantially eliminated by the addition of a half-sampling-frequency dither signal.

With regard to the U and V colour components of YUV video signals, tests have shown that quantising noise is slightly less visible than for the luminance components encoded using the same number of bits per sample [50]. Thus the use of 8 bits per sample should be sufficient for these U and V component signals.

The results described above apply to 625-line, 50 fields per second video signals having a bandwidth of 5.5 MHz. For any future higher definition television signal, the visibility of the random quantising noise should be less visible for the same number of bits per sample because the same noise power will be spread over a wide range of spatial frequencies on the display and the visibility of noise falls as its spatial frequency increases.

Sampling frequency

The subjective tests indicated that very satisfactory PCM encoding of 5.5 MHz composite video signals could be achieved with a sampling frequency only 9% greater than the

Nyquist limit, i.e. at 12 MHz. On critical test signals, however, some picture distortion caused by alias components and non-flatness of the frequency characteristics near 5.5 MHz could be observed.

Scaling of the characteristics of the filters discussed in this article indicated that a sampling frequency of about 12.7 MHz, i.e. 15% above the Nyquist limit, is required if aliasing impairment is to be eliminated and at the same time flat amplitude and group-delay frequency responses are to be maintained up to 5.5 MHz.

There are, of course, other factors to be considered in the determination of the most suitable sampling frequency for the digital coding of video signals. These other factors include the desirability of locking to line or colour-subcarrier frequencies and the ease of performing digital processing operations.

A frequency which is an integer multiple of the colour-subcarrier frequency, i.e. 13.3 MHz or 17.7 MHz for PAL signals with a 4.43 MHz colour subcarrier, has generally been found to be the most suitable sampling frequency for encoding composite colour video signals [47]. Where transmitted bit-rate is a restriction, a sub-Nyquist frequency of twice the colour-subcarrier frequency can also be used [51, 52].

For the luminance component of *YUV* video signals with a luminance bandwidth no greater than 6 MHz, a line-locked sampling frequency of 13.5 MHz has been selected as a worldwide standard (CCIR Recommendation 601).

Apart from being reasonably but not excessively close to the Nyquist limit, this frequency was selected on the basis that, to within a tolerance of a few hertz, it is an integer multiple of the line-scan frequency for both the 625-line, 50 fields per second and 525-line, 60 fields per second broadcast television standards, thus providing a degree of compatibility between digital video signals and equipment operating on these two standards.

For the next generation of higher definition television, sampling frequencies in the range 50 to 100 MHz are likely to be required.

MEASUREMENT OF PERFORMANCE OF A–D AND D–A CONVERTERS

Measurement of signal-to-quantising-noise ratios

As far as video signal processing is concerned, the most useful specification of the coding accuracy of A–D and D–A converters is given by the difference between measured and theoretical values of signal-to-quantising noise ratios. (Theoretical values of signal-to-quantising-noise ratio are given by equation 2.4.) To obtain a useful assessment of signal-to-quantising-noise ratio, it is essential that the video signal being quantised should be able to excite all the important types of quantising error. A form of video test signal which has been found to be very satisfactory is illustrated in Figure 2.25. The presence of the high frequency component in this test signal ensures that dynamic inaccuracies are measured in addition to non-linearities in the static transfer characteristics.

Methods of measuring signal-to-quantising-noise ratios and typical results obtained will now be described in relation to tests made on the codecs used in the work previously described. In these tests, the high-frequency component of the test signal was at the PAL colour-subcarrier frequency of 4.43 MHz.

MEASUREMENT OF PERFORMANCE OF A–D AND D–A CONVERTERS

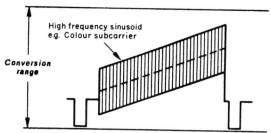

Figure 2.25 Test signal for signal-to-quantising-noise ratio measurements

In addition to measurements on combined A–D and D–A units supplied with an analogue test signal, the accuracy of D–A units on their own was measured using a digital source of data providing a similar video waveform to that shown in Figure 2.25.

Signal-to-quantising-noise ratios were measured by means of a standard analogue video noise meter which was preceded by a gating unit whose purpose was to remove video synchronising and blanking information. Within the gating unit, low frequency video information was removed by a 0.25 MHz high-pass filter and the high frequency component of the test signal was removed by a 3 MHz low-pass filter.

Measured values of the signal-to-noise ratio are shown in Figures 2.26, 2.27 and 2.28. In these figures, signal-to-noise ratios are given relative to a signal magnitude of 140 LSB (8-bit), i.e. relative to the normal spacing between black and white levels when composite video signals are digitally encoded. For these coding levels, the theoretical signal-to-quantising-noise ratio of an ideal 8-bit codec with sampling at 12 MHz is 57.2 dB (see equation 2.4) assuming a noise measurement bandwidth of 2.75 MHz. The results shown in these figures illustrate the usefulness of signal-to-noise measurements as a means of analysing the high frequency performance of A–D and D–A converters.

Curve (a) in Figure 2.26 shows results obtained for one complete codec. It can be seen that these results were quite close to the ideal value at low subcarrier amplitudes, but a

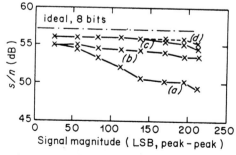

Figure 2.26 Signal-to-noise ratios (s/n) for one codec and one D–A converter plotted against colour subcarrier magnitude. (a) One codec, no sample and hold. (b) One codec, with sample and hold. (c) D–A converter alone, original deglitcher. (d) D–A converter alone, improved deglitcher

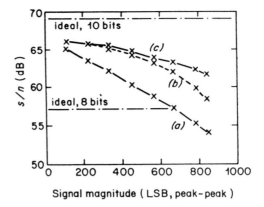

Figure 2.27 Signal-to-noise ratios (s/n) for one 10 bit D–A converter plotted against colour subcarrier magnitude. (Full Conversion range = 1024 LSB.) (a) Initial results. (b) Results for reduced clock jitter. (c) As (b) but with improved deglitcher

significant decrease to below the theoretical value for 7 bits per sample (51.1 dB) was obtained at high subcarrier amplitudes. Investigations into possible improvements in the codec performance included the addition of a sample-and-hold circuit in one of the A–D units, although the data sheet for the A–D integrated circuit stated that such a circuit was unnecessary. Curve (b) shows that this circuit gave a noticeable increase in signal-to-noise ratio for high amplitudes of the colour subcarrier, thus indicating that the internal circuitry of the A–D unit had a limited ability to handle high slew-rate video signals.

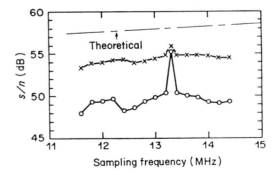

Figure 2.28 Signal-to-noise ratio (s/n) plotted against sampling frequency for 4.43 MHz, 180 LSB peak-peak video signal after one 8-bit codec
—·—·— theoretical signal-to-noise ratio,
∘—∘—∘ measured signal-to-noise ratio; no sample and hold in A–D,
×—×—× measured signal-to-noise ratio; with sample and hold in A–D converter

Results of signal-to-noise ratio measurements obtained for a D–A unit alone are given by curve (c) in Figure 2.26. Curve (d) indicates a slight improvement in performance as a result of a modification to the deglitching circuit used in the D–A unit. (This modification reduced slope-overload distortion in a buffer amplifier handling the sampled video signal prior to the video low-pass filter.)

A more accurate assessment of the performance of the D–A unit was possible because both this unit and the generator supplying the digital test signal were capable of operating at 10 bits per sample. In the first series of measurements on a 10-bit D–A unit, the signal-to-noise ratios obtained at different amplitudes of subcarrier in the test signal were as shown by curve (a) in Figure 2.27. This curve indicates that the quantising noise obtained for a peak-to-peak subcarrier amplitude equal to the full conversion range corresponded approximately to only 7 bits per sample. However, investigations revealed that a very small amount of timing jitter in the clock pulses fed from the digital generator was causing a substantial increase in the measured quantising noise. After this timing jitter had been reduced as far as possible, the improved results given by curve (b) in Figure 2.27 were obtained.

As discussed previously, the magnitude of the noise introduced by clock pulse jitter is directly proportional to the frequency and to the amplitude of sinusoidal video signals, i.e. jitter has most effect on high-frequency, high-amplitude video signals. Both theoretical arguments and practical measurements have indicated that random clock jitter with a peak-to-peak magnitude of 1 ns will introduce about the same amount of noise as 8-bit quantisation when the video signal consists of colour subcarrier with a peak-to-peak magnitude equal to one half the conversion range. It can thus be seen that considerable care is required in digital video equipment in order to ensure that the timing jitter in clock pulses fed to D–A or A–D units is kept to a very low level.

Curve (c) in Figure 2.27 shows a further improvement in coding accuracy achieved by the modified deglitching circuit previously mentioned in connection with Figure 2.26.

The above discussion shows that the use of high quality single chip D–A or A–D integrated circuits does not automatically ensure satisfactory PCM coding and decoding; considerable care is also required in the design of associated circuitry.

Further measurements of signal-to-noise ratio were performed on a complete 8-bit codec to examine the effect of altering the sampling frequency. The results obtained are given in Figure 2.28. This figure shows that with no sample-and-hold circuit, the sampling frequency had comparatively little effect on the signal-to-noise measurements apart from a significant increase for frequencies very close to 13.3 MHz i.e. near $3f_{sc}$ where f_{sc} is the colour-subcarrier frequency. Figure 2.28 also shows that the addition of the sample-and-hold circuit gave significantly higher signal-to-noise ratios at all sampling frequencies except those close to $3f_{sc}$. However, the results obtained near $3f_{sc}$ are misleading as discussed below.

Figure 2.29 shows the spectrum of the quantising noise obtained from one codec with sampling at 12 MHz for two different amplitudes of the subcarrier in the test signal. It can be seen that the spectrum obtained with a low subcarrier amplitude of 30 LSB peak-to-peak was virtually random and flat over the measured frequency range, but an increase of the amplitude to 140 LSB caused the addition of significant peaks into the spectrum. Since the fundamental quantising noise given by an ideal codec should become more random as the amplitude of the subcarrier is increased, the results shown in Figure 2.29 indicate that the peaks obtained at the higher subcarrier amplitude were caused by

56 ANALOGUE-TO-DIGITAL AND DIGITAL-TO-ANALOGUE CONVERSION

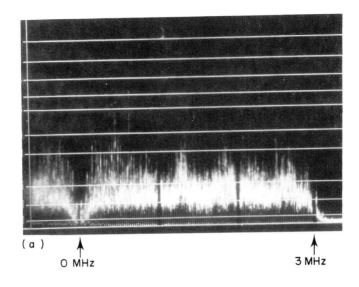

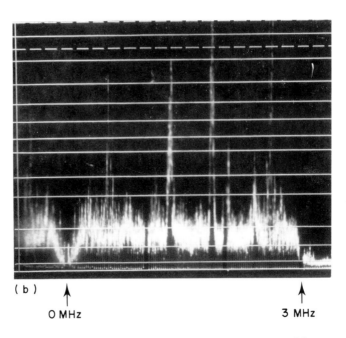

Figure 2.29 Spectrum of quantising noise generated by one codec using video signals containing different magnitudes of colour subcarrier. Horizontal scale: markers at 1 MHz intervals. Vertical scale: linearly related to signal magnitude. (a) Amplitude of subcarrier = 30 LSB peak–peak. (b) Amplitude of subcarrier = 140 LSB peak–peak

instrumental deficiencies in the codec rather than by fundamental quantising errors. This was proved by the fact that the addition of a sample-and-hold circuit in the A–D unit considerably reduced the magnitude of these spectral peaks.

Analysis of the frequency of the spectral peaks appearing in Figure 2.29b has shown that they are given by frequencies of the form $nf_s \pm mf_{sc}$ where f_s is the sampling frequency, f_{sc} is the colour subcarrier frequency and n and m are integers. Thus, when f_s is close to an integer multiple of f_{sc}, the frequencies of the peaks also occur close to integer multiples of f_{sc}. It follows that, for sampling close to $3f_{sc}$, the frequencies of the most significant instrumental quantising errors were close to 0, f_{sc}, $2f_{sc}$, etc., and, therefore, these errors were rejected by the 0.25 to 3 MHz band-pass filter employed in the noise meter. This rejection explains the misleadingly high signal-to-noise ratios shown in Figure 2.28 for sampling frequencies near $3f_{sc}$.

For measuring video codecs situated in equipment having a sampling frequency which is necessarily locked to f_{sc}, a video test frequency other than f_{sc} could be employed. A test frequency of about 5.5 MHz would have an advantage compared to f_{sc} in that quantising noise could then be measured in a bandwidth extending from low frequencies, e.g. 0.25 MHz up to 5 MHz, as in normal analogue signal-to-noise measurements.

Note that the above discussions indicate that any distinct unwanted beat patterns in large plain coloured areas of a display of a composite signal encoded using 8 or more bits per sample and with sampling not locked to the colour subcarrier, are mainly caused by instrumental deficiencies in the codec rather than by fundamental quantising errors.

In tests on codecs in tandem using the subcarrier plus sawtooth video signal, it was found that the noise power increased by 3 dB when the number of codecs was doubled, i.e. it increased in the same manner as for random noise. However, when the test signal was changed to a 1 kHz sinusoidal waveform, the signal-to-noise ratio for eight codecs in tandem was only about 7 dB lower than for one codec as compared to the 9 dB decrease obtained with the colour-subcarrier test signal. This was because the generation of exactly the same quantum levels in successive codecs is more likely with slowly changing video signals than with rapidly changing signals.

Linearity measurements

In the tests described on eight codecs in tandem, linearity errors in the codecs produced a noticeable curvature of a line–sawtooth waveform as shown in Figure 2.30. The maximum error was about 2 LSB compared to a straight line drawn through the points where the sawtooth crossed black and white levels, i.e. quantum levels 64 and 204. The corresponding value of the independent integral linearity errors is 1 LSB i.e. $+0.7\%$ of the difference between black and white levels. For picture containing areas of constant grey level, this non-linearity was sufficient to produce noticeable but not disturbing differences in brightness at mid-grey compared to the input picture when the both displays were set-up to have the same brightness at black and white levels. Thus this degree of non-linearity may be considered to be acceptable but an improved performance is desirable.

An alternative method of examining non-linearities is provided by differential gain measurements using a video test waveform similar to that used for signal-to-noise measurements as shown in Figure 2.25. These measurements indicate changes in slope

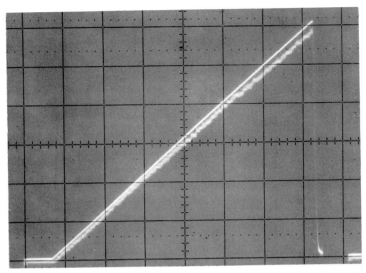

Figure 2.30 Response of eight codecs in tandem to a line–sawtooth video signal extending between quantum levels 64 and 204. Upper waveform given by input signal. Lower waveform given by output signal

of the input/output characteristic, the slope being averaged over the number of quantum levels given by the peak-to-peak amplitude of the colour subcarrier. Results obtained are illustrated in Figure 2.31. It can be seen that the differential gains of one codec and eight codecs in tandem were about 2% and 6% respectively for a luminance component varying between black and white levels. These values are the variations in the envelope of the traces shown in Figure 2.31. The thickness of the oscilloscope traces is caused by fundamental quantising errors as given by an ideally linear codec and should, therefore, be ignored as far as differential gain is concerned.

It should be noted that misleading results will be obtained from differential gain measurements if (a) the video test signal has a luminance component formed from a staircase waveform, as normally employed for analogue differential gain tests, instead of a linear sawtooth waveform and (b) the sampling frequency is locked to the colour-subcarrier frequency. For these conditions, the basic 'ideal' quantising errors are the same for each cycle of the colour subcarrier on a given step of the staircase and can cause changes in gain of over 4% for 8-bit coding of a 280 mV colour subcarrier added to a 700 mV (black-to-white) luminance waveform [53]. In these circumstances, instrumental quantising non-linearity cannot be distinguished from the basic quantising effects of an ideal codec. It is, therefore, important to randomise the basic quantising errors by using a sawtooth rather than a staircase waveform and, if possible, to employ a sampling frequency which is not locked to the colour-subcarrier frequency. The effects discussed above are illustrated in Figure 2.32.

Although an improved display of the integral linearity error could be obtained from a luminance line–sawtooth waveform, as shown in Figure 2.30, by subtracting the input signal form the output signal of a codec, it was found that measurements of differential gain could be made much more accurately than direct measurements of linearity error

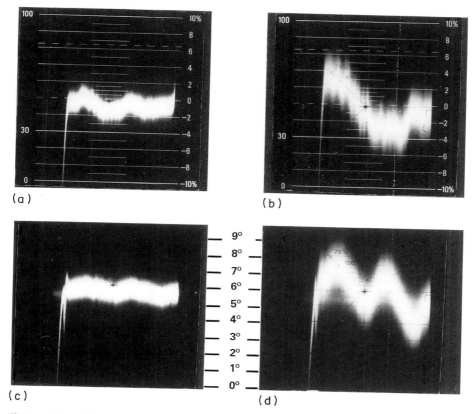

Figure 2.31 Differential phase and gain responses for one and eight codecs. Video test signal: 280 mV (56 LSB) peak–peak colour subcarrier added to 700 mV (140 LSB) line-sawtooth signal. Coding parameters: 8 bits per sample; 12 MHz sampling. (a) Differential gain of 1 codec. (b) Differential gain of eight codecs. (c) Differential phase of one codec. (d) Differential phase of eight codecs

for codecs already installed in video processing equipment. For this reason, and also because of the averaging effect of differential gain measurements, it has been concluded that differential gain specifications give a better guide to the overall linearity of codecs than integral linearity specifications. However, a limit on the maximum allowable integral linearity error is desirable in order to ensure that errors on single isolated quantising steps are not excessive.

TOLERANCES FOR CODING INACCURACIES

As already stated, it has been the author's experience that the most useful indication of coding accuracy of video A–D and D–A converters is given by signal-to-quantising-noise measurements. Other useful specifications are those for differential gain and independent integral linearity errors. Desirable tolerances for broadcast applications based on tests

60 ANALOGUE-TO-DIGITAL AND DIGITAL-TO-ANALOGUE CONVERSION

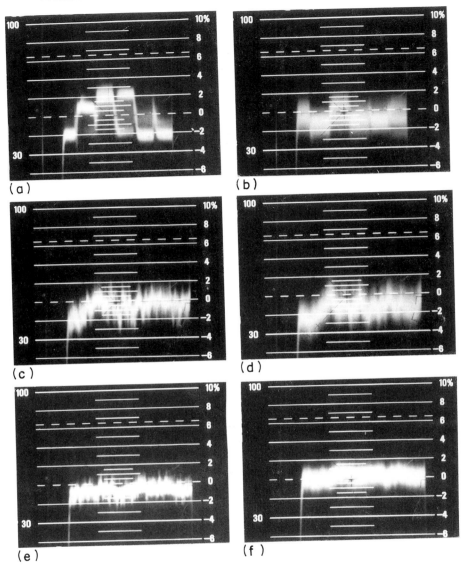

Figure 2.32 Effect of different video test waveforms and different sampling frequencies on measurements of differential gain. (a) 140 mV subcarrier on 6-step staircase, $3f_{sc}$ sampling. (b) 140 mV subcarrier on 6-step staircase, 12 MHz sampling (c) 140 mV subcarrier on sawtooth, $3f_{sc}$ sampling. (d) 140 mV subcarrier on sawtooth, 12 MHz sampling. (e) 280 mV subcarrier on sawtooth, $3f_{sc}$ sampling. (f) 280 mV subcarrier on sawtooth, 12 MHz sampling

described in this chapter and on experience of the best that can be economically achieved in practice are as follows.

The r.m.s. quantising errors generated by A–D units giving 8 bits per sample should not be more than 2 dB greater in excess of the noise generated by an ideal converter for all types of video input. A–D units in which the quantising noise increases up to 6 dB greater than ideal with increasing amplitudes of a high-frequency test signal are acceptable for most pictures but an improvement is desirable for coding high-amplitude colour subcarrier in composite video signals and critical electronically generated video signals. For 8-bit D–A converters, r.m.s. quantising errors should not be more than 1 dB greater than the theoretical value of an ideal D–A converter.

A suitable video test signal for these mesurements consists of a high-frequency component (e.g. colour subcarrier) added to a line-frequency sawtooth. It should be noted however that misleading results are obtained if the sampling frequency used in the conversion processes is an integer multiple of the high-frequency video component.

Where possible, measurements should be made on A–D and D–A units situated in final video processing equipment in order that the effects of clock pulse jitter and digital to analogue crosstalk can be included.

As already discussed, random clock jitter of 1 ns peak-to-peak causes significant quantising errors when high-frequency, high-amplitude video signals are being encoded and therefore this jitter should preferably not exceed 250 ps peak-to-peak.

The differential gain of A–D and D–A converters should preferably not exceed 1 %. This is a useful specification for assessing the overall linearity of codecs handling component (*YUV* or *RGB*) video signals as well as those handling composite colour video signals.

The independent integral linearity error should not exceed 0.2 %. This limit is mainly to restrict the magnitude of errors on isolated single quantising levels.

SPECIFICATION OF FILTER CHARACTERISTICS

The importance of filter characteristics in the achievement of high quality A–D and D–A conversion of video signals has been recognised by the CCIR by the inclusion in Recommendation 601 of very tight tolerances in the specifications of filters to be used with digital coding of *YUV* component video signals. The specifications for luminance filtering are shown in Figure 2.33. The specifications for filtering the *U* and *V* chrominance signals sampled at 6.75 MHz are similar but with frequencies halved and with pass-band tolerances approximately doubled.

CIRCUIT LAYOUT TECHNIQUES

In order to obtain satisfactory A–D and D–A conversion of video signals, it is essential that every effort is made to avoid crosstalk between digital video, clock pulse and analogue video circuits [54, 55]. As previously discussed, crosstalk from the digital video circuits to clock pulse circuitry can easily cause sufficient timing jitter in the analogue samples to give very noticeable picture impairment when high-frequency, high-amplitude signals such as colour subcarrier are being encoded. Crosstalk from digital to analogue circuitry in an A–D unit will also cause coding inaccuracies; one effect often noticed in slowly changing areas of a picture is that the edges of contours oscillate for a number of clock cycles between adjacent quantum levels as a result of positive feedback from digital circuits to analogue circuitry in an A–D converter.

The most usual cause of crosstalk is the generation of unwanted voltages in earth paths. A simple and effective method of avoiding this problem with a printed circuit board

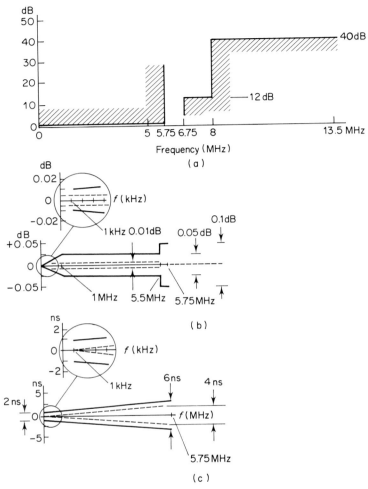

Figure 2.33 Specification for a luminance or *RGB* signal filter used when sampling at 13.5 MHz. (a) Template for insertion loss characteristic. (b) Pass-band ripple tolerance. (c) Pass-band group-delay tolerance

is to employ a low-impedance earth plane throughout the circuitry, thus making use of the fact that no voltages can be generated across zero impedance. In the author's experience, it is unlikely that any advantage will be gained by splitting the earth plane in an attempt to keep digital and analogue earth currents in separate areas of a circuit board. In fact, attempts to separate these earths can lead to voltage differences between them which worsen coding accuracy.

A common reason for crosstalk in earth paths is poor layout around PCB edge connections. On the PCB side of these connections, the earth plane should preferably be taken right up to the PCB connector; alternatively, the signal and associated earth lines leading up to the connector should be opposite one another on the two sides of the PCB thus forming a transmission line. Similar precautions should be taken on the back-plane

FACTORS AFFECTING PERFORMANCE OF VIDEO A–D AND D–A UNITS

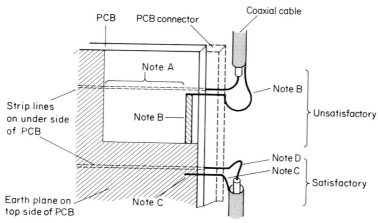

Figure 2.34 Diagram illustrating satisfactory and unsatisfactory construction techniques for analogue video signal connections to PCBs of A–D and D–A converters. Notes: A—Signal wire not below earth plane. Crosstalk from digital signals likely, via electromagnetic radiation. B—Long earth path between coaxial cable outer and PCB earth plane is likely cause of crosstalk via earth currents. C—Improved earth path compared to B. D—After splitting of inner and outer of coaxial cable, shortness of the unscreened inner is less important than shortness of the split-off outer earth connection

side of edge connectors. Also, the edge connector pins used for a signal and its associated earth should be very close so as to form part of a transmission line as far as is possible. It should be noted that where signal connections are made via a coaxial cable, experimental work has shown that it is much more important to minimise the length of the outer connection after it has been split away from the inner of the cable than it is to minimise the length of the unscreened inner lead (see Figure 2.34).

Poor earthing arrangements are usually the cause if the filtered analogue output of a D–A converter contains unwanted high-frequency components at clock pulse frequency and above rather than poor stop-band performance in the video low-pass filter. Note, however, that the filter can have a poor stop-band performance if it is not constructed using earth-plane techniques.

Crosstalk between digital and analogue circuits can also occur via electromagnetic radiation. Coupling of this type is most likely to occur within the coils of low-pass filters and these should preferably have a surrounding earthed screen. It can be reduced to negligible level in remaining parts of printed board circuitry by ensuring that signal connections are run opposite earth conductors as far as possible.

SUMMARY OF FACTORS AFFECTING PERFORMANCE OF VIDEO A–D AND D–A UNITS

The main factors affecting the overall performance of A–D and D–A units are as follows:

   (a) Basic coding accuracy of A–D and D–A circuitry (usually single chip integrated circuits).

(b) Jitter in clock pulses supplied to A–D and D–A circuitry.

(c) Crosstalk between digital and analogue circuitry.

(d) Frequency characteristics of associated analogue circuitry, particularly that of the low-pass filters.

(e) Performance of black-level stabilisation (clamp) circuitry.

It is important that A–D and D–A converter units used for broadcast television should have good performance with respect to all the above factors. It is not sufficient simply to employ high-quality integrated circuits for the conversion processes. The suitability of PCM codecs for broadcast television use in respect of factors (a), (b) and (c) can be largely measured by means of signal-to-quantising-noise measurements. Differential gain measurements supply useful supplementary information regarding overall linearity.

3 Semiconductor Storage of TV Signals

J. L. Riley

INTRODUCTION

The need to store television signals

The ability to store television signals has become essential for a wide range of applications. These include the construction of delay elements for standards conversion, PAL decoding, digital filters, television synchronisers and noise reducers. It also includes the storage of television pictures as randomly accessible 'stills' in a mass store. In all these cases about half a million data samples are required, sufficient to store a complete picture. However, there are other applications which require considerably less storage. For example, the filters based on television line delays or the stores in teletext decoders require only a few thousand samples. Except for the example cited in Chapter 1 and shown in Figure 1.1 it was not until the availability of semiconductor memory devices that even the applications requiring the more modest amounts of storage became viable propositions.

The developing semiconductor memory technology

The computer industry has, over the last few decades, grown around, and derived its momentum from, the developing memory technology. At first there were vacuum tubes and then ferrites in the 1950s and 1960s. Semiconductor memory devices followed and continue to the present day. So far, over this period there has been a tremendous rate of development which has seen chips grow from tiny 64-bit devices to a massive 1 Mbit capacity, from devices which were difficult to use to ones which are comparatively easy.

The whole range of memory devices which are now available are categorised by cost and speed performance in the graph shown in Figure 3.1. Semiconductor memory

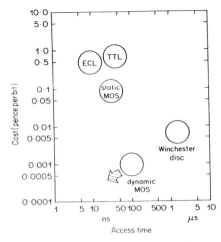

Figure 3.1 Cost–performance characteristic of several types of memory devices

encompasses the fastest, and yet costliest, devices (ECL, I²L and TTL) and the medium speed, medium cost devices with sub-100 ns access times costing about 1000 bits a penny. Slower but larger devices based on charge-coupled device (CCD) technology were for a time considered to be the future high-density, low-cost memory medium but early attempts to build reliable devices foundered. Bubble memory promised to fill this need but access times are presently very slow. At the cheapest, but certainly the slowest, end of the range lie the moveable-head magnetic devices although these are now facing stiff competition from semiconductor memory as the technology develops.

Applications within the computer industry have diversified in the 1980s with the widespread use of microprocessors because, being used in more portable machines, they demand even lower power consumption whilst their graphic support requirements demand faster access. This has led to the continued impetus for semiconductor memory development.

The main technology used in this chapter to describe the principles of storage is that of the metal–oxide–silicon (MOS) semiconductor memory device. This can be broadly divided into two groups referred to as 'static' and 'dynamic' because of their inherent cell structure. In this first section, the important milestones in the development of MOS technology are described in order to explain the trends in memory density, access time, power consumption and cost. Attention is drawn to the problems encountered in constructing such devices and how they were overcome. Each new generation has shown some improvement over its predecessor which makes for faster access, lower power-consumption and lower cost per bit.

Some of the first semiconductor memory devices to emerge were 64-bit and 256-bit ones from manufacturers like Intel which contained memory cells arranged in an orthogonal two-dimentional matrix as shown in Figure 3.2. Each cell was identified by a particular row and column address number and contained a digital binary digit (bit) of information which was either a '1' or a '0'. The individual cells of these devices were based on a cross–coupled flip-flop comprising, in most cases, six MOS transistors as shown

INTRODUCTION 67

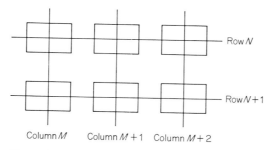

Figure 3.2 Memory cell matrix, showing memory cells addressed by a ROW and COLUMN address count

in Figure 3.3. Data are stored twice in their true and complement forms and this gives rise to pairs of column (bit) lines. Each cell is accessed by applying a voltage to a selected row (word) line so that the appropriate switch transistors, A and D, are turned on to connect one cell to each column. One column is selected by applying a voltage to the column switch transistors, G and H, to connect a single cell to the write and read transistor switches, I and J. In this particular design the data are written in true form and read from the complemented data line. Transistors B and C form the loads for the flip-flop. This type of cell forms the basis of static memory devices.

The features of these early devices were the simplicity of their structure but they were of comparatively large size and had a high power consumption. There was no decoding of the addresses on the chip. The MOS transistors were built using P-channel technology (P-MOS) which was more easily controlled in the early days. The resulting high cost promoted few serious applications but by the end of the 1960s a new generation 1024-bit device became available from Intel.

To overcome the large size and power-consumption disadvantages of the 256-bit devices, the 'dynamic' memory cell was devised. The static cell was replaced by a capacitor holding a stored charge and transistor switches to connect it to the memory matrix. In the 1 kbit Intel 1103, each memory cell consisted of three P-MOS transistors,

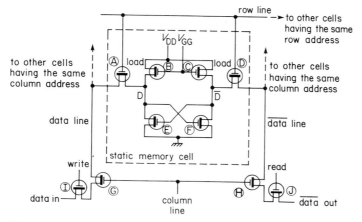

Figure 3.3 Basic static read–write memory cell

the gate-to-source capacitance of one transistor forming the cell capacitor. The problem, of course, with this arrangement is that the stored charge will eventually leak away through the finite resistance associated with the gate of this transistor to ground. This would typically occur within 2 ms. Special arrangments have to be made to sense the charge on each cell capacitor within this time and 'refresh' it to the original condition. This is normally performed row-by-row so that to refresh the entire matrix array, a refresh operation is applied to each row within 2 ms. The benefits of smaller cell size and lower power consumption more than outweighed the necessity to refresh.

The device was not without its drawbacks, however, and another 1K dynamic memory (DRAM), the MK4006 from Mostek, improved upon it and allowed the user to supply TTL level clocks with minimal timing considerations. Also, for the first time the address decoding circuits were incorporated within the chip. Memory access times were in the order of 350 ns from the supplied address clock.

The introduction of the next generation, 4K × 1, dynamic memory chips produced a variety of designs which competed with one another. There were at least five major designs comprising two differently arranged 22-pin dual-in-line packaged (DIL) devices, two differently arranged 18-pin DIL devices and a revolutionary 16-pin DIL device! Access times, power dissipation and chip size varied over a range of more than 2:1. The smaller 16-pin design was achieved by time-multiplexing the 12-bit address onto six signal pins so that the row-address and column-address components had to be supplied separately with independent clocks, referred to as the row address strobe (RAS) and column address strobe (CAS) respectively. The advantage of greater packing density compensated for the more awkward addressing arrangements and slightly inferior access times. On the later MK 4027 devices, internal circuitry allowed the complex timing of the row and column addresses to be handled on the chip making it relatively easy for the user to drive. Access times came down to about 150 ns.

The memory cell was eventually reduced to a capacitor and a single transistor as shown in Figure 3.4a and N-MOS technology was adopted as techniques developed. N-MOS is better because the threshold voltages are lower — suiting TTL compatibility — and the greater mobility of the electron carriers provides an inherently faster device. A low-resistance polysilicon material (POLY) for the gates and row lines replaced the aluminium used previously. The cross-section view shown in Figure 3.4b is accompanied by a plan view of the 'POLY' I process in Figure 3.4c to illustrate the compact cell arrangement. The smaller cell size and hence capacitor size (about 0.07 pF) meant that the voltages required to be sensed in the refresh operation were also correspondingly lower. These voltages are an attenuated version of the signal from the cell because of the capacitative divider action of the cell capacitance and the stray capacitance of the column lines — the latter may be measured in picofarads. One of the technological hurdles overcome, was to develop an amplifier design which could cope with these low voltage levels. Other problems remaining included the high power dissipation — largely due to the extra circuitry required to support the memory — inadequate noise margins and unexplained (at that time) 'soft' errors which are those produced randomly from other than physical defects on the chip but which can be recovered by reprogramming the data.

The next generation DRAM which appeared in 1976 — the 16K × 1 — attempted to overcome the drawbacks of the 4K × 1 device. The new technology, which permitted a greater memory density, incorporated two separate layers of polysilicon (POLY II) in the

INTRODUCTION 69

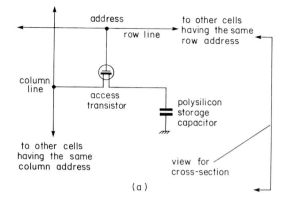

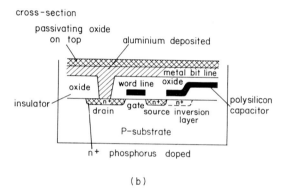

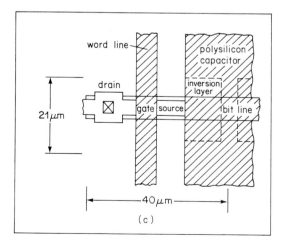

Figure 3.4 The single polysilicon for 4K × 1 devices: (a) circuit, (b) cross-section, (c) plan view

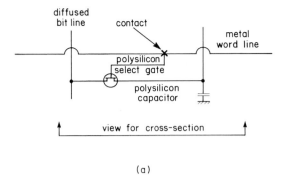

(a)

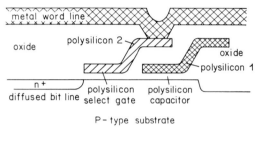

(b)

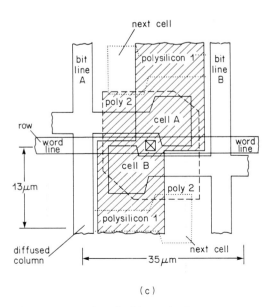

(c)

Figure 3.5 The double polysilicon process: (a) circuit, (b) cross-section, (c) plan view

cell construction (shown in Figure 3.5) and located the capacitor below the transistor. This technology was originally developed for CCD chips with minimum line widths down to 5 μm and resulted in cell sizes of less than 500 μm².

At this time there was considerable standardisation in the device package and performance so that to a large extent devices were interchangeable and effectively multi-sourced—a fact which played an important part in assisting costs to fall. The 16-pin DIL package became universal, the extra seventh address pin found by taking over a pin previously used for a chip select function. TTL compatibility was also important and the problems of high-power dissipation and inadequate noise margins were largely overcome by adopting balanced amplifier designs. The output driving specification was increased to permit two TTL loads and up to 100 pF to be handled [56]. The soft errors were found to be caused by alpha-particles emanating from the chip carrier casing and bombarding the chip—the result being to cause spurious electron–hole pairs to be generated and upset the stored charge [57]. Special coatings over the chip helped to reduce this effect to an acceptable level.

Once the single transistor cell had arrived, further advances in the technology relied on reducing the cell size. There are several ways to effect this reduction without altering the basic design. Throughout the 4K and 16K development period, cell sizes were progressively reduced in two dimensions in what is termed a 'shrinking' operation. At the transistor level, the length-to-width ratio of the channel determines its characteristics including resistive properties, gain, speed performance and relative size. Photographic size reduction produces a smaller device which has similar properties to the original provided the length-to-width ratio is kept constant.

A size reduction involving all three dimensions is called scaling and it is the technique which is responsible for the move to 64K DRAMs and the later generation 16K devices. With reference to Figure 3.6 and Table 3.1, an example is given of a transistor fabricated in the POLY II technology and scaled by a factor K. Since the field strength must be kept constant the voltage also scales by K: the device area is reduced by a factor of K^2 and the stored charge by a factor K. Both these reduce the transit time and increase the speed performance. The cell current is reduced by a factor K and hence power dissipation by a factor K^2. As both power and voltage are lower the reliability is improved. Scaling techniques are limited by the tolerance of the photolithographic equipment employed to make the masks.

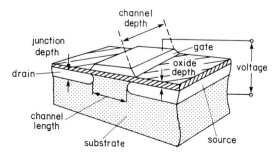

Figure 3.6 Simplified three-dimensional diagram of MOS transistor

Table 3.1
Example of scaled technology

	Scale factor	POLY II	Scaled POLY
Channel length (μm)	K	5	2.1
Power supply voltage	K	12	5
Junction depth (μm)	K	1.2	0.5
Oxide depth (Å)	K	850	354
Cell area (μm^2)	K^2	600	105
Capacitance (pF)	K	0.07	0.03
Power dissipation (μW)	K^2	40	7

The 64K DRAM was achieved by the results of several technological developments coming together [58]. The scaled N-MOS mechanism made TTL compatibility (and hence the provision of 5 V only) easier to achieve and allowed the +12 V supply to be dispensed with. Substrate bias circuits which generate the negative voltage required by the substrate, using the output of a ring oscillator capacitively coupled to the substrate, make it unnecessary to supply −5 V. The removal of the need to supply three separate voltages (±5 V, 12 V), as in most previous devices, released two of the three package pins for further address lines and opened the way to extend the now standard 16-pin DIL package right the way up to 256K DRAM devices. In many cases pin 1 was not connected to anticipate this—see Figure 3.7. The relative sizes of an Inmos 64K DRAM chip and its DIL package are illustrated in Figure 3.8.

As the device dimensions are reduced the stored charge is correspondingly reduced and this has two effects. The signal voltage available to the sense amplifiers is decreased making the sense amplifier design more critical. A method of folding the column lines back on themselves helped to reduce the stray capacitance and reduce the charge attenuation factor. Secondly, the difference between the number of electrons sensed as '1' and those sensed as a '0' is smaller. Since the number of electron–hole pairs produced within the silicon by an incident alpha-particle remains substantially constant, the probability of an alpha-particle error is increased. Package materials were improved and protective topcoats increased to reduce this problem. Fabricating the column lines in

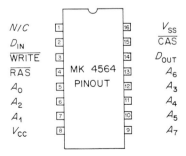

Figure 3.7 Pin layout for MK4564 64K dynamic memory chip

metal rather than by diffusion reduced the exposed silicon area and also helped to increase alpha-immunity.

Another problem arises when the channel length is less than about 3 μm: the source and drain regions are so close together that their respective depletion regions within the silicon may overlap causing an unwanted current path between source and drain which is out of control of the gate. This problem was tackled by a method of ion implantation whereby the diffusions of source and drain are graded to minimise the effects of the unwanted path.

The scaled N-MOS technology continued through several levels of scaling until line widths were down to about 2 μm and this was used for the first 256K DRAMs. At this density, one of the outstanding difficulties is achieving a satisfactory yield in the manufacturing process. To this end a number of spare columns of cells and row decoding circuity are normally integrated into the chip and subsequently selected and readdressed to replace defective ones prior to the final interconnection.

Figure 3.8 An Inmos 64K DRAM showing the relative sizes of DIL package and chip area

Recently there has been a move away from N-MOS to a complementary metal–oxide–silicon (CMOS) technology with its inherently lower power consumption resulting from balanced circuit design where no steady direct current is drawn [59–62]. The advanced CMOS technology combines all the improvements obtained with scaled N-MOS technology with the benefits of the complementary circuit design. The P-MOS transistors and memory cells are contained within $n+$ wells within the P-type substrate—see Figure 3.9: this 'burying' of the memory cell within the silicon improves the alpha-immunity. The smaller scale achieved (1.2 μm line widths) increases the device speed so that sub-100 ns access times are common.

Several manufacturers are now involved in 1 Mbit DRAM development with volume production expected in the 1990s. The technology depends on deeper entrenched cell capacitors where the surface area of the capacitor is effectively increased by utilising the side-walls of the trench as well as its base. This approach also helps to improve alpha-immunity. The problems yet to be overcome include the ability to achieve trenches of reasonable depth and profile and to make oxide layers thin enough to match the scaling.

Looking even further ahead, 4 Mbit and 16 Mbit devices are in the pipeline and are expected to require trench depths down to 7 μm deep with 0.5 μm design rules.

Developing trends of the technology

From the last section it is clear that there are several trends in semiconductor memory development which are worth summarising. These are essentially based on historical evidence and can be projected into the future with some degree of confidence. The

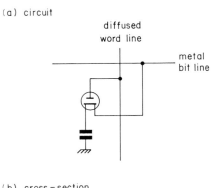

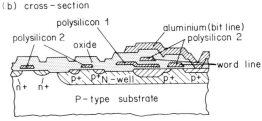

Figure 3.9 The advanced CMOS process: (a) circuit, (b) cross-section

important technological features of minimum line width, cell area, chip size, access time and power consumption are considered individually. A performance trend can be well illustrated in the speed–power product. Finally the costs are discussed in terms of per unit cell. The statistics presented relate to dynamic memory primarily although some parameters relate also to static devices.

One distinct trend which has characterised the evolution of dynamic memory chips is the quadrupling of storage capacity which has continued at the rate of a new generation about every four years since the 1970s and looks set to continue at an even greater rate for the rest of the century. The rapid build-up of the quantity of devices shipped has been fuelled by the ongoing success of the previous generations and stiff competition between manufacturers. The total number of bits taken worldwide is very nearly equivalent to a doubling every year.

The most advanced technologies tend to appear first in dynamic memory devices—see Figure 3.10—although there have been exceptions such as the CCD development which spawned the POLY II process and several programmable read-only memory devices (PROMs) which some manufacturers treated as guinea pigs for the scaled POLY technology. Each generation of device is often produced by more than one technology: some manufacturers are content to use the existing technology pushed to its limits to produce an early device ahead of the competition while others are keen to develop the new technology and beat the competition on performance later. In this way the technology is constantly under review and as each device is introduced, another is in the pipeline.

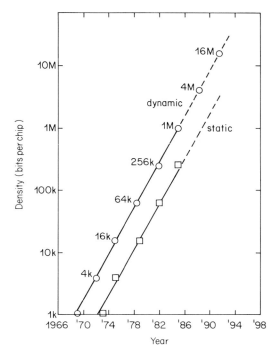

Figure 3.10 Semiconductor memory development

The minimum line width — see Figure 3.11 — has already been reduced by a factor of 10 since the early memory devices appeared. With line widths approaching 1 μm the limits of conventional photolithographic processes are reached and the future lies in electron beam and, later X-ray techniques. Associated with line width is the overall cell area — see Figure 3.12. The 1 Mbit chips are expected to achieve a cell area of 32 μm². The overall chip or die size has varied, generally between about 30 mm² down to 10 mm² depending on the memory capacity and technology used. The chip size affects system costs because the smaller the chip, the cheaper it is to produce and sell and also the device yield decreases exponentially with chip area.

The memory access time has decreased but not as rapidly as other factors. The multiplexed addressing arrangement has always penalised the access times of dynamic memories but smaller scaled cells and lower-resistance column-lines and row-lines have helped to reduce on-chip signal delays. With submicrometre geometries and now CMOS circuitry, power consumption levels are set to fall below 1 μW per bit for the first time, making the 256K DRAM chip run several times cooler than a 1K DRAM chip fifteen years ago. A useful overall figure-of-merit, combining a measure of speed and power performance is the speed–power product shown in Figure 3.13, measured in pJ.

The trend of dynamic memory development is such that as the package capacity increases the initial cost per bit is more than its immediate predecessor but will, at some time in its life, become competitive and eventually 'cross-over' and, therefore, cost less per bit until it is succeeded by the following generation. The cost of implementing a

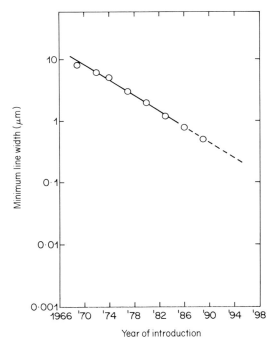

Figure 3.11 The trend in dynamic memory minimum line width

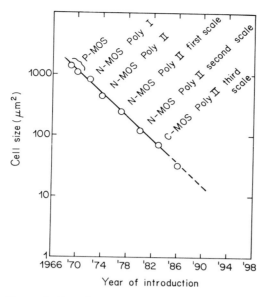

Figure 3.12 The trend in dynamic memory cell size

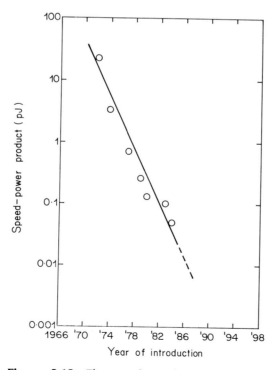

Figure 3.13 The trend in dynamic memory speed–power product

digital system based on these devices becomes competitive before the devices themselves because the higher density allows a reduced package count and reduced power-supply requirements. The figures presented in Figure 3.14 ignore the trends of individual devices and concentrate on the best value which was available at any time. The costs are based on the market rates pertaining to the UK in quantities of more than 1000 units. The steady fall in unit costs has continued by about a factor of ten every five years.

Memory device choices

The main choice, for the majority of applications to television engineering, is that between the use of static or dynamic semiconductor memory. Compared with other technologies, dynamic memories represent a very attractive and cost-effective solution for large random-access, mass storage applications where speed is not so important and where cost and overall power consumption dominate. This is assuming that the refresh mechanism poses no undue problems.

Static memory devices are suitable for smaller storage units (say up to about 64 Kbits) where the speed advantage is beneficial and the higher power consumption and cost can be tolerated. Thus, broadly speaking, dynamic memories tend to be used almost exclusively for building stores of television picture and multipicture capacity whereas static devices are more appropriate for television line stores and microprocessor memory.

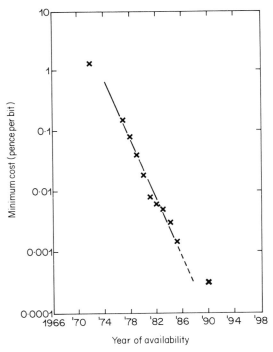

Figure 3.14 The trend in dynamic memory costs

FEATURES OF DYNAMIC SEMICONDUCTOR MEMORY

For very small stores, such as those required for delaying a video signal by a few sample periods, for example, the even faster bipolar devices are useful.

The rest of this chapter illustrates the way such memory devices may be used for a wide variety of television applications. The basic store design principles are discussed and, for each application, particular features are described in detail.

FEATURES OF DYNAMIC SEMICONDUCTOR MEMORY

Multiplexed addressing and basic read cycle

To avoid the memory chip package requiring an excessive number of pins the memory address has traditionally been multiplexed into its row and column components. Each component is latched into the memory device by independent address strobes and the precise timing of these is critical if the memory access time is to be minimised. The chip designers have attempted to simplify this timing problem by defining the timing relationships, as far as possible, on the chip itself.

A simplified functional block diagram of a typical dynamic read–write memory device is presented in Figure 3.15 which shows the applied memory address, as an n-bit wide

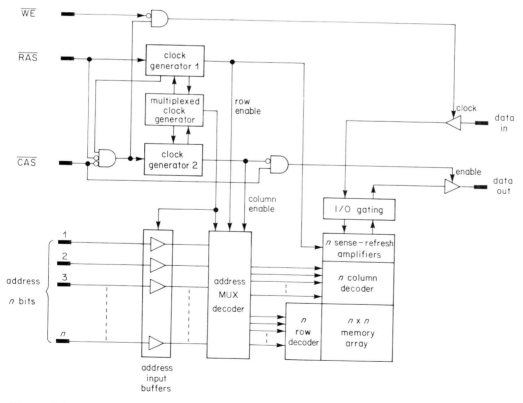

Figure 3.15 A simplified functional block diagram of a dynamic read–write memory device (signals only shown)

signal which is time-multiplexed to be alternately the row component and the column component of the address. For example, a 16K device, requiring 14 data bits to define it, has an *n* value of seven. The first action required to access the device is to apply the row address—see Figure 3.16a—and as soon as the row address inputs are valid (after a period t_{ASR}) the first address strobe may be activated. This strobe is referred to as the row address strobe ($\overline{RAS}$) and is an active low signal. It is responsible for initiating a variety of memory cycles which, once begun, must not be aborted.

The falling edge of $\overline{RAS}$ triggers an internally generated clock which performs three further functions. The first of these is to latch the row address into the chip and decode it. Secondly, the selected row is enabled and data are destructively read from each cell in the

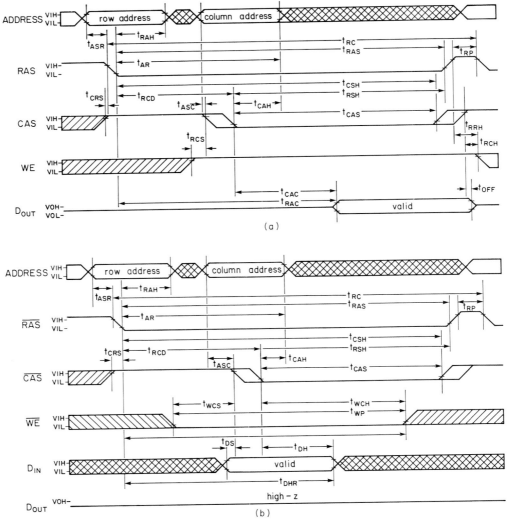

Figure 3.16 Dynamic read–write memory signal waveforms (a) read cycle, (b) write cycle

selected row by dumping its charge onto its respective column sense line. A sense amplifier for each column detects the change in voltage level on the column line as a result of the deposited charge and the signal is amplified. The third function is to latch the data into these sense amplifiers. The amplified signals are fed back onto the column sense lines, thus restoring (refreshing) the cells to their original voltages. At this time the sense amplifiers contain the same data as in the selected row — this remains so until $\overline{RAS}$ is deactivated. The minimum active period for $\overline{RAS}$ is necessary to allow the sense amplifiers time to restore the data (t_{RAS}).

Once the row address hold time (t_{RAH}) has been met the column address may be applied and as soon as this is valid (after a period t_{ASC}) the second address strobe may be activated. As soon as this column address strobe ($\overline{CAS}$) is applied the data output buffer is immediately disabled and the data output assumes a high impedance state. A delayed signal from the $\overline{RAS}$ generated clock and the $\overline{RAS}$ signal itself is gated with $\overline{CAS}$ to ensure that the $\overline{CAS}$ generated clocks do not commence until the optimum time and while $\overline{RAS}$ is active. The $\overline{CAS}$ generated clock latches the column address which selects the appropriate column of the memory array. The data from the selected sense amplifier are transferred to the output buffer within an access time (t_{CAC}) after $\overline{CAS}$ has become active. If the $\overline{CAS}$ is applied beyond the t_{RCD} (max) limit the access time is exclusively determined by $\overline{CAS}$ (t_{CAC}) rather than $\overline{RAS}$ (t_{RAC}). The output buffer is enabled by the $\overline{CAS}$ generated clock and this effectively completes the read access of the memory device. This buffer remains enabled until $\overline{CAS}$ becomes inactive. Before another access can occur the $\overline{RAS}$ must be held inactive for a prescribed precharge period (t_{RP}). The total time taken by the active and inactive period of the $\overline{RAS}$ represents the memory cycle which is often referred to as the random read cycle time.

Normal write cycle

During a normal read cycle, the write enable ($\overline{WE}$) signal is inactive, but for write operation it is activated at some time (t_{WCS}) prior to the $\overline{CAS}$ active edge — see Figure 3.16b. Until this point the access cycle is exactly as the read cycle already described. Once the $\overline{WE}$ signal is activated the data set up at the data input pin (t_{DS} applies) is latched into the chip on the $\overline{CAS}$ active edge. The data are held valid for a period (t_{DH}) after this. The data are written into the selected sense amplifier and the selected cell. During a normal write cycle the data output buffer is disabled.

This latter property together with the absence of a data output latch makes it possible to connect the data input and output pins together provided that only normal read and write cycles are used. This is referred to again later.

Other forms of write cycle

Another form of write cycle is common to all the most recent generations of dynamic read–write memory devices. It is the 'read–modify–write' cycle whereby an addressed cell can be accessed to read its data followed by different data written to the same address. Typical waveforms associated with the read–modify–write cycle are shown in Figure

3.17. When $\overline{WE}$ is delayed beyond the falling edge of $\overline{CAS}$ by a prescribed minimum period (t_{CWD}) the data output will contain data read from the selected cell in the same way as for a read cycle. Data to be written into this cell are set up and held relative to the falling $\overline{WE}$ edge which now directly performs the write function.

If it is not necessary to make use of the read data during this cycle, another alternative to the normal write cycle is one where writing can occur 'late'. A cell is addressed in the same manner as a read–modify–write cycle and the read data appear on the data output pin. However, the $\overline{WE}$ signal may now be applied before the data output is valid thus shortening the cycle length (t_{RW}). Input data are still applied referenced to the $\overline{WE}$ falling edge and the waveforms of Figure 3.17 are still valid. It is sometimes more convenient, in a particular application, to use the 'late write' form rather than 'early' (normal) write. For instance, in high-speed shift registers when data for writing does not become valid until after the memory address has been applied.

Attempts to speed up the cycle

The penalty of slower access time caused by multiplexed addressing has caused manufacturers to find ways of effectively increasing the speed of operation in some circumstances. One means of increasing speed without increasing operating power is possible provided that successive memory operations occur at multiple column locations of the same row address. This is known as 'page mode' and a typical read and write cycle is shown in Figure 3.18. The row and column components of an address are applied in the normal way and, depending on the polarity of $\overline{WE}$, data is either written to or read from the selected cell. However, if, when $\overline{CAS}$ is made inactive, $\overline{RAS}$ is maintained active, the data for the whole of the addressed row remain available on the sense amplifiers for

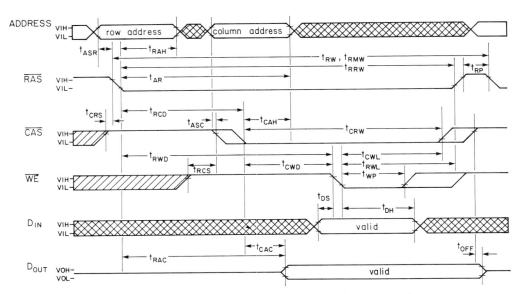

Figure 3.17 Dynamic read–write memory late-write/read–modify–write cycle

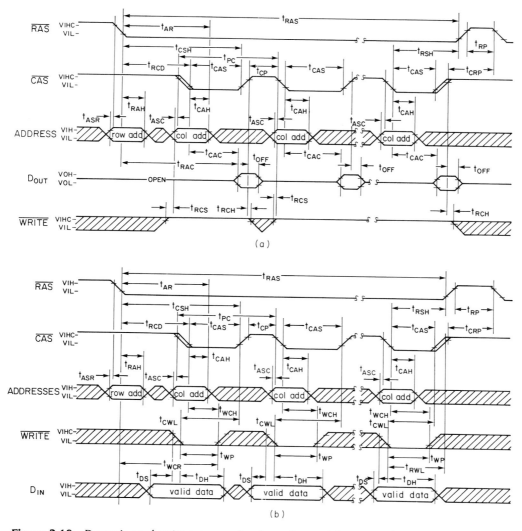

Figure 3.18 Dynamic read–write memory signal waveforms: (a) page mode read cycle, (b) page mode write cycle

that row. By applying a second column address and a second $\overline{CAS}$, in the case of read operation, another sense amplifier can be selected and its data transferred to the output buffer without having to readdress the row again. The whole of the cells in a row may be accessed in the same way and similarly for a repeated write operation. Successive accesses, after the first, can, therefore, be repeated at very much shorter intervals than is the case for normal read or write cycles. Typically a speed increase of the order of 30% is possible. Besides normal read and write cycles, read–modify–write cycles can also be used in page mode.

An improvement over page mode operation for random access requirements is achieved in some devices by accessing more than one cell at a time from a single applied address. In the form developed, four cells at successive column addresses are accessed in what has become known as 'nibble' mode. (The term 'nibble' is borrowed from computer terminology where it generally means half-a-byte, namely half of eight bits.) The resulting waveforms for a 64K DRAM (on which the nibble mode first became available) are given in Figure 3.19 which describes the read and write cycles. The first row and column address supplied determine the address of the first cell in the sequence of four. Toggling $\overline{CAS}$ causes the next three successive cells to be accessed. If a fourth active $\overline{CAS}$ is applied in the same sequence, the sequence of selected cells repeats. The nibble mode read and write minimum cycle time (t_{NC}) for 64K DRAMs is typically 55 ns.

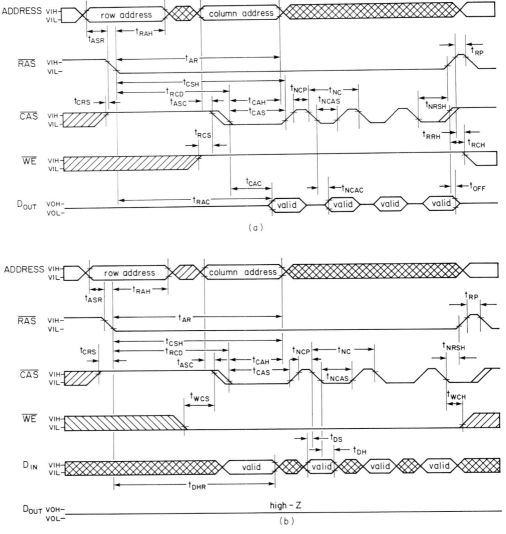

Figure 3.19 Dynamic read–write memory signal waveforms: (a) nibble mode read cycle, (b) nibble mode write cycle

Two new features were introduced on the later 256K CMOS devices. Transparent row address latches allow a much shorter row address capture window which reduces the row address hold time (t_{RAH}). Also the column address decoding is static so that no address strobe is required to select an individual cell in any row. Once a row has been selected, the column addresses may be freely changed and the output data follow them.

Power consumption

The dynamic circuitry causes most operating current to be drawn on address strobe edges. Thus, the operating power is primarily a function of operating frequency, namely the speed at which consecutive memory cycles occur. To a secondary extent the operating power also depends on the electrical loading of the data output connection. Typical current waveforms for a 16K DRAM (which employs three voltage levels) operating with different cycles are shown in Figure 3.20. It has been estimated that about 60% of the operational power is due to $\overline{RAS}$ and the remainder to $\overline{CAS}$. The reduction of operating current with reduced operating frequency is shown in Figure 3.21. At its minimum operating random read–write cycle length of 375 ns the maximum current drawn at 12 V for the mid-speed option (suffix-3) measures 35 mA but increasing the cycle length to 1 μs, it is reduced to 20 mA — a significant saving! The minimum overall system power consumption is achieved if $\overline{RAS}$ is used to chip select devices because unslected chips then revert to operation in the low-power standby mode regardless of $\overline{CAS}$.

Power distribution

The transient operating currents shown in Figure 3.20 can cause significant power rail and ground noise unless precautions are taken with the distribution of power to

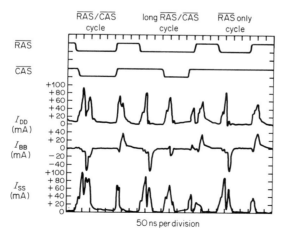

Figure 3.20 Dynamic read–write memory: typical current waveforms

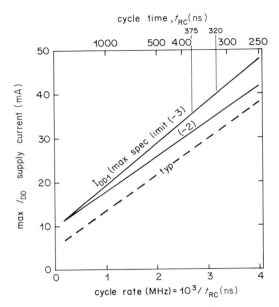

Figure 3.21 Current consumption characteristics of dynamic read–write memories

individual devices and adequate decoupling is provided. The power conductors and ground conductors should ideally be fully 'gridded' to minimise their impedance and reduce the amplitude of noise on these lines which can otherwise erode signal margins. The term 'gridding' means using power conductors interconnected orthogonally in the form of a lattice. Adequate decoupling may be provided by a 0.1 μF ceramic capacitor, connected as directly as possible between the power and ground pins of each device, to suppress high-frequency transients. Also, a larger tantalum capacitor, say 47 μF, should be placed near the edge connector of the memory board where the power lines connect to the motherboard. This provides the bulk energy storage required to prevent an unacceptable voltage drop due to the main power supply being remote from the memory board at the end of a relatively long inductive path. In the earlier triple voltage level devices it was also important for the substrate bias supply to be applied first and removed last: otherwise it was possible to cause catastrophic failure of some parts of the device by semiconductor junctions becoming effectively shorted.

Data output control

It has been common practice for the data output buffer of dynamic read–write memory devices to be controlled by signals derived from the applied $\overline{CAS}$. The block diagram of Figure 3.15 illustrates the hardware implementation of this which has also applied to the succeeding generations of device from the 16K onwards. In these cases, whenever $\overline{CAS}$ is high the data output are unconditionally high impedance—see Figure 3.16. When $\overline{CAS}$

is activated and $\overline{WR}$ is held high the data output pin becomes active after the appropriate access period and contains the data read from the selected cell. This applies equally to the read, late–write and read–modify–write cycles. In a normal write cycle the data output remains high impedance because the active $\overline{WR}$ signal which precedes the active $\overline{CAS}$ edge overrides the enabling action of $\overline{CAS}$. If the device operation is restricted to normal read and write modes it is possible to connect the separate data input and output pins together to form a common data input–output bus.

The data appearing on the data output pin during a read cycle are not normally latched within the device but remain valid until the end of the active $\overline{CAS}$ pulse. During this time there is the opportunity to latch the data using external circuitry. This method of operation, which has again been common practice, allows devices from the 16K generation onwards to have their output pins interconnected to form a common output data bus without the need for the data from unselected devices sharing the data bus to be turned off.

A more recent innovation has appeared in the 16K × 4 arrangement of 64K DRAMs offered by a number of manufacturers. An output enable, $\overline{OE}$, function provides as extra level of output control which allows a common input/output bus even in the read–modify–write mode and in this device the data input and output pins are connected internally.

Refresh

The volatile nature of dynamic read–write memory devices makes it necessary to refresh the data stored within the capacitative cells before the charge dwindles to such a degree that data are lost. Each succeeding generation of DRAM has tried to maintain some level of compatibility with its predecessor as far as the refresh requirements are concerned. Thus, for example, when 64K devices were introduced, some manufacturers offered a 128-cycle every 2 ms refresh requirement to match the 16K devices and others, seeking to halve the number of sense amplifiers within the chip, provided a 256-cycle every 4 ms refresh. No manufacturer provided both options on one chip because that would have increased the die size to include the extra sensing amplifiers and option selection mechanism.

Apart from any special provision that may be made to meet refresh requirements, any type of memory cycle which accesses a row of the memory matrix causes all the cells within that row to be refreshed. There are other ways in which refresh can be achieved. It is sufficient to operate a $\overline{RAS}$ only cycle to perform a refresh operation and because no $\overline{CAS}$ is required there is a significant power saving in this method and it is used in all generations from 4K through to the latest 256K DRAMs. Memory addresses are supplied externally. A $\overline{CAS}$ before $\overline{RAS}$ method, used by some manufacturers for the 64K and 256K devices, avoids the need to provide external memory addresses with consequent savings of power and board space. The $\overline{CAS}$ is brought low before $\overline{RAS}$ and this triggers an internal counter to source the refresh address: the address pins of the device are ignored. Refresh only is available in this mode and no data can be written or read. Subsequent $\overline{CAS}$ before $\overline{RAS}$ cycles auto-increment the refresh address counter to access all rows in turn. No device selection occurs and the data output pins of each device remain unchanged. This means that if this type of refresh is applied directly after a read cycle, for instance, the data output is maintained and the refresh action is effectively 'hidden'. On some 64K DRAMs, where one of the pins of the dual-in-line package is otherwise unallocated, this internal refresh function can be initiated by applying a signal to this pin. No separate $\overline{CAS}$ is, therefore, required.

Interfacing

As the operating frequency of memory devices increases it becomes important to reduce the propagation delay and 'ringing' of applied addresses and the other signals because of the capacitative loading which the devices present. In practice the situation is far from that of an ideal transmission line. Typically a driving buffer supplying signals to a string of memory devices along a printed-circuit board (PCB) track might introduce large overshoots by the time the signal reaches the last device. The signal waveform can be improved by either including a small series resistor between the driver and first device in an attempt to match the source impedance of the driver to the PCB track impedance or by terminating the transmission line directly in a simple RC network.

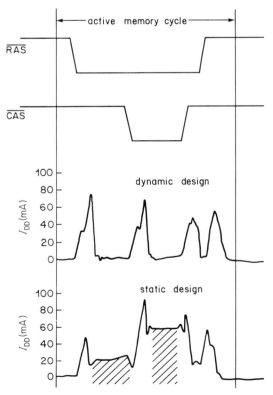

Figure 3.22 A comparison of the current consumption characteristic in static and dynamic read–write memory devices

Reliability

The results of tests carried out by Hitachi and published in their semiconductor memory data book indicate that MOS memories are very reliable devices. At elevated temperatures (up to 150 °C ambient) the failure rate was measured at less than 1 in 10^5 out of a batch of devices tested over a period of over a million component hours. Tests at high humidity (RH = 85%) indicate a similar failure rate. No failures were detected due to thermal cycling (between -55 °C and $+150$ °C), soldering heat (260 °C for 10 s), mechanical shock (1500 G for 0.5 ms), variable frequency (20 Hz to 2 KHz) or constant acceleration (20 000 G).

Comparison with static memory devices

The faster access time of static memory devices owes much to the direct addressing of the memory cell matrix. Read cycles operate in a completely static mode in that no external clocks are required to access the stored data. This is accomplished by a sense address transition circuit which initiates an internal clock, wherever a change occurs in the logical state of the address lines. The static loads associated with the static sense amplifier circuitry account for a steady current drain in these devices. A comparison of current drawn in dynamic and static devices is presented in Figure 3.22. For static devices the current waveform is more dependent on the active duty cycle and a greater overall power is dissipated.

DESIGN PHILOSOPHY

The use of semiconductor memory generally falls into one of two categories, namely, that which serves as a delay and that which offers random access of a storage block. The first requires a relatively simple means of control in which the memory address is continually incremented for a period defining the delay and then reset to the start address. For random access a means must be provided to generate the store address and supply the necessary WE polarity for either read or write cycles.

The main questions to answer when designing a large dynamic semiconductor-memory-based random access store are

    (a) What total store size is required (in terms of storage capacity and the number of bits to define each data sample)?

    (b) Which type of memory chips are available? Are there special features e.g. page mode, nibble mode?

    (c) What multiplexing arrangements are required to accommodate the fastest data transfer rate?

    (d) How many independent read or write access ports are required?

    (e) How many memory chips and their support chips can be satisfactorily housed on a single printed circuit board?

90 SEMICONDUCTOR STORAGE OF TV SIGNALS

Another question concerns the refresh requirements of dynamic memories. In television picture storage applications, the memory chips can be arranged to be accessed sufficiently frequently during normal video read cycles to service the refresh function and no special precautions are generally necessary.

Some of the important design aspects raised by the questions above are now considered in more detail before illustrating the way the design guidelines have been applied and developed as more sophisticated devices are used.

Store size

The number of data samples required to support one television picture depends on the digital video sampling frequency and the appropriate television standard. Figure 3.23 shows the number of data samples applicable to two television standards, System I (UK) and System M (USA), for a range of sampling frequencies between 12 MHz and 20 MHz. It is often sufficient to store the active picture area only and to omit those samples which occur during the field- and line-blanking intervals and both cases are presented in the figure. It is worth noting that at the sampling frequency of 13.5 MHz, shown by a broken line, which has since become an international digital television standard [9], more than 512K samples are required to hold a single picture using System I. This is slightly inconvenient in terms of the number of memory devices required to store this many samples because of the modulo-two factor inherent in memory device package sizes.

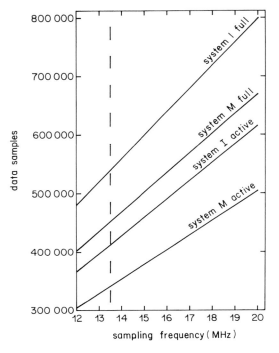

Figure 3.23 The number of data samples generated, in television pictures based on the System I and System M television standards, as a function of digital sampling frequency

Multiplex factor

The relatively slow speed of dynamic memory devices can be matched to video data rates by demultiplexing the data by a factor depending on the ratio of these quantities. There are basically two ways of doing this. Firstly, the data may be sequentially distributed like the action of a commutator as in Figure 3.24a and retrieved from the memory devices in similar fashion. The advantage of this method is that a minimum delay can be achieved, but the disadvantage is that multi-phase clocks and addresses are required. Alternatively the incoming data can be assembled into blocks and presented to the stores simultaneously as in Figure 3.24b and on reading, the retrieved blocks dispersed. The advantage

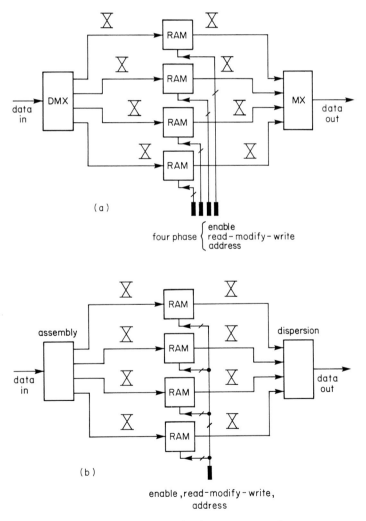

Figure 3.24 Two methods of multiplexing semiconductor memory devices

of this method is that only one clock and address phase is required. With either method the amount of delay is quantised into units of F clock periods where F is the demultiplex factor. Finer delay trimming can be obtained using a small buffer store.

Store configuration

A number of television picture stores have been constructed based on the schematic shown in Figure 3.25a for one bit of data. This arrangement splits the storage into two separate sections, labelled A and B, which are operated on an alternate write and read

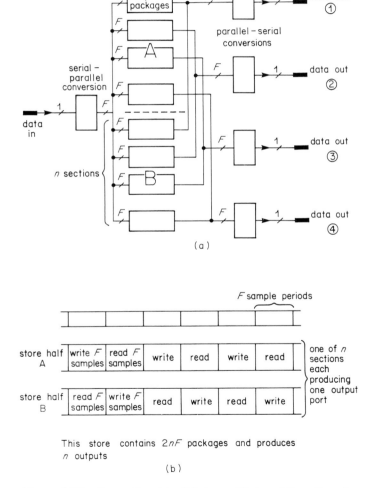

Figure 3.25 Conventional 'split' store with A and B sections for alternate reading and writing: (a) block schematic, (b) alternate read–write sequence

cycle — see Figure 3.25b — (dynamic memory devices cannot accept simultaneous write and read addresses). The input data is demultiplexed by a convenient factor denoted by F. The minimum value for F depends on the data sampling rate and the minimum memory cycle time. The maximum value for F depends on the overall size of the store which, in turn, determines the maximum package count. In general there are n subsections in each half of the store and each subsection contains F packages. In the example shown, $n = 4$ and the store can be considered as four separate stores 'in parallel' sharing a common

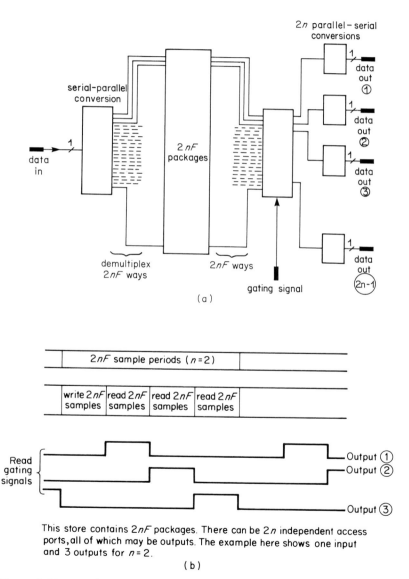

Figure 3.26 An alternative store arrangement with gated outputs: (a) block schematic, (b) read–write sequence and gating waveforms

data input port but each feeding a separate data output port. In this case, four separate output ports can be provided, one from each pair of subsections taken from A and B as shown. For standards conversion, for example, four separate outputs containing data on four successive television lines can be provided by such a store arrangement. Moreover, the store control is relatively straightforward requiring an input F-way serial–parallel converter, an F-way parallel–serial converter and simple address generators driving each section independently. The maximum number of access ports for this arrangement is $2n$ but because of the alternate write–read operation, there can be only n independent ports used for data output. In this configuration the store cycle length is defined to be $2F$ clock periods and so the store delay is quantised into units of the same amount.

An alternative store arrangement is shown in Figure 3.26a. The demultiplexing factor is now much greater than the previous case and can take a value between F and $2nF$; the figure is drawn for the maximum value of $2nF$. The store cycle is now $2nF$ sample periods long. The timing diagram (Figure 3.26b) is given for a store with $n = 2$. Because F sample periods are required for each device read or write cycle it is possible to independently access the store four times within the store cycle. The first is used, in this example, to write a block of $4F$ samples, one into each package. In the second to fourth cycles $4F$ samples of data are read from a different address each cycle. The output data is off-loaded onto a common data bus and gating signals must be applied to separate the data destined for separate output ports. The control is, therefore, more complex than in the previous store arrangement, requiring larger serial–parallel and parallel–serial converters and also the generation of output gating signals to separate the output data. The advantage of this arrangement, however, is that three outputs can be provided or, in general, a maximum of $2n$ (with no writing cycle), which is double that of the previous arrangement.

EARLY DESIGNS

The 'Wardrobe stores'

The earliest random access store of any size was that constructed from 1 kbit shift registers in the mid 1970s before the simplest read–write memory devices became cost effective [63]. The store was required to have a word capacity at least equivalent to the active area of one television picture, when sampled at the then common sampling standard of three times the colour subcarrier frequency which is for System I, 576 lines of 692 samples and each sample was to be 8 bits. The maximum data transfer rate had to exceed 13.3 Mword per second. To meet the requirements of a large variety of applications, it was necessary to incorporate a wide range of facilities in the final design. The operation and control of the store also had to be largely self-contained in order to avoid the need for different user groups to design and construct complex store-control equipment. This feature made the use of the store by such groups relatively economical, as each group had only to concentrate on the problems peculiar to its own work and not on the internal design of the store.

The store was constructed in two autonomous halves so that two different groups could each use one half, that is one field (288 lines), of storage. The inputs and outputs were asynchronous, and any line of data stored was randomly accessible. Two independent 'read' outputs were available from each half of the store making a total of four outputs when the two halves were used together. This arrangement was convenient for work such as standards conversion where interpolation between several lines is necessary.

In addition to these facilities, four requirements were imposed on the design. These were that the stores should be transportable, they should incorporate a comprehensive self-checking arrangement, they should operate at lower sampling frequencies than the maximum and that they should perform statically although employing 'dynamic devices'.

The stores as constructed are shown in Figure 3.27, displaying their single picture. Each field store consumes about 1 kW!

Figure 3.27 The 1975 field store bays

Experiments with early read–write memories

A number of experiments were made following the construction of the stores just described to study the use of the emerging read–write memory devices. The major advantage to be gained in constructing a memory store from such devices over shift-registers is the flexibility of write-in and read-out addressing. Read–write memories of 1 kbit capacity were available constructed from both bipolar (TTL, ECL) and static MOS technology. The bipolar products were fast (~ 100 ns cycle time) but had very high dissipation (~ 1 mW per bit) and were expensive. MOS devices, on the other hand, were relatively cheap, had a lower power dissipation (~ 0.2 mW per bit) but suffered from a slow cycle time (~ 300 ns). At digital video sampling rates of three times the System I colour subcarrier frequency a convenient demultiplex for these latter devices was found to be sixteen giving a store cycle time of 1.2 μs and allowing two video accesses. If the devices which stored just 1 bit per address were chosen a minimum number of 128 (8×16) devices were necessary for an 8-bit video signal giving a total minimum capacity of 16 kbytes or about twenty television lines. To reduce the package count for a small experimental store a device with parallel storage of several bits at each address was preferred and one device with an organisation of 256×4 bit was produced by several manufacturers. 32 of these gave the necessary demultiplex of 16 and provided storage of more than four television lines. AMD produced the fastest version of this device and its timing requirements were less damanding than some of the others. A small store was, therefore, constructed using 32 devices to test the feasibility of the techniques proposed and the results were encouraging.

When the $4K \times 1$ devices first became available they were dynamic and required thought to be given to arranging for their periodic refresh. They were, however, lower in power dissipation and, for the same store cycle time proposed earlier, the minimum size of store possible with these devices was about 77 television lines. A printed circuit board design was eventually produced, containing 32 devices and this was capable of holding a quarter of a television field, 4 bits wide on a single 4U BMM card. (This is a binary-metric-modular racking system (BMM) based on a '19 inch' wide rack and rack heights in units of 1U which is approximately 45 mm high. A 4U PCB measures approximately 156 mm × 256 mm.) It was supplied with three different d.c. voltages ($+5$ V, $+12$ V and -5 V) and incorporated special circuitry to ensure that the substrate voltage (-5 V) was supplied before the other and removed after them. A television picture store was constructed using sixteen of these boards and controlled by a special logic personality card which was intended to provide flexible store control for different potential users. Several units were built and used for early work on standards conversion [64] for which the four independent outputs were essential.

A digital field delay using 64K CCDs

During 1978 a charge-coupled device, capable of storing 64 kbit became available from Fairchild—the CCD 464. By considering the active field period only the storage can be accommodated in 32 64K devices, which gives a convenient organisation, for an 8-bit signal, of four 64K stores for each bitstream. These can be assembled on a single 4U board

EARLY DESIGNS   97

and such an economy of space was thought to be well worth achieving. In fact this work by A. Oliphant was somewhat speculative as the CCD devices were not being manufactured in volume at the time and those used were pre-production samples. However, in other respects they appeared to be ideally suited for applications such as the noise reducer described in Chapter 5 since their shift register organisation, with low power consumption, was exactly what was needed.

The maximum clocking frequency of the device was quoted as 4 MHz. This implied a four-way demultiplex at the chosen sampling frequency, giving 32 parallel bit streams, one device per stream. As a result, serial-to-parallel and parallel-to-serial converters operating on the input and output data, respectively, were required. The charge-coupled device itself, is organised as a serial–parallel–serial structure with 16 parallel registers, each of capacity 4 bits. Data are stored by filling each register successively, using a 4-bit address to determine which register is used at any instant. Data are shifted in all 16 registers simultaneously on receipt of clocks and reading takes place at the same location as writing. The complex clocking arrangements meant that the 32 storage devices required an overhead of 27 ancillary devices. Nevertheless, it proved feasible to mount all

Figure 3.28 General view of the compact field store board based on 64K CCDs

the components on a single 4U board with a power consumption of about 14 W. A general view of the board is shown in Figure 3.28.

The clock waveforms and addresses for both field store boards are generated on a separate housekeeping board. Trimming of the data delay is accomplished by omitting clock pulses during the line blanking interval and, for NTSC operation, which requires significantly less storage than PAL, omitting CCD addresses. These operations are again performed on the housekeeping board.

In operation, the store was found to be fairly reliable but several CCDs failed in the course of two or three years and had to be replaced. Eventually, it became impossible to obtain further devices and failed devices were allocated to least significant bits of the store where their isolated errors would cause least damage.

THE 16K GENERATION

An experimental picture store using 18 MHz sampling frequency

Introduction

This section describes the design principles and construction of an experimental digital television picture store which was developed in early 1979 for work on television synchronisers and PAL decoding experiments at sampling frequencies up to 18 MHz.

Before the establishment of the 13.5 MHz sampling frequency of Recommendation 601 [9] there were several proposals based on frequencies locked to parameters of the composite signal as discussed in Chapter 4. At this time one contender for a standard video sampling frequency was around 18 MHz and specifically four times the frequency of the colour subcarrier ($4f_{sc}$) for a System I television standard. The 1975 digital picture store based on shift registers and described earlier was adequate for many applications but suffered from a long access time (more than four television lines). A more recent television field store based on 4-kbit dynamic read–write memory devices and occupying about one quarter of the volume of the 1975 store was not capable of operating at 18 MHz. The introduction of 16-kbit dynamic memory devices made it worth considering the development of a new television picture store for experimental purposes. The solution proposed was to use the existing 4K-based store board design with a modified personality control board and replace the 4K devices directly with 16K devices. The mechanical arrangement of the 4K-based store, however, contained many hand-wired looms which made the construction time-consuming. A speedier solution was achieved by redesigning the control circuit to operate at the faster rate and building these circuits on conventional 4U BMM boards with a printed circuit motherboard. The resulting picture store consists of a total of 12 PCBs which together with power supplies fits into the same case as the 4K-based field stores (0.5 m × 0.5 m × 0.44 m) with room to spare — and the new picture store is illustrated in Figure 3.29. The total cost at 1979 prices was about £3000 and the total power consumption is 225 W. Television data are stored on a line-by-line basis where each line is identified by a line label (10 bit) and the beginning of a line by a line trigger and these together with a data output clock are user supplied. An asynchronous buffer allows data to be written into the store at a rate which is different from that at which it is read out. The buffer is made of sufficient capacity (48 samples in this case) to

Figure 3.29 An experimental $4f_{sc}$ picture store based on 16K dynamic memory devices (1979)

allow the delay of data through the entire store to be continuously variable provided the input and output clock frequencies are within ±0.1%.

The picture store is built as two separate field stores sharing a common control system. Each field store—see Figure 3.30—is divided into two halves, A and B, so that bunches of F samples of data can be written into A while F samples are read from B and vice versa. Data are output from A and B via tristate buffers which are enabled alternately every F sample periods to form one continuous stream of serial output data. In the main part of the store the writing and reading operations are synchronous with the clock rate of the output data and the delay is variable in steps of 16 output clock periods.

The initial asynchronous buffer is required to serve two functions. The overall store delay is quantised in steps of $2F$ samples—the length of the store cycle. A buffer, of minimum capacity $2F$ samples, can, therefore, make the overall delay variable in discrete single sample steps. Further, for the buffer to be made asynchronous (i.e. the clock rates associated with data into and out of the buffer are allowed to be different), the overall

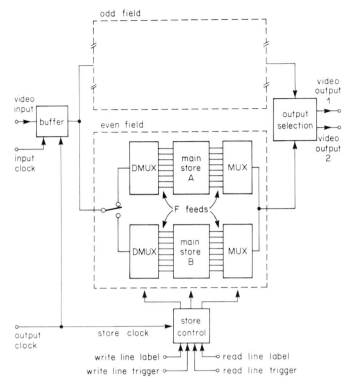

Figure 3.30 A basic schematic of the experimental $4f_{sc}$ picture store

picture store delay can be made continuously variable up to a maximum of one television picture period. For asynchronous operation data are applied at an input data rate and removed at another clock rate which is user supplied and subsequently used to synchronise the main store operation.

Store board organisation

A block diagram of the storage arrangement, for one of the four bits, available on the store board designed originally for 4-kbit devices, but modified for 16K devices (type 4116–3) and for operation at a higher frequency is shown in Figure 3.31. The board holds 32 devices each arranged as four parallel 1-bit data stores with a demultiplex factor of eight. Two of these boards are required to make up the 8-bit total required in each field block A or B. Taking the worst-case memory cycle time of 375 ns and the period between data samples as 55 ns the minimum de-multiplexing factor possible, F, is 375/55 which is ~ 7. The maximum value of F which can be used depends on the total number of samples which each half of a field store, A or B, is required to hold. At a $4f_{sc}$ sampling rate the total number of samples generated each half-field (see Figure 3.23) is 180 000. These can only be accommodated in 11 devices per bit: no multiplexing beyond 11 is, therefore, possible without under-using the available capacity of the memory devices. A summary

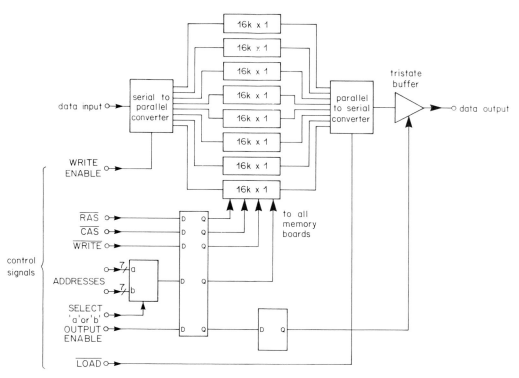

Figure 3.31 Block diagram of one-quarter of a store board which handles one bit of storage in one-half of a field

of the number of samples stored for values of F between 8 and 11 is given in Table 3.2 where the cases (a) to (c) are defined as

   (a) all samples generated at $4f_{sc}$
   (b) all samples generated at $4f_{sc}$ excluding field blanking
   (c) all samples generated at $4f_{sc}$ excluding field and line blanking.

With a demultiplex factor of eight the total store capacity is equivalent to about 99 % of a $4f_{sc}$ sampled picture excluding both field and line blanking.

Table 3.2

Demultiplex factor F	Total number of devices for picture store	Number of samples stored in field block A or B	% samples stored		
			(a)	(b)	(c)
8	256	131 000	73	79	99
9	288	147 000	82	89	111
10	320	163 000	91	99	123
11	352	180 000	100	—	—

Store control

The main waveforms associated with the memory control signals required for both A and B sections of the field stores over a complete store cycle are shown in Figure 3.32: all are generated synchronously from a counter whose successive clock periods are numbered 0 to 15. Consider the operation of a read cycle in the A section of the store. A read address, A_m, is set up during counts 1 to 8; the row part of the address occurring in the first half of this period (1 to 4) and the column part of the address in the second half (5 to 8). (The same address, A_m, is used to read from the other half of the store, B, during counts 9 to 0.) The negative-going edges of the $\overline{RAS}$ and $\overline{CAS}$ waveforms latch the row and column parts of the address respectively into the memory devices. Because the $\overline{WRITE}$ A waveform is in the 'high' state after this time a read cycle is initiated and the data associated with the addressed memory locations are available after the appropriate access time which is approximately during counts 9 to 10 as shown. The $\overline{LOAD}$ pulse at count 9 loads eight data bits at a time into parallel-to-serial shift registers and the data exits serially from the store during counts 10 to 1. An identical series of operations occurs for the B section of the store 8 clock periods later.

Next consider the operation of a write cycle in the A section of the store. Serial data is input to the store boards during count 4 to 11 and fed into serial-to-parallel converters so that during the period 12 to 3 the parallel data is held steady for writing into the memory

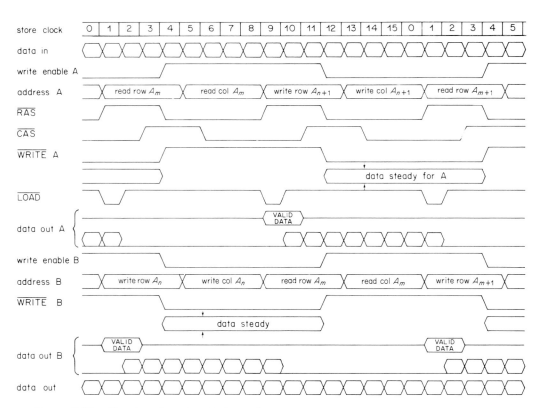

Figure 3.32 The store board control waveforms

devices. A write address A_{n+1} is set up during count 9 to 0; the row part on counts 9 to 12 and the column part on counts 13 to 0. (The same address is used to write into the other half of the store B during counts 1 to 8.) The address is latched into the memories in the same manner as before but because the $\overline{\text{WRITE}}$ A signal is now in a 'low' state the available data is written into the memory devices.

In general the main store clock frequency gives rise to a non-integer number of store cycles over a television line period and arrangements are made for the last memory cycle in a line to be completed whilst at the same time resynchronising the start of a new memory cycle to the next line trigger.

Microcomputer-controlled stores

Introduction

By the end of the 1970s a new factor had appeared in the application of television picture stores. This was the increasing availability of microcomputers which lifted the use of picture storage into a new dimension. Now it was possible to interactively communicate with the stored video data, albeit at a slow rate (~ 1 Mbit/s), and process it in some way. The development of an electronic graphics generator produced a new picture store design which combined on a single PCB, the semiconductor storage, video multiplexing arrangements and an independent slow-rate data access port to connect it to a microprocessor, for a single data bit. (This design was also later adopted for Enhanced Teletext equipment [65] which required 24 bits for each picture sample and allocated as 8 bits each for the red (R), green (G), and blue (B) components.) A video data sampling rate of 13.5 MHz was chosen for the System I television standard because by this time the discussion on the proposed international standard had demonstrated the attractions of this proposal. Thus the 'active' picture area contains 720×576 pixel locations making for a minimum picture store capacity of just over 1.2 Mbytes (one byte is equivalent to 8 bits). The new picture store design contained 32, $16K \times 1$ read–write memory devices, which is sufficient to hold 512K samples, and which is 25% more capacity than is required for one video component. Three such stores with 8 bit-planes each are, therefore, required to hold the R, G, B samples and the extra capacity is used as auxiliary storage to the computer port only.

Each group of 8 bit-plane boards is interfaced to the microcomputer bus through control circuitry arranged as shown in Figure 3.33. The block which connects directly to the computer bus (the X–Y controller) allows the picture store to be addressed asynchronously at the operating rate of the microcomputer by supplying two 10-bit numbers which identify a particular horizontal offset (X) and vertical offset (Y) and define X, Y coordinates of a displayed pixel. Slow-rate video data can be written to or read from the store via the computer data port independently of the full-rate video port; access to data and video ports is arranged to alternate approximately every microsecond. The store can also be used simply as an extension to the computer memory, in which case a single 19-bit address is used and data can be retrieved within about 2 μs. The timing of control signals and address generation is handled on the store control board which, in common with the other two picture stores, receives 13.5 MHz clock pulses and video synchronisation pulses from a common clock-pulse generator board.

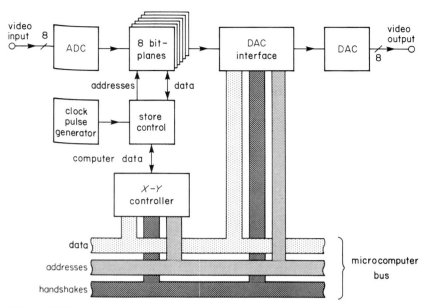

Figure 3.33 The picture store arrangement for one component of the *RGB* store (1981). The clock pulse generator is shared between the three components

In the design there is only one full-rate video port which is shared for video input and output. Video output generally takes precedence but single frames of video can be grabbed into the store by momentarily interrupting the video output for two television field periods. Each video input port requires its own analogue-to-digital converter (ADC). The video outputs are taken to their respective digital-to-analogue converters (DACs) through additional boards which also interface with the microcomputer bus. These boards contain an array of small read–write memories which can be programmed to allow simple video post-processing such as picture blanking or gamma correction. The store is contained in an 8U BMM rack together with the associated ADCs and DACs and is shown in Figure 3.34. The three X–Y controller functions are housed in a separate microcomputer rack. The complete R, G, B store dissipates about 250 W.

Store board organisation

At the time of design the type 4116–3, 16K × 1 dynamic read–write memory devices were the most cost effective mass memory building block although the next generation 64K devices were beginning to appear. With the lower video sampling frequency of 13.5 MHz the minimum F is reduced to about five and for a total of 32 devices the maximum F could be made 32. A value of 16 was chosen as a compromise for minimising circuit complexity. This also meant that the picture storage could be divided into two independently accessible field store sections as shown in the block diagram of Figure 3.35 and addressed by separate $\overline{CAS}$ control signals. When a computer access takes place one section of 16 devices is activated and one of these is singled out by the lower significant four bits of the computer address.

Figure 3.34 A general view of the *RGB* picture store together with the analogue-to-digital and digital-to-analogue converters

Store control

Each store cycle of 16 video sampling clock periods is divided into two independent store accesses which are allocated alternately to video and computer use. The video access is normally used for monitoring the contents of the picture store at video rate and at the same time fulfilling a refresh function. The video addresses are generated in counters clocked at the video data sampling rate and locked to supplied mixed synchronising pulses. Addresses are supplied for the active field period only so that the store is blanked in the vertical interval. When required, the video access can be used for writing data to the store at video rate and in this case the reading of video data is interrupted for one picture period.

During a computer data access, the address supplied by the computer command must be synchronised to the store cycle. To accomplish this, the data to be written from the computer memory into the picture store are effectively delayed until they can be handled in the next available computer access slot. Data to be read from the picture store are held ready on the store output latches until the computer issues a read instruction to take them.

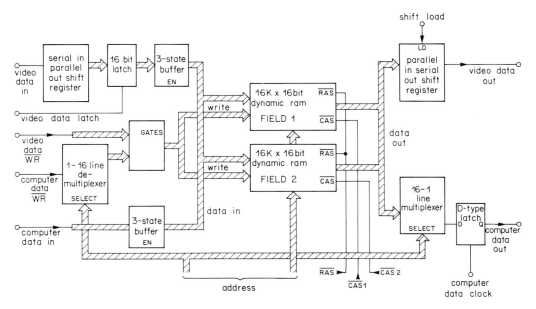

Figure 3.35 Block diagram of the 1-bit, single component, picture store board

Animation store buffers

Introduction

In 1979 the digital television Animation Store system was conceived [66, 67] to provide a mass store for still pictures and the sophisticated range of picture processing facilities described in Chapter 7. The design was based on an Ampex magnetic disc drive as the main storage element with sophisticated head control and multiplexing to achieve video data rates and real-time sequence replay. The total capacity achieved was equivalent to 815 pictures and access was controlled by means of a local dedicated computer. Besides the storage and retrieval of stills, several other features were implemented. These included the ability to grab pictures and sequences into the store at video rate and an interface to a video rostrum camera system whereby stills and sequences could be generated manually and animated sequences assembled. Chroma-key facilities were provided to allow the electronic mixing of pictures to create others.

At an early stage in the development it was decided that the digital video signal would be stored in its component form (rather than as a digitised composite signal) in an attempt to produce a better overall picture quality. Had the signal been stored in its composite form, signal processing would have been required to regenerate the PAL eight-field sequence from one or two stored fields. Moreover, where pictures are derived from a local television camera or other sources available in component form, the processes of PAL decoding and encoding could be avoided.

To support the PAL decoding function and to provide temporary buffer storage and display storage facilities, a number of semiconductor memory stores were required. The proposed EBU video component coding standard, at the time that the choice for this system had to be made, was based on a sampling frequency of 12 MHz for the luminance (Y) and 4 MHz for each of the chrominance vectors (C_B and C_R). In the absence of an international standard and, for reasons discussed in Chapter 7 in connection with the maximum available data rate, the 12:4:4 parameters were selected for this application. The buffer stores are required to operate asynchronously so that data may be moved into or out of them without relation to the phase of the signals maintaining the store cycle. The storage is addressed by line labels in the same manner as that described previously and one control function of the computer is to keep track of the line numbers allocated during writing and reading operations.

A block diagram of the store arrangement is shown in Figure 3.36: the video data paths are near the top and the control functions below. The digital video components are multiplexed to be alternately Y and $C_B C_R$ and in this case are allocated 8 bits per sample. The main storage boards are arranged to hold 1-bit each of one picture of $YC_B C_R$—a total of 655 kbit. Separate video input and output buffers provide the opportunity to synchronise the writing and reading sampling phase to the store cycle and adjust the data delay in single sampling period steps to overcome the finite minimum block size demanded by the store multiplex. Independent Y and $C_B C_R$ output processing blocks allow simple operations to be performed on the video data prior to being read out: examples are picture inversion, linear movement in the horizontal or vertical directions or a mixture of each and also the addition of video blanking. The video data are transferred to and from the store using ECL logic levels because of the greater ruggedness they give to signals over the long video data paths involved between racks of equipment. The function of the computer in this design is purely to control the store: there is no transfer of computer data possible into the main storage blocks.

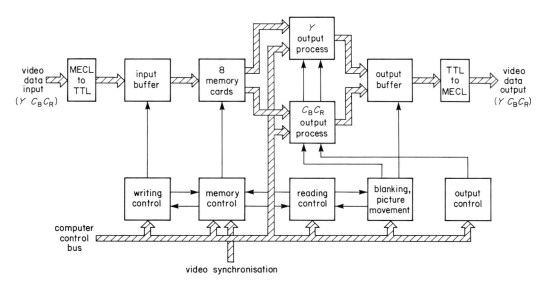

Figure 3.36 Block diagram of a $YC_B C_R$ picture store with post-storage processing facilities

Store board organisation

The semiconductor bit-plane board is based on 16K × 1 DRAMs with a 375 ns random read–write cycle time. At a 12 MHz sampling frequency there are less than 370 000 data samples generated in a picture period: this requires a minimum of 22 devices to support them. To give some margin and allow the storage to be divided four ways in a similar manner to the $4f_{sc}$ store described previously a total of 24 devices are allocated to the main Y store. This means that there are six devices per block and a video demultiplex factor of six is acceptable to match the devices to video rate. The chrominance, C_B and C_R,

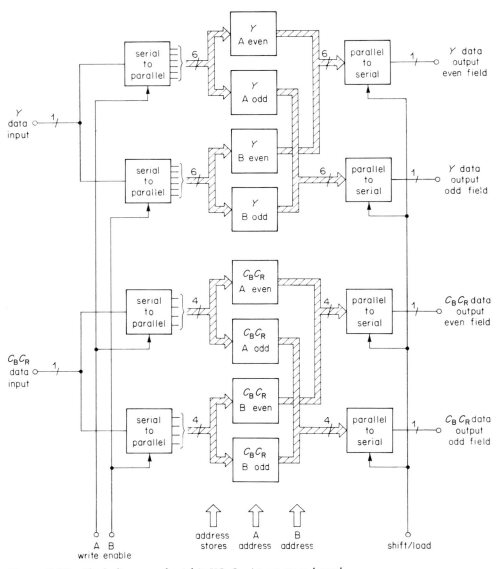

Figure 3.37 Block diagram of a 1-bit $YC_B C_R$ picture store board

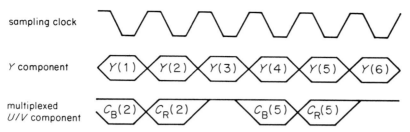

Figure 3.38 The Y and multiplexed $C_B C_R$ digital sampling arrangements for the $Y C_B C_R$ picture store

components can be supported in eight devices each and are arranged time-multiplexed in a block of 16 with a similar store multiplex of six. A block diagram of the store board is given in Figure 3.37 and shows the video multiplexing and demultiplexing functions resident on-board. The time-multiplexed phasing of the luminance and chrominance components is shown in Figure 3.38.

Store control

Apart from the main store board control signals, such as $\overline{RAS}$, $\overline{CAS}$, write enable and memory addresses, which are similar to stores already described, this store design is different in its post-processing facilities and only these will be referred to here. Their control is handled in the 'blanking and picture movement block' shown in Figure 3.36 and allows pictures to be repositioned on replay.

The computer supplies X and Y offsets to define the amount of movement required horizontally and vertically. To accommodate a total shift range of one picture width and one picture height the normal propagation delay of the video data through the Y and $C_B C_R$ output processing blocks is made equal to one line and one picture period. The horizontal shift value is added to or subtracted from a value which is loaded into a horizontal counter incremented at the video sampling rate so as to produce a suitably delayed 'start of line' pulse and appropriately phased start and end triggers to generate video blanking. The vertical shift value is used to appropriately delay the time at which the first stored line is read out and when reading ends. From both the horizontal and vertical offset information a mixed video blanking signal can be generated and also a 'background' control which is responsible for filling in the part of the display area from which the picture has moved with either blanking level or some other colour specified by the computer (possibly intended for chroma-key application). In addition to this 'dynamic' picture movement there is a further 'static' offset facility which allows retiming of pictures within the animation store as the internal configuration changes.

THE 64K GENERATION

A general purpose digital storage card

Introduction

During 1981 a versatile digital storage system was developed by N. E. Tanton based on 64K × 1 bit dynamic memory devices. Previous general purpose store designs had tended to be over-complex in an attempt to accommodate all possible users. The new system design philosophy aimed to limit circuit complexity by making it dependent upon the user's system requirement. Figure 3.39 shows a typical organisation of the component parts: it includes one computer port, four video ports (which may be input or output), a refresh port to supplement the default refresh provision (if necessary) and a priority arbiter to oversee the system management.

As a compromise between complexity and versatility it was proposed that the new store should consist of a common set of identical storage cards together with a set of user-specified access ports. Communication between the modules would be via parallel address, data and control buses, all under the control of the priority arbiter. The external system behaviour is determined by the access ports and priority scheme, which is specified by the user to suit his application. For instance, a simple picture delay with identical input and output data rates requires one write and one read access port with an elementary priority scheme. A more complicated application with, say, three read operations at one data rate, a write operation at a second (higher) data rate and a microprocessor random access facility would need three read access ports, a microprocessor access port and a write access port with additional buffering, together with a more complicated priority scheme. The system complexity is directly related to the application, with no unnecessary overheads.

Store board organisation

The store board is built on an 8U BMM card and contains 64 64K × 1 devices, as shown in the block diagram of Figure 3.40. Data are presented to the data bus as a 64 parallel word (1 bit per device) allowing the memory to be configured to suit the user's requirements. The data capacity of a single card is 524 288 bytes which for System 1, when sampled at 13.5 Mbyte/s, corresponds to just over a complete picture or alternatively, $2\frac{1}{2}$ fields if only the active line is stored. Each storage location is identified by a 16-bit address A_0–A_{15}. The remaining four address bits A_{16}–A_{19} serve to select one of sixteen cards sharing the same bus. The store card has an automatic refresh facility

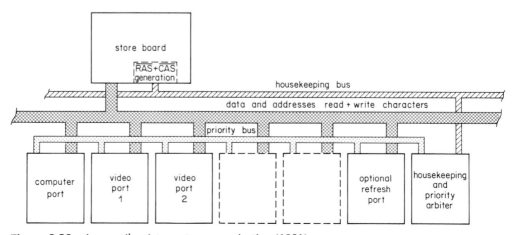

Figure 3.39 A versatile picture store organisation (1981)

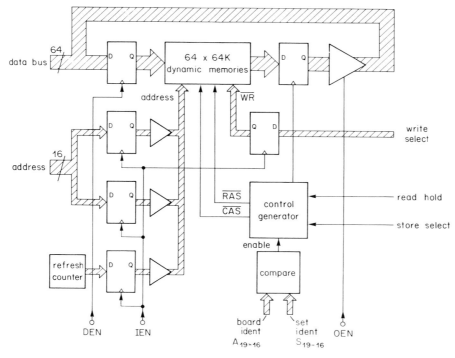

Figure 3.40 Block diagram of a 64K-based memory store board

which operates whenever an idle store cycle occurs. When a refresh cycle is initiated an internal counter is incremented so that each of the 256 pages of dynamic memory are eventually refreshed.

The memory devices can be programmed to operate in a normal read or write cycle mode or alternatively in a page mode to shorten the cycle. The maximum data rate of the store can therefore be increased if it is possible to pre-configure the addresses to allow page mode access. The storage card is intended to operate asynchronously at a clock rate which is either chosen to maximise data throughput, in which case an appropriate mix of normal and page-mode access cycles are arrived at, or is chosen to minimise interface complexity by synchronising the store to the external system clock, for example. A convenient frequency for the first case is 22.22 MHz allowing a page mode access in four clock periods and a normal access in six.

Timings of data, address and control signals on their respective buses are relatively flexible as shown in Figure 3.41. The passage of information along the various buses is determined by three signals — input control enable (IEN), data enable (DEN) and output enable (OEN). Address, write select and store address signals are accepted by the store on the store clock pulse following the signal IEN, preparing the store fully for its next access cycle. Data are accepted into the store following the signal DEN, and are stored in the address specified under IEN, provided that a write access was specified and the store selected. In the event of a read access, output data become available some time after IEN, t_{ia}, and generally after the IEN of the following cycle. Output data are allowed onto the data bus by the signal OEN, which should be timed with respect to their *own* IEN.

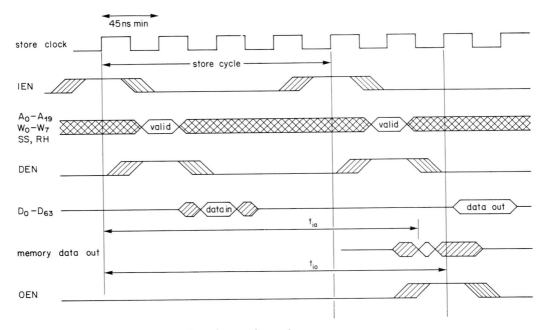

Figure 3.41 Memory store board control signal requirements

Film dirt concealment application

This application requires a storage system capable of providing four simultaneous digital video outputs while accepting one digital video input, all at 13.5 Mbyte/s [68]. The four outputs correspond to points in four consecutive fields scanning in unison. Outputs 2 and 3 are the even and odd fields of a picture and fields 1 and 4 are the nearest odd and even fields respectively from the two adjacent pictures. Outputs 2 and 3 move to the next picture once every two field periods. Output two is required only during the first field of the output frame.

The first decision is to establish what data demultiplex factor is needed to allow five consecutive store accesses per store cycle. An 8-way demultiplex has a store cycle length of 592 ns, which is sufficient only to allow 2 normal or 3 page mode accesses at the full store speed of 22.22 MHz. A 16-way demultiplex has a store cycle length of 1184 ns which is long enough for $4N$, $2N + 3P$, or $6P$ accesses at full speed where N = Normal and P = Page. Also with a synchronous store clock of 13.5 MHz it is possible to accommodate $4N$ or $1N + 4P$ accesses. Of these possibilities, $2N + 3P$ and $1N + 4P$ look most promising. The $1N + 4P$ scheme is attractive because it uses a 13.5 MHz clock and can be made synchronously. Unfortunately, it is difficult to implement because the requirement for an adjustable propagation delay implies a non-static address sequence and it is not always possible to guarantee that four out of five accesses can be made with the same memory row address. The $2N + 3P$ option, with its attendant problems of asynchronous operation, is, therefore, chosen.

This system has been tested using a computer model to investigate the maximum buffer sizes and worst case access sequences imposed by the rest of the application.

Parameters of the model are store access time, write block length and read length (sixteen times the write and read clock periods respectively) and whether the addresses are singly or doubly buffered. Two additional parameters called 'write address down time' and 'read address down time' are included to model the time taken for addresses to change. Assuming that the four read ports are singly buffered and operating at 13.5 MHz each, the results of the simulation were able to indicate what level of buffering is required for the write port, for a given input data rate, and to indicate whether additional refresh provision was required.

The application can be satisfied with five, albeit more complex, ports and two store cards giving a store efficiency of 98.7% and refresh requirements are handled entirely by the default mechanism resident on the store boards.

A 2 Mbyte storage module

For a random access application the basic architecture of Figure 3.39 was modified so that the total video input and output data rate capability could be increased, particularly with HDTV applications in mind. Four of the general purpose store cards were interconnected via a common address and control bus, as shown in Figure 3.42, but the four video access ports are provided on video interface boards with independent local data bus connections to each store card. Other control functions which are handled on separate cards include the generation of video addresses, control interface and computer

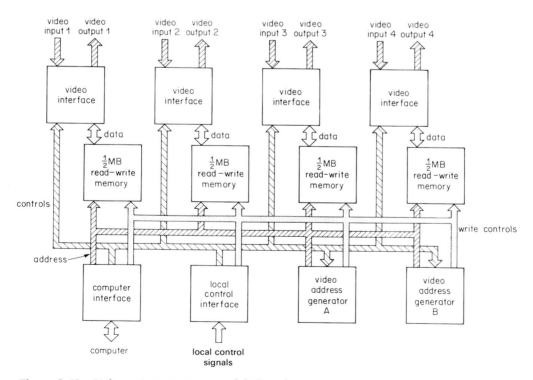

Figure 3.42 2Mbyte picture storage module based on 8U store boards

interface. The total storage amounts to 2 Mbyte with a potential maximum video data rate of more than 54 MHz.

The storage boards are operated on a synchronous (13.5 MHz) eight-way video demultiplex cycle whose timing diagram, presented in Figure 3.43, is based on the general waveforms of Figure 3.41. The eight clock-period store cycle is split into two independent access slots, referred to as A and B. The A slot is used exclusively for video reading which provides a monitor display of the store contents and fulfils refresh requirements (independently of the default refresh function incorporated within the store boards). The B slot is used either for a second video access or a computer access allowing both writing or reading. In the case of video writing the video input is stripped of its line codes and the synchronising information is used to lock the scanning control for slot B accesses. Video addresses, which are independently generated for the A and B access slots, can be modified, partly by firmware and partly by software, to alter the scanning ranges of the addressable X, Y, Z coordinates which specify the horizontal (picture width) dimension, vertical (picture height) dimension and the depth (frame number) dimension respectively.

The computer interface (D11W) is based on a 16-bit data bus together with control signals for RECEIVE, SEND and DATA FUNCTION CONTROL. There are eight types of data function provided and these include computer data input or output, loading an X or Y coordinate address, post-auto incrementing those addresses by one, two or four units and store module number. The remote coupling of the store to the computer, in this way, means that it is difficult to use the storage as 'working memory' because of the penalty in access time. Thus, in normal operation, a data area from the store is read into the

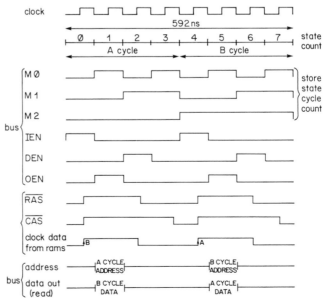

Figure 3.43 The 2Mbyte picture store control signal waveforms

computer for processes to be performed on it before transferring it back again for direct viewing in real time.

High-definition television application

The potentially high data rate of the 2 Mbyte storage module makes it suitable for an HDTV display store. Three modules have been used to hold three components of a video signal, R, G, and B as shown in Figure 3.44. The four video output signals from each 2 Mbyte module are fed into a multiplexing unit which combine them into a 54 MHz serial data stream before being passed on to high-frequency DACs.

The 54 MHz clock is generated locally and drives a multiplex unit associated with each module. Also, counters generate the video synchronising information to control each module and provide a mixed synchronisation signal to accompany the video output. Independent computer access, at a slow rate, is available via the DR11W computer interface so that with suitable demultiplexing units, an HDTV real-time signal input could be accommodated.

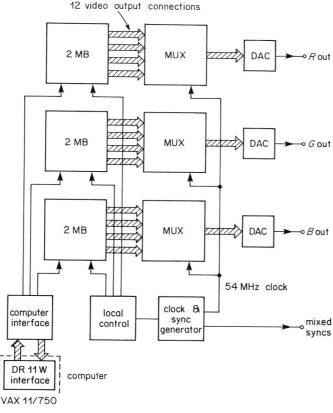

Figure 3.44 Arrangement of three 2 Mbyte modules for an HDTV display application

Sequence store application

To permit the subjective evaluation of digitally encoded short video sequences, three of the 2 Mbyte modules have been arranged with sequencer units shown in Figure 3.45. The sequencer units allow the video signals sampled at 13.5 MHz to be directed to and from the individual store boards according to the sequence control which supplies the necessary video synchronisation and board decoding signals.

The control of the sequence store is sufficiently flexible to allow the following configurations:

(a) Video data coding can be *RGB*, $C_B C_R$, *Y* or PAL, etc, at 8 bits for each component.

(b) The architecture can be effectively changed to accommodate four *RGB* stores, one HDTV store, six $C_B C_R$ stores or twelve *Y* only stores.

(c) Partial picture sizes can be used to increase the sequence length at the expense of picture area. Fixed fractions of one-half and one-quarter are provided.

This allows a number of optional sequence arrangements which include, for a 625-line System I standard:

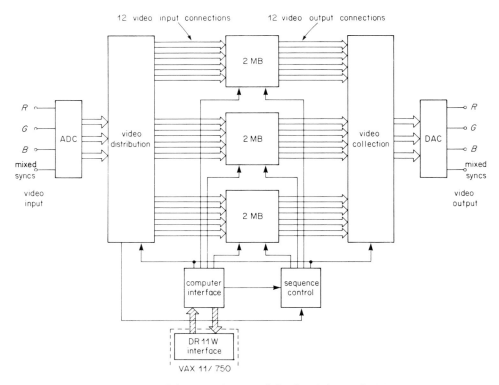

Figure 3.45 Arrangement of three 2 Mbyte modules for a short television sequence store

(a) 12 Y only pictures (about half a second)
(b) 24 Y pictures, half area (about 1 second)
(c) 48 Y pictures, quarter-area (about 2 seconds)
(d) 24 $YC_B C_R$ pictures, quarter-area (about 1 second).

The sequences can either be repeated cyclically, which will cause a 'hiccup' to occur in between sequence transitions, or palindromically so that a continuous display can be formed. Although it is bulky in terms of hardware and suffers from the limitations of the DR11W interface this equipment has provided a beginning for the image processing of short sequences. A general view of the equipment comprising the short sequence is given in Figure 3.46.

A compact random access picture store

Introduction

A television semiconductor picture store of 1.5 Mbyte capacity and capable of holding an R, G, B picture when sampled at 13.5 MHz has been described as part of equipment built

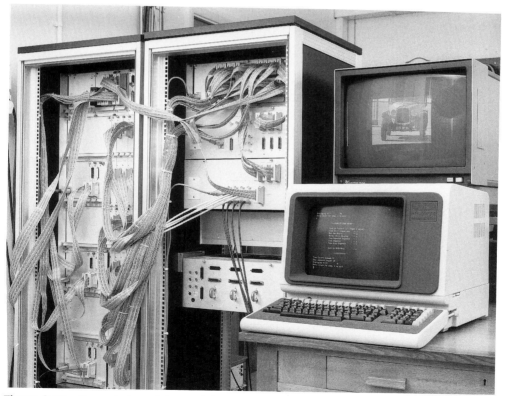

Figure 3.46 General view of three of the 2 Mbyte picture stores (at far left) together with the video distribution and collection hardware, computer interface, picture display monitor and computer terminal arranged as a short television sequence store

to study enhanced teletext [65]. In addition to the ability to read or write pictures at the full video rate, it is possible to read or write data at a slower rate under the control of a local microcomputer. When duplicate equipment was required three years after the original design the pace of technology made it worth considering a redesign of the semiconductor store using the latest available generation of dynamic memories — namely 64K devices. (256K dynamic memories were becoming available but at prices which would not become cost effective until 1985.) A new design held out the promise to be more compact, cheaper to build, cheaper to operate because of lower power consumption and lower heat dissipation — and also, to operate with an improved performance margin.

The original store design contained one 4U PCB for each bit of each R, G and B component — making 24 memory boards in all — see Figure 3.34. Each R, G and B component required one 4U control PCB and a further three interface boards to connect the store to the microcomputer, housed in a separate rack. The power supplies (5 V, 12 V and -5 V) were contained in an additional rack making a total of 30 PCBs plus power supplies spread over four 19 inch equipment racks.

A new design made it possible to reduce the number of 4U memory boards from 24 to 6. By combining a number of the store control functions of the R, G and B stores, the three original control boards can be replaced by one. Also, the considerably fewer number of store boards can be connected directly to the computer bus itself, thus dispensing with the three interface boards. The whole R, G, B store, therefore can be built on seven boards occupying just over a quarter of one 4U rack. The 24-bit picture stores dissipates less than 50 W. A block diagram of the new arrangement is shown in Figure 3.47.

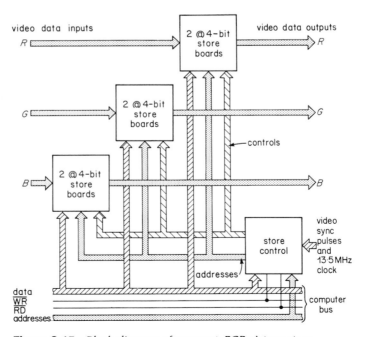

Figure 3.47 Block diagram of compact *RGB* picture store

Store board organisation

Using memories with 64K address capability a video multiplex factor of eight permits the required dual access port capability with relatively little extra capacity over the minimum required for an active picture when sampled at 13.5 MHz. On this basis a picture store built with 64K dynamic memories requires just 8 chips per bit. It is possible, in practice, to accommodate about 30–40 memory chips in 16 or 18 pin DIL packages together with multiplexing and demultiplexing chips on a 4U PCB. This suggests that four bits per board is a reasonable aim for a convenient storage unit. There is little to choose between the 64K × 1 or 16K × 4 options as either is suitable but the latter was chosen (type IMS 2620-12) and because of its common data input–output connection, the PCB layout was simplified.

A block diagram of the memory is shown in Figure 3.48. The 32 memory chips are arranged in four groups of eight. They all receive a common address input but each group is fed with a different version of $\overline{RAS}$ and $\overline{CAS}$ which in effect act as a further two address bits to identify one group out of the four. Data can be written to and read from the board in two ways each. Video input data are clocked into a serial–parallel shift register and every eight clock periods, eight data samples are stored in one of the four groups of memories. Similarly video output data are formed by reading out data from one of the four groups of memories and supplying it to the eight inputs of a parallel–serial shift register. Data can also be transferred into and out of the board at a slower rate of

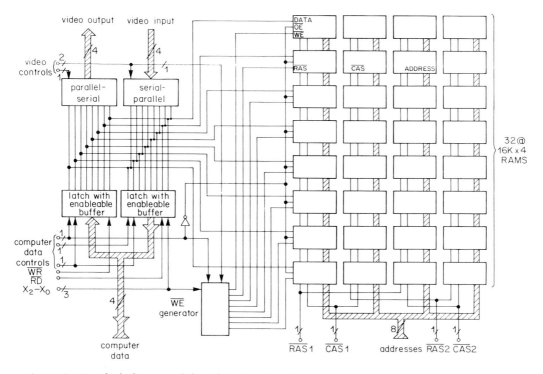

Figure 3.48 Block diagram of the 4-bit store board

($13.5 \div 8$) Mbyte/s by means of the computer bus. The computer address associated with a computer command is converted into a store board address on a separate control board and supplied to the store board along with the other control signals, including a 3-bit address referred to as X_2–X_0 which selects one of eight memory devices in the multiplex block.

Store control

The store signal waveforms associated with a read and a write cycle for the Inmos $16K \times 4$ memory chip IMS 2620-12, as used in the store, are given in Figure 3.49. $\overline{CAS}$ is fed into an active delay unit, with stated accuracy of ± 2 ns, before being applied to a pair of fast buffers and the memory chips. The effect of this is to offset the $\overline{CAS}$ waveform from the row–column address transition to meet the column address set-up criterion and ensure that the slower mid-speed option of the memory device can be used. Alternate memory cycles are used exclusively for video reading to continually monitor the contents of the store and provide refresh. The other memory cycle is used for a video write access or a computer data access, either reading or writing.

The video 'grab' function is provided in addition to, and not 'instead of', the video read function; when the store is grabbing a video picture the computer data transfer process is interrupted. This allows the video input to the store to be monitored continually while the grab operation of a single picture is executed. In addition to a picture grab, a field grab may be requested and also a field repeated in both halves of the store to achieve, crudely, a non-flickering field grab of moving picture material.

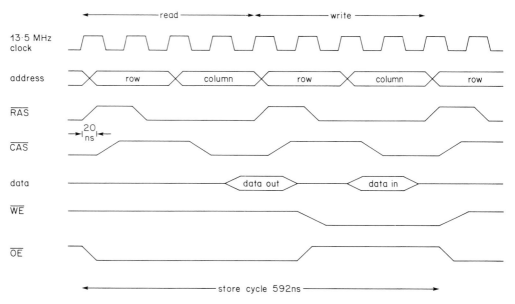

Figure 3.49 Control signal waveforms for a READ and WRITE CYCLE of a type IMS 2620–12 read–write memory

THE 64K GENERATION   121

Table 3.3
I-O address list for the *RGB* store control

I-O	Port address	Data	Function
0	1C4	00 hex	Video write a picture
0	1C4	01 hex	Video write a field.
0	1C4	03 hex	Video write a field and duplicate in both field halves of store.
0	1D0	Xmsbyte	Load Xmsb's address.
0	1D1	Xlsbyte	Load Xlsbyte address.
0	1D2	Ymsbyte	Load Ymsb's address.
0	1D3	Ylsbyte	Load Ylsbyte address.
0	1D8	data	Shift LEFT & write to all *R, G, B* stores together.
0	1D9	data	Shift LEFT & UP & write to all *R, G, B* stores together.
0	1DA	data	Shift UP & write to all *R, G, B* stores together.
0	1DB	data	Shift RIGHT & UP & write to all *R, G, B* stores together.
0	1DC	data	Shift RIGHT & write to all *R, G, B* stores together.
0	1DD	data	Shift RIGHT & DOWN & write to all *R, G, B* stores together.
0	1DE	data	Shift DOWN & write to all *R, G, B* stores together.
0	1DF	data	Shift LEFT & DOWN & write to all *R, G, B* stores together.
I-O	1E0	data	Access GREEN store only, no XY change.
I-O	1E1	data	Access RED store only, no XY change.
I-O	1E2	data	Access BLUE store only, no XY change
0	1E3	data	Leave XY alone & write to all *R, G, B* stores together.
I-O	1E8–1EF	data	Access GREEN store only but with autoincremented addresses as detailed in 1D8–1DF.
I-O	1F0–1F7	data	Access RED store only but with autoincremented addresses as detailed in 1D8–1DF.
I-O	1F8–1FF	data	Access BLUE store only but with autoincremented addresses as detailed in 1D8–1DF

Computer access serves three independent blocks of storage containing 524 288 bytes of data addressed on an X–Y coordinate basis where X, Y = 0, 0 represents the top-left corner of a display of the store contents and X increases horizontally to the right and Y increases in a downwards direction. An auto-increment of X, Y address can be invoked by nominating different computer input–output (I–O) ports in the computer data transfer process. Nine ports identify no shift and one of eight directional shifts along either axis in either direction or any of the four intermediate diagonal axes. This has proved essential for high-speed data transfer between the store and an external storage device as a Winchester disc drive.

A full list of the computer commands associated with the store is given in Table 3.3. The first column indicates the input–output direction of the computer data transfer — input means into the memory and output means out of the computer memory. For example, when a computer input instruction is given to the computer port address 1E0, data are read from the store holding the green video component at the current X, Y pixel address and the X, Y counters are not auto-incremented. The instruction list includes commands to instigate the video 'grab' function and load a new X and Y pixel address.

DELAY APPLICATIONS

Short video delay

A 64 bit static memory chip, with an access time of 25 ns (type Am 29701), provides a means of producing short, variable video data delays. To give a comfortable operating margin the devices are demultiplexed two ways and operated in A and B sections as discussed previously. Each device is arranged as 16 × 4 bits so that a minimum of four are needed to service eight bits. Data are read from address N in section A while data are written into address N-1 in B. During the next clock period, data are written into address N in A and read from N in the second. The delay is governed by the length of the address sequence and is quantised into multiples of two sample periods. Delays of odd numbers of periods are achieved by selecting the input or output of an additional D-type latch at the delay output. Using a 4-bit address generator, the range of delay possible with this configuration is three to 33 clock periods. A simplified block diagram of the arrangement is shown in Figure 3.50.

Television line delays

Several versions of television line delay units have been designed and constructed on 4U BMM cards. All require a two-way demultiplex to match video rates and their delay setting control is similar to that already described. The first — see Figure 3.51a — is a dual line delay organised as two completely independent blocks of up to 1028 8-bit words. The delay for each block can have one of two values, each of which is preset on a switch mounted on the PCB and selected by separate control inputs. The second (Figure 3.51b) is a quadruple line delay organised as four blocks of up to 1028 8-bit words, each block having separate data inputs and outputs but sharing a common delay setting which is preset by an on-board switch. Both these units are based on Fairchild 256 × 4

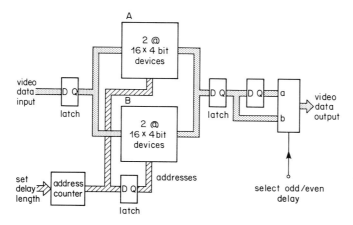

Figure 3.50 Programmable 8-bit short video delay

bit static read–write memories type 93L422 or 93422, depending upon the required operating speed. These devices have worst case read access times of 60 ns and 45 ns respectively.

The third unit contains four 10-bit delays connected in tandem with a common delay setting as shown in Figure 3.51c and was designed at a later time when Intel type 2148 HL-3 1K × 4, static devices became available. The access time for these in 55 ns. Four and two half-devices are required for each delay making a total delay range of between 1 and 2049 clock periods which corresponds to more than two television lines of a digital video signal sampled at 13.5 MHz. The single and dual clock periods of delay are achieved by deselecting the memory devices and operating them in a power-down fashion.

Television picture delay

A programmable picture delay board which can be set in the range 20 to 589 852 clock-periods delay, is constructed on a 4U BMM board, using 64K semiconductor dynamic memory devices and dissipates less than 10 W at 5 V. Each board handles four bits of data. The delay can be directly set by a 20-bit control word or, alternatively, up to 32 delays can be preprogrammed in on-board programmable read-only memory devices and remotely selected by a 5-bit control word.

A primary aim when designing this board was to produce a standard 4U BMM unit which would be self-contained and combine a complete picture store and delay setting control. A further consideration was to make the delay continuously variable and independent of any multiplexing of storage devices. The first solution devised was based on a television field delay with 8-bit wide capacity. There were two reasons why this solution was not adopted. Firstly, the number of chips required to realise the design was far in excess of the number generally accepted to be reasonable for a 4U board. Secondly it was realised that an 8-bit wide unit might cause an unnecessary restriction in data resolution. In many video processing applications 9- or 10-bit resolution is preferred, if not obligatory. This led to the development of a design based on a television picture with

4-bit wide capacity. The reduced amount of multiplexing due to halving the word-width made a significant difference to the overall chip count which made the design feasible. A real advantage of this arrangement is gained in the board count when more than 8-bits of data are handled.

A video demultiplex of nine allows up to 589 824 bits to be stored which is sufficient to hold a full television picture when sampled at 13.5 MHz. A straight 9-way demultiplex

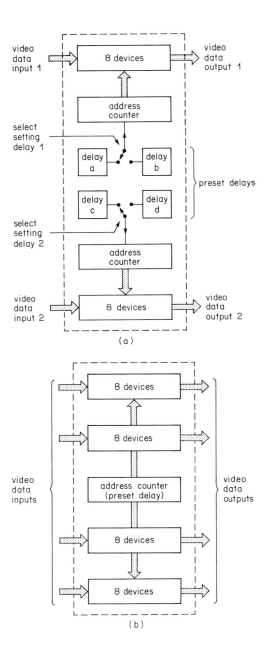

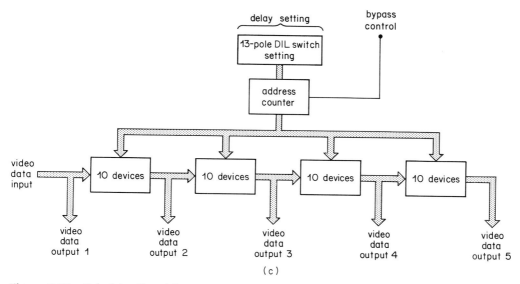

Figure 3.51 Television line delay arrangements: (a) dual line delay with independent control, (b) quadruple line delay with common control, (c) quadruple two-line delay with common control

has been chosen for three reasons. The minimum memory cycle time of 270 ns requires a demultiplexing factor of at least 4 which does not engender a convenient submultiple of 9. A nine clock-period cycle allows the devices to be accessed at a rate which is three to four times slower than their design capability. This makes for a lower overall power consumption because the chip dissipation increases proportionally with cycle rate. Also the demultiplexing and remultiplexing functions for a factor of 9 can be realised with a reasonably economical chip count. The total number of devices required is, therefore, 36 per board. A block diagram of the delay board is shown in Figure 3.52.

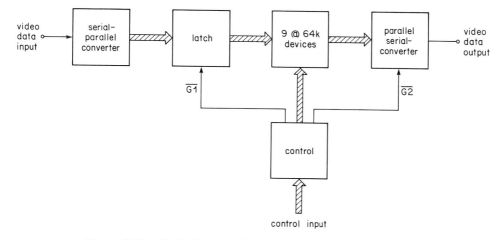

Figure 3.52 Block diagram of a programmable picture delay

126 SEMICONDUCTOR STORAGE OF TV SIGNALS

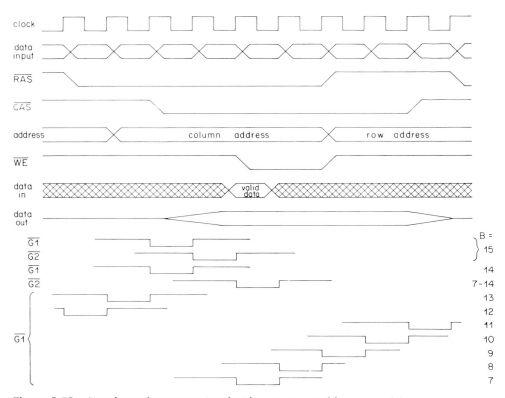

Figure 3.53 Signal waveforms associated with a programmable picture delay

With this arrangement all devices take the same 16-bit address. The delay is altered by allowing a certain range of addresses only to be generated but instead of the A/B alternate read–write control approach the memories are arranged to operate in a read–modify–write cycle. Data, which was written on the previous access of an address, are read out and subsequently shifted out through the output registers. During the same cycle and at the same address, new currrent data are written in. Thus the number of addresses generated determines the total delay through the store.

The waveforms associated with one store cycle are shown in Figure 3.53 and illustrate the way in which the delay can be continuously altered within the 9-way demultiplex range by changing the precise clock-edge on which the data are enabled into the input latch and shifted out from the devices.

THE 256K GENERATION

Introduction

Semiconductor dynamic memory chips offer an alternative solution to Winchester disks operating in parallel for the bulk storage of television pictures required for picture processing by computer. Television picture stores built to date have proved reliable and repeatable — there are no moving parts to go wrong and once a satisfactory PCB design has been achieved it can be replicated in large quantities. A number of other advantages can be identified. The whole store behaves like a random access memory — each pixel is equally accessible — and there is no need for separate buffer picture stores: a very short buffer of less than a microsecond is sufficient to deal with the multiplexing of memory chips to match video data rates. The full standard television *RGB* data rate of 324 Mbit/s is easily obtained by adding extra store boards. Full-size HDTV pictures can be accommodated by accessing further stores in parallel. There is no restriction on the phasing of the store cycle making it possible to lock easily to an external source of mixed synchronising pulses. A 'clean' environment is not essential. Another advantage is that a bulk store based on semiconductor memory chips can be built in several units, each of which is useful in its own right and can be operated by several users independently. One unit could be built initially with relatively little capital outlay and the individual store boards of a unit are interchangeable.

There are three main disadvantages. The first is that the memory must be preloaded, at power-on, from a back-up device, for example, a streaming tape cartridge or, for faster operation, a Winchester disc. Secondly, dynamic memory chips require frequent 'refreshing' to maintain their memory. If accidental power loss is to be safeguarded, special provision must be made to generate sufficient refresh cycles to allow the memory to be backed up immediately. This may require some form of battery facility. Thirdly, the length of picture sequence which can be housed in a standard 19 inch bay is probably limited to about five or six seconds of *RGB* full rate. Longer sequences can only be achieved by reducing the data rate to say, quarter-picture size only or one component only — or by building more bays of equipment.

This section outlines some of the design considerations when using semiconductor memory chips in this application and discusses the specification for a 'package' of storage which can be fabricated on a single PCB.

Choice of memory devices

The 256K generation of dynamic memory devices is available with a minimum normal read or write cycle time 200 ns — which is somewhat faster than the 64K device. A memory cycle of four clock periods at the 13.5 MHz sampling rate (296 ns) is still appropriate, however, and allows the slower options of the device, which are generally cheaper, to be used with comfortable operating margins and operate at less than their stated maximum current. To permit simultaneous video read and write or computer read or write at a slower rate, the store must be accessed twice which results in an eight-way demultiplex and a store cycle similar to that used for the 64K device implementation. With 256K devices, therefore, the smallest store possible is $8 \times 256K = 2$ Mbit made from eight chips — this is equivalent to four (1 bit) active television pictures.

Another device is configured as $32K \times 8$ bit. The addressing is direct (15 bit) which results in a faster cycle time of 180 ns minimum and a saving of power — the maximum active current consumption is 55 mA. The byte-wide data bus is shared for input and output data and the device is available either in a 28-pin DIL package (which is larger) or a leadless chip carrier which occupies a board area of about one third of a square inch.

The same arguments as described above, for multiplexing chips to match video data rates, apply here but there is the possibility of reducing the memory cycle time to three video clock-periods (222 ns). This would allow the video demultiplex factor to be decreased to six (for the same dual-port access) or increased to nine to achieve a triple-port access. In the former case it would be necessary to provide more memory addresses than the equivalent store based on a multiplex factor of eight. It may, however, be possible to reduce the total number of chips required to store a convenient number of active television pictures on a single store board. The faster cycle rate is likely to result in a greater power consumption and decreased operating margin. In the latter case it would be possible conveniently to store four *full* television pictures in 72 chips but with the same power consumption and operating margin penalties although it is not generally necessary to store the full field and line blanking periods for picture processing purposes.

A 64K × 4 vision is available in an 18-pin DIL package and with similar performance to the 256K × 1. This device makes it possible to upgrade the compact picture store board previously described and quadruple its storage to four pictures of 4 bit per board.

STORE BOARD ORGANISATION

There have been previous attempts to make a general memory storage board so that individual users can match them to their requirements by adding dedicated boards of their own. Experience gained from these attempts, leads to two particular conclusions. The first is that the video demultiplexing of memory chips is best performed on the same board as the chips themselves to avoid carrying an excessive number of data connections from one board to another. Secondly, the memory board should not be too large physically — large board area can result in signal waveform distortions when signals are conveyed from one side of a board to another and this is particularly critical in the case of distributing the 13.5 MHz sampling clock.

A reasonable package of memory storage is, therefore, one which can group an array of memory chips with the required video multiplexing and demultiplexing and computer decoding chips on a board which is not too large. A greater physical package efficiency may be achieved by moving from 4U memory boards to 6U. A 4U board is capable of supporting 32 memory chips comfortably (the compact *RGB* picture store memory board is an example of this): it may be possible to support 64 memory chips on a board only 50% larger because there is very little increase in the number of ancillary chips needed to support them. This forms the basis of a new memory store design which has both video read and video write ports as well as a computer data port and has a capacity of 4 Mnibble (4 bits wide).

A rack of storage and beyond

An arrangement of 24 such boards, grouped in pairs suitable for mounting in a single rack of hardware is shown in the block schematic of Figure 3.54. They are configured in this case as eight boards each for *RGB* making a total storage capacity per rack of almost 50 Mbyte. The video input processing required to strip off line codes and interface video to the store boards can be handled on one board. Similarly video output processing requires

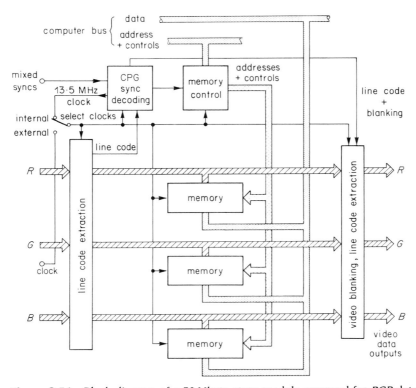

Figure 3.54 Block diagram of a 50 Mbyte store module arranged for *RGB* data

one board to apply video blanking and reinsert the line codes. A clock pulse generator, capable of being locked to supplied mixed synchronising pulses or free-running, provides an internal source of clocks. Alternatively clocks may be provided from video input data. Each store board requires a 21-bit address (15 for each chip and 6 to distinguish 1 of 64 chips) and two further bits are required in each rack to distinguish one of four boards for *RGB* — this makes a total of 23-bit address. This, together with all the usual memory control signals, are generated on a separate control board.

An image-processing tool incorporating a large television picture sequence store is presented in Figure 3.55. Based on a commercial 16-bit microcomputer, it incorporates a Winchester disc and streaming tape drive for picture back-up and an interface to a fast local area network (LAN) for interconnection to a more powerful processor, for example, a VAX 11/750. The sequence store may be used for:

(a) the random access of individual pictures, part pictures or pixels within a picture.

(b) the output of a sequence of pictures at video rate. One bay would provide about 5 s of *RGB* active picture area. Alternatively the addressing could be arranged to give, say, an *RGB* quarter picture area sequence for 20 s or a single component (*Y*) sequence, quarter-picture area for one minute.

As the speed–power product performance improves and the cost per bit falls there will be a tendency towards including a greater amount of storage in professional equipment.

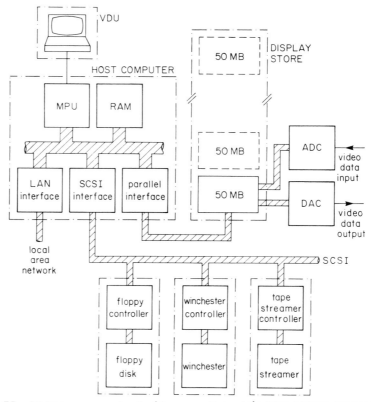

Figure 3.55 An image-processing tool incorporating a television picture sequence store

This could take the form of digital filters, with improved performance, employing picture and field delay taps, for example, to achieve a sophisticated PAL decoder, making a better job of separating the chrominance and luminance information in a composite signal. Greater mass storage could herald stills stores of even greater capacity, capable of holding the equivalent of hundreds of digital television pictures for almost instant retrieval and display.

The technology trend also brings improvements to the domestic consumer. The reductions in cost, size and power consumption of stores for studio equipment in the period covered by this chapter support the view that they can now be used to enhance the performance of TV receivers.

4 Digital Encoding and Decoding of Composite Television Signals

C. K. P. Clarke

INTRODUCTION

Digital signal coding, in which the television waveform is represented as a series of samples encoded as binary numbers, has considerable advantages for signal processing. In addition to protecting the signals against waveform distortion, binary coding allows digital arithmetic and storage techniques to be used. This greatly extends the range of processes which can be applied and ensures stable and predictable performance.

These attributes on their own have been sufficient to promote the application of digital techniques to composite coders and decoders using analogue inputs and outputs. With the opportunity for large scale integration, this approach has become increasingly attractive for domestic receivers, although in this case the techniques are applied more for improved stability and reliability than to provide more advanced methods. However, digital implementation has become even more attractive at the composite signal interfaces to digital studios working on component signals. Then the relatively high cost of the digital equipment compares more favourably with that of analogue equipment and an associated digital interface unit.

Digital signal coding for use in studios

Many of the parameters of digital signals for use in studios have been standardised in CCIR Recommendation 601 [9]. This defines a form of coding based on line-locked sampling of the separate component luminance and colour difference signals Y, $R - Y$ and $B - Y$. An important property of locking the sampling to a harmonic of line frequency is that it produces an orthogonal grid of sample points on the picture. This means that filtering in each of the horizontal, vertical or temporal dimensions becomes a simple one-dimensional process. The use of separate component signals simplifies processing by avoiding the subcarrier phase variations which occur in composite encoded signals. Notwithstanding these advantages, it is only relatively recently that line-locked sampled component signals have been considered suitable as the basis for a studio signal standard.

In early digital television experiments, the PAL composite signals were sampled either at three times the colour subcarrier frequency or at a nearby harmonic of line frequency [32, 69]. Exactly which reference frequency was used depended on the application. If line-repetitive processing was involved such as in vertical interpolation [70], line-locked sampling was favoured, while if similarly phased samples were required, such as in DCPM prediction [71], then subcarrier-locked sampling was used.

A refinement of subcarrier-locked sampling was to relate the sampling phases to the subcarrier reference phase [72], in particular to specify that one sample should fall on the $+U$ axis. This had two advantages. First, it became possible to synchronise two digital colour signals for mixing by shifting the signals horizontally by one or two samples to match one another, and secondly, the phase of samples on adjacent lines was then related, in spite of the action of the PAL V-axis switch. For example, if samples OABOAB ... were taken on one line (O being the 'on-axis' sample), then the samples would be in the order OBAOBA... on the adjacent lines. This feature simplified the synthesis of digital colour waveforms using a three-times subcarrier frequency clock, because each colour could then be generated by repeating a sequence of three sample values [73]. Also, two-dimensional prediction became possible for DPCM because samples of similar subcarrier phase were available from the previous line [74].

For a short period, three-times subcarrier frequency phase-locked sampling was widely used and regarded as a possible standard until a method of sampling PAL at a sub-Nyquist frequency of twice the colour subcarrier [75] was invented. In this method samples were taken at phases of $45°$ and $225°$ to the $+U$ axis. The attraction of the reduced data rate resulting from twice subcarrier sampling was judged to outweigh the small impairment involved, especially as this was found to be, in part, non-cumulative. Shortly afterwards another system of twice subcarrier frequency sampling was found [76] in which samples were taken at phases of $135°$ and $315°$. This system produced digital Y and $(U + V/U - V)$ signals which could be converted back to PAL with almost no impairment.

The development of improved analogue-to-digital and digital-to-analogue converters, as reviewed in Chapter 2, introduced the possibility of sampling at four-times subcarrier frequency. By sampling with phases of $45°$, $135°$, $225°$ and $315°$ to the $+U$ axis, it was easy to produce both the twice subcarrier systems of sampling. The digital comb filters used in the conversions from four-times subcarrier had better defined, more stable characteristics, so the resulting twice subcarrier sampled signals were improved in quality. The two systems of twice subcarrier sampling were thought to satisfy a wide range of requirements, including low data rate transmission [7] and reversible PAL coding and decoding [77], with four-times subcarrier sampling providing a link between the two. In view of this, the two and four-times subcarrier sampling methods were suggested as the basis for a possible standard [78].

However, further investigations revealed that the two systems of twice subcarrier sampling were not compatible [79]. When used in cascade, the two systems caused a serious loss of vertical chrominance and diagonal luminance resolution and largely negated the special features of the systems when used individually. For this reason, more emphasis was placed on four-times subcarrier frequency sampling. For studio processing, this sampling rate combines the advantages of subcarrier phase-related sampling with many of the features of line-locked sampling, notably the nearly orthogonal structure and the static grid of samples on the picture. Nevertheless, the extra storage capacity requirements and the reduced processing time in a clock period remained significant drawbacks.

PAL decoding with subcarrier-locked sampling

Throughout this period of development, a critical factor working in favour of subcarrier-locked sampling was the realisation that in some cases it would be too complicated to process the composite signal directly. Because of this, the signals would need to be decoded to component form. With samples taken at specific phases of the subcarrier, the content of each sample is known. For example, at four-times subcarrier with the samples locked to the 45° phases (preferred for the derivation of twice subcarrier samples), the four different phases contain $(U \pm V)/\sqrt{2}$, $(-U \pm V)/\sqrt{2}$, $(-U \mp V)/\sqrt{2}$ and $(U \mp V)/\sqrt{2}$, with the sign of the V component alternating with the PAL switch. At three-times subcarrier the weighting factors are somewhat more complicated with the three phases containing U, $-\frac{1}{2}U \pm \sqrt{3}V/2$ and $-\frac{1}{2}U \mp \sqrt{3}V/2$. Summing weighted proportions of the individual samples over a cycle of the subcarrier waveform can be used to extract both the colour difference signals and the luminance content. Thus digital PAL decoding with subcarrier-locked sampling was considered to be a relatively simple matrixing operation. This concealed the underlying contradiction that the heart of the 'digital' decoder was, in fact, the analogue phase-locked sampling process.

Subcarrier-locked decoding, however, was not without its drawbacks. It perpetuated the complications of digital PAL processing by carrying the subcarrier relationship into the *YUV* component samples. Twice- and three-times subcarrier frequency sampling produced moving sample grids which were much less convenient for component signal processing than the static orthogonal samples of a line-locked system. With the alternative of four-times subcarrier sampling, although the sampling grid was static and nearly orthogonal, there was uncertainty in its position relative to the picture information, since this depended on the undefined subcarrier-to-line phase relationship at the original PAL coder.

Further problems were caused if non-mathematical PAL signals were encountered, since these signals do not have the correct subcarrier frequency to line frequency relationship. In this case, because the sampling structure follows the subcarrier, the picture information drifts sideways until reset with a jerk when sufficiently out to step with the line pulses. Although the horizontal movement amounts to only one sample width, the repeated drift and jerk is very disturbing. Also, with monochrome signals, in which the colour burst is omitted to prevent cross-colour on colour receivers, there would be no subcarrier available as a reference for the sampling loop.

The absence of any satisfactory solutions to the problems of subcarrier-locked decoding finally weighed against having a studio standard based on digitised PAL signals. Also it was recognised that so many processes would benefit from using digital component signals, preferably in line-locked form, that it was better that the signals should remain in component form rather than being converted to and from PAL. The task of PAL decoding was therefore relegated to one of interfacing the existing analogue PAL system to the new line-locked digital component signals during a changeover period.

PAL decoding with line-locked sampling

With line-locked sampling there is no simple relationship between individual samples and the subcarrier reference phase, so the simple matrixing method used for subcarrier-locked sampling cannot be applied. Indeed, because of this it was widely held that digital decoding was impracticable unless subcarrier-locked sampling was used [78, 80]. Digitising the outputs of an analogue PAL decoder with three A–D converters using line-locked sampling seemed the only economically viable method of obtaining line-locked component signals from PAL, but necessarily lost the performance benefits of digital processing techniques in the decoder. However, the problems of digital decoding in a line-locked sampling system were confronted during the development of an experimental converter working between the 625/50 PAL and 525/60 NTSC television standards [81] (see Chapter 6).

The converter itself [64] used two closely related line-locked sampling frequencies for the input and output standards. The frequencies were chosen to make the number of active-line samples the same on the two standards, so avoiding the need for any horizontal interpolation in the converter.

The use of analogue decoding during these experiments was not favoured because quite complicated comb filtering methods were to be compared and analogue parameter variations between the methods could have affected the results. The possibility of encountering non-mathematical composite signals also ruled out the use of subcarrier-locked sampling, even if two appropriately related frequencies had been available. The convenience of operating digital decoders and coders directly at the line-locked sampling frequencies of the converter was, therefore, very attractive.

The method of decoding developed for this application differed from the prevailing understanding of digital decoding in that the analogue sampling process made no contribution to demodulating the colour signals. Instead all the processes of generating a local reference subcarrier, locking it to the incoming burst phase with a feedback loop and extracting the colour signals using product demodulators were carried out by digital circuitry [82, 83]. It was found that similar techniques could be used for PAL and NTSC, and could be applied to composite encoders as well. The confirmation that these line-locked techniques provided a viable and cost-effective solution for coding and decoding proved to be a significant step in support of standardisation on the use of line-locked sampled digital component signals for studios [84]. Since then, techniques have been used in more general investigations aimed at optimising the luminance–chrominance separation filters for digital studio inputs. The development of these separation filters is described later.

The technique used to generate subcarrier signals from a line-locked sampling frequency is fundamental both to digital coders and to decoders. In the following description, therefore, this is treated first, followed by the explanation of modulation and its application in coders. In decoders, it is advantageous to apply digital techniques to control the sampling loop for analogue-to-digital conversion of the composite signal. The process of demodulation, although similar to modulation in many respects, has the additional requirement to lock the local reference subcarrier to the incoming burst phase. The description of this process includes an explanation of the way in which a line-locked sampled demodulator is able to accommodate non-mathematical PAL signals. Then the application of special techniques such as PAL modifiers in decoders is described.

While the early sections deal with the mainly practical aspects of applying digital techniques to coders and decoders, the final section describes problems of the deployment of coders and decoders in the context of digital studios. The main problem is that of recoding signals which have residual impairments from a previous decoding process. In some circumstances 'clean' composite coding systems, which provide a degree of reversibility could be used to avoid the problem. However, these systems are not always ideal and as an alternative, a way of minimising the problem by using a particular layout of studio centre is described. Throughout the text, although the main description deals with the PAL system, the differences necessary for NTSC operation are also included.

SUBCARRIER GENERATION FROM A LINE-LOCKED SAMPLING FREQUENCY

Colour coders and decoders both use quadrature subcarrier signals locked to the subcarrier reference phase. In a digital circuit, a train of samples representing a subcarrier waveform can be generated by selecting an appropriate sinusoidal sample value corresponding to the current phase of the reference. If subcarrier phase-locked sampling is used, then each sample is associated with a defined value of the reference phase. This can be monitored by a simple counter, for example, with three-times subcarrier sampling, the phases $0°$, $120°$ and $240°$ could be identified by counter states 0, 1 and 2. Sample values corresponding to 0, $\sqrt{3}/2$ and $-\sqrt{3}/2$ for a sine wave or 1, $-\frac{1}{2}$ and $-\frac{1}{2}$ for a cosine wave could then be selected, as shown in Figure 4.1.

With line-locked sampling there is no direct relationship defined between individual samples and the subcarrier reference phase. However, the subcarrier and sampling frequencies are related, although in a complicated way.

Subcarrier phase generation

Frequency relationships

In the PAL (System I) and NTSC (System M) colour standards there are prescribed relationships [85] between the subcarrier frequency (f_{sc}) and the line frequency (f_h). For PAL

$$\frac{f_{sc}}{f_h} = \frac{709\ 379}{2500}$$

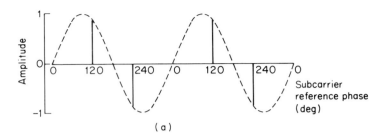
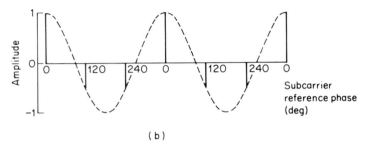

Figure 4.1 Three-times subcarrier phase-locked sampling: samples of subcarrier frequency waveforms generated from the subcarrier reference phase producing (a) a sine wave and (b) a cosine wave

Also, the 13.5 MHz sampling rate (f_s) of the digital studio standard is related to the line frequency [9], as shown

$$f_s = 864 f_h$$

Combining these equations produces the following relationship between the subcarrier frequency and the line-locked sampling frequency of the digital studio standard

$$\frac{f_{sc}}{f_s} = \frac{709\,379}{2\,160\,000}$$

which can also be expressed in terms of the sampling period (t_s) and the subcarrier period (t_{sc})

$$\frac{t_s}{t_{sc}} = \frac{709\,379}{2\,160\,000}$$

In each clock period, therefore, the subcarrier phase advances by this fraction of a subcarrier cycle.

The following somewhat simpler relationships are obtained for NTSC

$$\frac{f_{sc}}{f_h} = \frac{455}{2}$$

SUBCARRIER GENERATION FROM A LINE-LOCKED SAMPING FREQUENCY

and [9]

$$f_s = 858 f_h$$

so that

$$\frac{f_{sc}}{f_s} = \frac{35}{132}$$

or

$$\frac{t_s}{t_{sc}} = \frac{35}{132}$$

Similar relationships could be obtained for any other line-locked sampling frequencies, although in most cases these would not provide a common sampling frequency (13.5 MHz) on the two standards.

Ratio counters

It is possible to design a counting system [73, 86] based on any ratio in the form $p:q$. This consists of an accumulator in which the smaller number p is added at each clock pulse modulo another number q. In hardware, the counter consists of an adder and a register as shown in Figure 4.2. The contents of the register are constrained so that, if they would exceed or equal q, q is subtracted from the contents.

In the present application, this overflow corresponds to the completion of a full cycle of subcarrier. Since only the remainder, that is, the subcarrier phase, is required, the number of whole cycles completed is of no interest. During each clock period, the output of the register shows the relative phase of a subcarrier frequency waveform in $q$ths of a subcarrier period. So, by using the register contents to address a read-only memory (ROM) containing a sine wave characteristic, a numerical representation of the sampled subcarrier sine wave can be produced.

While the 132 phases required for NTSC are perfectly manageable, the PAL ratio would require a ROM with 2 160 000 words capacity. This can be avoided by partitioning the ratio into two fractions, the more significant of which provides the subcarrier reference phase. In these circumstances it is advantageous to make the denominator of

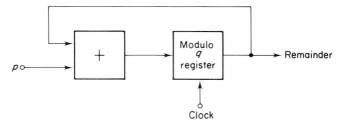

Figure 4.2 A $p:q$ ratio counter

the more significant fraction a power of two, partly because the overflow of the ratio counter is then automatic and partly to use the full capacity of the ROM.

If the subcarrier period is divided into 2048 phase steps, then the partitioned ratio for PAL becomes

$$\frac{t_s}{t_{sc}} = \frac{672\frac{10\,064}{16\,875}}{2048}$$

or, if a 16-bit accumulator is used for the less significant ratio

$$\frac{t_s}{t_{sc}} = \frac{672\frac{20\,128}{33\,750}}{2048}$$

The partitioning can also be applied to NTSC to ensure automatic overflow in the counter feeding the ROM and to achieve greater commonality of processing for the two systems. Then

$$\frac{t_s}{t_{sc}} = \frac{543\frac{1}{33}}{2048}$$

which, with a 16-bit accumulator, becomes

$$\frac{t_s}{t_{sc}} = \frac{543\frac{1024}{33\,792}}{2048}$$

In this form, the corresponding ratio counter has two stages as shown in Figure 4.3. The less significant stage produces a sequence of carry bits which correct the approxi-

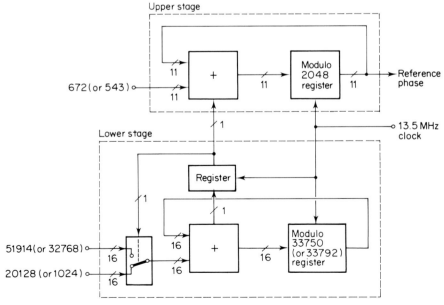

Figure 4.3 A two-stage ratio counter for generating PAL or NTSC subcarrier reference phase from a 13.5 MHz clock frequency. Values for NTSC operation are shown in parentheses

mate ratio of the upper stage by altering the counting step from 672 to 673 (543 to 544 for NTSC). The upper stage then produces an accurate 11-bit reference phase output to address a ROM.

While the upper stage adder automatically overflows to provide modulo 2048 operation, the lower stage requires additional circuitry because 33 750 (33 792 for NTSC) is not an integer power of two. In this case, the 16-bit register has a maximum capacity of 65 535 and the adder generates a carry for any value greater than this. To produce the correct carry sequence, it is necessary, each time the adder overflows, to adjust the next number added to make up the difference between 65 536 and 33 750 (33 792). This requires

$$65\,536 - 33\,750 + 20\,128 = 51\,914 \text{ for PAL}$$

or

$$65\,536 - 33\,792 + 1024 = 32\,768 \text{ for NTSC}$$

Although this changes the contents of the lower stage register, the sequence of carry bits is unchanged, so ensuring that the correct phase values are generated.

Quadrature subcarrier generation

Each value of the 11-bit reference phase signal corresponds to one of 2048 waveform values taken at a particular point in the subcarrier cycle period and stored in a ROM. The maximum phase error produced by this quantisation is $\pm 0.09°$ which causes a maximum amplitude error of $\pm 0.075\%$ (relative to the peak-to-peak amplitude) at the steepest part of the sine wave signal. In comparison, quantisation of a standard level composite signal to 8 bits in an A–D converter causes $\pm 0.28\%$ error in the largest possible peak-to-peak amplitude of subcarrier. Therefore, in a demodulator, quantisation of the reference phase signal is a minor contribution to the total quantisation distortion. Even so, a further reduction of distortion could be obtained by using a 12-bit phase signal, although twice the ROM capacity would then be required to store the subcarrier wave-shape.

The ROM requirements can be minimised by storing only one quadrant of the subcarrier wave-shape for each of the quadrature signals. The values for the other quadrants can then be produced using the symmetrical properties of the sinusoidal waveshape. This requires the 512 values used to store a quadrant to be positioned symmetrically throughout a 90° section of the characteristic. Thus, rather than taking the sample values at the 2048ths points on the waveform, they are taken at odd multiples of one 4096th of the total period, as shown in Figure 4.4. This avoids end-effects when the sample values are read out in the reverse order.

Figure 4.5 shows a circuit arrangement for generating quadrature subcarriers from an 11-bit subcarrier reference phase signal. This uses two ROM units with 9-bit addresses to store quadrants of the sine and cosine waveforms. Exclusive-OR gates are used to invert the addresses for generating time-reversed portions of the waveforms and to invert the output polarity to make negative portions of the waveforms. An additional gate is included in the sign bit for the V subcarrier to allow injection of a PAL square wave signal to provide phase inversion of the V signal on alternate lines. The output word length of

140 DIGITAL ENCODING AND DECODING OF COMPOSITE TELEVISION SIGNALS

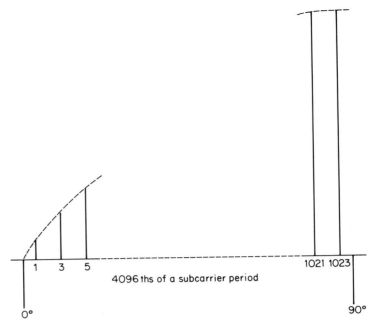

Figure 4.4 Positions of stored sample values for one quadrant of a subcarrier cycle. Samples for other quadrants can be generated by inverting the addresses and/or the sample values

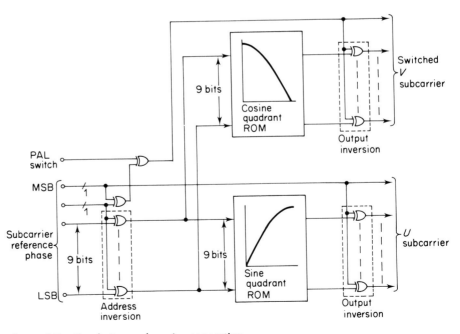

Figure 4.5 Quadrature subcarrier generation

COLOUR ENCODING

the ROMs used to generate the subcarrier samples tends to be limited more by the following circuitry than by constraints imposed by the ROM size.

A further feature that is often conveniently included in the ROM is a gain change. By storing a scaled version of the waveform, the gain factors converting from, for example, $B - Y$ to U or $R - Y$ to V can be accommodated. This is discussed further, later in this chapter.

COLOUR ENCODING

In addition to the quadrature subcarrier generation described in the previous section, the other basic process of a conventional PAL coder is chrominance modulation. The weighted colour-difference signals are modulated onto the high frequency subcarrier signals and added to the luminance as shown in Figure 4.6. Although no compensating delay is shown, the luminance and modulated chrominance signals should be co-timed at the point of addition. The same basic arrangement is used for NTSC although there are minor differences as described later in this section. 'Clean' coding methods for PAL and NTSC which involve additional filtering in the coder are dealt with later.

Features of the PAL signal

The PAL system uses double sideband suppressed-carrier amplitude modulation of two subcarriers in phase quadrature to carry the two weighted colour-difference signals, U and V. Because the subcarriers are orthogonal, the U and V signals can be separated perfectly from each other. However, if the modulated chrominance signal is distorted, either through asymmetrical attenuation of the sidebands or by differential phase distortion, the orthogonality is degraded, resulting in crosstalk between the U and V signals. As this is quite likely to occur, the PAL system incorporates alternate line switching of the V signal component. This is an additional form of modulation which provides a frequency offset between the U and V subcarriers in addition to the phase offset. Thus, when decoded, any crosstalk components appear modulated onto the

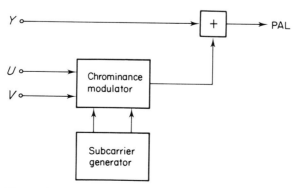

Figure 4.6 Basic processes of a PAL coder

alternate line carrier frequency, in plain coloured areas producing the moving pattern known as Hanover bars. This pattern can then be suppressed in the decoder by a comb filter averaging equal contributions from switched and unswitched lines.

The PAL chrominance signal can be represented mathematically as a function of time (t) by the expression

$$U \sin \omega t \pm V \cos \omega t$$

where

$$\omega = 2\pi f_{sc}$$

and for system I PAL signals

$$f_{sc} = 4.43361875 \text{ MHz}$$

The sign of the V component alternates from one line to the next. This signal is formed by the product modulators of Figure 4.7 which multiply the orthogonal subcarrier waveforms by the instantaneous values of low-pass filtered baseband U and V signals. The low-pass filters use a frequency characteristic approximating to the Gaussian template shown in Figure 4.8.

The modulation process is shown in spectral terms in Figure 4.9. Figure 4.9a represents the baseband spectrum of a full bandwidth colour-difference signal, the high frequency components of which are attenuated by low-pass filtering to produce the spectrum of Figure 4.9b. When convolved with the line spectrum of the subcarrier signal, Figure 4.9c, this produces the modulated chrominance signal spectrum of Figure 4.9d.

A Gaussian frequency characteristic (Figure 4.8) is used for the colour-difference low-pass filters for optimum compatibility with monochrome receivers. This wide-band, slow roll-off characteristic is needed to minimise the disturbance visible at the edges of coloured objects because, with a monochrome display, the signal receives no further filtering. A narrower, sharper-cut characteristic would emphasise the subcarrier signal at these edges, widening the transitions and introducing ringing.

Because the coder uses wide-band U and V filters, the modulated chrominance spectra can overlap to cause aliasing, as shown in the zero frequency region of Figure 4.9d. Also, since the PAL coder does not a contain 5.5 MHz low-pass filter at the output, the upper chrominance sidebands extend well beyond normal video frequencies. In practice, the PAL signal may sometimes retain these out-of-band components up to the broadcast transmitter, at which point the portion of the upper sideband above 5.5 MHz is removed.

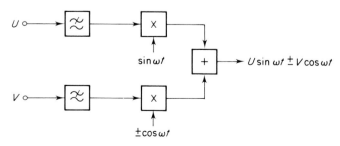

Figure 4.7 PAL chrominance modulation

COLOUR ENCODING 143

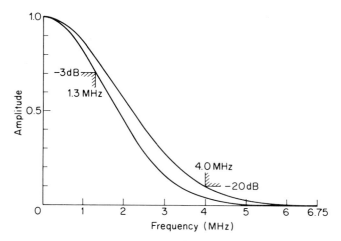

Figure 4.8 Gaussian low-pass filter characteristics for use in PAL coders

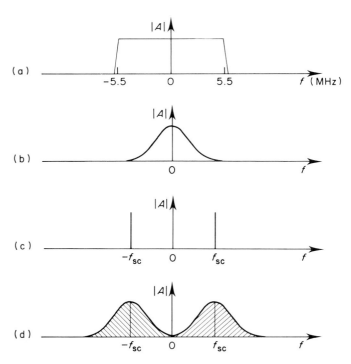

Figure 4.9 Frequency spectra in PAL chrominance modulation: (a) the baseband colour-difference signal, (b) the Gaussian filtered colour-difference signal, (c) the subcarrier sine wave and (d) the modulated chrominance spectrum produced by convolving (b) and (c)

144 DIGITAL ENCODING AND DECODING OF COMPOSITE TELEVISION SIGNALS

Although the coder maintains a wide chrominance bandwidth, the colour-difference signal bandwidth reproduced in the decoder is usually much narrower than this. If the decoder produces colour signal frequencies above 1.07 MHz, the loss of the upper sidebands above 5.5 MHz leads to ringing and U–V crosstalk on colour transitions. These effects are caused, respectively, by the sharp-cut characteristic of the 5.5 MHz filter and the sideband asymmetry. Also, any increase in the decoder bandwidth causes a proportionate increase in cross-colour, whilst the additional chrominance resolution diminishes rapidly as the bandwidth is increased. Therefore, if compatibility with monochrome displays were to be ignored, the encoded chrominance bandwidth could be reduced without losing any resolution for the colour viewer. Moreover, this would have a beneficial effect for the colour viewer because the spread of the lower chrominance sidebands into the low frequency luminance region would be reduced. This, in turn, would limit the amount of low frequency cross-luminance, which is difficult to suppress in a decoder.

Digital PAL encoding

Whereas an analogue PAL coder depends on several waveforms from an associated synchronising pulse generator, it is usually convenient to incorporate this function as part of a digital coder. Two alternative arrangements are shown in Figure 4.10. Figure 4.10a, for analogue *RGB* input signals, contains analogue-to-digital converters and a

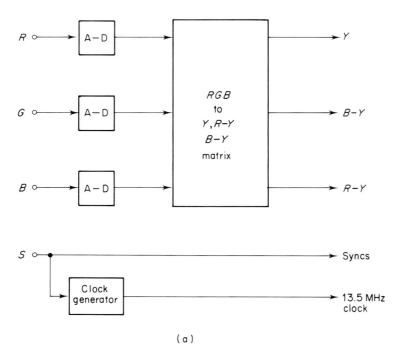

(a)

COLOUR ENCODING

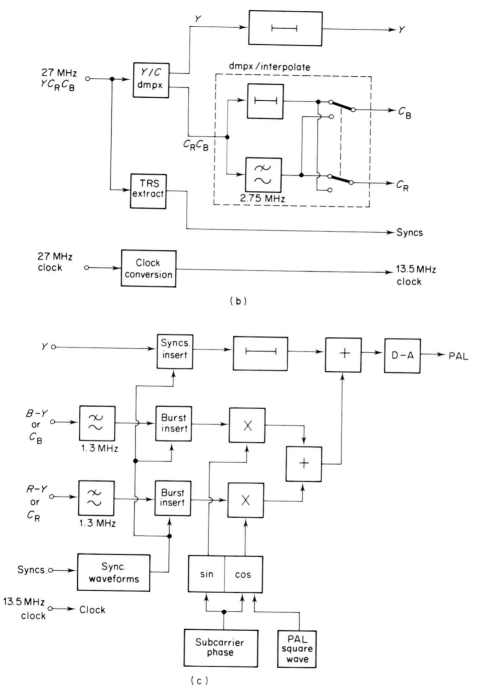

Figure 4.10 Alternative input circuit arrangements for digital PAL encoding: (a) for analogue *RGB* and (b) for digital component signals. The common modulator circuitry is shown in (c)

13.5 MHz clock pulse generator locked to the line frequency of the incoming synchronising pulses. The digitised *RGB* signals are then matrixed to form luminance and colour-difference components. Alternatively, Figure 4.10b shows the demultiplexing and sample interpolation circuitry [87] needed to convert from a 27 MHz multiplexed digital component signal. In this case, synchronising information is obtained from the timing reference signal (TRS) codes [10] and 13.5 MHz clocks are produced from division of the 27 MHz clock carried with the digital component signal. The remaining circuitry shown in Figure 4.10c, including sync and burst insertion, the digital transversal low-pass filters and the four-quadrant digital multipliers, is common to both arrangements.

Although in most respects the alternatives of Figure 4.10 would produce substantially similar results, the *RGB* input arrangement of Figure 4.10a has a significant advantage. The PAL System I specification requires that the composite signal should represent only signals within the normal ranges of *R*, *G* and *B* [88]. The *RGB* inputs effectively ensure this because any signals extending significantly beyond the normal range will be clipped by the A–D converters. Alternatively, digital *RGB* inputs could be clipped in a similar manner. The digital YC_RC_B input, however, does not have this facility, since the only way of finding whether the YC_RC_B combinations will produce valid *RGB* values is to convert to *RGB*. The responsibility for ensuring that the combinations are valid is, therefore, passed to the YC_RC_B system as a whole. In practice, this may be difficult to control unless processes such as electronic graphics generation, which could produce unreal colours when operating in YC_RC_B, always operate with *RGB* signals.

Modulation

When a coder is implemented with digital circuitry, each signal is represented by a stream of binary coded samples. Because of this, the frequency spectra of the modulation process are those of the analogue coder (Figure 4.9) repeated at harmonics of the sample rate as shown in Figure 4.11. With a sample rate of 13.5 MHz, the upper sidebands of the modulated signal extend beyond the Nyquist limit and can cause high frequency chrominance to be aliased to fall near to the subcarrier frequency. For a digital coder, then, the use of a rather narrower pre-modulation low-pass filter would be more suitable to reduce the alias caused by 13.5 MHz sampling.

When the 13.5 MHz modulating signals are obtained from a digital 4:2:2 YC_RC_B input as shown in Figure 4.10b, the colour-difference signals will have already been attenuated above 2.75 MHz, with alias components from the 6.75 MHz sampling frequency present between 2.75 and 4 MHz. The spectrum of these signals is shown in Figure 4.12a. The aliased region is then further attenuated by the Gaussian filters of the modulator. However, when modulated, this region is aliased again by the 13.5 MHz sampling frequency to produce components between 5.07 and 6.32 MHz. This is shown in Figure 4.12b with the effect of the Gaussian filters omitted for clarity. As many of these components are above 5.5 MHz, the characteristic of the D–A converter low-pass filter shown in Figure 4.12c will provide some additional attenuation at the coder output. Overall, therefore, the effect of the alias components due to 6.75 MHz sampling is negligible. However, the loss of the upper sidebands above 5.5 MHz results in ringing and *U–V* crosstalk as described in a later section.

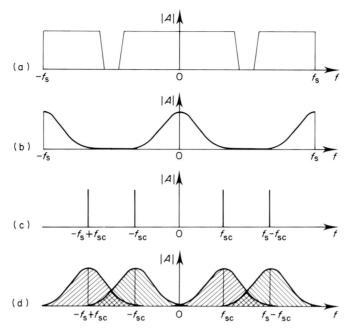

Figure 4.11 Frequency spectra in digital PAL chrominance modulation: (a) the baseband colour difference signal, (b) the Gaussian filtered colour-difference signal, (c) the subcarrier sine wave and (d) the modulated chrominance spectrum produced by convolving (b) and (c)

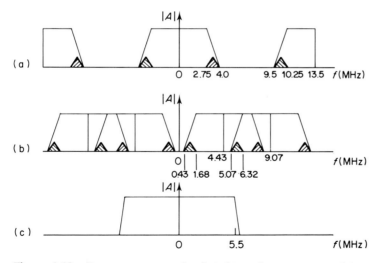

Figure 4.12 Frequency spectra for digital PAL chrominance modulation with signals derived from C_R and C_B signals sampled at 6.75 MHz: (a) the spectrum of a 13.5 MHz sampled colour-difference signal derived from 6.75 MHz samples showing the region of attenuation and aliasing (shaded); (b) the positions of the spectral components after modulation, ignoring the effect of the coder Gaussian filters and (c) the video passband region producing additional attenuation above 5.5 MHz

Signal relationships in coders

The PAL encoding process involves a number of signal conversion and scaling operations to produce the composite signal from either RGB or digital YC_RC_B component signals. If converting from analogue RGB, it is convenient to make the coding range at the A–D converters 16 to 235 for a standard level signal, to accord with CCIR Recommendation 601 [9]. The Y signal can be obtained from RGB using the following relationship [85]

$$Y = 0.299R + 0.587G + 0.114B$$

However, Y can be calculated more conveniently from the rearranged form [89]

$$Y = 0.299(R - G) + G + 0.114(B - G)$$

with the colour-difference signals $R - Y$ and $B - Y$ being obtained subsequently by subtraction. The matrix arrangement based on this relationship is shown in Figure 4.13.

Alternatively, $R - Y$ and $B - Y$ can be obtained as shown in Figure 4.10b from the digital colour-difference signals C_R and C_B by the following relationships [9]

$$B - Y = \frac{0.886}{0.5} \frac{219}{224} C_B$$

$$R - Y = \frac{0.701}{0.5} \frac{219}{224} C_R$$

Further gain changes are required to convert the $R - Y$ and $B - Y$ signals to the weighted colour-difference signals U and V [85]

$$U = 0.493(B - Y)$$

and

$$V = 0.877(R - Y)$$

and an additional factor of 0.639, that is 140/219, must be included in Y, U and V to optimise the coding range of the composite PAL signal for D–A conversion.

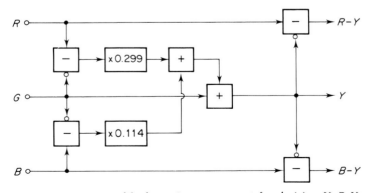

Figure 4.13 A simplified matrix arrangement for deriving Y, R–Y, and B–Y signals from R, G and B

For the colour-difference signals it is advantageous to combine these two sets of scaling factors into a single factor for each signal. These can be applied conveniently by weighting the subcarrier sine and cosine values stored in each ROM, although the colour burst amplitudes have to be adjusted to take account of the subsequent gain factor. The luminance path, however, requires a dedicated ROM for gain conversion and to alter the black level code from 16 to 64 in the composite output signal. The output levels for a composite PAL 100% colour bars signal are then as shown in Figure 4.14 with sync. bottom at level 4 and chrominance at the top of the yellow bar reaching level 251. Although Figure 4.14 shows codes appropriate for an 8-bit signal, it is advantageous to carry additional lower significance bits through from the modulators. Then a 10-bit D–A converter can be used with only a small increase in cost.

Synchronising pulse and colour burst generation

With line-locked sampling, although the generation of subcarrier waveforms is more complicated than it would be with subcarrier-locked sampling, the generation of substantially line-repetitive waveforms such as synchronising pulses, blanking edges and the colour burst envelope is greatly simplified. These waveforms are generated from the 13.5 MHz clock frequency by a combination of sample-rate and line-rate counters, for PAL, dividing by 864 and 625 respectively. The two counters, shown in Figure 4.15a, are locked to line and picture pulses derived from either mixed syncs, in the case of an analogue *RGBS* signal, or the TRS code, for a digital component input. In the absence of an input signal, the counters can free-wheel so that standard sync. and burst waveforms are maintained at the output.

The method used to generate the mixed sync., burst gate and mixed blanking waveforms was originally developed (and subsequently enhanced by P. W. Fraser) for use in an electronic zone plate generator based on line-locked sampling [90]. The two-level control waveforms are produced by the system of read-only memories shown in Figure 4.15b driven by the sample and line-rate counters. One ROM converts the sample numbers into codes representing the periods between all possible transition points of the

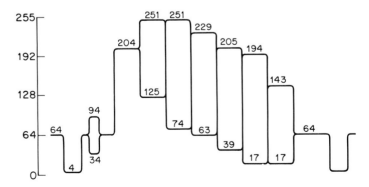

Figure 4.14 8-bit coding levels in a 100% colour bar composite PAL signal

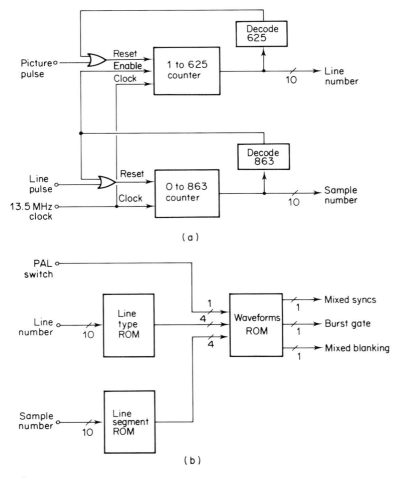

Figure 4.15 Line-repetitive waveform generation for the 625/50 PAL system: (a) sample and line-rate counters and (b) the arrangement of ROMs used to generate sync., burst and blanking waveforms

three waveforms over the line period. These are shown in Figure 4.16 with the thirteen line segments between the transition points labelled 0 to C as hexadecimal numbers. A second ROM is used to convert the line number from the line-rate counter into codes 0 to E, as hexadecimal numbers, identifying fifteen different types of composite lines from which the complete set of waveforms can be assembled: for example, broad pulses, broad pulse followed by equalising pulse, equalising pulses, etc. The output codes identifying the line segment and the line type are used in a third ROM which defines the state of each control waveform, high or low, in each segment on each type of line. For some line types, the presence of the colour burst depends on whether the first and second or third and fourth fields are being generated. Because of this, the V-axis which has the opposite sense on corresponding lines of alternate pictures, is used in the third ROM to control the generation of the burst blanking sequence.

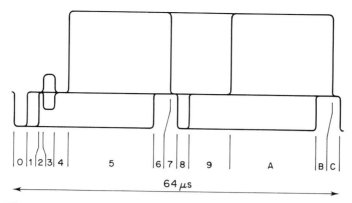

Figure 4.16 Line segment positions for mixed sync., mixed blanking and burst gate in 625/50 PAL

While the two-level control waveforms generated by the third ROM in Figure 4.15b identify the positions of the edges to the nearest sample period, the sample values representing each edge have to be chosen to give an appropriate risetime and shape. A suitable shape of edge characteristic is obtained by integrating a raised-cosine impulse such as the '*T*' pulse shown in Figure 4.17a. This impulse and the edge produced from it,

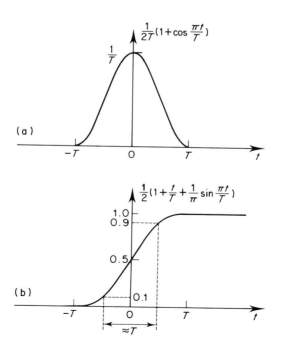

Figure 4.17 Band-limited edge generation: (a) the '*T*' pulse and (b) the edge formed by integrating a '*T*' pulse

Figure 4.17b, have the property [91] that there is virtually no signal energy beyond a frequency f where

$$f = \frac{1}{T}$$

Therefore, this characteristic provides a fast risetime, without ringing, within a well defined bandwidth. The risetime of the edge between the 10 and 90% points is also approximately equal to T (more accurately, the edge duration, 10 to 90% is equal to 0.964 T). So, appropriate sample values can be chosen for the synchronising pulse edges (nominal risetime 0.25 μs) and for the burst envelope (0.3 μs). These values can then be read from a small ROM by a counter triggered by edges in the two-level control waveforms produced by the circuit of Figure 4.15b. Similarly, values from a 0.3 μs edge can be used as coefficients in a ROM multiplier for the insertion of blanking edges.

Filter implementation

The Gaussian U and V low-pass filters used in a PAL coder can be produced conveniently using symmetrical transversal filters with an odd number of taps. A suitably accurate response can be obtained with a nine-tap filter using the five coefficients shown in Table 4.1. These coefficients were derived by choosing a characteristic in the middle of the template of Figure 4.8 and calculating the response value at five equispaced frequencies from 0 to 6.75 MHz. These values were used to solve five simultaneous equations to yield the coefficients [92].

The ladder structure of the digital transversal filters used is shown in Figure 4.18. All the weighted values of a single input sample are produced simultaneously by the coefficient multipliers a_0 to a_4. These values are then added into the delay line registers so that the taps are arranged in the correct sequence. This arrangement has the advantage that the delay line elements also act as pipeline delays for the adders. Also, if the multipliers are implemented as ROM look-up tables, the coefficient values are not subject to quantising distortion (only the products are quantised so that the filter accurately reproduces the designed filter characteristic). Filters with a more rapid rate of cut require more coefficients and can be produced by adding more of the modular stages. Quantising noise can be maintained at a suitably low level by using 10 bits for the input words (1024-word ROMs) and twelve bits throughout the adder ladder. The use of twos

Table 4.1
Coefficient values for a nine-tap transversal filter giving a Gaussian characteristic

$a_0 = 0.3153$
$a_1 = 0.2308$
$a_2 = 0.0903$
$a_3 = 0.0191$
$a_4 = 0.0021$

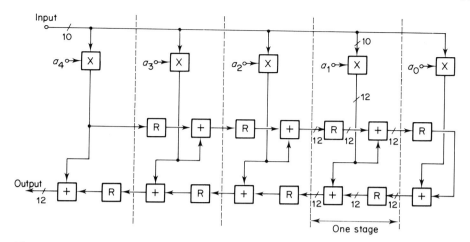

Figure 4.18 A modular transversal filter arrangement consisting of ROM multipliers, adders and registers suitable for use as a low-pass filter in a digital PAL coder

complement signal coding avoids the need to adjust signal offsets during the filtering process.

NTSC encoding

The NTSC system, although based on suppressed-carrier amplitude modulation of two orthogonal subcarriers, has two main areas of difference from PAL. First, there is no frequency offset between the subcarriers and secondly I (In-phase) and Q (Quadrature) colour-difference signals are used, instead of the U and V signals of PAL. Thus the NTSC chrominance signal can be represented mathematically by the expression

$$Q \sin \omega t + I \cos \omega t$$

where $\omega = 2\pi f_{sc}$ and for NTSC signals $f_{sc} = 3.579\,545$ MHz.

Because there is no frequency offset between the subcarriers, differential phase distortion produces static hue errors which cannot be removed by filtering. Also, sideband asymmetry results in hue errors on colour transitions. This makes the NTSC signal much less robust than PAL.

The NTSC system is based on the principle that significantly lower resolution is acceptable for some colours, notably for blue–magenta and yellow–green. So the Q and I colour-difference signal axes are rotated by $33°$ from U and V as shown in Figure 4.19 and have the following relationships to U and V

$$Q = U \cos 33° + V \sin 33°$$

and

$$I = V \cos 33° - U \sin 33°$$

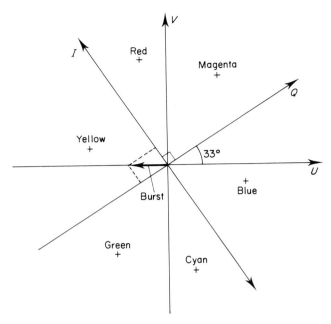

Figure 4.19 Positions of the *I* and *Q* modulation axes of the NTSC system relative to *U* and *V*, the colour primaries and their complements, and the NTSC colour burst

This rotation aligns the Q signal with the axis of colours least affected by a loss of resolution, so minimising the impairment caused by encoding the Q signal with a low bandwidth.

In practice, with the arrangements of Figure 4.10, the I and Q signals are produced either by matrixing directly from $B - Y$ and $R - Y$ signals, or from C_R and C_B signals, using the additional factors given previously. These matrix arrangements are shown in Figure 4.20.

As the signal bandwidth of the NTSC system is only 4.2 MHz, the spectrum space available for the chrominance components is relatively limited. In particular, making the subcarrier a fine, high frequency pattern to reduce its visibility results in a significant loss of the upper chrominance sidebands. If both colour-difference components were wideband, the asymmetry would cause crosstalk between the two signals. However, as the Q signal is low bandwidth, it remains a double sideband signal and causes no crosstalk. The higher frequencies from the wideband I signal, on the other hand, do cross into the Q signals, but at higher frequencies than the real Q frequencies. The I–Q crosstalk components can therefore be removed by low-pass filtering the demodulated Q signal.

The specifications [85] for I and Q coder low-pass filters are shown in Figure 4.21. While the I characteristic is similar to that of the Gaussian U and V filters used for PAL, the Q filter bandwidth is only about 0.5 MHz and has a much sharper roll-off. Although a fair approximation to the Q filter specification can be produced with a 23-tap filter using the coefficients shown in Table 4.2, possibly twice as many taps as this would be needed

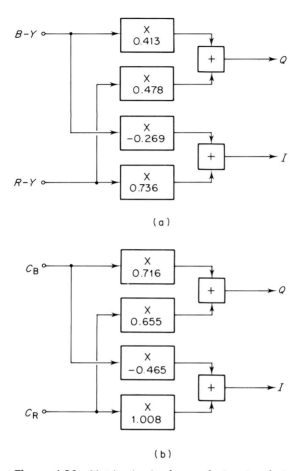

Figure 4.20 Matrix circuits for producing *I* and *Q* colour-difference signals: (a) from *B–Y* and *R–Y* and (b) from C_R and C_B

to match the template accurately. Additional delays are needed in the *I* and *Y* signal channels to compensate for the extra delay of the *Q* filter.

As the NTSC colour burst lies on the $-U$ axis, as shown in Figure 4.19, this could be generated as proportions of *I* and *Q*. However, in a digital coder, it is perhaps more convenient to insert the envelope purely in the *Q* signal and to alter the subcarrier reference phase by the addition of a 33° offset during the burst. Also phase switching of the *V* subcarrier is not required for NTSC.

The final feature of the NTSC signal is included by low-pass filtering the encoded signal to 4.2 MHz to remove part of the upper sideband of the *I* signal. Where access to a digital NTSC signal is not required, this low-pass action can be produced by using a 4.2 MHz filter in the digital-to-analogue converter.

A considerable simplification of a digital NTSC coder can be made by ignoring the need for a narrow bandwidth *Q* filter. Instead, wideband Gaussian filters can be used in both

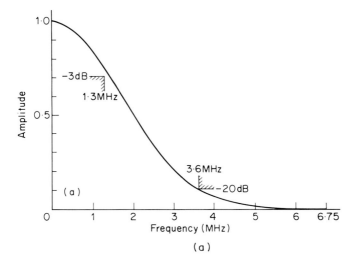

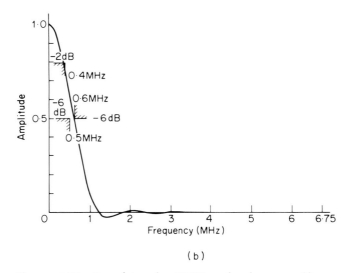

Figure 4.21 Templates for NTSC coder low-pass filters: (a) the template for the *I* channel showing, as an example, the characteristic produced by the coefficients of Table 4.1 and (b) the template for the *Q* channel with the characteristic produced by a 23-tap approximation using the coefficients of Table 4.2

channels, as in PAL. If both filters are the same, the matrix used to rotate the modulation axes to *I* and *Q* is not needed, so that the *U* and *V* signals can be modulated directly. Also, the colour burst can be generated directly in the *U* channel without the need to offset the subcarrier by 33° and there is no need for a 4.2 MHz output low-pass filter.

In the studio this so-called 'broadband' NTSC remains compatible with normal NTSC and provides improved quality on colour transitions. When bandlimited to 4.2 MHz as

CHROMINANCE DEMODULATION

Table 4.2
Coefficient values for a 23-tap approximation to the NTSC Q-filter specification

$$a_0 = 0.0889$$
$$a_1 = 0.0872$$
$$a_2 = 0.0821$$
$$a_3 = 0.0742$$
$$a_4 = 0.0639$$
$$a_5 = 0.0522$$
$$a_6 = 0.0398$$
$$a_7 = 0.0278$$
$$a_8 = 0.0171$$
$$a_9 = 0.0085$$
$$a_{10} = 0.0027$$
$$a_{11} = 0.0002$$

may occur in a transmitter, the upper sideband is lost. This results in crosstalk between the Q and I channels in the decoder on the high frequency colour components. This would show up mostly as static hue errors on edges, although the effect is unlikely to be very serious. This is because the hue is incorrect on edges in normal NTSC due to the loss of high frequency Q signals. The broadband approach could, therefore, be included as a simple modification to a digital PAL coder in circumstances where the full extra complexity of NTSC coding cannot be justified.

CHROMINANCE DEMODULATION

Decoder configurations

The main processes of a colour decoder consist of separating the luminance and chrominance, and demodulating the chrominance to produce two colour-difference signals. However, these functions need not be carried out precisely in this order. When comb filters are used, it is often convenient to obtain the luminance components by first separating a chrominance signal and using this to cancel appropriately delayed chrominance components from the input signal, as shown in Figure 4.22a. Although the use of different chrominance pass characteristics is often advantageous, it is sometimes possible to simplify the circuit to use only one filter, as in Figure 4.22b. A further rearrangement which can produce equivalent results is shown in Figure 4.22c. In this case the chrominance is demodulated first and then the unwanted luminance components (cross-colour) are suppressed with baseband colour-difference filters. Chrominance components in the input signal (cross-luminance) are cancelled using a remodulated version of the filtered colour-difference signals. Whichever arrangement is used, the demodulation process can be performed using the techniques described in this section. Also described here are techniques such as PAL modification which is used in chrominance–luminance separation filters, but has much in common with synchronous demodulation.

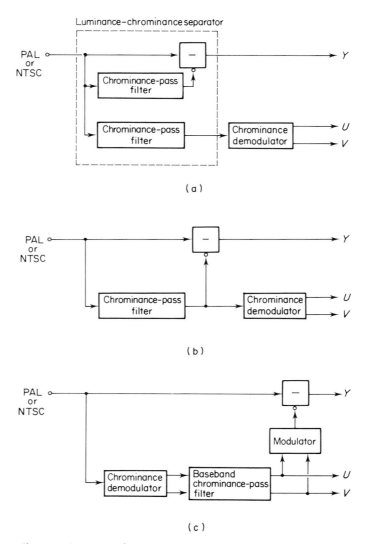

Figure 4.22 Decoder configurations: (a) with non-complementary luminance and chrominance separation filters, (b) with complementary filtering and (c) with complementary filtering of the baseband colour-difference signals

Line-locked sampling of composite signals

Effects of quantising distortion

When a television signal is encoded as a stream of binary numbers, the process of quantisation distorts the signal to produce low-level harmonics of the signal frequencies. The quantisation errors are partly the result of the fundamental accuracy of coding (usually at least 8 bits are used) and partly the result of instrumental errors in A–D and

D–A conversion equipment. When coupled with the effect of sampling, the harmonic components can be converted to low frequencies which may be visible as beat patterns on the picture. These are more noticeable on composite colour signals when the high frequency chrominance components are demodulated to form baseband colour-difference signals.

The form of the beat patterns depends on both the signals and the sampling frequency, but special cases occur with subcarrier-locked sampling and signals representing areas of constant colour, for example, colour bars. Then the distortion results in a constant hue error which is much less noticeable. With the same signal and line-locked samples, the beat pattern visibility is not dissimilar to that at other frequencies (non-line-locked). For this reason, subcarrier-locked sampling was regarded as superior to line-locked sampling [93]. However, this is the result of the exactly constant levels of the colour bar signal and is not representative when other signals are considered. With gradually changing signals, such as a constant hue on a luminance sawtooth signal, the beat patterns produced by subcarrier-locked sampling are somewhat more noticeable than those produced by line-locked sampling. In general, therefore, there is little to choose between the two.

Fortunately with modern A–D conversion equipment and 8-bit quantisation, the level of beat patterns introduced by digital encoding is below the level of perception [69, 94]. Also the use of low-level dither signals [69] in the A–D conversion process can disguise the beat patterns for any sampling frequency by converting them to random noise. Similar results are obtained when digitising NTSC signals.

Sampling frequency stabilisation

If the comb filters of a digital decoder contain field or picture delays, then the pulses used to sample the composite colour signal have to be particularly stable. With such a long delay between storing and reading the samples, even a small clock frequency error can produce a variation of delay which is a significant proportion of a subcarrier period. Such delay variations would destroy the effectiveness of a comb filter for separating chrominance and luminance signals.

It has proved much easier, in practice, to obtain a stable sampling frequency with subcarrier-locked sampling, in which the colour burst is used as the reference, than with line-locked sampling, which uses the line pulse falling edge as the reference. This might be ascribed to the greater timing information content of the burst, although this has been calculated by T. A. Moore as being between only two and seven times that of the sync. edge, depending on whether the noise spectrum of the video signal is assumed to be triangular or flat. A more significant factor is the use of analogue slicing techniques, which are particularly susceptible to noise amplification, in obtaining the line pulse reference. These are less effective than the synchronous demodulators used in phase comparators for subcarrier-locked sampling.

A more effective phase comparator for line-locked sampling can be produced using digital circuitry immediately following the A–D converter, as shown in Figure 4.23. The digital sample values describing the sync. pulse falling edge are compared in Figure 4.23(a) with a line pulse produced by division from the clock frequency. The resulting control signal, changing at line rate, is converted to analogue form to adjust the frequency of a 13.5 MHz crystal oscillator using a conventional loop filter [95].

The phase comparator, shown in more detail in Figure 4.23b, requires the black level of the video signal to be clamped in the A–D converter, usually at level 64 for PAL. Then with a standard level signal and the converter gain set to 5 mV per step, the coding level corresponding to the line timing reference point is 34 as shown in Figure 4.24a. Because of this, 34 is subtracted from each video sample so that the waveform value at the timing reference point is zero. A control waveform could be produced by sampling this waveform with the line pulse from the 864 divider. This would give zero error when the divider line pulse was coincident with the timing reference point. However, a more effective control signal can be obtained by summing a series of weighted samples from the region of the sync. edge using the multiplier-accumulator shown in Figure 4.23b. This arrangement uses more of the timing information from the sync. edge and can be used to suppress noise. Accumulating the weighted samples is equivalent to a transversal filter with an impulse response given by the series of weighting factors. The optimum suppression of noise can be obtained by relating the frequency characteristic of this filter to the spectral content of the edge. Also the filter should have zero response at subcarrier frequency to ensure that chrominance components extending below black level are suppressed. The frequency characteristic of such a filter is shown in Figure 4.25, which is produced by the coefficients of Table 4.3.

To perform the filtering, the series of weighting factors are read from a read-only memory by a counter triggered from the 864 divider line pulse. This produces the

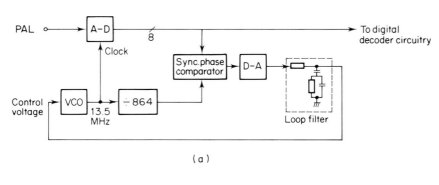

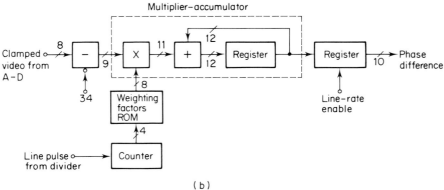

Figure 4.23 (a) A digital loop for stabilisation of a line-locked sampling oscillator. (b) The sync. phase comparator shown in more detail

CHROMINANCE DEMODULATION

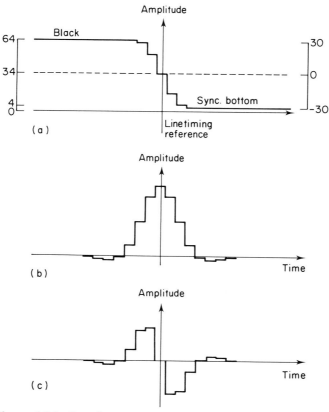

Figure 4.24 Waveforms in the digital line phase comparator: (a) the line sync. edge, (b) the series of weighting factors and (c) the weighted sync. edge samples

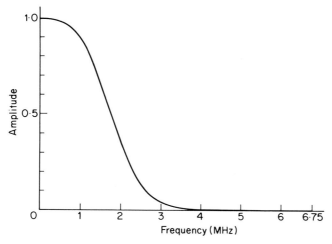

Figure 4.25 Low-pass characteristic corresponding to the filtering action of the line-phase comparator

Table 4.3
Weighting factors used in the line phase comparator

$a_0 =$	0.265
$a_1 =$	0.224
$a_2 =$	0.132
$a_3 =$	0.044
$a_4 =$	-0.004
$a_5 =$	-0.016
$a_6 =$	-0.011
$a_7 =$	-0.003

waveform of Figure 4.24b at the multiplier input which, when multiplied by the sync. edge, produces the waveform of Figure 4.24c. When the central weighting factor is coincident with the line timing reference, this waveform integrates to zero.

Although the loop will lock with the phase detector alone, the lock-up time can be rather slower than necessary if the 864 divider phase has a large offset from the incoming sync. phase. This can be avoided by initially using a conventional sync. separator to force the divider into approximately the correct phase before transferring control to the phase detector.

The phase detector of Figure 4.23b uses the assumption that the input signal will be at standard level. If the signal level is not standard, the locking position will vary with the sync. height until at half amplitude the circuit will fail to lock. This can be avoided by a small modification which involves subtracting 64 from the incoming signal instead of 34. The filter, therefore, accumulates a negative value at the half amplitude position of the sync. edge which is then offset by subsequently accumulating additional values taken from the sync. bottom. Locking at half the sync. height is then assured by weighting the samples taken at the sync. bottom by coefficients with half the amplitude and the opposite sign to those used on the sync. edge.

Implementation of the line phase detection circuits after the A-D converter eliminates the possibility of noise pick-up associated with analogue sync. slicing techniques. This provides sufficient sampling clock stability to avoid any significant variations in field and picture delay comb filters. Also, because the A-D converter is included in the control loop, a specified sampling pulse to sync. phase is ensured in the digital signal. Thus registration of the picture information on the sampling grid is reproducible.

Although controlled by the phase-locked loop, the stability of the sampling pulse is in part dependent on the basic stability of the sampling clock oscillator. For this reason, a crystal oscillator is normally used. However, the locking range of a simple crystal oscillator is insufficient to accommodate the full range of non-standard line frequencies. While sufficient range for all non-standard signals could be obtained with an L–C oscillator, adequate basic stability might be difficult to achieve.

PAL chrominance demodulation

As described in the previous section the PAL chrominance signal can be represented by the expression

$$U \sin \omega t \pm V \cos \omega t$$

CHROMINANCE DEMODULATION

with the sign of the V component alternating from one line to the next. The colour information can be demodulated by multiplying the signal by appropriately phased subcarrier frequency sine waves and low-pass filtering the resulting products, as shown in Figure 4.26a. This is known as synchronous demodulation. U is obtained by multiplying by $2 \sin \omega t$ and V is obtained by multiplying by $\pm 2 \cos \omega t$, so that

$$(U \sin \omega t \pm V \cos \omega t) 2 \sin \omega t = U - U \cos 2\omega t \pm V \sin 2\omega t$$

and

$$(U \sin \omega t \pm V \cos \omega t)(\pm 2 \cos \omega t) = V \pm U \sin 2\omega t + V \cos 2\omega t$$

In each case, the twice subcarrier frequency (2ω) terms are removed by the low-pass filters, leaving only the U and V signals at the demodulator outputs. Using a switched subcarrier waveform in the V channel removes the PAL switch modulation as part of the same process.

If the local reference subcarrier phase were incorrect, as could result from differential phase distortion, then the line-to-line pattern known as Hanover bars would result. This

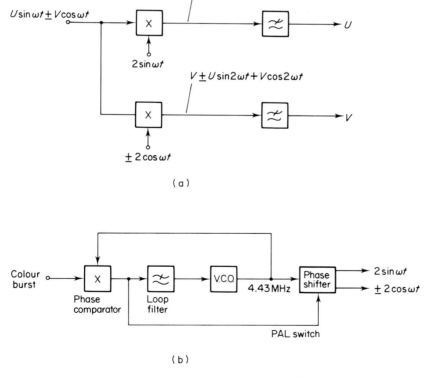

Figure 4.26 PAL chrominance demodulation: (a) signals in a PAL demodulator and (b) the separate phase detector arrangement used in a conventional analogue demodulator

can be shown by including a reference phase error θ making the demodulating subcarriers $2 \sin(\omega t - \theta)$ and $\pm 2 \cos(\omega t - \theta)$, so that now

$$(U \sin \omega t \pm V \cos \omega t) \, 2 \sin (\omega t - \theta)$$

and

$$(U \sin \omega t \pm V \cos \omega t)[\pm 2 \cos (\omega t - \theta)]$$

produce

$$U \cos \omega \mp V \sin \theta$$

and

$$V \cos \theta \pm U \sin \theta$$

after low-pass filtering. In areas of constant colour, averaging equal contributions from odd and even lines by, for example, averaging across a line delay, in each case cancels the alternating crosstalk component, leaving only a desaturation of the true component by the factor $\cos \theta$.

In a conventional analogue demodulator, the incoming reference burst is used to lock a subcarrier frequency crystal oscillator which includes its own phase detector, as shown in Figure 4.26b. Alternate bursts pull the oscillator towards 135° and 225° so that, by including a long time constant in the loop filter, the reference oscillator takes up the mean phase of 180°. The error signal of the phase detector then provides the alternating V switch square wave directly.

The demodulation process is shown in spectral terms in Figure 4.27. Figure 4.27a represents the baseband spectrum of an analogue video signal with chrominance

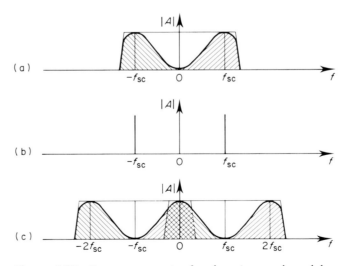

Figure 4.27 Frequency spectra for chrominance demodulation: (a) the spectrum of an analogue video signal with chrominance information in the shaded region around the colour subcarrier frequency, (b) the spectrum of a subcarrier frequency sine wave and (c) the result of convolving (a) and (b)

information in the shaded region around the colour subcarrier frequency, whilst Figure 4.27b represents the spectrum of the subcarrier frequency sinewave used for demodulation. Convolution of these two spectra (equivalent to multiplication in the time domain) produces the spectrum of Figure 4.27c in which the baseband video spectrum has been shifted to be centred on f_{sc} and $-f_{sc}$. Therefore, the chrominance information which now appears as a baseband signal can be separated from the low frequency luminance centred on f_{sc} and the chrominance centred on $2f_{sc}$ by a low-pass filter.

The frequency characteristic of the demodulation low-pass filters is a compromise between several factors. To extract all the transmitted chrominance resolution requires a wide passband, which would increase the amount of cross-colour because a greater range of luminance frequencies would be included. Also, above 1.07 MHz, the loss of the upper chrominance sidebands would introduce ringing and U–V crosstalk. Filters with a sharp-cut at 1.07 MHz would maximise the retention of chrominance while avoiding U–V crosstalk on edges and much of the cross-colour. However, such filters would accentuate ringing on chrominance edges and for this reason slow roll-off filters are normally used. For professional decoders, these have a relatively wide passband, similar to the colour-difference filters used in PAL coders (Figure 4.8). Decoders using in domestic receivers, however, tend to have a narrow bandwidth, thus producing poorer resolution, but reducing cross-colour, ringing and U–V crosstalk on edges.

Digital demodulation

In a sampled system, the demodulation process is similar to that for continuous signals except that the baseband spectra of Figure 4.27 are repeated around harmonics of the sampling frequency. This is shown in Figure 4.28 for a sampling frequency of 13.5 MHz which is slightly greater than three times the PAL subcarrier frequency. While the sampled video and subcarrier spectra, Figure 4.28a and 4.28b, are identical to those of Figure 4.27 up to half the sampling frequency, the repeated spectra of the demodulated signal, shown in Figure 4.28c, overlap to produce alias components which were not present in the corresponding analogue signal spectrum. However, with sampling frequencies above the Nyquist rate for the video signal, these alias components do not overlap with the main components of the baseband chrominance spectrum and, therefore, can still be removed by low-pass filtering.

In a digital decoder, the synchronous demodulators of Figure 4.26a consist of four-quadrant digital multipliers followed by digital transversal filters. A similar filter design method can be used to that described earlier for the coder. However, the decoder requires a slightly sharper roll-off to ensure suppression of the sampling alias components which, with a 13.5 MHz sampling frequency, start to become significant above the 3 MHz region. A suitable characteristic is shown in Figure 4.29 which can be obtained with a 15-tap filter using the coefficients of Table 4.4.

Subcarrier-to-burst synchronisation

The subcarrier generation process in a digital demodulator is based on the same ratio counting method already described. However, the local phase signal generated by the ratio counters must be adjusted initially to match the reference phase of the incoming

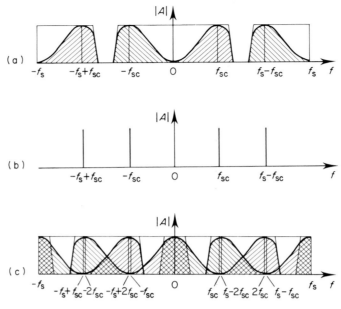

Figure 4.28 Frequency spectra for digital PAL chrominance demodulation at $f_s = 13.5$ MHz: (a) the spectrum of sampled composite video, (b) the spectrum of a sampled subcarrier sine wave and (c) the result of convolving (a) and (b)

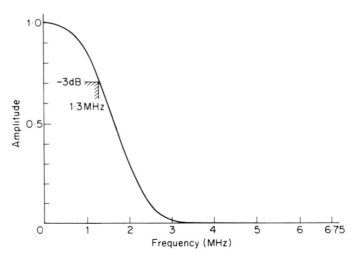

Figure 4.29 The frequency characteristic of a low-pass filter for use in a digital PAL demodulator. This is obtained from the coefficients of Table 4.4

CHROMINANCE DEMODULATION

Table 4.4
Coefficient values for the demodulator low-pass filter characteristic of Figure 4.29.

$a_0 =$	0.2460
$a_1 =$	0.2136
$a_2 =$	0.1356
$a_3 =$	0.0540
$a_4 =$	0.0019
$a_5 =$	-0.0146
$a_6 =$	-0.0103
$a_7 =$	-0.0031

signal. Moreover, if the subcarrier and line frequencies are not in the normal relationship, then continuing adjustments of the local phase are necessary.

These adjustments can be made automatically by a digital phase-locked loop, the main units of which are shown in Figure 4.30. The digital loop derives a phase error signal by examining the demodulated colour burst signals produced at the V output of the demodulator. This avoids the separate multiplier used for burst phase detection in the conventional analogue demodulator shown in Figure 4.26b. Because of this, any comb filtering preceding the digital demodulation must be bypassed for the duration of the burst.

Before the loop is locked, the arbitrarily phased subcarriers demodulate the incoming signals, including the burst. This produces signals related to the difference between incoming burst and the local subcarrier phase. The phase comparator, shown in Figure 4.31a, is triggered by a pulse derived from the line timing reference to accumulate sixteen consecutive sample values from the nominally flat, middle region of the demodulated V-channel burst waveform. Although more samples could be used, this is not necessarily

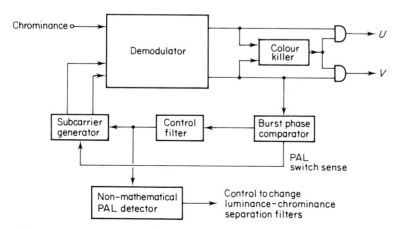

Figure 4.30 The subcarrier phase synchronisation loop in a digital decoder

168   DIGITAL ENCODING AND DECODING OF COMPOSITE TELEVISION SIGNALS

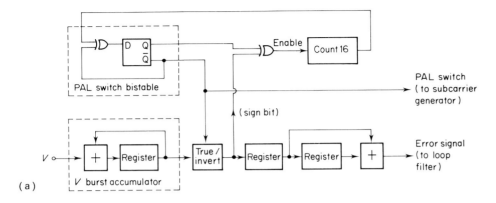

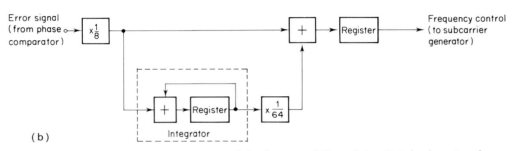

Figure 4.31 (a) The phase comparator and (b) the control filter of the digital subcarrier phase-locked loop

beneficial. This is because the edges of the burst sometimes suffer transient distortions which would tend to reduce the accuracy of phase measurement.

The phase detector also contains the local PAL switch bistable which provides the line alternating phase inversion used in generating the switched V subcarrier. The same signal is used to reintroduce the PAL switch into the control loop by inverting alternate values produced by the V-burst accumulator. The local PAL switch bistable is synchronised to the incoming signal by comparing the local switch sense with the sign of the accumulated switched burst values. If the sense is consistently incorrect for several lines, then the bistable is reset as shown in Figure 4.31a. The accumulated values from two adjacent lines are then averaged to produce the error signal. When the local subcarrier is correctly phased, the accumulated values from alternate lines cancel, so that the phase error signal is zero.

The line rate samples of the error signal produced by the phase detector are filtered to give the loop suitable stability and locking characteristics. The control filter shown in Figure 4.31b produces a subcarrier frequency control signal consisting of proportional and integrated contributions. The gain factors shown were chosen using a computer model of the loop to give a rapid, slightly underdamped lock-up with a time constant of about ten lines. A rather slower response might be preferable for very noisy signals. Also, at the beginning and end of the burst blanking period (the nine lines without bursts in each field interval) there is a failure of cancellation which causes transients in the error signal. The rapid response of the loop ensures that this is not a problem without the need to blank out these spurious error signals.

For non-mathematical operation, the control filter has to maintain a non-zero output to provide the appropriate frequency offset in the locally generated subcarrier. It should be noted that, when the line frequency is incorrect, the subcarrier frequency generated by the ratio counters will also require correction, even if the incoming subcarrier frequency is correct. This is because the sampling frequency, being line-locked, is also incorrect. With proportional control alone this would cause the loop to lock with a steady phase error. The inclusion of the integrator allows the filter to maintain an output without an input, so allowing the loop to be locked to non-mathematical signals in both frequency and phase. As designed, the loop can accommodate subcarrier frequency offsets of approximately ± 100 Hz. Although adequate for most purposes, this limitation is caused by number overflow and could be improved by extending the arithmetic to more bits.

The filtered error signal is added to the increments used in the lower section of the ratio counters to alter the local subcarrier frequency by a small amount. With the 16-bit lower-stage counter shown in Figure 4.32a, the least significant bit corresponds to approximately 0.1 Hz change in subcarrier frequency. Whereas the correct ratios must be used in a coder, the subcarrier phase generator of a decoder could be simplified by substituting an approximate fraction with a binary denominator. Although the same number of bits would be required in the ratio counter, the modulo-q overflow would be automatic, thus eliminating the data selector and one of the frequency control adders, so producing the arrangement of Figure 4.32b. The error in frequency would be corrected automatically by the phase-locked loop adding a small offset at the frequency control input.

A non-mathematical PAL input signal can be detected as shown in Figure 4.33a by integrating the signal at the frequency control input of the subcarrier phase generator for a picture period. When the correct ratios are used, this signal integrates to zero for mathematical signals, or to a known value if an approximate ratio is used. Thus, a level comparator can be used to detect if the value is significantly different from that for a mathematical signal. Integration for a picture period is used because picture content and noise can produce small line-to-line variations in the control signal. With a non-mathematical error of more than about 1 Hz in the subcarrier frequency, some field and picture-based comb filters do not completely suppress the subcarrier components. The result is cross-luminance and hue errors in coloured areas of the picture. Under these circumstances, it is possible to use the non-mathematical detection signal to switch to a line-based comb filter method, which is substantially unaffected by the frequency error.

The demodulation system of Figure 4.30 also includes a means of detecting the absence of the colour burst in monochrome signals. This circuit, shown in detail in Figure 4.33b, samples the demodulated U and V signals in the middle of each colour burst period. If neither channel includes a burst for more than the nine lines of burst blanking, the signal is recognised as being monochrome. The control signal produced can then be used to suppress the colour-difference signals, thus avoiding cross-colour. Also

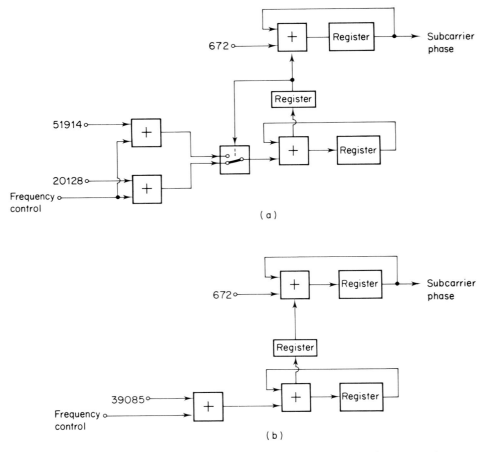

Figure 4.32 Frequency adjustment of ratio counters: (a) generating the exact subcarrier frequency for zero control input and (b) a simplified circuit which has a small frequency error for zero control input

since there will be no subcarrier to suppress, all luminance filtering can be removed so that the full signal bandwidth is displayed.

These features of the line-locked sampling demodulator method allow the full range of colour signals to be accommodated without undue complication. Also the method avoids passing on any signal frequency defects to cause complications in a subsequent digital components studio system.

Signal relationships in decoders

The conversion and scaling operations required in a digital PAL decoder are, in general, the reverse of those described for PAL encoding in an earlier section. Figure 4.34 shows the options of converting either to analogue *RGB* and syncs, or digital 4:2:2 multiplexed component signals.

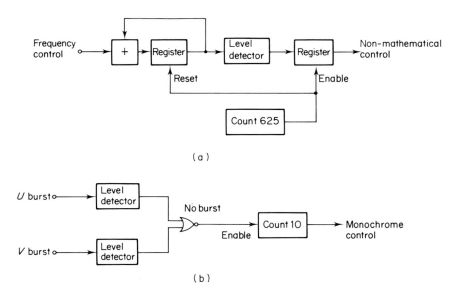

Figure 4.33 Circuit arrangements for detecting (a) non-mathematical PAL and (b) monochrome signals in a digital PAL demodulator

Unweighted demodulation of chrominance produces U and V colour-difference signals at the demodulator outputs. These can then be matrixed to produce RGB according to

$$\begin{bmatrix} R \\ G \\ B \end{bmatrix} = \begin{bmatrix} 1 & 0 & \dfrac{1}{0.877} \\ 1 & \dfrac{-0.114}{0.587 \times 0.493} & \dfrac{-0.299}{0.587 \times 0.877} \\ 1 & \dfrac{1}{0.493} & 0 \end{bmatrix} \begin{bmatrix} Y \\ U \\ V \end{bmatrix}$$

The circuit arrangement needed to produce these relationships is shown in Figure 4.35. Alternatively, the digital component signals, Y, C_R and C_B can be produced by scaling U and V according to

$$C_B = \frac{224}{219} \times \frac{0.5}{0.886} \times \frac{1}{0.493} U$$

and

$$C_R = \frac{224}{219} \times \frac{0.5}{0.701} \times \frac{1}{0.877} V$$

with the factor 224/219 arising because the Y coding range is slightly less than that used for C_R and C_B.

172   DIGITAL ENCODING AND DECODING OF COMPOSITE TELEVISION SIGNALS

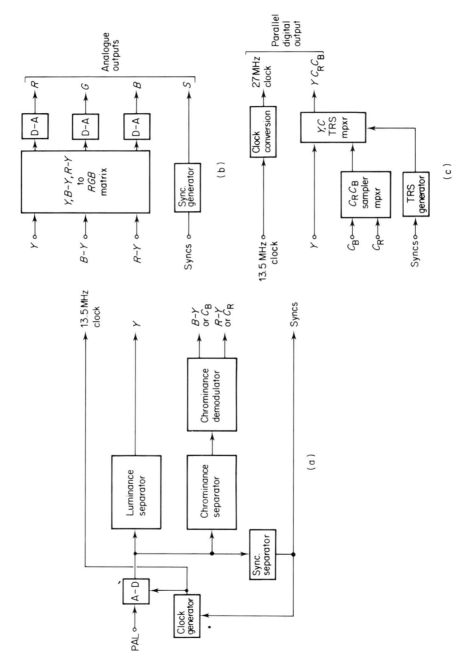

Figure 4.34 Alternative output configurations for digital colour decoding: (a) common circuitry, (b) an analogue *RGB* output interface and (c) a parallel digital output interface

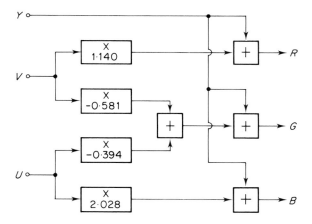

Figure 4.35 A YUV to RGB matrix

However, C_R and C_B can be produced directly by including these gain factors in the demodulator using scaled sine and cosine waveforms. In this case, the matrix to produce RGB becomes

$$\begin{bmatrix} R \\ G \\ B \end{bmatrix} = \begin{bmatrix} 1 & 0 & \frac{219}{224} \times \frac{0.701}{0.5} \\ 1 & \left(\frac{-219}{224} \times \frac{0.886}{0.5} \times \frac{0.114}{0.587}\right) & \left(\frac{-219}{224} \times \frac{0.701}{0.5} \times \frac{0.299}{0.587}\right) \\ 1 & \frac{219}{224} \times \frac{0.886}{0.5} & 0 \end{bmatrix} \begin{bmatrix} Y \\ C_B \\ C_R \end{bmatrix}$$

Another alternative is to simplify the matrix by using scaling factors of $1/0.493$ and $1/0.877$ in the demodulators to produce $B-Y$ and $R-Y$ signals directly. The relationship for RGB then becomes

$$\begin{bmatrix} R \\ G \\ B \end{bmatrix} = \begin{bmatrix} 1 & 0 & 1 \\ 1 & \frac{-0.114}{0.587} & \frac{-0.299}{0.587} \\ 1 & 1 & 0 \end{bmatrix} \begin{bmatrix} Y \\ C_B \\ C_R \end{bmatrix}$$

which requires only two non-unity coefficients. Scaling is then required in the digital component outputs according to

$$C_B = \frac{224}{219} \times \frac{0.5}{0.886} \times (B - Y)$$

and

$$C_R = \frac{224}{219} \times \frac{0.5}{0.701} \times (R - Y)$$

Additional factors are sometimes required to take account of the coding ranges used in the A–D and D–A converters. Ideally, a 9-bit A–D converter would be used with black-and-white corresponding to levels 16 and 235 in a coding range from -128 to $+383$. The standard [9] 8-bit coding ranges for Y, R, G and B would then be produced directly, with the additional bit providing range for the positive and negative excursions of chrominance and syncs. At present, however, the high cost of a 9-bit converter makes its use unattractive. Because of this, it is usual to use an 8-bit converter with black stabilised at level 64 and white set at level 204. The composite signal then extends from level 4 at the sync. bottom to level 251 at the most positive chrominance excursion. This almost minimises the quantising distortion, but leaves virtually no headroom for highly coloured overloads. With the 8-bit converter, an additional scaling factor of 219/140 is required in all three channels. While this can be combined with other factors in the colour-difference channels, this requires an additional code conversion process in the luminance channel to provide the correct gain and black level.

Because of the large gain factors involved, cross-colour signals produced by the demodulators can go well beyond the normal range of RGB signals, particularly in the B channel. Some reduction of the worst cross-colour can, therefore, be produced by limiting the signals at the matrix output to the normal R, G and B ranges. While this is generally beneficial, it should be noted that this reduces the reversibility of the decoding process and, if carried out in the sampled domain, leads to aliasing of the harmonic components produced by limiting. The latter effect can be avoided by using extra headroom in the D–A converters and limiting in the analogue domain.

Synchronising information required for the generation and insertion of timing reference signals can be obtained from the divider circuit of the line-locked sampling loop previously described. Analogue mixed synchronising pulses to accompany the RGB outputs can be regenerated using the method previously described.

Modulated chrominance processing techniques

The PAL demodulation process is inherently phase-sensitive. Because of this, some post-demodulation chrominance–luminance separation methods that can be used in the arrangement of Figure 4.22c cannot be reproduced before the demodulators, as in Figure 4.22b, using conventional filtering techniques. Instead PAL modifiers and phase-shift filters have to be included to alter the chrominance contributions from different lines to all have the same reference phase.

PAL modifiers

The alteration of subcarrier phase and V-switch sense is obtained using a PAL modifier. This consists of a multiplier fed with a twice subcarrier frequency sine wave, followed by a chrominance band-pass filter, as shown in Figure 4.36. The phase of the twice subcarrier frequency sine wave relative to the reference phase of the PAL signal determines the phase shift of the modified signal. As drawn, with a PAL chrominance signal represented by $U \sin \omega t \pm V \cos \omega t$ and multiplied by $-2 \cos 2\omega t$, there is no shift of the reference phase, but the V-switch sense is inverted.

CHROMINANCE DEMODULATION

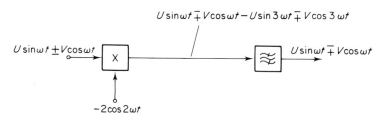

Figure 4.36 A PAL modifier

Alternatively, if $2 \sin 2\omega t$ is fed to the multiplier, then a 90° phase shift is produced as well as the V-switch inversion

$$(U \sin \omega t \pm V \cos \omega t)(2 \sin 2\omega t)$$
$$= U \cos \omega t \pm V \sin \omega t - U \cos 3\omega t \pm V \sin 3\omega t$$
$$= U \sin(\omega t + 90°) \mp V \cos (\omega t + 90°)$$
$$- U \cos 3\omega t \pm V \sin 3\omega t$$

and the terms at three-times subcarrier frequency are removed by the chrominance band-pass filter.

The action of a modifier is shown in spectral terms in Figure 4.37. Figure 4.37a represents the baseband spectrum of an analogue video signal with chrominance in the shaded region around the colour subcarrier frequency and Figure 4.37b represents the spectrum of the twice subcarrier frequency sine wave used in the modifier. Convolving these spectra produces the results shown in Figure 4.37c in which the chrominance information from f_{sc} is shifted to $-f_{sc}$ and that from $-f_{sc}$ is shifted to f_{sc}.

If the same operation were performed in a sampled system, the resulting spectrum would be that of Figure 4.37c folded back for frequencies above half the sampling frequency. The positions of the half sampling frequency are shown relative to the modified spectrum in Figure 4.37d. Although the required chrominance modification would have been achieved, aliased low frequency luminance signals centred on $2f_{sc}$ in Figure 4.37c would occupy the same band as chrominance at f_{sc}. For sampling frequencies at or near to $3f_{sc}$, for example 13.5 MHz, this can be avoided by inserting a chrominance band-pass filter before the multiplier. Figure 4.38 shows the repeated spectra of a modifier system, including prefiltering, for a sampling frequency slightly greater than $3f_{sc}$. Figure 4.38a shows the spectrum of the sampled video signal, whilst in Figure 4.38b the low frequency luminance has been removed by a chrominance band-pass filter. When the spectrum of Figure 4.38b is convolved with the sampled twice subcarrier frequency sine wave, Figure 4.38c, the modified spectrum of Figure 4.38d is produced. Finally, the unwanted products can be removed by a second chrominance band-pass filter to leave the modified spectrum as shown in Figure 4.38e.

The implementation of a PAL modifier in digital circuitry uses the same four-quadrant multiplier techniques used in the demodulator, except that the read-only memory contains an appropriately phased sine wave of twice the frequency. Although the subcarrier phase information is again derived from the demodulator reference phase, a constant phase offset is added to take account of processing delays between the modifier and the demodulators.

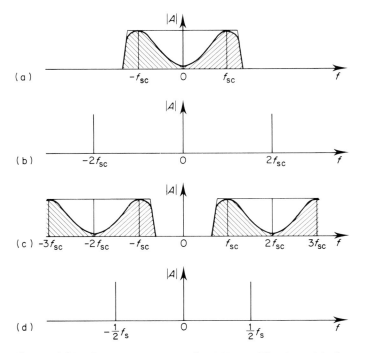

Figure 4.37 Frequency spectra for PAL modification: (a) the spectrum of a composite video signal, (b) the spectrum of the twice subcarrier frequency sine wave used in the modifier and (c) the result of convolving (a) and (b). The half sampling frequency positions of 13.5 MHz sampling are shown in (d) relative to the modified components in (c)

By considering the processes involved it can be shown that the combination of a demodulator and a remodulator is sometimes equivalent to a PAL modifier. If the colour-difference signals are remodulated in the same phase as they were demodulated, the circuit has the effect of a band-pass filter. However, if the V signal is inverted at some stage in the demodulation and remodulation process, then the effect is that of a PAL modifier. These relationships, shown in Figure 4.39, allow an equivalent filtering action to be produced either before or after demodulation.

A considerable disadvantage in the use of PAL modifiers is that their operation is adversely affected by differential phase distortion. This is because the phase shift introduced by a modifier is a function of the subcarrier reference phase in the demodulator. As differential phase distortion alters the relationship between the reference burst and the active-line chrominance signals, modified signals no longer match the phase of those from unmodified lines. In luminance separation comb filters, the cancellation of the subcarrier components depends on the phases of the individual contributions from different lines being accurately matched. Consequently, if a modifier is used, phase distortion leads to incomplete subcarrier suppression, even in plain areas. This is particularly serious for decoding applications where subsequently the signals are to be recoded to PAL, such as at the inputs to a digital component studio. Then the

CHROMINANCE DEMODULATION

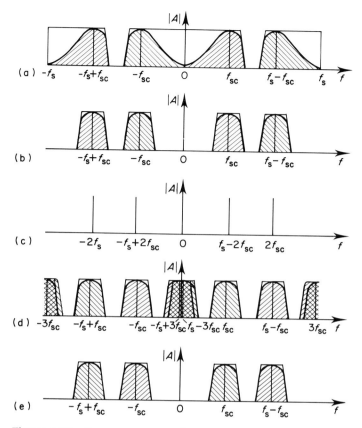

Figure 4.38 Frequency spectra for a PAL modifier operating at a sampling frequency of 13.5 MHz: (a) the spectrum of sampled video signals, (b) sampled chrominance signals obtained by band-pass filtering, (c) the spectrum of a sampled twice subcarrier frequency sine wave, (d) the result of convolving (b) and (c), and (e) the spectrum of PAL-modified chrominance obtained by band-pass filtering (d)

residual subcarrier interferes with the subcarrier in the following coder to cause a hue error when the signal is finally decoded.

Phase-shift filters

While the effect of differential phase distortion makes modifiers unattractive for luminance separation, there is no deleterious effect for chrominance. On the contrary, combining oppositely switched lines together to produce the chrominance signal for demodulation ensures suppression of Hanover bars. However, when oppositely switched lines are only used for obtaining chrominance, the complication of a modifier is unnecessary because an equivalent effect can be produced more simply. By using separate inputs to the U and V demodulators, the V-phase contributions can be inverted, where necessary, without

178 DIGITAL ENCODING AND DECODING OF COMPOSITE TELEVISION SIGNALS

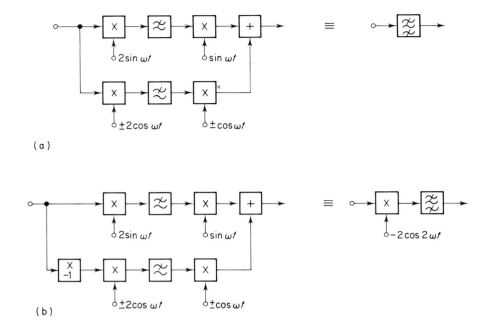

Figure 4.39 The effect of decoding and recoding: (a) with remodulation in the same phase as demodulation and (b) with the V signal remodulated with the opposite phase to demodulation

affecting the U-phase signals. Also the phase shift of 90° required in some contributions for both U and V signals can be obtained by a special form of band-pass transversal filter.

The symmetrical transversal filters previously described produce no phase shift of one frequency relative to another. However, if a filter with antisymmetrical taps and no contribution from the centre term is used, all frequencies are shifted by 90°. As the amplitude characteristic is a summation of harmonic sine wave components weighted by the filter coefficients, the response at zero frequency is necessarily zero. Thus, only band-pass or high-pass characteristics are possible. Appropriate coefficient values for the desired response can be obtained by solving a set of simultaneous equations [92], except that in this case each equation is a summation of sine terms instead of cosines.

A digital filter with antisymmetrical taps can be implemented using a similar ladder structure to that for a filter with symmetrical taps shown in Figure 4.18. Antisymmetry is obtained by altering the centre-term stage to give zero contribution and to invert the contributions added in the upper delay line. This arrangement is shown in Figure 4.40.

NTSC demodulation

Because of the similarity between the two systems, many of the features described for PAL demodulation can also be used for the NTSC system. Also many of the differences in NTSC signals have been included in the section on NTSC encoding. However, there are some additional features which are specific to NTSC demodulation.

CHROMINANCE DEMODULATION

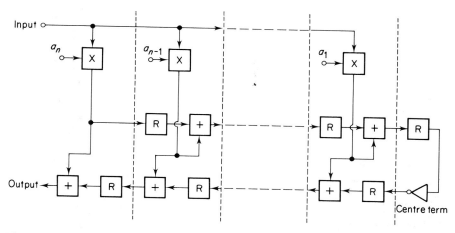

Figure 4.40 Phase shift filtering: using the alternative centre term stage shown here produces a characteristic which has a 90° phase shift at all frequencies with the same modular filter structure as in Figure 4.18

The NTSC chrominance signal, represented by the expression

$$Q \sin \omega t + I \cos \omega t$$

can be demodulated by multiplying by $2 \sin \omega t$ and $2 \cos \omega t$ and low-pass filtering the resulting products. As Q is a low bandwidth (0.5 MHz) double sideband signal and I is a wideband (1.3 MHz) asymmetrical sideband signal, different Q and I low-pass filters are required. For Q, a filter similar to that used in the coder can be used (Figure 4.21b) with the sharp cut-off ensuring that crosstalk from high frequency I signals is removed. For I, a slow roll-off filter with a somewhat narrower bandwidth than the wideband I filter used in the coder is preferred. Some gain in this filter in the range 0.5 to 1.3 MHz could be included to compensate partially for the sideband asymmetry of the I signal. However, this is of doubtful benefit because of the increase in cross-colour that would result.

After demodulation, the axis rotation of the I and Q signals has to be removed using the relationships

$$U = Q \cos 33° - I \sin 33°$$

and

$$V = Q \sin 33° + I \cos 33°$$

These relationships could be implemented with the matrix arrangement of Figure 4.41a or for *RGB* outputs, could be incorporated in the *YUV* to *RGB* matrix of Figure 4.35. Alternatively, for digital component signals, the additional relationship between U, V and C_B, C_R signals could be incorporated into the matrix coefficients producing the values shown in Figure 4.41b.

In the digital phase-locked loop only minor modifications are required to change from PAL to NTSC operation. As the NTSC signal includes no *V*-switch modulation, the PAL switch bistable must be disabled. The loop can then lock directly to the reference phase

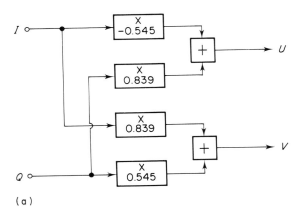
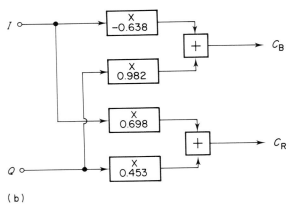

Figure 4.41 Matrix circuits for producing (a) U and V or (b) C_B and C_R colour-difference signals from I and Q

(the $+U$ axis) using as the error signal the amount of the $-U$ burst signal demodulated in the V channel. The 33° phase rotation required to generate I and Q subcarriers is then provided by a constant offset added to the reference phase signal during the active-line period. It should be noted that, although the counting ratios described earlier are relatively simple for NTSC signals, the full accuracy of the 16-bit lower stage counter is required in the demodulator. This is necessary in order to provide sufficient precision in the frequency control of the digital loop.

Similar 'broad-band' techniques to those described for NTSC coders can also provide considerable simplifications in an NTSC demodulator. As in the coder, ignoring the sideband asymmetry allows the multipliers to demodulate U and V directly and to use identical low-pass filters, thus avoiding the need for an I, Q matrix and the 33° phase offset from the burst. This approach provides much greater similarity to the PAL demodulator.

CHROMINANCE–LUMINANCE SEPARATION

Conventional decoding

While the luminance and chrominance components in a colour encoder are combined by simply adding the two signals together, separating the signals again in a decoder is much more difficult. Because of this, filtering forms a large part of the decoding process and is the most important factor in determining decoder performance.

The operation of a conventional decoder, as used in domestic television receivers, is shown in spectral terms in Figure 4.42. A notch filter is used in the luminance channel to remove the main subcarrier components as shown in Figure 4.42b. In the chrominance channel, the modulated chrominance signals are separated from low frequency luminance by a band-pass filter, as shown in Figure 4.42c and then demodulated to form the baseband spectrum of Figure 4.42d. In practice, because the spectra overlap, inevitably the signals are decoded with some chrominance in the luminance channel causing cross-luminance (unsuppressed subcarrier) and some luminance is demodulated as chrominance to produce cross-colour. Similarly, supression of the main chrominance components loses some luminance and suppression of unwanted colour effects loses some

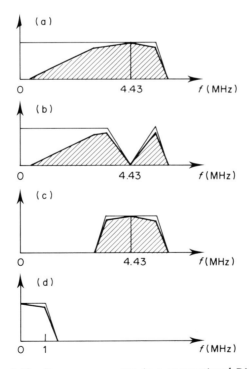

Figure 4.42 Frequency spectra in a conventional PAL decoder: (a) the composite signal spectrum (b) the luminance signal spectrum obtained using a notch filter, (c) the separated chrominance spectrum and (d) the demodulated chrominance spectrum

chrominance resolution. So differences between decoders essentially reduce to different compromises in the filters between the loss of wanted frequencies and the suppression of unwanted components.

In addition to cross-luminance and cross-colour impairments, there is the possibility of phase distortion destroying the orthogonality of the subcarriers. This results in U–V or I–Q crosstalk when the signals are demodulated. While in NTSC this results in a static hue error which cannot be removed, in PAL, the frequency offset of the V-switch modulation allows the crosstalk to be filtered out by averaging contributions from adjacent lines.

In the conventional circuit, Figure 4.43a, the 'line' delay is not exactly one line period, but is, instead, a whole number of half subcarrier periods, commonly $283\frac{1}{2}$ or $284T_{sc}$. This small change in delay acts as a 90° phase shift at subcarrier frequency and so provides a simple phase relationship between the two ends of the delay. For a $284T_{sc}$ delay, the U subcarrier signals are cophased, while V subcarrier signals are anti-phased because of the inversion of the V-switch sense from one line to the next. Thus, the low vertical frequency components of U and V can be separated by using an adder and a subtractor to feed the U and V demodulators, respectively. The high vertical frequency components of V, which pass through the adder with the low frequency U signals, are rejected by the phase discriminating property of the U channel sychronous demodulator. Similarly, high vertical frequency U components are suppressed by the V channel demodulator.

Although this use of a delay as a phase shift has adequate performance in practice, it is actually imperfect. This is because the phase shift of the delay is a function of frequency

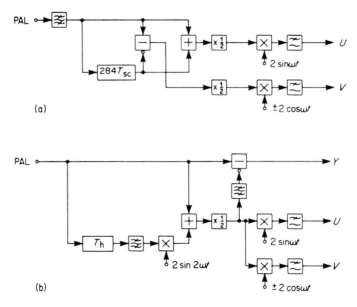

Figure 4.43 (a) The conventional form of a delay line PAL chrominance circuit and (b) a generalised circuit which is functionally equivalent

so that the subcarrier sidebands are not phase shifted by exactly 90°. This results in hue errors on vertical chrominance transitions, for example, on the edges of conventional vertical colour bars. An additional effect is that the frequency characteristic of the comb filter is diagonal, that is, partly vertical and partly horizontal, instead of being a purely vertical filter. This causes diagonal chrominance detail of one slope to be attenuated more than that of equal, but oppositely signed, slope. The correct method, therefore, would be to substitute a full line delay and a 90° phase shifter for the $284T_{sc}$ delay of Figure 4.43a. A method of designing phase shift filters has already been described.

The generalised circuit, Figure 4.43b accordingly uses a full line delay, but instead of just using the 90° phase shifter, substitutes a PAL modifier. This has the dual purpose of providing an overall 90° phase shift and inverting the V subcarrier, so that the chrominance from the modifier is identically in phase with the input chrominance. The two signals can be combined, therefore, in a common averager and share a common path to the demodulators.

In this generalised form, based on the block diagram of Figure 4.22b, the luminance and chrominance separation filters can be improved using the techniques described in the remaining parts of this section.

The three-dimensional spectrum of composite signals

Some of the deficiencies of conventional composite decoders can be overcome by using comb filtering techniques to provide additional degrees of freedom for the optimisation of filter responses. Although comb filter action can be explained by detailed examination of the comb 'teeth' in the conventional one-dimensional horizontal frequency spectrum, a much better understanding can be obtained by using the vertical and temporal frequency dimensions as well. In these dimensions the sampling action of the television raster scan has a considerable effect as described below.

Effects of scanning

The brightness of an image falling onto a television camera tube is a function varying in three dimensions: horizontal position (x), vertical position (y) and time (t). The action of scanning the image converts this three-dimensional function into a one-dimensional signal.

In the scanned signals the horizontal dimension is still continuous, but the image brightness is now only defined at discrete intervals in the vertical and temporal dimensions. Thus, the image brightness has been sampled in the two dimensions of vertical position and time.

The action of sampling repeats the frequency spectrum of the image at harmonics of the sampling frequency. The image spectrum consists of horizontal frequency (m), vertical frequency (n) and temporal frequency (f) components. Horizontal frequencies in the image are not affected by scanning because the horizontal dimension of the signal remains continuous. Therefore, only the vertical and temporal components of the image spectrum, represented diagrammatically in Figure 4.44, are affected by scanning.

If the scene from which the image is formed is completely still, then the frequencies which make up the image spectrum have no temporal component and all the spectral

184   DIGITAL ENCODING AND DECODING OF COMPOSITE TELEVISION SIGNALS

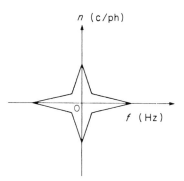

Figure 4.44 A diagrammatical representation of the vertical/temporal frequency (n–f) spectrum of an image

energy lies along the vertical frequency axis. If an object in the scene appears and disappears at a regular rate, such as a light flashing, this results in temporal frequency components which are positioned away from the n axis by an amount depending on the rate of flashing. However, most temporal frequencies are produced by the movement of objects in the scene. The effect of movement is more complicated because the temporal frequencies produced depend on both the rate of movement and the spatial frequencies which make up the object.

For purely horizontal movement, all the spectral energy is concentrated along the f axis of the n–f spectrum. If, instead, an object is moving vertically, this produces a series of temporal components along a diagonal line passing through the origin; the angle made between this line and the n axis increases with the speed of movement of the object.

There is virtually no limit to the spatial and temporal frequencies that can occur in a scene, but some filtering is introduced by the action of the camera. The camera tube integrates the light falling on it at each point for a field period, thus attenuating the higher temporal frequencies. (It is generally assumed that interlaced line positions are discharged on each field scan.) Also, the width of the scanning spot is not infinitely small, so that the charge from adjacent areas of the tube surface contributes to the output current, thereby filtering the image vertically.

The effect of scanning is to repeat the spectrum of the image of the original scene, Figure 4.44, filtered by the camera characteristic, at harmonics of the scanning rates. A 312 lines per picture, 50 fields per second sequential scan produces the spectrum shown in Figure 4.45a, while interlaced scanning, with 625 lines per picture and 50 fields per second, produces the spectrum shown in Figure 4.45b. Although only the first quadrant is shown, the spectra extend into the other quadrants in the same pattern.

As components of the spectra can extend well beyond the outlines shown in Figure 4.45, aliasing forms an accepted part of normal television pictures. For example, with interlaced scanning, flicker arises from vertical frequency components in the region around $(0, 312\frac{1}{2})$ in the image spectrum. In the scanned spectrum these components are repeated around $(25, 0)$, arising from the spectra centred on $(25, \pm 312\frac{1}{2})$. So the vertical detail flashes with a temporal frequency of 25 Hz.

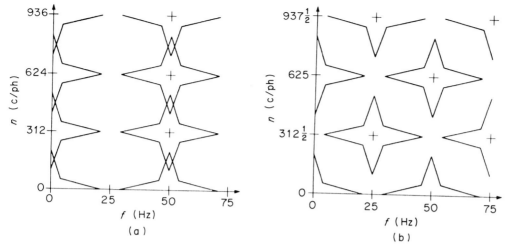

Figure 4.45 The effect of television scanning in the frequency domain: the spectrum of the image (Figure 4.44) is repeated at harmonics of the scanning rates (a) by a 312 lines per picture, 50 fields per second sequential scan and (b) by a 625 lines per picture, 50 fields per second interlaced scan

Spectra for other scanning standards would, in general, be similar to those of Figure 4.45 with appropriate scaling of the axes. In particular, for the 525 lines per picture, 60 fields per second, interlaced standard, the repeated spectra would be centred on (0, 525), (30, 262½), (60, 0), (60, 525), etc.

Colour subcarrier in the vertical/temporal spectrum

The colour subcarrier frequencies used in the PAL and NTSC systems were chosen to minimise the visibility of the subcarrier dot pattern produced on monochrome receivers. Because of this, besides being a high horizontal frequency, the subcarrier has a high vertical frequency component to ensure that there is a phase change from line to line and a high temporal frequency component to produce a phase change from picture to picture. Thus, the centres of the modulated chrominance components in the vertical/temporal spectrum are well separated from the main luminance frequencies as shown in Figure 4.46. In Figure 4.46a the offset between the U and V components results from the inversion of the V signal on alternate lines, known as the PAL switch.

The colour-difference signals produced from the image contain vertical and temporal frequency components similar to those described in the previous section. In the PAL encoded signal, these appear as sidebands extending from the colour subcarrier frequencies, Figure 4.46a. For vertical frequencies, the sidebands extend from the subcarrier positions along lines parallel to the n axis, while for temporal frequencies, as might be produced by a coloured object moving horizontally, the sidebands extend parallel to the f axis. However, the colour-difference signals usually have a horizontal bandwidth of about 1 MHz, much lower than that of the luminance signals. Because of this, horizontal movement produces much less high temporal frequency energy in the colour-difference signals. For this reason the extent of the temporal chrominance sidebands shown in the diagram is less than that shown for luminance.

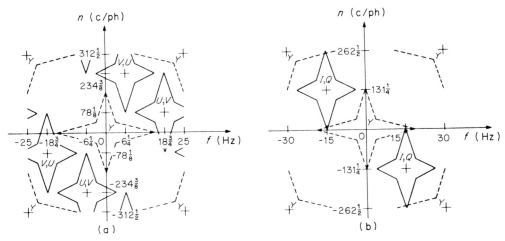

Figure 4.46 Positions of the main chrominance and luminance components in the vertical/temporal spectrum of encoded colour signals: (a) for 625/50 PAL and (b) for 525/60 NTSC. Luminance (Y) regions are marked with a broken outline

The positions of the U and V subcarriers are interchanged for positive and negative values of m. Thus, the U subcarrier is located at $(18\frac{3}{4}, 78\frac{1}{8})$, $(-6\frac{1}{4}, -234\frac{3}{8})$, etc., for positive values of m, and at $(6\frac{1}{4}, 234\frac{3}{8})$, $(-18\frac{3}{4}, -78\frac{1}{8})$, etc., for negative values of m. As well as being orthogonally phased, the V subcarrier has offsets of $-12\frac{1}{2}$ Hz and $156\frac{1}{4}$ c/ph (cycles per picture height) from the U subcarrier position due to the modulating action of the V-axis switch. The positions shown in Figure 4.46a minimise the visibility of the subcarriers by giving the highest combinations of temporal and vertical frequency consistent with the offset resulting from the V-axis switch.

For 525/60 NTSC signals, the I and Q subcarriers are only separated by their orthogonal phase relationship. Therefore, the subcarriers are both located at the same positions in the n–f spectrum, as shown in Figure 4.46b. As for PAL, this location combines the highest possible values of temporal and vertical frequency to give the minimum visibility of subcarrier. Because there is no offset to accommodate between the subcarriers, NTSC subcarriers are less visible than the equivalent PAL subcarriers for the same scanning standard. Vertical detail and horizontal movement result in modulating sidebands similar to those described for PAL.

Chrominance demodulation in the vertical/temporal spectrum

Chrominance demodulation consists of multiplying the modulated chrominance signals by subcarrier frequency sine and cosine waves. Accordingly, in the frequency domain, the modulated chrominance spectrum is convolved with an impulse function at each subcarrier position. In the n–f plane, this shifts the U, I and Q components to be centred on the origin, as shown in Figure 4.47. The spectrum for demodulated V signals is similar to Figure 4.47a with both the U and V positions, and the cross-colour $+m$ and $-m$ positions interchanged. Any luminance components entering the demodulators to produce cross-colour are similarly shifted in frequency to be centred on the positions

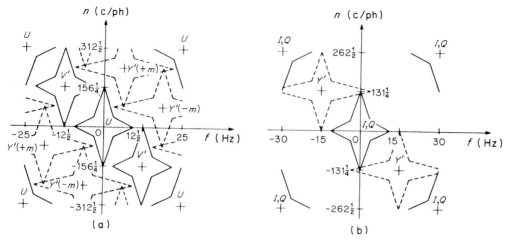

Figure 4.47 Positions of the main chrominance and luminance components in the vertical/temporal spectrum of encoded signals after chrominance demodulation: (a) for 625/50 PAL (U-channel) and (b) for 525/60 NTSC. In (a), Y' $(+m)$ and Y' $(-m)$ denote positions of the luminance (cross-colour) components for positive and negative horizontal frequencies, respectively

previously occupied by the subcarriers. Thus, the cross-colour components Y' $(\pm m)$ are centred on $(6\frac{1}{4}, 234\frac{3}{8})$ and $(18\frac{3}{4}, 78\frac{1}{8})$ for PAL and $(-15, 131\frac{1}{4})$ for NTSC. The demodulation process also shifts chrominance information to twice subcarrier frequency but these components are removed by the horizontal low-pass filters of the demodulators.

If the orthogonal phase relationship of the subcarriers has been maintained, there will be no crosstalk between the two chrominance channels. However, if there is differential phase distortion or if the alignment of either the coder or decoder is inaccurate, then some crosstalk will occur. In the PAL system, V components entering the U channel will be centred on $(\mp 12\frac{1}{2}, \pm 156\frac{1}{4})$ as shown in Figure 4.47a. This accounts for the moving pattern of horizontal lines, known as Hanover bars, which is present in plain coloured areas when a phase distorted PAL signal is decoded using a simple demodulator. In the NTSC system, when phase distortion produces crosstalk between the I and Q signals, this results in a static hue error in plain coloured areas.

Multi-dimensional luminance and chrominance filters

The conventional methods of PAL and NTSC decoding, as used in domestic television receivers, make little use of the vertical and temporal offsets of the subcarrier frequency. However, comb filters can fully exploit the offsets between the true and interfering frequency components shown in Figure 4.46.

In most cases the circuitry is simpler if chrominance and luminance are separated first, followed by chrominance demodulation as shown in Figure 4.22a. Then the chrominance separation filter needs to retain the areas centred on the colour subcarriers shown in Figure 4.46. Luminance separation is also achieved by retaining broadly the same areas which are then subtracted from the delayed input signal, thus cancelling the chrominance components and leaving luminance. Because of this, the comb filter defines a region which is 'not luminance' (that is, predominantly chrominance) and is consequently sometimes referred to as an $\bar{L}$ filter. For luminance separation, the comb filtering effect is only applied to the high horizontal frequencies in the composite signal where chrominance energy predominates, thus avoiding distortion of low frequency luminance.

Line-delay comb filters

In general, virtually all the simple, symmetrical comb filters fall into a relatively small number of categories, mostly depending on whether the length of the delays is an odd or even number of line periods. The four principal arrangements are shown in Figure 4.48, in this case based on line-delay comb filters. In each case, to achieve cancellation of chrominance components at the luminance subtractor, the main contributions from the comb filter have to match the phase of the subcarriers at the centre tap of the delays. The circuits of Figure 4.48 represent different methods of combining the delayed signals to ensure that this phasing requirement is met.

With delays of one line period, the PAL-switch sense is inverted and the phase difference is approximately 90° or 270° relative to that at the centre tap. Thus the phase of both U and V subcarriers is almost exactly inverted across the two delays. The approach taken by Roe [96] was to combine the signals in a subtractor as shown in Figure 4.48a which cancels this phase inversion and averages the two signals while making them co-phased. The remaining 90° phase difference and inversion of the PAL-switch sense is removed by the PAL modifier circuit.

In contrast, the comb filter of Figure 4.48b uses an adder to combine the signals from the extreme ends of the delays. Because the subcarrier signals from these points are in anti-phase, most of the chrominance information is cancelled, leaving luminance. This predominantly luminance signal is then subtracted from the signal at the mid-point of the delays to leave chrominance. Since most of the incorrectly phased chrominance is suppressed by cancellation, the circuit requires no modifier nor further phase shifts even though the original contributions are not co-phased. The overall effect is to produce a cosine-shaped response in the high frequency luminance region [97].

A further important method of decoding [77] using one-line delays is shown in Figure 4.48c. As shown, the Weston circuit amounts to a combination of the two methods in Figures 4.48a and 4.48b with half the chrominance being contributed by each. However, in its earliest form, rather different circuitry was used [76].

While the earlier circuits demonstrate the approaches used for one-line delays, a substantially simpler method can be used when each delay is equal to two line periods. Then all the contributions in the filter have a common PAL switch sense and are approximately cophased or in anti-phase. This allows the contributions to be combined directly as shown in Figure 4.48d to produce a raised cosine response [98] for high frequency luminance.

The chrominance–luminance separation performance of each circuit can be summarised as a set of vertical frequency characteristics. There are eight curves describing each circuit, four for chrominance, shown in Figure 4.49, and four for luminance, shown in Figure 4.50. The abscissa of each curve represents a band of frequencies extending from one line harmonic to the next. Negative portions of the characteristics are shown as

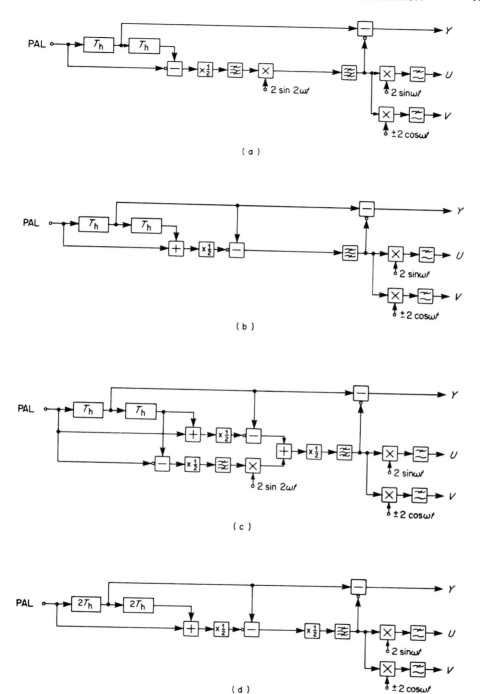

Figure 4.48 Basic line-based comb filter arrangements for PAL decoding. (a) Two-line Roe, (b) two-line cosine, (c) two-line Weston and (d) four-line cosine

190 DIGITAL ENCODING AND DECODING OF COMPOSITE TELEVISION SIGNALS

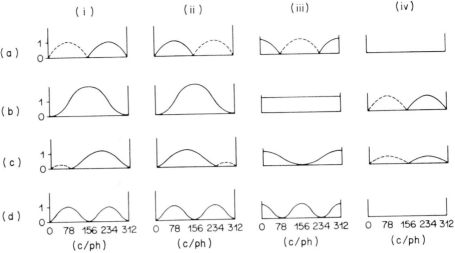

Figure 4.49 Chrominance vertical frequency characteristics of the decoders of Figure 4.48. The four curves for each decoder show: (i) U-channel modulated chrominance and cross-colour, (ii) V-channel modulated chrominance and cross-colour, (iii) U- and V-channel demodulated chrominance, and (iv) U–V crosstalk in demodulated chrominance. (a) Odd lines (direct) (b) Odd lines (cancellation), (c) Odd lines (Weston), (d) Even lines

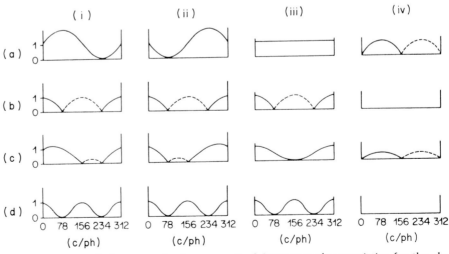

Figure 4.50 High frequency luminance vertical frequency characteristics for the decoders of Figure 4.48. The four curves for each decoder show (i) U-phase luminance and cross-luminance, (ii) V-phase luminance and cross-luminance, (iii) true luminance (components unaffected by modifier), and (iv) aliased luminance (components passing through modifier)

dashed lines. The general circuit of Figure 4.22b is shown again in Figure 4.51 marked with the positions at which the various response curves are taken.

Columns (i) and (ii) in Figure 4.49 show the vertical frequency characteristics of each chrominance filter up to the demodulator input. Two sets of curves are shown because the response of the filter for signals in the U and V phases may differ if the filter includes a modifier. In the vertical frequency spectrum of modulated chrominance, the U and V subcarrier signals are centred on the $\frac{3}{4}$ and $\frac{1}{4}$ line-offset positions respectively. Therefore, in the first column of characteristics (Figure 4.49i), for U phase signals, each curve has unity gain at the $\frac{3}{4}$ line-offset position (234 c/ph). Similarly, the second column of curves (Figure 4.49ii), for V phase signals, all have unity gain at $\frac{1}{4}$ line-offset (78 c/ph). In addition, these curves show the cross-colour performance of each circuit. Any luminance in the chrominance frequency band will be attenuated by these characteristics before being demodulated to produce cross-colour in the U or V outputs.

When the chrominance signals are demodulated, the spectrum is shifted from the chrominance band to the baseband and from the $\frac{3}{4}$ line-offset position (for U) or the $\frac{1}{4}$ line-offset position (for V) to the line harmonic position. Thus, the chrominance filtering and demodulation has the resultant effect on baseband chrominance signals shown in the third column of curves (Figure 4.49iii), which apply similarly for U and V.

In some cases, this third curve is just a shifted version of the modulated chrominance characteristics, but in Figures 4.49iiib and 4.49iiic, this is not so. For these filters, the modulated chrominance curves are asymmetrical about the U and V subcarrier positions. As a result, the demodulated output curve, produced by averaging contributions from the upper and lower sidebands, has a different shape from the modulated chrominance curve. Additionally, because of the asymmetry, the demodulators are unable to separate U and V phase signals perfectly and there is crosstalk between the U and V signals, proportionate to the degree of asymmetry. Accordingly, the fourth column, Figure 4.49iv, shows the vertical frequency U–V crosstalk characteristics for the demodulated signals.

The curves of Figure 4.50 show the vertical frequency characteristics of each luminance filter over the chrominance band, that is, at frequencies near to 4.43 MHz. As with the chrominance curves, the inclusion of a modifier can lead to different characteristics for signals in the U and V phases. The first column of curves (Figure 4.50i) shows the effect of the luminance filter for signals in the U phase, whilst the second column of curves (Figure 4.50ii) applies for signals in the V phase. Therefore, these curves show directly the degree of suppression of cross-luminance from U and V chrominance signals.

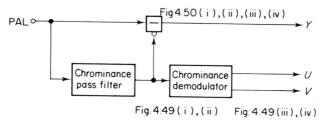

Figure 4.51 A general decoder circuit showing the positions of the chrominance and luminance characteristics of Figure 4.49 and 4.50

In particular, all the curves of Figure 4.50i have zero response at the U subcarrier ($\frac{3}{4}$ line-offset, 234 c/ph) position, whilst the curves of Figure 4.50ii all have zero response for the V subcarrier ($\frac{1}{4}$ line-offset, 78 c/ph) position. The complementary relationship between the luminance and chrominance filters is readily apparent, as the curves of Figures 4.50i and 4.50ii can be obtained by subtracting the corresponding curves of Figures 4.49i and 4.49ii from unity.

As an alternative to the characteristics for luminance signals in phase with the U and V subcarriers, Figures 4.50i and 4.50ii respectively, the effect of the luminance filter can be expressed as a combination of true and aliased luminance characteristics. Thus, the curves in the third column, Figure 4.50iii, show the action of the filter for signals not affected by the modifier, whilst the characteristics for signals that pass through the modifier and are, therefore, aliased about twice subcarrier frequency, are shown in Figure 4.50iv. Although the true and aliased luminance characteristics are shown separately for clarity, both the filtered true luminance and aliased luminance signals would be present in the output. As the aliasing adds a temporal component, the aliased luminance alternately reinforces and cancels the true luminance.

Comparison of the vertical frequency characteristics shown in Figures 4.49 and 4.50 for the different comb filter arrangements of Figure 4.48 shows that each represents a different compromise between retaining resolution and suppressing cross-effects. With the Roe circuit, Figure 4.48a, there is a substantial loss of vertical chrominance resolution as shown in Figure 4.49iiia, and chrominance at 156 c/ph is inverted, so there is no suppression of Hanover bars. Cross-colour is reduced, in particular there is no cross-colour from luminance resolution bars, as shown in Figures 4.49ia and 4.49iia. The circuit retains full luminance resolution, Figures 4.50iiia, and the full amplitude alias falls to zero at line and half-line frequency harmonics as shown in Figure 4.50iva. Luminance resolution bars are, therefore, retained at full amplitude and are free from aliasing. The possibility of overloading is present for the $\frac{1}{4}$ and $\frac{3}{4}$ line-offset luminance positions where the addition of true and aliased components can produce double amplitude signals. Cross-luminance extends over two lines on each field at horizontal chrominance transitions.

For Figure 4.48b, the vertical frequency characteristic, Figures 4.49ib and 4.49iib, is a raised cosine with nulls at the line-offset points and a gain of two at 156 c/ph. The gain is unity at both the U and V subcarrier positions. Accordingly, line-repetitive luminance gives no cross-colour, whilst diagonal luminance at 156 c/ph gives 6 dB more cross-colour than that from a simple decoder. As the circuit contains no modifier, U and V signals have the same frequency characteristic.

The demodulation process averages the responses from above and below the subcarrier positions and, in this case, the modulated chrominance characteristic is antisymmetric about the subcarrier frequencies. Because of this, the characteristics for true chrominance after demodulation are flat, since the falling response from one side-band is neutralised by an equal increase from the other sideband. However, the antisymmetry introduces crosstalk between the U and V channels. The U–V crosstalk characteristic, shown in Figure 4.49ivb, has a sinusoidal form with nulls at line and half-line offsets. The mechanism for the crosstalk is that, if the chrominance information changes from line to line, chrominance signals do not cancel in the adder. Then, because these signals are obtained from different lines, without being phase-shifted or modified, the phase of the subcarriers used for the subsequent demodulation is inappropriate. The crosstalk appears as flicker at 12.5 Hz on horizontal chrominance edges, such as those in horizontal colour bars.

For high frequency luminance, the circuit produces a cosine characteristic with nulls to suppress subcarrier signals at the $\frac{1}{4}$ and $\frac{3}{4}$ line-offset points (Figure 4.50b); diagonal luminance between these points is inverted. Because there is no modifier, the luminance signal is free of aliasing. However, cross-luminance at horizontal chrominance transitions extends over two fields lines, in common with other methods using two line delays.

The Weston decoder shown in Figure 4.48c can be seen as a combination of the two-line Roe circuit, Figure 4.48a and the two-line cosine circuit, Figure 4.48b. Because of this, the chrominance and luminance vertical frequency characteristics shown in Figures 4.49c and 4.50c are the average of those in Figures 4.49a and 4.49b and in Figures 4.50a and 4.50b, respectively.

For chrominance, the Weston circuit introduces a loss of vertical resolution falling to zero at 156 c/ph so that Hanover bars are suppressed, Figure 4.49iiic. The U–V crosstalk characteristic Figure 4.49ivc has the same form as that for the two-line cosine circuit, but the amplitude is halved. Cross-colour reaches a peak value 1.21 times that for a simple decoder at $\frac{5}{8}$ths line offset for U and $\frac{3}{8}$ths line offset for V. There is no cross-colour for line-repetitive luminance, such as test card resolution bars, Figures 4.49ic and 4.49iic.

Similarly, the Weston luminance circuit retains line-repetitive luminance, whilst diagonal luminance at 156 c/ph is lost, Figure 4.50iiic. The luminance alias, Figure 4.50ivc, is half that produced by the two-line Roe circuit and cross-luminance on horizontal chrominance transitions is present on two lines on each field. Phase distorted chrominance signals, which would lead to Hanover bars in a simple PAL decoder, appear at the luminance output of the Weston decoder as a residue of unsuppressed subcarrier signals. This occurs in other methods using PAL modifiers.

Using two two-line delays as shown in Figure 4.48d avoids both the need for a modifier and the introduction of U–V crosstalk. However, there is a serious loss of vertical chrominance and diagonal luminance resolution, Figures 4.49d and 4.50d. In addition, Hanover bars are unsuppressed and cross-luminance at horizontal colour bar transitions extends to four lines on each field.

Field- and picture-delay comb filters

As the circuits of Figure 4.48 depend primarily on the subcarrier phase relationship at the ends of the line delays, other delay lengths can be used provided that any phase differences that this incurs are taken into account. For example, the phase and PAL switch sense across a 625-line delay is almost exactly the same as that for a 1-line delay. Thus, picture delays can be substituted directly in the comb filters of Figures 4.48a, 4.48b and 4.48c. This produces temporal characteristics similar to the vertical characteristics of Figures 4.49 and 4.50, except that the abscissa would have a scale of 0 to 25 Hz. Also, with 313-line delays, the phase is inverted relative to that with a one-line delay. This can be compensated for by inverting the contributions from the ends of the delays, for example, by altering the modifier phase in Figure 4.48a or by changing the lower subtractor to an adder in Figure 4.48b. With 312-line delays, the phase and PAL-switch sense is the same as with two-line delays, so the circuit of Figure 4.48d can be used directly.

194 DIGITAL ENCODING AND DECODING OF COMPOSITE TELEVISION SIGNALS

Whereas the one-dimensional frequency characteristics of Figures 4.49 and 4.50 can be interpreted for the diagonal response of 312- and 313-line delay comb filters, these are more easily visualised in the two-dimensional vertical–temporal frequency plane. The effects of line-, field- and picture-delay comb filters in this plane are shown diagrammatically in Figure 4.52. The shaded areas in each diagram show the main frequencies rejected by the chrominance separation and $\bar{L}$ filters. For 312-line delays, the curves corresponding to Figure 4.49 and 4.50 would extend along a diagonal line from the origin to the 50 Hz, 625 c/ph point, while for 313-line delays, the path would extend from the origin to -25 Hz, $312\frac{1}{2}$ c/ph.

Of the characteristics shown in Figure 4.52, the 312- and 625-line comb filters are of most interest. Using 312-line delays best exploits the relative positions of the luminance and chrominance energy centres. Accordingly this filter gives a good balance of impairments, keeping cross-effects at relatively low levels for both still and moving pictures. With 625-line delays the response is only a function of temporal frequency. For still pictures, all the signal energy is concentrated along lines parallel to the vertical axis, located at $\pm 6\frac{1}{4}$, and $\pm 18\frac{3}{4}$ Hz. In this case, therefore, the 625-line comb filter gives perfect separation of luminance and chrominance. However, for moving pictures, the

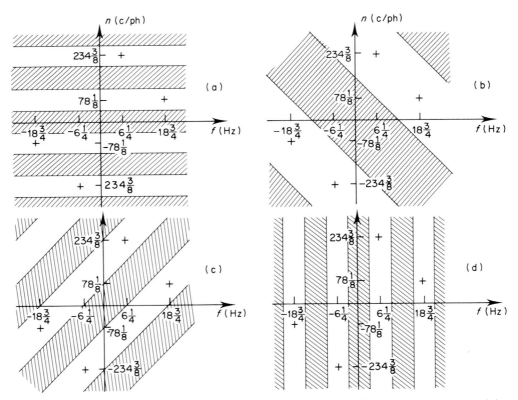

Figure 4.52 The vertical/temporal (n–f)frequency characteristics of line-, field- and picture-delay comb filters for chrominance–luminance separation. The frequencies retained are shown as hatched areas. (a) 1-line, (b) 312-line, (c) 313-line and (d) 625-line delays

temporal response is very limited, causing movement to become blurred and cross-luminance and cross-colour to return.

A further disadvantage of the 625-line Weston method, in common with other methods using PAL modifiers, is its performance on signals affected by differential phase distortion. This causes the signals derived by phase shifting chrominance from adjacent lines to be no longer in the correct phase to cancel the subcarrier on oppositely switched lines. As the distortion increases, there is a rapid increase in the level of unsuppressed subcarrier in uniformly coloured areas. When the signals are subsequently recoded, the unsuppressed subcarrier could produce hue errors at the final decoder.

Although the 312-line field-based filtering of Figure 4.52b gives a good compromise in most aspects of performance, the filter provides no suppression of U–V crosstalk components (Hanover bars). In addition it provides more resolution for one sense of vertical movement than for the other. As far as the wanted components are concerned, there is no justification for this and characteristics retaining only the lower temporal and vertical frequencies with more response along the axes, would be preferable.

Improved comb filters

The most obvious drawback of the 312-line decoder is its failure to suppress Hanover bars. This could be avoided by combining additional contributions from oppositely switched lines, so that the total of contributions from odd- and even-numbered lines was equal. Use of the 313-line contributions suppresses Hanover bars, but leads to an unsatisfactory characteristic with large negative responses. Using both 1-line and 313-line contributions in conjunction with the 312-line taps produces a more satisfactory response as shown in Figure 4.53 which has four quadrant symmetry. Alternatively, using 625-line contributions as well, as shown in Figure 4.54, improves the vertical chrominance resolution for still pictures while still suppressing Hanover bars.

The use of these additional lines requires that the contribution weights take account of the subcarrier phase on each line relative to the demodulator reference phase. As this advances approximately in multiples of 90°, this requires only that some contributions are inverted and some are passed through a 90° phase shift network operating over the chrominance band. This phase shift can be obtained using a transversal filter with an antisymmetrical impulse response as previously described. To take account of the PAL switch, some contributions in the V channel must be inverted, so separate feeds are needed to the U and V demodulators. Alternatively, if the same filter is to be used to define the luminance response, a PAL modifier could be used.

The luminance signals produced would then be complementary to the chrominance signals. The disadvantage of this is that, in circumstances where the PAL signal is affected by differential phase distortion, the impairment will appear in the luminance output as unsuppressed subcarrier.

An alternative is to use the pseudo-complementary filter obtained by inverting all the contribution weights except the centre term which is subtracted from unity. This is used to filter the input signal directly and, when high-pass filtered to define the region of combing, is subtracted from an appropriately delayed version of the composite input signal to produce luminance. This method of filtering does not require a PAL modifier and so avoids leaving areas of unsuppressed subcarrier for signals with differential phase distortion. The contributions for dual 1, 312, 313-line and dual 1, 312, 313, 625-line

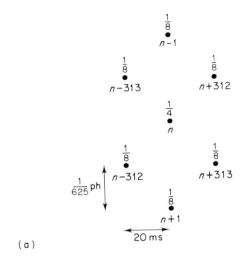

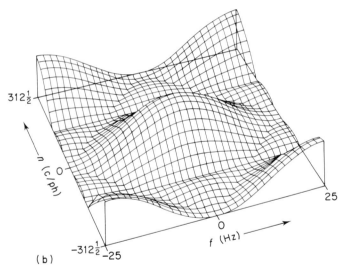

Figure 4.53 (a) Contribution weights and (b) frequency characteristics for a dual 1, 312, 313-line decoder

luminance filters are shown in Figure 4.55. This maintains the advantage of using binary fractions for most of the contributions so that the weighting factors can be accommodated by shifting the binary digits.

When chrominance filtering before demodulation is used with 'pseudo-complementary' luminance, the delays, the weighting circuits and some of the adders can be made common to both filters. Such an arrangement is shown in Figure 4.56 for the 1, 312, 313, 625-line decoder and a similar circuit could be used for the 1, 312, 313-line decoder as well. In some circumstances, the luminance circuit of a dual 312-line decoder might

CHROMINANCE–LUMINANCE SEPARATION

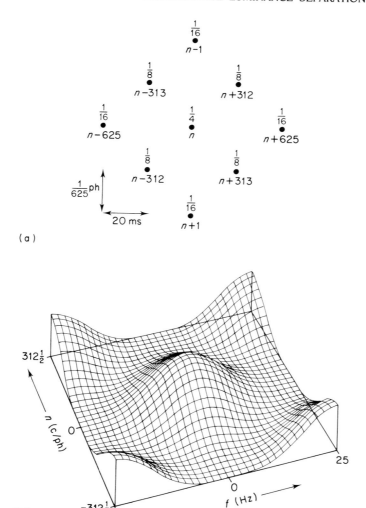

Figure 4.54 (a) Contribution weights and (b) frequency characteristics for a dual 1, 312, 313, 625-line decoder

be preferable and the modifications to provide this instead are straightforward. Particularly in the case of the dual 1, 312, 313-line decoder, this would reduce the loss of diagonal luminance and would make the cross-luminance less noticeable.

Complex comb filter design

The synthesis of two-dimensional frequency characteristics conforming to the constraints of Figure 4.57 is particularly difficult because the diamond shape required is non-variables-separable. Also, the non-rectangular array of line positions in the interlaced scan is an added complication. If the array of lines was rectangular and a rectangular

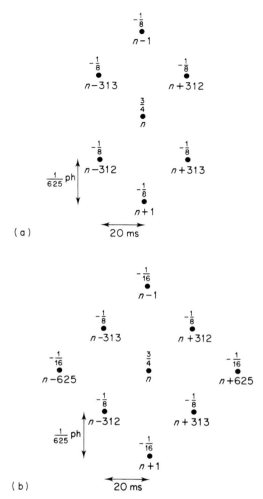

Figure 4.55 Contribution weights for 'pseudo-complementary' luminance in (a) a dual 1, 312, 313-line decoder and (b) a dual 1, 312, 313, 625-line decoder

response was required, the characteristic could be calculated as two variables-separable one-dimensional responses.

It is possible to proceed with this approach by rotating the axes of the frequency characteristic by 45° to make the response rectangular and variables separable. Also this would make the array of line positions in the interlaced scan correspond to the orthogonal structure of a sequential scan. However, from a practical viewpoint, the resulting array of contributions required would not be a good match to the array of storage available. Because line delays are cheaper than field delays and a repetitive structure is easier to construct, the contributions provided by the store tend to produce a rectangular array of contributions on the original axes, as shown in Figure 4.58. Also

CHROMINANCE–LUMINANCE SEPARATION

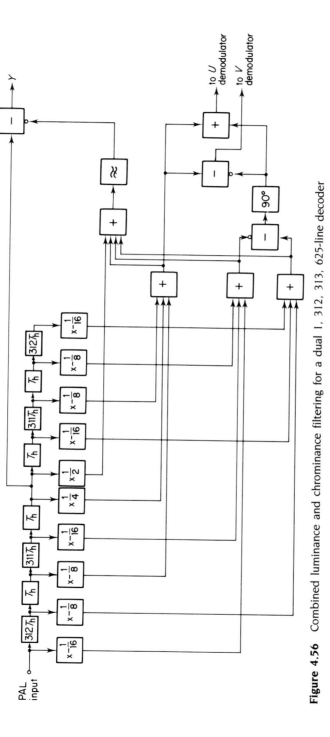

Figure 4.56 Combined luminance and chrominance filtering for a dual 1, 312, 313, 625-line decoder

200 DIGITAL ENCODING AND DECODING OF COMPOSITE TELEVISION SIGNALS

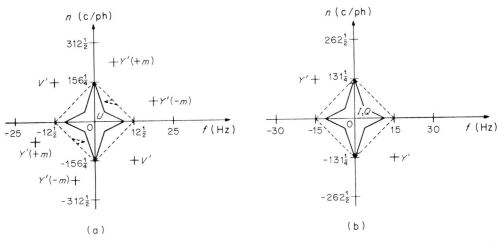

Figure 4.57 Suppression of unwanted components in the demodulated chrominance signals using a filter characteristic with reduced response to vertical and temporal frequencies: (a) For 625/50 PAL (U-channel and (b) For 525/60 NTSC. In (a), Y' ($+m$) and Y' ($-m$) denote the positions of the luminance (cross-colour) components for positive and negative horizontal frequencies, respectively

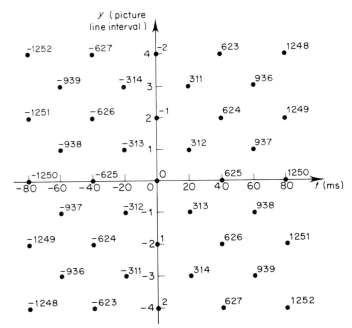

Figure 4.58 The array of contributions available in the vertical/temporal filters

there is no justification for treating vertical and temporal frequencies in the same way, which would be a necessary consequence of this approach.

A preferable alternative is to treat the available array of lines as being rectangular by adding the missing line positions of the corresponding sequential scan. This results in no loss of generality. If the frequency characteristic is chosen to ensure that coefficient values at the added line positions are always zero, the remaining coefficients can be immediately applied to an interlaced signal application.

A commonly used method of filter design is to calculate the coefficients for an ideal characteristic and to use an approximation by 'windowing' the coefficients to match the available contributions. In this case, because the number of taps in each dimension is relatively small, this method is not very suitable because the choice of the window function virtually determines the shape of the response. Instead it is much better to calculate the frequency characteristic by a method which automatically includes the limitations of the filter size. Specifically, the overall dimensions of the filter aperture determine the spacing of points at which the frequency characteristics can be specified.

The basis for such a method is demonstrated for a one-dimensional case in Figures 4.59 and 4.60. The discrete, aperiodic impulse response of the filter shown in Figure

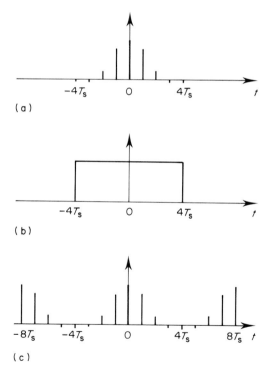

Figure 4.59 The constraint of limited filter width in the time domain: (a) the discrete, aperiodic impulse response of a transversal filter may be considered as the product of (b) a block pulse equal to the aperture width of the filter and (c) a discrete, periodic impulse response

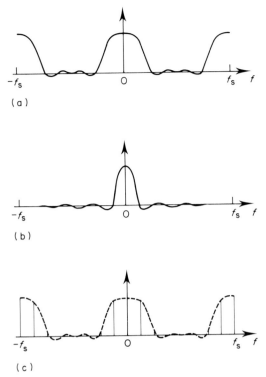

Figure 4.60 Spectra corresponding to the waveforms of Figure 4.14: (a) the continuous, periodic frequency characteristic of a transversal filter can be obtained by convolving (b) the sinc function and (c) the series of weighted impulse values representing points on the response of (a)

4.59a corresponds to the continuous, periodic frequency characteristic shown in Figure 4.60a. However, Figure 4.59a can be considered as the result of multiplying together the discrete periodic impulse response of Figure 4.59c and the rectangular pulse of width equal to the filter aperture, shown in Figure 4.59b. The corresponding operation in the frequency domain consists of convolving the discrete, periodic spectrum of Figure 4.60c with the sinc function of Figure 4.60b. As the sinc function value is zero at all other impulse positions, the continuous characteristic (Figure 4.60a) produced by this convolution has the same values as the discrete impulses at the positions shown in Figure 4.60c. Thus, the impulse values of Figure 4.60c can be set independently to produce the desired continuous characteristics, consistent with the resolution of the set frequency points. If a wider filter aperture is used, the spacing of the set points is reduced, so allowing a faster rate of cut to be achieved.

As Figures 4.59c and 4.60c are both discrete and periodic, they are related in the two directions by the discrete Fourier transform and its inverse. So the impulse response values of Figure 4.59c can be calculated from the set values of Figure 4.60c. If there are shared values at the end of the repeat period in Figure 4.60c, these have to be halved to take account of their effect in each cycle period. Similarly, shared values at the edge of the cycle period in Figure 4.59c have to be halved to produce the coefficient pattern of the filter shown in Figure 4.59a.

The two-dimensional case can be simplified by assuming four-quadrant symmetry so that the pattern of set points then appears as in Figure 4.61a. The array of contributions shown in Figure 4.58 allows response values to be set at intervals of $6\frac{1}{4}$ Hz and $78\frac{1}{8}$ c/ph, thus providing points at the centres of the unwanted spectra. Any smaller aperture filter than Figure 4.58 would not allow this important feature to be obtained. However, because the set points relate to a sequential standard, only approximately half the points can be set independently. To take account of interlaced scanning, all the pairs of points centred on $(12\frac{1}{2}, 156\frac{1}{4})$ have to be equal values. Then, when the frequency points are transformed to produce the coefficient pattern of Figure 4.61b, zero coefficients are produced at all the positions where no lines exist. As in the one-dimensional case, edge values have to be shared, so values on the 25 Hz or 312 c/ph lines are halved and the $(25, 312\frac{1}{2})$ value reduced to one quarter. A similar weighting pattern is applied to the converted values to obtain the correct filter coefficients.

The symmetry requirement of interlaced scanning reduces the number of independently set frequencies from 25 to 13 in the quadrant. Of these, the origin must have a value of unity and the main centre of cross-effects $(18\frac{3}{4}, 78\frac{1}{8})$ zero response. In addition, to suppress U–V crosstalk in chrominance signals requires a zero at $(12\frac{1}{2}, 156\frac{1}{4})$. Also, it should be noted that the response value cannot change abruptly from one set point to the adjacent ones without causing large overshoots in the continuous characteristic between the set frequencies. A reasonably slow roll-off characteristic is required anyway to avoid strong multiple images of moving objects, which result from ringing in the temporal dimension.

Because of these constraints, the range of different filters that can be produced is relatively small and the design becomes a simple compromise between resolution loss and suppression of cross-effects. However, because the temporal components of colour signals

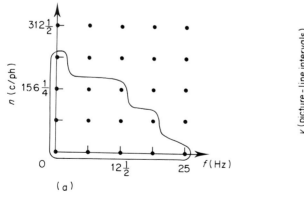
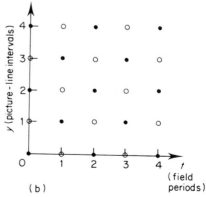

Figure 4.61 (a) The pattern of specification points in one quadrant of the vertical/temporal spectrum and (b) the corresponding array of contribution points in the time domain. Choosing independently only the 13 values in the outline in (a) obtaining the remainder by symmetry about $(12\frac{1}{2}, 156\frac{1}{4})$, makes alternate values zero in (b) as required for an interlaced scan

are limited because of their lower horizontal bandwidth, quite narrow temporal bandwidths can be used without serious impairment. Also, as the filtering effect in luminance is restricted to the high horizontal frequencies, again narrow temporal bandwidths can be used with effects not unlike extended camera integration. This results in improved suppression of cross-effects over the simpler filters previously discussed. Whether the improved performance is sufficient to justify the substantially increased cost of these complex filters needs a detailed assessment of the performance requirements of individual applications.

Performance comparison

With the very large number of different PAL decoding methods that have been developed, it has become increasingly difficult to keep track of their individual strengths and weaknesses and, therefore, to understand the progression towards improved performance. As an aid to visualising this, a chart has been prepared (Table 4.5) which tabulates the performance of nine different decoding methods in eight categories of impairment. Each impairment is graded on a five-point scale.

In the chart, the impairments produced vary considerably in their nature, even within the same category. With this variation it is difficult to maintain consistency from one part of the table to another in all respects. Therefore, the emphasis has been on maintaining consistency within each category of impairment rather than from category to category for each decoder. Accordingly, the cross-colour performance of a two-line Weston decoder is broadly equivalent to that of a conventional decoder, but the impairment resulting in each of these cases is not necessarily equivalent to that produced by the vertical chrominance resolution loss in a dual 1, 312, 313-line decoder. Also, with the five-point grading system used, it is not possible to bring out every nuance of performance. Thus the cross-colour performance of the dual 1, 312, 313, 625-line decoder is noticeably better than that of the dual 312-line decoder, but is not sufficiently so to make it comparable to that of the dual 1, 2, 625, 1250-line decoder.

The details of the nine decoding methods are given in Table 4.6. They range from the simplest possible methods, through elementary comb filters, to increasingly more sophisticated comb filters, in an order roughly corresponding to their chronological implementation.

The eight categories of impairment are resolution loss, cross-effects and self-effects, each for luminance and chrominance, followed by their response to differential phase and to movement. The assessment for movement is a composite judgement mainly based on the response to moving saturated chrominance, that is, chrominance resolution and cross-luminance, but also takes account of cross-colour, self-effects and luminance resolution.

The following brief description brings out the main features of the chart. Successful decoders are a compromise, avoiding the worst impairments, but perhaps not achieving spectacular performance in any respect. Thus, while the simple decoder has extremes, such as 'very good' cross-luminance suppression at the expense of 'bad' h.f. luminance resolution, the conventional decoder maintains 'fair to poor' performance in all categories, avoiding the most serious impairments.

The two-line Weston decoder was seen as a means of obtaining improved luminance resolution and of reducing cross-colour, in particular, retaining resolution bars on test

Table 4.5
Comparison chart for PAL decoding methods

	Simple	Conventional	2-line Weston	2-picture Weston	Dual 312-line	Dual 1, 312 313-line	Dual 1, 312 313, 625-line	Dual 1, 2 625, 1250-line	Improved field, picture-based
h.f. luminance resolution	●	○[6]		**	*		*	*	*
vertical chrominance resolution	**	○[7]	○	**	*	○[10]		**	*
cross-colour	●	○	○	*[9]		○		*	*
cross-luminance	**[3]	○	○	*[9]					*
luminance aliasing[1]			○	○					
U–V crosstalk[1]	●[4]		●	●	●[4]				
differential phase[2]			●[8]	●[8]					
movement				●				○	

** — very good
* — good
○ — fair
● — poor
● — bad

1. As luminance aliasing and U–V crosstalk are impairments introduced unnecessarily by some decoding methods, it gives a misleading picture to grade their absence other than as 'fair'.
2. As the effects of differential phase distortion can be suppressed easily, it is misleading to grade their absence other than as 'fair'.
3. While the performance of a 3 MHz low-pass filter is 'very good' for reducing cross-luminance, the suppression is not complete because some chrominance components in the encoded signal extend below 3 MHz.
4. Differential phase distortion (which is rare in studio centres) would result in Hanover bars.
5. Although decoders without field or picture delays cannot introduce any temporal filtering impairments, movement has a strobing effect on cross-luminance and cross-colour components. Because such components are often present with high amplitudes with these decoders, their movement performance can only be described as 'fair'.
6. A proportion of this impairment is due to the group delay errors which result from using a non-phase-corrected notch filter.
7. A proportion of this impairment is due to chrominance being shifted down the picture by one picture-line pitch relative to luminance.
8. Differential phase distortion resulting in unsuppressed subcarrier.
9. Although with this decoder there are virtually no cross-effects on still pictures, the cross-effects are exaggerated on movement.
10. Better luminance and cross-luminance performance and a simpler implementation can be obtained by using the dual 312-line luminance circuit instead.

Table 4.6
Decoder details of methods compared in Table 4.5

	Luminance	Chrominance
Simple:	3 Mhz l.p.f	Direct demodulation
Conventional:	Non-phase-corrected notch	Single delay line (Figure 4.43)
2-line Weston:	Figure 4.48c	Figure 4.48c
2-picture Weston:	Figure 4.48c with 625 T_h delays	Figure 4.48c with 625 T_h delays
Dual 312-line:	Figure 4.48d with 312 T_h delays	Figure 4.48d with 312 T_h delays
Dual 1, 312, 313-line:	Figure 4.55a	Figure 4.53a
Dual 1, 312, 313, 625-line:	Figure 4.55b	Figure 4.54a
Dual 1, 2, 625, 1250-line:	$C_0 = 0.75$	$C_0 = 0.3125$
	$C_1 = -0.125$	$C_{625} = 0.2375$
	$C_2 = -0.0625$	$C_{1250} = 0.10625$
	$C_{625} = -0.125$	
	$C_{1250} = -0.0625$	
Improved field, picture-based (main coefficients):	$C_0 = 0.72$	$C_0 = 0.16$
	$C_1 = -0.10$	$C_1 = 0.01$
	$C_{312} = -0.08$	$C_{312} = 0.08$
	$C_{313} = -0.08$	$C_{313} = 0.08$
	$C_{625} = -0.07$	$C_{625} = 0.13$
	$C_{1250} = -0.03$	$C_{937} = 0.04$
		$C_{938} = 0.04$
		$C_{1250} = 0.04$

card F at full amplitude and with no cross-colour. However, the extra impairments introduced for other picture material were not fully recognised. The most serious in the context of studio decoding is the failure to suppress plain area subcarrier in signals affected by differential phase distortion. The next development, the two-picture Weston decoder, made spectacular improvements to still pictures, but introduced serious impairments to movement and retained the differential phase problems of the line-based method.

Figure 4.62 (see colour plates between pp 282 and 283) compares the performance of the two-picture Weston decoder (column 4 of Table 4.5) with a conventional decoder (column 2) using the *RGB* signal as the reference. In this case the zone plate test pattern generator [90] was used to produce a pattern of concentric circles of increasing spatial frequency towards the edges of the picture. Thus in the case of the *RGB* signal, Figure 4.62a, the resolution is limited along the horizontal axis by the base bandwidth and at the top of the picture by aliasing due to the line structure.

The zone plate is a very severe test of the conventional decoder, Figure 4.62b, showing up the cross-colour which produces a moving pattern of coloured circles radiating from spatial frequency positions of 4.43 MHz, the frequency of the subcarrier. The effect of the notch-filter can also be seen in Figure 4.62b as a loss of resolution around the subcarrier frequency before it increases again towards the edge.

Figure 4.62c shows the results obtained with the two-picture Weston decoder which are indistinguishable from the *RGB* results for this test. The impairments are significant with moving images which can of course not be demonstrated on the photograph. These can be largely overcome by the dual 312-line method. Although this decoder is still sensitive to different phase distortions this is rarely present in studio centres, thus Hanover bars are unlikely to appear. The static performance of this decoder is shown in Figure 4.63c (see colour plates between pp 282 and 283) with test card in comparison with the *RGB* picture, Figure 4.63a, and the conventional decoder, Figure 4.63b. The resolution performance of (c) appears similar to that of (a) when compared on the test card but in fact there is some loss of diagonal resolution, hence the grading of 'good' in Row 1.

The improvement compared with the conventional decoder, Figure 4.63b, can most readily be seen in the resolution bars immediately to the right of the picture in the circle. The other area to study is around the colour bars at the top of the picture where improvements in cross-luminance and vertical chrominance resolution can be seen at the transition between the colour bars and their lower edges respectively. Figure 4.63c is much closer to Figure 4.63a in these respects than the conventional decoder, Figure 4.63b.

A decoder using the dual 312-line method has been produced in small quantities by the BBC Design and Equipment Department for use in areas where the interface between composite and component signals is particularly critical such as electronic graphics preparation for News and Current Affairs programmes using moving or still video inserts. Figure 4.64 is a photograph of the equipment.

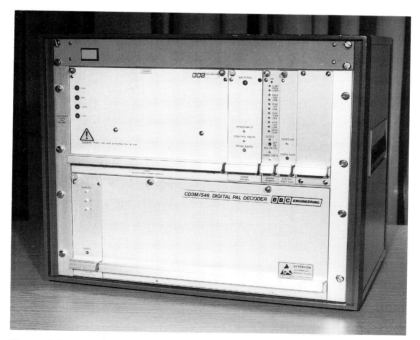

Figure 4.64 Dual 312-line decoder equipment

The dual 1, 312, 313-line and dual 1, 312, 313, 625-line decoders are developments of this, devised to remove Hanover bars. In the first case this is obtained by sacrificing vertical chrominance resolution directly, whilst adding the 625-line contributions reduces the vertical chrominance resolution loss. This then represents a reasonably satisfactory decoder, being judged 'fair' in all respects, but comes nowhere near the strengths of the two-picture Weston method.

To obtain any significant improvement in still picture performance over the dual 1, 312, 313, 625-line method, it is necessary to use narrower temporal filters, thus requiring more field and picture delays. The dual 1, 2, 625, 1250-line decoder allows spatial resolution to be increased and cross-effects reduced at the same time. The disadvantage is that the blurring and appearance of cross-luminance that occurs with moving chrominance is of greater extent, although the impairment is less than might be expected because it has a reduced amplitude. Nevertheless, significant improvements in resolution and reduction of cross-effects can be obtained as a development of the dual 1, 2, 625, 1250-line coder by using a smoother and more gradual temporal response to maintain reasonable movement performance. The resulting increase in cross-colour is kept in check by sacrificing some vertical chrominance resolution to be comparable to that of the dual 312-line method. Luminance resolution is increased and cross-luminance reduced by replacing intra-field vertical filtering (1- and 2-line taps) by inter-field vertical filtering (312, 313, and 1-line taps). Although the temporal aperture required for these filters is doubled, the number of taps required may not increase very greatly. The 1, 312, 313, 625 and 1250-line taps are the most important, but access to the 937 and 938-line taps is also beneficial.

Adaptive techniques

While the complex filters developed in the previous section represent one approach to improving on the performance of simple comb filters, an alternative is to use adaptive techniques. Many of the simple comb filter methods give good performance for some signals, but produce quite noticeable impairments for others. In their simplest forms, adaptive techniques detect the signal conditions under which an impairment might occur and change to a different decoding method to avoid the impairment. For example, the 625-line comb filter is very good for still or slowly moving areas of picture, providing near perfect separation, but produces more serious impairments on rapid movement. Therefore, in principle, it would be possible to detect moving areas of picture and to change to a different comb filter (for example, based on one-line delays) to avoid movement impairments. This would combine the good still-picture performance of the picture-delay filter with the acceptable movement performance of the line-delay filter.

The detection of movement in PAL signals is complicated by the subcarrier itself being a moving pattern, even for still pictures. Thus, the movement detection circuit must have a temporal characteristic with zero response for stationary luminance and for both the modulated chrominance energy centre positions. Even then, the chrominance detection circuits can be confused by high frequency luminance into changing modes when this is not desirable. In addition, the movement detector may be confused by the presence of noise in the signal.

For many signals, therefore, the decoder may remain in the line-based movement mode for the majority of the time. The same detection problems are likely to be met with other movement-adaptive decoding methods. Therefore, although adaptive decoding may achieve good results in test signals which are free from noise and distortion, and give well defined signals in the movement detector circuits, on practical signals the decoder may remain almost entirely in the poorer performance movement mode. The advantages of picture delay comb filtering for still pictures would then be lost.

Accepting that movement detection will be imperfect under some signal conditions, it is prudent to ensure that the overall performance of both the still and moving picture modes is reasonably good. In practical terms, this means making use of the improved comb filter methods described in the previous section. In view of this, the complexity of adaptive methods needs to be considered carefully in relation to their benefits. Besides the extra circuitry required for movement detection, it is necessary to provide separate comb filter circuits for the still mode and the movement mode. If for either comb filter the luminance and chrominance circuits are non-complementary, this leads to an additional increase in complexity. Therefore, the circuitry required for an effective adaptive method may be up to three times that required for a non-adaptive decoder.

Although the benefit of these simple adaptive techniques is in some doubt, more sophisticated methods based on movement measurement and compensation are currently being considered. While these methods hold out the possibility for substantially improved performance, the presence of the moving subcarrier pattern may still prove to be a confusing influence in the movement adaptive circuitry. Eventually, if the DATV techniques mentioned in Chapter 1 can be used in conjunction with composite signals, data for adaptive processes derived before encoding can be provided for the decoder without the confusing influence of the subcarrier or noise.

DECODERS IN DIGITAL STUDIOS

Multiple decoding and recoding

The changeover to digital component studios will inevitably involve a period of mixed PAL and component working. Analogue PAL signals will be decoded at the inputs to a digital studio, processed in the same way as directly sourced component signals and then recoded to PAL at the output, thus introducing an extra PAL decode and recode into the signal chain. In adverse circumstances, several such decode–recode operations could be cascaded.

If the signals from a decoder are recoded to PAL, true luminance and chrominance components are unaffected, but the crosstalk components are incorrectly treated, so causing impairments in the final PAL decoder. In the recoding process, cross-colour components are remodulated to form high frequency luminance again, but these may be shifted in phase to reinforce or cancel true luminance components or, if the PAL switch sense is different from that in the original coder, may be aliased about twice subcarrier frequency to become a moving diagonal pattern. With a conventional final decoder these effects are partly suppressed by the luminance notch and the overall effect is to produce cross-colour which may be either greater than or less than its normal amplitude.

Cross-luminance, however, passes straight through the second coder, but is decoded subsequently by the reference subcarrier of the second coding process. This subcarrier may be different in phase, in PAL switch sense or even in frequency from that used in the original coder, so these signals, when decoded in the final decoder, produce hue errors. In practice, this is a much more serious impairment than those resulting from cross-colour, so the suppression of cross-luminance in the first decoder is of prime importance for recoding.

These problems can be exacerbated by processing in the component signal domain; for example, grabbing a still frame derived from PAL freezes any residual subcarrier which is then demodulated in the final decoder to produce a flashing area of colour. Also, if multiple decoding and recoding occurs, then any impairments are rapidly compounded.

Under some special circumstances, the decoding–recoding processes can be made transparent, so that the recoded PAL signals are identical to the original PAL. This requires the luminance and chrominance filters in the decoder to be complementary and requires the subcarrier phase and PAL switch sense in the second coder to be the same as those in the first coder. However, this is of limited usefulness because any processing of the component signals, including editing of video tapes, is likely to destroy the transparency of the decode–recode process. Also the system could not be used widely because of the difficulty of maintaining a standardised subcarrier phasing and PAL switch sense at all PAL coders. In view of this, the only satisfactory solution to the decode–recode problem is to devise a decoding method for use in the first decoder position which provides a minimum of cross-luminance and cross-colour, but without adversely affecting the wanted luminance and chrominance components.

Clean coding systems

The impairments introduced by conventional colour coding and decoding methods could be avoided by substituting a clean coding system. As the clean signals can be decoded without cross-effects, this prevents the most serious impairments occurring when the signals are recoded. There would, however, be a cumulative loss of resolution.

The only filters present in a conventional PAL coder are the Gaussian low-pass filters used to shape the spectrum of the colour-difference signals. Because of this, there is considerable spectral overlap between the luminance and chrominance components, which, in general, makes it impossible to separate the signals completely by filtering at the decoder. However, in both PAL and NTSC, additional filtering can be included at the coder to reduce the severity of subsequent cross-colour and cross-luminance impairments.

Cross-colour in the decoder can be reduced by removing high frequency luminance in the coder as shown in Figure 4.65a. For PAL, the simplest example of this is the inclusion of a luminance notch filter which, although reasonably effective when conventional decoders are used [99], reduces the potential for improvements in the decoder.

With the stability of digital circuitry, more complicated vertical and temporal comb filtering techniques [81, 100] can be employed in the coder. Such techniques allow fine, moving luminance, which produces the more noticeable coarse, stationary cross-colour, to be removed while retaining the full static luminance resolution. Similar techniques can be applied in the chrominance channel, as in Figure 4.65b, to reduce cross-luminance. This can be achieved by restricting the vertical and temporal chrominance

DECODERS IN DIGITAL STUDIOS 211

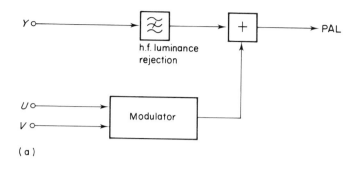

(a)

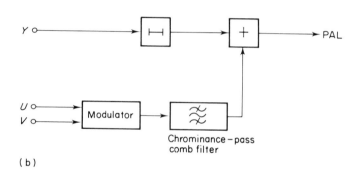

(b)

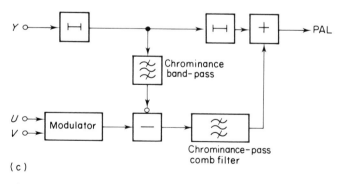

(c)

Figure 4.65 Clean PAL coding systems: (a) the removal of h.f. luminance to reduce cross-colour, (b) the removal of vertical/temporal chrominance detail to reduce cross-luminance and (c) a complementary clean PAL coder arrangement

sidebands, although if too severely applied, the chrominance information can become blurred.

Many rearrangements of these clean coder circuits are possible, in some of which the colour-difference signals are filtered before modulation. A particularly efficient rearrangement is shown in Figure 4.65c, in which a single chrominance-pass comb filter is used to provide cross-colour and cross-luminance suppression. With 'ideal' rectangular filters, such an arrangement would ensure that luminance and chrominance were kept perfectly separate in the spectrum. However, the use of rectangular filters is ruled out by the severe ringing which would occur at abrupt transitions, as well as by the high cost of sharp-cut filters.

A more satisfactory method of ensuring perfect separation of luminance and chrominance is to use the Weston clean coding and decoding technique [76, 77, 79]. As alternatives to separation in frequency, the constituent parts of a colour television signal can be encoded either separately in time, or by separation in phase, modulated onto orthogonal subcarriers. In a PAL signal, the U and V spectra overlap in frequency space but the signals are kept separate by placing them on quadrature phases of subcarrier. The chrominance signals cannot be kept separate from the luminance signal if the luminance spectrum overlaps the two chrominance spectra. However, if we only wish to transmit, along with the luminance signal, *one* chrominance signal such as $U + V$ (or U) on one line and $U - V$ (or V) on the next, we can place this single chrominance component on one phase of subcarrier as before and use the luminance signal to modulate the quadrature phase of the subcarrier.

The method of luminance modulation, which gives a modulated signal which is 'compatible' with the original luminance component, is to multiply the luminance baseband signal by a $2f_{sc}$ carrier and add the resulting lower sideband back to the baseband component. This process is equivalent, within the basebandwidth, to sampling at $2f_{sc}$. The phase of the luminance 'subcarrier' will be determined by the phase of the $2f_{sc}$ sampling.

A simple, clean composite signal could therefore be formed by sampling the luminance at $2f_{sc}$ and adding to this the single chrominance component modulated at f_{sc} (with the relative phases of the $2f_{sc}$ sampling and chrominance subcarrier chosen appropriately). The luminance component can be separated from this composite signal by resampling at $2f_{sc}$ and arranging that the sampling instants lie at the zero crossings of the chrominance subcarrier. Combining these signals in appropriate pre- and post-sampling filters, such as adding and subtracting across a line-, field- or picture-delay, produces a compatible variant of PAL in which the two chrominance signals are carried on one phase of subcarrier, whilst high frequency luminance is carried on the orthogonal phase. Because of this, the luminance and chrominance signals can be separated perfectly by synchronous demodulation. (The Weston system was first explained in these terms by N. D. Wells.)

Although both additional coder filtering and matched coder–decoder combinations can improve the picture quality obtained from PAL signals, the introduction of such systems into a conventional PAL environment does present problems. With systems that require both modified coders and decoders, such as Weston clean PAL, the improvements for viewers with conventional decoders are small and are sometimes offset by the introduction of additional impairments. Also such systems require expensive receivers. On the other hand, if modifications are confined to the coder, then the improvements are less spectacular, but still beneficial from a picture quality viewpoint. However, the large number of PAL coders used and the studio timing difficulties caused by the introduction of extra delays in the comb filters currently remain as considerable drawbacks. In the future, though, when large areas have been converted to digital component signal operation, the use of complicated clean coders for the network outputs could be worthwhile.

The changeover to digital studios

A necessary outcome of achieving greater suppression of cross-effects is that the complementary nature of chrominance and luminance filtering is lost. In a complementary decoder, all the input signal is preserved, either in the chrominance or the luminance channel. However, if the luminance and chrominance responses are both narrowed to achieve better suppression of cross-effects, the separate signals have irretrievably lost some information which results in gaps in the spectrum of the recombined signals. These non-complementary decoders are therefore unlikely to retain adequate performance through multiple decoding and recoding operations. This is because with practical filters, the spectrum gaps will broaden as the number of cascaded filters increases.

In view of this, it is necessary to consider the positioning of decoders in the television system with a view to minimising the number of decoding and recoding processes. For example, it would be unreasonable to include the decoder in a loop with a digital video tape recorder, as the resolution loss would be unacceptable after multiple generations.

If digital *YUV* equipment is introduced by spreading initially from the signal origination equipment (cameras, telecines, etc.) then no extra decode–recode processes are required. Also, *YUV* equipment can spread back from the broadcast transmitter. Since the links would not alter the *YUV* component signals, it would be possible to phase-lock the PAL coder at the transmitter to a complementary PAL decoder at the studio and thus achieve no additional loss of quality at the home receiver. However, the introduction of most *YUV* equipment so far has been for post-production processing. Also it is widely accepted that the development of *YUV* areas will centre around the introduction of digital video tape recorders. This approach tends to magnify the decoding problem.

One approach that tends to minimise the number of cascaded decode–recode operations over the transition period is to keep the analogue composite and digital component areas separate. Connections within analogue composite areas would be made through a large centralised routing matrix as at present. As the digital component areas develop, these would be interconnected through a comparable digital matrix. All connections between composite and component signals would then be made between the analogue and digital routing matrices. Centralising the interface in this way should ensure that no unnecessary conversions take place and also minimises the number of coders and decoders required. This aspect is particularly important because of the high cost of the complex decoding methods needed for adequate suppression of cross-effects. It also ensures that full system flexibility is retained throughout the changeover period.

As the conversion to component signals proceeds, the number of PAL sources will reduce. However, decoding to *YUV* will still be needed, even after the completion of the conversion to digital components systems, to provide access to the large amount of archival material in analogue PAL form.

5 Digital Filtering of Television Signals

J. O. Drewery

INTRODUCTION

Digital technology allows us to deal with signals as time series and perform mathematical operations on them to produce further time series. Thus digital filters, unlike their analogue counterparts which involve capacitors, inductors and resistors, are based on precise mathematical relationships. As such, their behaviour is independent of hardware imperfections and their performance is predictable (within limits) and, if necessary, can be simulated using a digital computer to model the design.

The term 'digital filter' is implicitly assumed to be a device which affects the spectral characteristic of a signal. Within this broad definition there is a large class of filters whose properties are independent of the signal values (linear) and of time (time-invariant) and this chapter will be mainly concerned with these. An interpolator is an example of a time-dependent filter and these will be dealt with in Chapter 6.

Over the past ten years the subject of digital filtering has been studied intensively and there is now an enormous body of literature concerned with it. This chapter can do no more than introduce the concepts and provide the barest outline of the theory before indicating some practical applications to video. For a comprehensive treatment of the theory, see [101, 102]. One particular application, that of video noise reduction, although based on a particular kind of filtering, is a good example of how many other signal processing techniques need to be drawn upon to solve certain problems. As this merits a fuller treatment it forms a substantial part of the chapter.

At the outset, a distinction must be made between the operation of sampling and its consequences and the operation of quantisation and its consequences. This chapter will be mostly concerned with the first aspect. As the spectral domain is the domain of interest we will begin by examining the Fourier transform and, in the context of video filtering, its particular relevance to image filtering.

THE TWO-DIMENSIONAL FOURIER TRANSFORM

The Fourier transform is one of the basic tools of communication engineering which allows us to transform a signal from the time domain into the frequency domain. In this domain the operation of filters then becomes far more tractable.

Just as a one-dimensional function, such as a sound signal, may be expressed as a superposition of one-dimensional sinusoids, so a two-dimensional function, representing the brightness of an image, may be expressed as a superposition of two-dimensional sinusoids or spatial frequencies [103]. These may be visualised as sloping sinusoidal gratings having integral numbers of cycles per picture width and height; Figure 5.1 shows the first few members of the set. The image brightness $E(x, y)$ is then related to the amplitudes and phases of the frequencies by the equation

$$E(x, y) = \sum_{m=0}^{\infty} \sum_{n=0}^{\infty} a_{mn} \cos [2\pi(mx/w + ny/h) + \phi_{mn}] \quad (5.1)$$

where a_{mn} and ϕ_{mn} are the amplitude and phase of the frequency with components m and n cycles per picture width and height, and the picture dimensions are $w \times h$. The

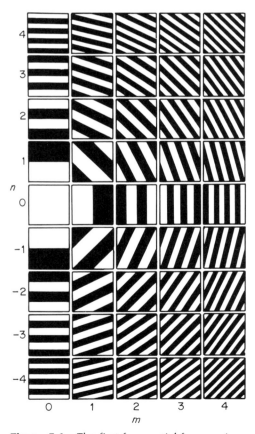

Figure 5.1 The first few spatial frequencies

function represented by equation 5.1 is two-dimensionally periodic in the picture dimensions as shown in Figure 5.2.

Just as with one-dimensional transforms, the physical frequency may be expressed in terms of conjugate exponential frequencies so that the amplitude and phase can be combined into a single complex number. Thus the relationship in equation 5.1 becomes

$$E(x, y) = \sum_{m=-\infty}^{\infty} \sum_{n=-\infty}^{\infty} A_{mn} \exp j2\pi(mx/w + ny/h) \quad (5.2)$$

where

$$A_{mn} = \tfrac{1}{2}a_{mn} \exp j\phi_{mn} \quad \text{and} \quad A_{-m,-n} = A_{mn}^*$$

The array A_{mn}, plotted as a function of m and n, constitutes the two-dimensional Fourier transform of the image.

Note that A_{mn} is a complex quantity so that both an amplitude and a phase spectrum exist, the former having a centre of symmetry and the latter, a centre of antisymmetry. Figure 5.3 shows examples of the amplitude spectrum of three different images and Figure 5.4 shows the corresponding phase spectra. As can be seen it is possible to pick out certain features on the amplitude spectra, i.e. they are somewhat coherent, whereas the phase spectra appear as an incoherent jumble.

The phase spectrum often gets forgotten although it is equally, if not more, important for images [104]. For example, if the phase transform of one image is paired with the amplitude transform of another and vice versa then the inverse transforms closely resemble the originals whose phase transforms were used but have little similarity to the originals whose amplitude transforms were used, as shown in Figure 5.5. This shows that most scenes have much the same amplitude characteristic and that most of the significant information is carried by the phase transform. Figure 5.6 shows the reconstruction of an image from the phase transform alone where the amplitude transform is set to unity.

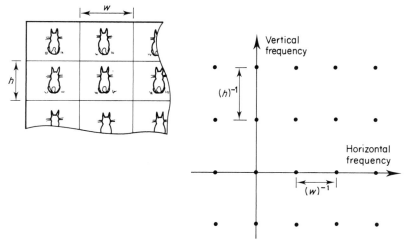

Figure 5.2 The two-dimensional image represented by a Fourier series

(a)

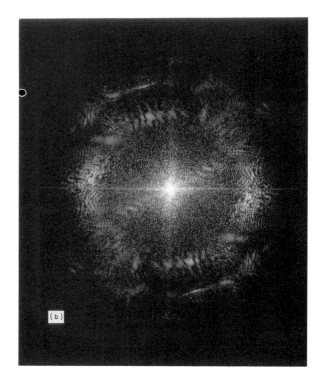

(b)

Figure 5.3 The amplitude spectrum of three different images

Typically, the amplitude characteristic peaks at the origin and dies away as the reciprocal of the spatial frequency magnitude in much the same way as the transform of an edge. There are exceptions, of course, when a scene contains a significant proportion of periodic detail such as test pictures or random detail such as a crowd scene, which tends to have a flatter spectrum. A further exception is where the three components of a coloured image are combined into a composite signal so that the colour information modulates a subcarrier. The subcarrier has a defined position in the two-dimensional transform space and develops sidebands having the same symmetry properties about the carrier as the overall spectrum.

Clearly, then, a filter designed to select or reject a particular region, or regions, of the space must have due regard to the phase characteristic, i.e. the phase transform must be unaffected or, at worst, suffer no more than the addition of a linear function, corresponding to a delay. Failure to observe this restriction results in undesirable effects on edges. Figure 5.7 shows, for example, the effect of a filter which notches out a region of the spectrum near the subcarrier and which has a poor phase response. Note the effects near vertical edges.

Another class of filters is designed to deliberately affect the phase transform whilst leaving the amplitude transform unaffected. This requirement usually arises in the context of signals modulated on a subcarrier.

In the case of filters designed to pass or reject certain bands of frequencies the requirements of sharpness of cut and step-response overshoot suppression always

220 DIGITAL FILTERING OF TELEVISION SIGNALS

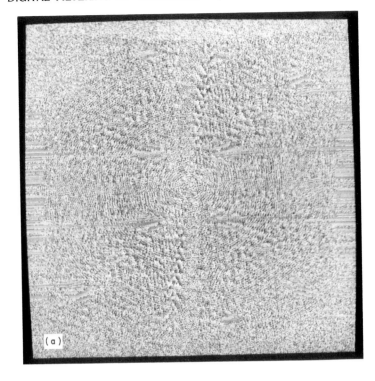

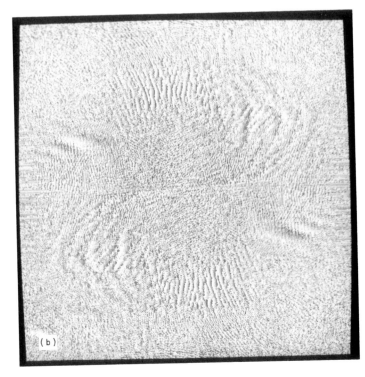

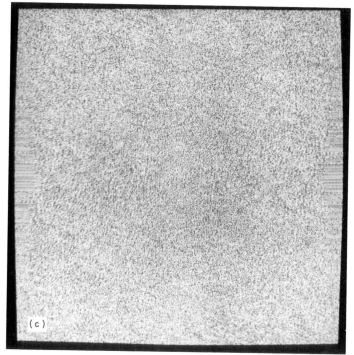

Figure 5.4 The corresponding phase spectra of the three images

conflict. Often the designer is faced with needing as sharp a cut as possible to satisfy some system requirement such as minimisation of crosstalk or loss due to repeated filter cascading. On the other hand, an infinitely sharp cut causes a step to develop a slowly decaying sinusoid at the cut frequency, an effect known as 'ringing'. For example, with a low-pass filter, the effect is for the edge to overshoot and undershoot by some 8.4% as shown in Figure 5.8.

THE THREE-DIMENSIONAL FOURIER TRANSFORM

Television is concerned with two-dimensional pictures that move, so it is natural to introduce time as a third dimension. Then, in a way entirely analogous to the two-dimensional case, the three-dimensional brightness function of a moving image, $E(x, y, t)$, may be expressed as

$$E(x, y, t) = \sum_{m=-\infty}^{\infty} \sum_{n=-\infty}^{\infty} \sum_{f=-\infty}^{\infty} A_{mnf} \exp j2\pi(mx/w + ny/h + ft) \qquad (5.3)$$

The physical function $\cos[2\pi(mx/w + ny/h + ft) + \phi_{mnf}]$ is a three-dimensional frequency and may be thought of as a sloping grating which is moving so that crests pass a fixed point at a temporal frequency of f Hz [98]. The array A_{mnf}, plotted as a function of m, n and f, constitutes the three-dimensional Fourier transform of the moving image whose

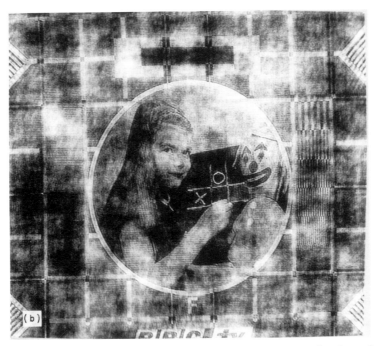

Figure 5.5 The images resulting from cross-pairing amplitude and phase spectra (a) amplitude Test Card, phase 'Couple', (b) amplitude 'Couple', phase Test Card

THE THREE-DIMENSIONAL FOURIER TRANSFORM

Figure 5.6 An image reconstructed from only the phase transform

Figure 5.7 The effect of a filter with a poor phase characteristic

Figure 5.8 Ringing at an edge caused by sharp-cut filtering

amplitude and phase parts have the same symmetry properties as in two dimensions.

When an image moves, the spatial frequencies of which it is composed must move in the same way. Supposing that the motion is pure translation with velocity components u and v, then the image brightness may be written, through equation 5.2 as

$$E(x - ut, y - vt) = \sum_{m=-\infty}^{\infty} \sum_{n=-\infty}^{\infty} A_{mn} \exp j2\pi(mx/w + ny/h - mut/w - nvt/h) \qquad (5.4)$$

Comparing equations 5.3 and 5.4, it is seen that the relationship between the temporal frequency and the velocity is given by

$$f = -(mu/w + nv/h) \qquad (5.5)$$

This equation is that of a plane in m, n, f space whose gradient is proportional to the magnitude of the velocity (if m and n are suitably scaled) and whose direction of slope is the direction of the velocity. Thus the transform of a translating image lies on a tilted plane in the three-dimensional transform space as shown in Figure 5.9.

A filter whose characteristic is a function of only the temporal frequency, i.e. a pure temporal filter, has, nevertheless, an effect on a moving image. For, supposing the characteristic to be $H(f)$, it can be seen from equation 5.5 that this becomes, for a moving object,

$$H(-mu/w - nv/h)$$

SAMPLING AND THE DISCRETE FOURIER TRANSFORM

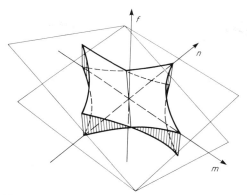

Figure 5.9 The transform of a translating image

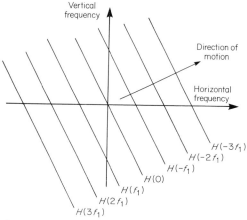

Figure 5.10 The contour mapping for a temporal filter acting on a moving object

Thus a point value of H at a temporal frequency f becomes a contour value in m, n space along the straight line contour

$$um/w + vn/h = -f$$

which is just equation 5.5 in another guise. These contours are perpendicular to the direction of motion as shown in Figure 5.10. In other words, the temporal filter, in combination with the velocity, gives rise to an equivalent spatial filter in the direction of motion. Inasmuch as a linear phase characteristic is a desirable property of spatial filters, it follows that this must also be true for temporal filters.

SAMPLING AND THE DISCRETE FOURIER TRANSFORM

So far, the discussion has been framed in terms of images described by continuous functions of space and time. However, the conversion of an image to a signal necessarily involves scanning which is a sampling process in time and one spatial dimension. Moreover, in a digital context, the signal itself is sampled, so creating a three-dimensional lattice of sample points.

Conventional television scanning samples the images vertically in an interlaced fashion so that, with European standards, the samples appear as in Figure 5.11. The quantities y and t take on discrete values given by

$$\left.\begin{array}{l} y = jY \\ t = kT \end{array}\right\} \quad j + k = 2p \tag{5.6a}$$

where Y is the picture line pitch, T the field period and j and k the line and field indices, respectively, or

$$y = (q + r)Y \tag{5.6b}$$
$$t = (q - r)T$$

where q and r are orthogonal vertical/temporal indices as shown in Figure 5.12.

The sampling involved in digital coding can correspond to any spatial structure but, after considerable discussion, as described in Chapter 1, it has been agreed that the sampling frequency will be such as to give an orthogonal pattern of samples, i.e. will be a multiple of the line frequency. Thus the values of x and y take on the discrete values

$$x = iX \tag{5.7}$$
$$y = jY$$

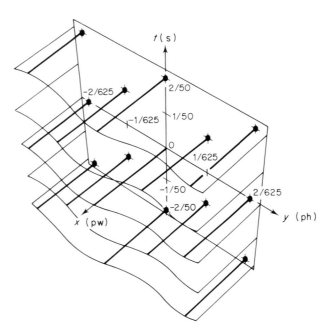

Figure 5.11 The vertical/temporal sampling of interlaced scanning

SAMPLING AND THE DISCRETE FOURIER TRANSFORM 227

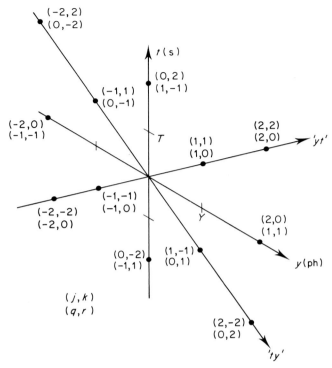

Figure 5.12 An alternative indexing for vertical/temporal sampling

Just as with one-dimensional sampling this two-dimensional sampling causes the spectrum to repeat in a two-dimensional fashion which sets a limit to the horizontal/vertical or vertical/temporal detail that can be handled by the system. Letting

$$M = w/X$$
$$N = h/Y \tag{5.8}$$

so that the image consists spatially of $M \times N$ pixels, Figure 5.13 shows the spectral effect of the horizontal/vertical sampling from which we see that spatial detail must be confined to the range given by

$$|m| < M/2$$
$$|n| < N/2 \tag{5.9}$$

In practice the horizontal limit is set by the analogue prefilter in the ADC as explained in Chapter 2. However the vertical limit is set only by the characteristics of the scanning system, the dominant part of which is the rather ill-defined scanning aperture.

In the vertical/temporal plane, letting

$$N = h/Y$$
$$F = 1/T \tag{5.10}$$

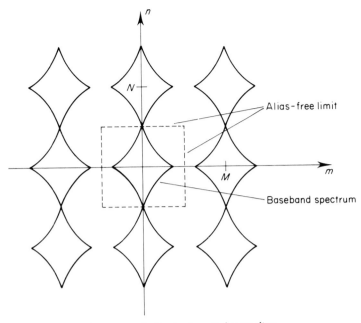

Figure 5.13 The spectral effect of spatial sampling

Figure 5.14 shows the effect of the vertical/temporal sampling from which we see that there is a variety of possible vertical/temporal bounds that avoid overlap. One bound might be

$$|n| < N/2$$
$$|f| < F/4 \qquad (5.11a)$$

another might be

$$|n| < N/4$$
$$|f| < F/2 \qquad (5.11b)$$

and yet another might be

$$|n/N + f/F| < 1/2$$
$$|n/N - f/F| < 1/2 \qquad (5.11c)$$

In practice, the bounds are set by the ill-defined vertical and temporal characteristics of the scanning system, the temporal part of which is provided by the integration time of the camera.

Given these effects of sampling, the transform relationships of equations 5.2 and 5.3 between continuous functions now become relationships between discrete functions. Substituting equations 5.7 and 5.8 into equation 5.2 and taking into account the bounds

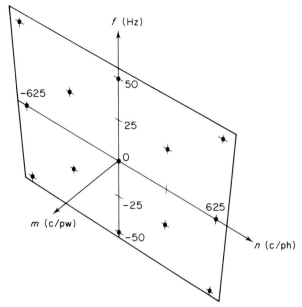

Figure 5.14 The spectral effect of vertical/temporal sampling

for m and n described by equations 5.9, we have

$$E(iX, jY) = E_{ij}$$

$$= \sum_{m=-M/2}^{M/2} \sum_{n=-N/2}^{N/2} A_{mn} \exp j2\pi(mi/M + nj/N)$$

or, since the addition of M to m and N to n leaves the expression unaltered,

$$E_{ij} = \sum_{m=0}^{M-1} \sum_{n=0}^{N-1} A_{mn} \exp j2\pi(mi/M + nj/N) \qquad (5.12)$$

This is the inverse discrete Fourier transform (IDFT). The inverse relationship is

$$A_{mn} = (MN)^{-1} \sum_{i=0}^{M-1} \sum_{j=0}^{N-1} E_{ij} \exp -j2\pi(im/M + jn/N) \qquad (5.13)$$

which is the discrete Fourier transform (DFT). Note that E_{ij} and A_{mn} are now both periodic with subscript periods M and N.

To obtain analogous relationships for three-dimensional images the concept of a temporal bound must be introduced as shown in Figure 5.15. The image is now regarded as a block of dimensions $w \times h \times \tau$, giving $M \times N \times L$ pixels (if all sites were populated) where

$$L = \tau/T \qquad (5.14)$$

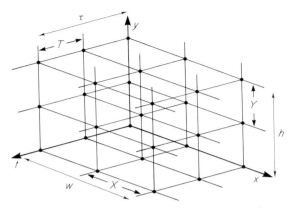

Figure 5.15 Definitions associated with a three-dimensionally bounded image

Defining a new temporal frequency variable by

$$l = f\tau \tag{5.15}$$

analogous to m and n and substituting equations 5.6a, 5.14 and 5.15 into equation 5.3, we have

$$E(iX, jY, kT) = E_{ijk}$$
$$= \sum_{m=0}^{M-1} \sum_{n=0}^{N-1} \sum_{l=0}^{L-1} A_{mnl} \exp j2\pi(mi/M + nj/N + lk/L), \quad j + k = 2p \tag{5.16}$$

This is the three-dimensional IDFT whose inverse, the DFT, is

$$A_{mnl} = (MNL)^{-1} \sum_{i=0}^{M-1} \sum_{j=0}^{N-1} \sum_{k=0}^{L-1} E_{ijk} \exp -j2\pi(im/M + jn/N + kl/L), \quad j + k = 2p \tag{5.17}$$

E_{ijk} and A_{mnl} are now both three-dimensionally periodic with subscript periods M, N and L.

These two- and three-dimensional DFT relationships are important in the derivation of filter characteristics as will be seen.

GENERAL CONSIDERATIONS OF DIGITAL FILTERS

From the foregoing it is clear that digital filters are concerned with the manipulation of sample values horizontally, vertically and temporally adjacent to the point being processed.

Working in just one dimension, for clarity, the input and output can be considered as time series, i.e. sequences of values occurring at regular instants of time, kT. The most general relationship between an output sample r_k and the input and output sequences $\{s\}$

and $\{r\}$ is then given by [105]

$$r_k = r(kT) = C_0 s_k + C_1 s_{k-1} + C_2 s_{k-2} + \cdots - D_1 r_{k-1} - D_2 r_{k-2} - \cdots$$

$$= \sum_{i=0}^{M-1} C_i s_{k-i} - \sum_{i=1}^{P-1} D_i r_{k-i} \tag{5.18}$$

where $\{C_i\}$ and $\{D_i\}$ are sets of constant coefficients of size M and $P-1$, respectively. In other words, the output sample at any given time is a linear combination of the present input sample and previous input and output samples up to defined limits. If the coefficients D_i are all zero then the filter is said to be transversal, otherwise it is said to be recursive.

Although the difference equation (5.18) helps to predict the pulse response, it does not give much insight into the frequency characteristic, which is usually of prime interest. However, just as the Laplace transform comes to the aid of solving the differential equations that arise with analogue filters, an analogous process can be used here to solve the difference equations.

If we define a variable z in terms of the Laplacian variable p and the sampling interval T as

$$z = \exp pT$$

then the Laplace transform of a sequence becomes the z transform. The z transform $S(z)$ of the sequence $\{s_k\}$ which is zero for $k < 0$ is derived as

$$S(z) = \sum_{k=0}^{\infty} s_k z^{-k}$$

Applying this transform to both sides of equation 5.18 and noting that the transform of a sequence delayed by i samples is simply z^{-i} times the original transform, we obtain

$$R(z) = \left(\sum_{i=0}^{M-1} C_i z^{-i} \right) S(z) - \left(\sum_{i=1}^{P-1} D_i z^{-i} \right) R(z)$$

or

$$R(z)/S(z) = H(z)$$

$$= \sum_{i=0}^{M-1} C_i z^{-i} \bigg/ \sum_{i=0}^{P-1} D_i z^{-i} \quad D_0 = 1$$

where $H(z)$ is a z transform transfer function analogous to the transfer function of a continuous system.

It can be seen that H is a ratio of two rational polynomials in the variable z^{-1}. Each polynomial has zeros, those of the denominator becoming the poles of H. Thus the poles and zeros of H may be plotted in the z plane to give insight into the filter behaviour in the same way as a p plane plot gives insight with continuous filters. Now it is well known that, in the design of continuous systems, a stable transfer function may not have any poles in the right-hand half of the p plane. As the imaginary axis of the p plane transforms into the unit circle of the z plane given by

$$|z| = 1$$

it is clear that a stable recursive filter may have poles only within the unit circle. Transversal filters, having no poles, are always stable.

If

$$z = \exp j\omega T$$

we obtain the frequency characteristic as

$$H(\exp j\omega T) = \sum_{i=0}^{M-1} C_i \exp - ji\omega T \bigg/ \sum_{i=0}^{P-1} D_i \exp - ji\omega T \qquad (5.19)$$

This is periodic in the frequency domain, $\omega/2\pi$, with a period of T^{-1}, i.e. the characteristic repeat unit is the sampling frequency, a property of sampled systems.

The foregoing treatment can be extended to two and three dimensions by introducing the concept of two- and three-dimensional z transforms and sequences (although the term 'sequence' in more than one dimension is somewhat questionable). Thus, in two dimensions, the 'sequences' and coefficients become two-dimensional and the z transform is defined as

$$S(z_1, z_2) = \sum_{i=0}^{\infty} \sum_{j=0}^{\infty} s_{ij} z_1^{-i} z_2^{-j}$$

This gives rise to a four-dimensional transfer function $H(z_1, z_2)$ since z_1 and z_2 are both complex. The general treatment of such filters is beyond the scope of this chapter. However, putting

$$z_1 = \exp j\omega_1 X$$

$$z_2 = \exp j\omega_2 Y$$

where

$$\omega_1 = 2\pi m/w$$

$$\omega_2 = 2\pi n/h$$

we obtain a two-dimensional frequency characteristic as

$$H(\exp j\omega_1 X, \exp j\omega_2 Y)$$

$$= \sum_{i=0}^{M-1} \sum_{j=0}^{N-1} C_{ij} \exp - j(i\omega_1 X + j\omega_2 Y) \bigg/ \sum_{i=0}^{P-1} \sum_{j=0}^{Q-1} D_{ij} \exp - j(i\omega_1 X + j\omega_2 Y) \quad (5.20)$$

which is periodic in the two-dimensional frequency domain $\omega_1/2\pi$, $\omega_2/2\pi$ with periods X^{-1} and Y^{-1}.

Analogous relationships can be derived for three dimensions.

TRANSVERSAL FILTERS

For a transversal filter, equation 5.19 becomes

$$H'(\omega) = \sum_{i=0}^{M-1} C_i \exp - ji\omega T \qquad (5.21)$$

while equation 5.20 becomes

$$H'(\omega_1, \omega_2) = \sum_{i=0}^{M-1} \sum_{j=0}^{N-1} C_{ij} \exp{-j(i\omega_1 X + j\omega_2 Y)} \quad (5.22)$$

which are Fourier transform relationships, the latter like that of equation 5.2. Thus we obtain the important result that *the frequency characteristic of a transversal filter is the Fourier transform of the coefficient pattern, in however many dimensions.*

Further, for a one-dimensional filter, if M equals $2K + 1$ so that there is an odd number of coefficients and redefining H, we may write

$$H(\omega) = \exp{-j\omega KT} \sum_{i=0}^{2K} C_i \exp{j(K-i)\omega T}$$

Now if we put

$$C'_i = C_{i+K}$$

i.e. we renumber the coefficients from the centre and choose

$$C'_i = C'_{-i}$$

i.e. make the coefficient pattern symmetrical about the new origin, then we have

$$H(\omega) = \exp{-j\omega KT} \sum_{i=-K}^{K} C'_i \exp{-ji\omega T}$$

$$= \exp{-j\omega KT}\left(C'_0 + 2\sum_{i=1}^{K} C'_i \cos i\omega T\right) \quad (5.23)$$

The term in brackets is purely real and is the amplitude characteristic, whilst the first term is simply a linear phase characteristic corresponding to a group delay of KT. Thus *we have achieved the goal of a linear phase filter* which is why so much attention has been paid to symmetrical transversal filters in video processing.

Alternatively, by choosing

$$C'_i = -C'_{-i}$$

with C'_0 equal to zero, i.e. making the coefficient pattern antisymmetrical about the origin, we have

$$H(\omega) = \exp{-j\omega KT}(2j^{-1}) \sum_{i=1}^{K} C'_i \sin i\omega T$$

As before, the first term corresponds to a group delay and the last term is the amplitude characteristic, but now there is a pure imaginary term which corresponds to a 90° phase shift for all frequencies. Thus we have the basis of a wideband phase shift network.

Similar considerations apply to two-dimensional transversal filters in that a linear phase characteristic is obtained by choosing the coefficient pattern to have a centre of symmetry, i.e.

$$C_{-i,-j} = C_{ij}$$

Moreover, it is often the case that the required frequency characteristic has four-quadrant symmetry, i.e.

$$H(-\omega_1, \omega_2) = H(\omega_1, \omega_2) = H(\omega_1, -\omega_2) = H(-\omega_1, -\omega_2)$$

This gives rise to a coefficient pattern with four-quadrant symmetry also.

A large class of filters are further required to have a variables-separable frequency characteristic, i.e.

$$H(\omega_1, \omega_2) = H_1(\omega_1)H_2(\omega_2)$$

In such cases the coefficient pattern is also variables-separable, i.e.

$$C_{ij} = C_i^h C_j^v$$

and this means that the filter can be realised by cascading one-dimensional filters with a consequent saving in hardware. If the dimensions of the filter are $(2M + 1) \times (2N + 1)$ then only $2(M + N + 1)$ coefficients are required which represents a considerable saving over $(2M + 1)(2N + 1)$ if M and N are large.

In general the remarks made above concerning linear phase, four quadrant symmetry and variables-separable behaviour are equally applicable to three-dimensional transversal filters.

DESIGN OF TRANSVERSAL FILTERS

The content of this section is really directed towards the problem of deriving the coefficient pattern, given the frequency characteristic. For a one-dimensional filter there are several ways of achieving this including

Frequency sampling
Windowing
Recursive frequency domain interpolation.

For a two-dimensional filter, techniques include:

Frequency sampling
Windowing
Two-dimensional mapping of a one-dimensional filter.

Frequency sampling

In the frequency sampling technique, the frequency characteristic is specified at discrete, equi-spaced points in the frequency domain and the coefficients computed using the resulting DFT relationship. The drawback of this method is that the characteristic may overshoot some template constraint in between the specification points but at least the behaviour is 'tied down' at the points in question.

If the characteristic of a one-dimensional filter is a function of M coefficient values, as in equation 5.21, it may be specified at M points which are multiples of T^{-1}/M. Thus if

$$H(\omega/2\pi) = \sum_{i=0}^{M-1} C_i \exp - ji\omega T$$

then

$$H(kT^{-1}/M) = H_k = \sum_{i=0}^{M-1} C_i \exp - j2\pi ik/M, \quad k = 0 \ldots M-1$$

This is the IDFT and thus the inverse relationship is

$$C_i = M^{-1} \sum_{k=0}^{M-1} H_k \exp j2\pi ki/M, \quad i = 0 \ldots M-1$$

with the restriction that

$$H_{M-k} = H_k^*$$

for the coefficients to be physically realisable, i.e. the frequency characteristic must obey the symmetry conditions for amplitude and phase.

In the same way, a two-dimensional frequency characteristic which is a function of MN coefficients, arranged in a rectangular array of spacing X and Y as shown in Figure 5.16, can be specified at MN points on a rectangular lattice of cell dimensions X^{-1}/M and Y^{-1}/N, as shown in Figure 5.17. Thus if

$$H(\omega_1/2\pi, \omega_2/2\pi) = \sum_{i=0}^{M-1} \sum_{j=0}^{N-1} C_{ij} \exp - j(i\omega_1 X + j\omega_2 Y)$$

then

$$H(kX^{-1}/M, lY^{-1}/N) = H_{kl}$$
$$= \sum_{i=0}^{M-1} \sum_{j=0}^{N-1} C_{ij} \exp - j2\pi(ik/M + jl/N), \quad k = 0 \ldots M-1, l = 0 \ldots N-1$$

This is, again, an IDFT relationship so that

$$C_{ij} = (MN)^{-1} \sum_{k=0}^{M-1} \sum_{l=0}^{N-1} H_{kl} \exp j2\pi(ki/M + lj/N), \quad i = 0 \ldots M-1, j = 0 \ldots N-1$$

with

$$H_{M-k, N-l} = H_{kl}^*$$

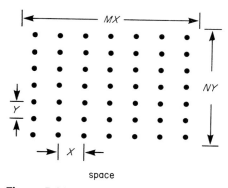

Figure 5.16 Definitions associated with a rectangular array of coefficients

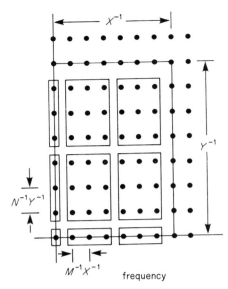

Figure 5.17 Specification of the frequency characteristic of a two-dimensional filter

As these are DFT relationships, the derived filter coefficient patterns are, themselves, periodic. Truncation of the patterns to one cycle to form a real filter convolves the discrete frequency characteristic with the transform of the cycle bound. This is a one- or two-dimensional 'sinc' function as the case may be. In the one-dimensional case the sinc function is

$$(\sin M\omega T/2)/(M\omega T/2)$$

and in the two-dimensional case it is

$$(\sin M\omega_1 X/2)/(M\omega_1 X/2) \times (\sin N\omega_2 Y/2)/(N\omega_2 Y/2)$$

It is the convolution which provides the interpolation between the points and may cause the characteristic to overshoot beyond the required specification.

Windowing

In this technique the filter is first specified as an ideal characteristic, usually involving sharp cuts between bands. From what has been said above it should be clear that this implies an infinite coefficient pattern. To overcome this difficulty the pattern is then truncated by multiplying by a 'window' function. This has the effect of smoothing the discontinuities in the frequency characteristic as it is convolved with the transform of the window so that sharp edges become integrals of the transform. Clearly, a desirable property of the window is that both it and its transform should be as compact as possible. It can be shown that the functions which possess this property are the prolate spheroidal wavefunctions, which are somewhat complex. A simpler approximation, however, based on Bessel functions is due to Kaiser and hence has become known as a Kaiser window

[106]. This window has a single shape parameter which, together with the width, controls the behaviour of the filter characteristic at the band edges. The resulting designs have pass- and stop-band ripple which die away from the band edges and which are approximately equal near the edge.

Kaiser derives an empirical relationship between the shape parameter α and the ripple attenuation in dB, ATT, as

$$\alpha = 0.1102(\text{ATT} - 8.7), \qquad \text{ATT} > 50$$
$$= 0.5842(\text{ATT} - 21)^{0.4} + 0.07886(\text{ATT} - 21), \quad 21 < \text{ATT} < 50$$

Likewise, the number of coefficients N as an empirical function of the ripple attenuation and the normalised transition bandwidth ΔF is given by

$$N - 1 = (\text{ATT} - 8)/(14.36\, \Delta F)$$

where

$$\Delta F = \Delta f / f_s$$

and f_s is the sampling frequency.

Assuming that the filter is symmetrical with an odd number of terms so that

$$N_p = (N - 1)/2$$

and that the coefficients of the idealised filter are a_n, obtained by Fourier analysis, the coefficients of the actual filter are given by

$$c_n = a_n I_0\{\alpha[1 - (n/N_p)^2]^{1/2}\}/I_0(\alpha), \quad n = 0 \ldots N_p$$

where I_0 is the modified Bessel function of the first kind and zeroth order and α is calculated from the above equation.

The advantage of this method is that it is fast by virtue of the explicit relationships between the parameters which are simple. The disadvantage is that it cannot be used to design equi-ripple characteristics. A more complete description is given by Kaiser [106].

Recursive interpolation (Remez exchange)

In this method a guess is made of the extremal frequencies of the characteristic of the filter that has the minimum equi-ripple for the specified number of terms. These frequencies are those where the deviation from the desired characteristic is greatest. It is then possible to estimate the value of the deviation at these points and, using these values, the complete characteristic by Lagrange interpolation. As a result, a new set of extremal frequencies is derived. The process is then repeated until there is no change in the extremal frequencies. The whole iterative process is carried out using a fine grid of discrete frequencies.

A full description is given by McClellan *et al.* [107] which includes a computer program. The inputs to the program are
  filter length
  number of frequency bands
  grid density

frequencies of lower and upper band edges (normalised to f_s)
ideal characteristic values in the bands
error weights in each band (inversely proportional to ripple magnitude).

The output of the program comprises the ripple magnitudes attained in each band, the extremal frequencies and the coefficient pattern. Design might typically involve running the program several times until a satisfactory ripple performance is obtained consistent with using the minimum number of terms.

A disadvantage of this method is that it is slow due to the iterative nature of the program. Execution time increases rapidly with filter order. An advantage is that the program can be modified to allow the idealised filter characteristic to be any arbitrary analytical function rather than just constant values in the bands.

Two-dimensional windowing

The one-dimensional method can be extended to two dimensions by repeated application horizontally and vertically [108]. This implies that the window is variables-separable and not, for example, circular. Transition bands in the two dimensions are specified, resulting in attainable ripple attenuation separately but, as the final frequency characteristic is a double convolution, it is difficult to predict how the two dimensions interact. This can only be deduced by inspection. For example, the characteristic can be searched to find the extrema or it can be output as a contour plot. This adds considerably to the design iteration time.

One- to two-dimensional mapping

This technique is based on the idea that values of a one-dimensional frequency variable can be mapped onto contours of a two-dimensional frequency space by a transformation [109–111]. It is then possible to map a symmetrical one-dimensional filter characteristic into a two-dimensional space. The structure of the realisation is a combination of a one-dimensional filter and a two-dimensional transformation filter and as, in general, the transformation is of low order, the structure complexity is predominantly that of the one-dimensional filter. Moreover, a rearrangement allows the coefficient values of the original one-dimensional filter to be used directly.

As the one-dimensional filter is assumed symmetrical its characteristic is given by

$$H(\omega) = \sum_{n=0}^{N-1} a_n \cos(n\omega)$$

where

$$a_0 = h(0)$$
$$a_n = 2h(n)$$

discounting the overall group delay. Using the Chebyshev polynomials $T_n(x)$, $\cos(n\omega)$ can be expressed as a polynomial of degree n in $\cos \omega$. So

$$\cos(n\omega) = T_n(\cos \omega)$$

and

$$H(\omega) = \sum_{n=0}^{N-1} a_n T_n(\cos \omega)$$

making the characteristic a function of the variable $\cos \omega$. This equivalence corresponds to the equivalence between the circuits of Figures 5.18a and 5.18b where the filter G is of the form

$$G(\omega) = (z + z^{-1})/2 = \cos \omega$$

A transformation filter characteristic, $G(\omega_1, \omega_2)$ can now be substituted for the one-dimensional filter G so that

$$H(\omega_1, \omega_2) = \sum_{n=0}^{N-1} a_n T_n[G(\omega_1, \omega_2)]$$

where

$$G(\omega_1, \omega_2) = \cos \omega$$

This gives a two-dimensional filter in which values of $H(\omega)$ at frequencies ω are mapped onto contours of ω_1, ω_2 given by G.

The attraction of the method is that standard techniques can be used to design the one-dimensional filter whilst the function G, being 'well behaved', allows the overall result to be reasonably well predicted.

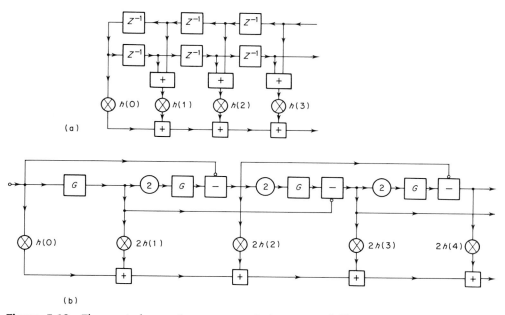

Figure 5.18 The equivalence of two symmetrical transversal filter circuits: (a) normal form (b) rearranged form embodying Chebyshev polynomials

HARDWARE CONSIDERATIONS OF TRANSVERSAL FILTERS

The hardware involved in digital filters must provide means of access to adjacent pixels, multiplication and accumulation. The means of access to adjacent pixels, remembering that 'adjacent' can also mean vertically or temporally, is provided by storage registers corresponding to one sample period, or one line period or one field period plus or minus half a line period. If the coefficient pattern calls for access to more than one dimension then the line and/or field periods become nominal, being shortened by the extent of the other dimension(s).

Multiplication, if fixed, is performed by programmable read-only memories (PROMs) containing the input–output relationship of a constant scale factor. This means that inaccuracies due to coefficient rounding are avoided and the accuracy is limited only by the number of output bits. Variable multiplication, such as might be needed in an adaptive filter, can be performed by random-access memories (RAMs) if the coefficient values change slowly enough for the contents of all the RAMs to be calculated and updated by an external computer. Rapidly varying multiplication is carried out using dedicated multi-bit multipliers.

Negative values of multiplier, multiplicand and product are best dealt with using two's complement number representation. This is especially true for the product for it then allows the following accumulation operation to proceed without regard to sign. Moreover, commercial four-quadrant multipliers are invariably in two's complement form. PROMs or RAMs may, however, assume other sign conventions for their inputs.

Accumulation of the products is conventionally performed using a parallel tree as shown in Figure 5.19. However, this tree is often difficult to build because it is too big and has too many inputs to fit onto a single board. Once built it is likely to be difficult to extend.

The parallel tree may be rearranged as a serial accumulation ladder as shown in Figure 5.20. This may be built in modular form and is, therefore, easy to extend by adding extra

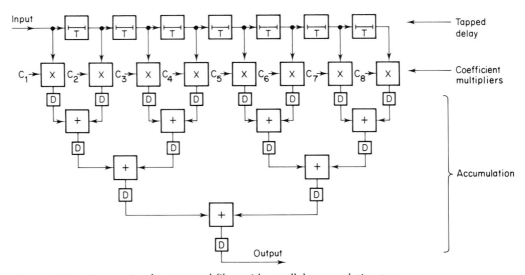

Figure 5.19 Conventional transversal filter with parallel accumulation tree

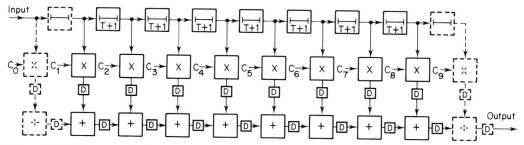

Figure 5.20 Transversal filter with serial accumulation ladder

modules, as shown by the broken lines. However, the delays in the accumulation ladder must be matched by increasing the delay of the tapped delay line elements by one clock period. The overall delay through the filter is also increased.

These problems can be overcome by reversing the direction of flow of the accumulation ladder as shown in Figure 5.21. This requires shorter tapped delay elements and has less overall delay than either Figure 5.20 or Figure 5.19. It is also even easier to extend than forward accumulation because extra modules can be added without breaking the main signal path.

Often the delay element T is only one clock period. In this case Figure 5.21 simplifies to Figure 5.22 which has no separate tapped delay, the delay being performed by the reverse flow of the accumulation ladder.

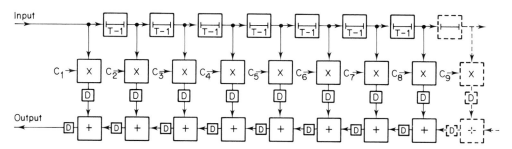

Figure 5.21 Transversal filter with reverse serial accumulation ladder

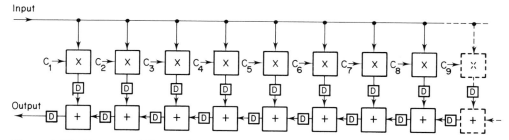

Figure 5.22 Transversal filter without separate delay line

If the transversal filter is symmetrical the number of modules may be reduced by adding the input samples together in pairs before multiplying by the weighting coefficients. This is shown in Figure 5.23 which is obtained by folding the accumulation ladder of Figure 5.21. Such a symmetrical filter may be constructed by cascading the module delineated by the broken lines.

Alternatively, starting with the arrangement of Figure 5.22, the generalised filter can be rearranged as in Figure 5.24 where the multipliers are placed before the delay elements. Note that the coefficients are applied in the reverse order. Compared with the arrangement of Figure 5.21 this needs one less storage element per stage. Further, because all the multipliers are working with the same sample at once, the multipliers could be simplified to a system in which common partial products are generated and the values required in each stage are selected according to the individual coefficient values.

The disadvantage of the arrangement is that more bits must be carried in the delay line if the accumulation of serious rounding errors is to be avoided. However, if T is small, as is usually the case, this is not important.

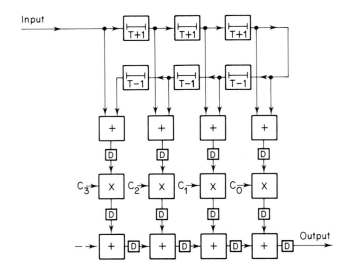

Figure 5.23 Symmetrical transversal filter

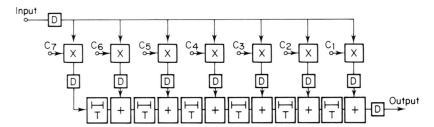

Figure 5.24 Transversal filter with multipliers before the delay line

For symmetrical filters the same simplification as before can be applied to this arrangement to yield the arrangement of Figure 5.25. Compared with Figure 5.23 each module now needs two less storage elements. Further, the arrangement of Figure 5.23 suffers the disadvantage that, as there is an addition before the multipliers, the signals entering them have one bit more than those in Figure 5.25. Therefore, either a more complex multiplier must be used or a loss of accuracy must be accepted.

Finally, the module of Figure 5.25 may be rearranged slightly as in Figure 5.26 so that both accumulation ladder inputs are latched. This avoids interfacing problems at high word-rates which may arise when modules are separated by appreciable distances. Moreover, the module can be modified as in Figure 5.27 to accommodate a central coefficient for symmetrical filters or as in Figure 5.28 to provide a zero-weighted tap and sign-reversal for anti-symmetrical filters.

Four basic modules, based on the form of Figure 5.26, can be designed to fit onto a single 4U BMM (defined in Chapter 3) printed circuit board, becoming a super module.

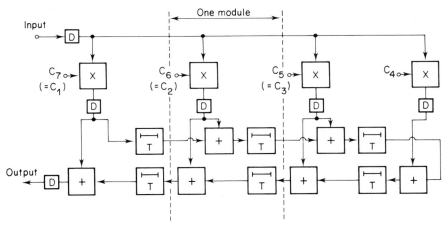

Figure 5.25 Symmetrical transversal filter with multipliers before the delay line

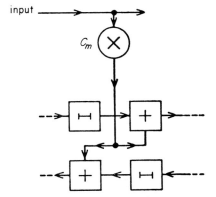

Figure 5.26 A rearranged version of the module in Figure 5.25

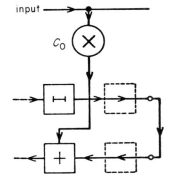

Figure 5.27 Centre term module for symmetrical filters

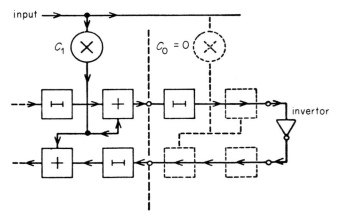

Figure 5.28 Centre term module for antisymmetric filters

Boards may then be cascaded indefinitely, subject to the driving capability of the hardware supplying the signal, to make a transversal filter of any required size. Each board can be designed so that the inputs to cascaded boards may be bussed on the backwiring and a similar interrupted bus structure allows the cascaded inputs and outputs to be connected between boards. Each supermodule thus contributes eight taps to a transversal filter, unless it provides the central term when it contributes seven. A seven-tap symmetrical filter can thus be accommodated on a single board, a fifteen-tap filter on two boards, etc.

The arrangements of Figures 5.19 to 5.28 imply that the delay elements between the coefficients are identical, giving a one-dimensional filter. Thus, elements of one clock period give a horizontal filter, elements of one line period give a vertical filter and elements of one picture period give a temporal filter. However, there is no reason, in principle, why modules with different delay elements may not be mixed to form a multi-dimensional filter, notwithstanding the fact that a module based on a line, field or picture delay will be bulkier than one based on a one-sample delay. Use of mixed modules based on the form of Figure 5.26 will still yield a symmetrical filter which will have a centre of symmetry in multi-dimensional space.

SOME APPLICATIONS

When considering video applications of digital filters it is necessary to distinguish between real-time hardware situations and computer simulation. The former are likely to involve rather rudimentary designs since implementation of really high-order filters would result in hardware of unacceptable proportions. There is thus, at present, an enormous gulf between the design methods outlined and practical situations. No such constraints exist, however, for computer-simulated filters where the complexity is limited only by the processing time so that the designer can indulge in 'overkill'. Indeed, such a technique is extremely valuable in helping the designer of a practical system to discover the relative importance of various factors.

Real-time aperture correction

The need for aperture correction, sometimes called 'enhancement', arises out of the frequency distortion introduced by the source (and display) scanning spots. It can be shown that the effect of source aperture distortion is to multiply the Fourier transform of the source by the Fourier transform of the aperture [103]. These transforms are three dimensional, as already shown, the temporal part of the aperture being caused by the integration time and lag of the camera which manifests itself as a blur on moving objects. This part is usually neglected so that the term 'aperture correction' is understood to refer to the spatial part. However, temporal aperture correction will also be discussed here. In addition, aperture correction can also compensate for loss of resolution due to other causes in the camera such as lens imperfections and image spreading in a camera tube or film emulsion but these effects are usually less important. Correction of display aperture, which should be carried out at the display, is usually omitted because, with a correctly adjusted display, the spot has a negligible effect on spatial resolution and the phosphor afterglow, which provides the temporal aperture, is a small fraction of a field period. As the Fourier transform of a real aperture is generally of low-pass form with a progressive roll-off, it follows that an aperture corrector is required to have a frequency characteristic which rises with spatial frequency.

Aperture correction has been in existence far longer than digital processing and so the traditional hardware to obtain this high-frequency boost is based on analogue techniques. Nevertheless, it simulates the operation of a digital filter as the circuit of a conventional corrector is an analogue transversal filter, taking contributions from the continuous signal delayed by various amounts. The delays give access to neighbouring points and whereas there is no choice vertically, being the signal on adjacent lines, the horizontal delay can be chosen to optimise the characteristic. One form of corrector comprises two cascaded circuits, each like that of Figure 5.29, one correcting the horizontal aperture and the other the vertical aperture. In the vertical corrector the delay, τ, would be one line period whereas in the horizontal corrector τ would be about 70 ns. This could be called a variables-separable design. The other form of corrector,

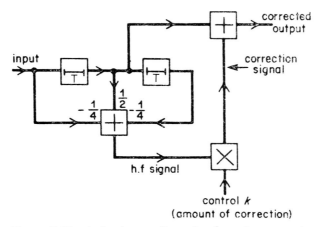

Figure 5.29 A simple one-dimensional aperture corrector

shown in Figure 5.30, treats both dimensions at the same time. This allows the frequency characteristic to be non-variables separable.

In either case the corrector works by first deriving an enhancement signal formed by weighted contributions from the neighbouring points and the central point such that the sum of the weighting coefficients is zero. This ensures that the zero-frequency gain of the corrector is unity. The enhancement signal is then weighted by a variable amount, k, to provide an adjustment and added to the value of the central point. It will be noted that in neither case does more than one term either side of the central point contribute. Thus, in the variables-separable case, each corrector has a frequency characteristic which consists of only a constant and a fundamental term, the polarity of the latter being such as to peak at the frequency $1/2\tau$. Figure 5.31 shows the effect of this characteristic on an assumed Gaussian aperture distortion for various values of τ where the amount of correction is adjusted to keep the overall characteristic maximally flat at zero frequency. It can be seen that such a corrector is nowhere near accurate enough to correct the aperture distortion perfectly but the result improves as τ decreases and k correspondingly increases. Unfortunately, this cannot be allowed to happen indefinitely because, in the limit, it corresponds to an infinite amount of correction at an infinitely high frequency. For the vertical aperture corrector there is no choice as τ must correspond to one vertical sample equivalent to the spacing of the field lines, giving a peak at the equivalent of about 3.7 MHz. For the horizontal corrector a compromise is usually struck with the peak at about 7 MHz and this is known as 'out of band' correction.

Figure 5.32 shows the two-dimensional frequency characteristics of both variables-separable nine-point correctors and non-variables-separable five-point correctors operating on a Gaussian aperture [112]. The amount of correction has again been adjusted to give a maximally flat response at zero frequency. The correctors have the same characteristic along the axes but, as can be seen, the nine-point corrector is better able to correct the diagonal frequencies although there is less than a dB difference between the

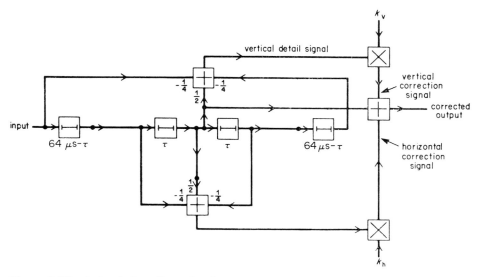

Figure 5.30 A simple two-dimensional aperture corrector

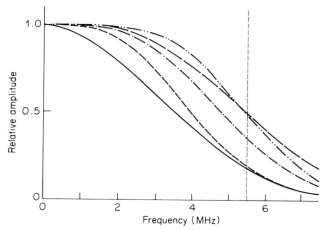

Figure 5.31 Aperture corrected Gaussian frequency characteristics: (———) no correction (– – –) $\tau = 150$ ns, $k = 0.25$, (–·–·–) $\tau = 75$ ns, $k = 1$, (— — —) $\tau = 7.5$ ns, $k = 100$, (–··–) second-order correction $\tau = 75$ ns, $k_1 = 11/6$, $k_2 = -5/24$

two. However, if the correction is increased, as it would be in practice, the nine-point corrector gives far more overcorrection on diagonals than the five-point corrector, increasing the likelihood of ringing. So for practical aperture correctors which are likely to be operated with slight amounts of overcorrection the five-point corrector is to be preferred as it gives a more isotropic response.

The block schematic of a digital version of such an aperture corrector would appear very similar except that the horizontal delay is forced to be the reciprocal of the digital sampling frequency. This has now been standardised at 13.5 MHz and so the delay is about 74 ns, giving a peak correction at 6.75 MHz. With digital techniques, however, it is possible to have a corrector of any order, either horizontally or vertically, without incurring unacceptable impairments due to delay imperfections. In practice, it is found that higher order vertical correction is not worthwhile because even first-order correction peaks at such a low frequency that it is well in band [113]. As a result, it is not able to correct the aperture very accurately and adequate subjective sharpness can only be obtained by accompanying 'ringing'. However, higher order horizontal correction is worthwhile because it can improve the accuracy of correction, as shown by the second-order curve in Figure 5.31. It can be seen that the weighting coefficients of the extra samples are much smaller than those of the nearest neighbours and reversed in sign. The extra contributions provide a spectral component which peaks at half the frequency of the first-order contribution, in-band, and the way in which the components build up the overall characteristic is shown in Figure 5.33. The block diagram of a second-order horizontal corrector is shown in Figure 5.34.

The use of such a corrector in conjunction with a vertical corrector can be justified where the highest quality of signal is required. One such corrector has been incorporated into an experimental digital channel, processing the signals obtained from a telecine sensor, either flying-spot or solid-state [114].

The inability of the vertical corrector to enhance vertical frequencies beyond the equivalent of 3.7 MHz can be overcome by taking contributions from neighbouring

248   DIGITAL FILTERING OF TELEVISION SIGNALS

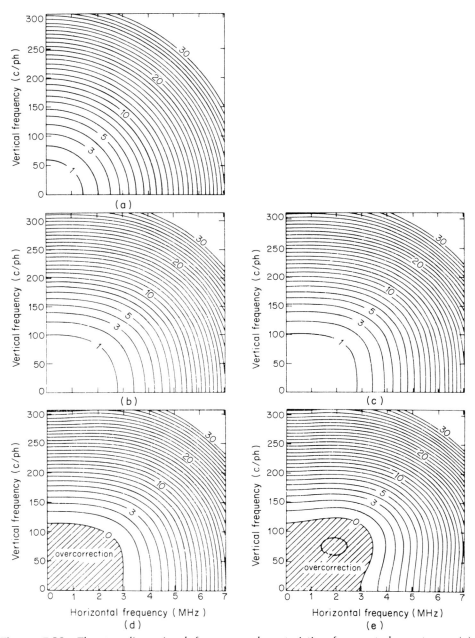

Figure 5.32 The two-dimensional frequency characteristic of corrected apertures. (a) No correction, (b) maximally flat five-point correction, (c) maximally flat nine-point correction, (d) overcorrected five-point correction, (e) overcorrected nine-point correction

SOME APPLICATIONS 249

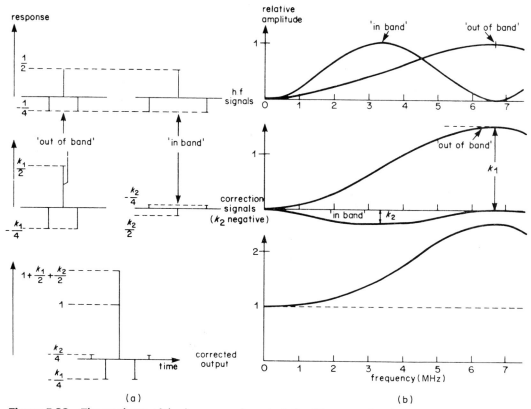

Figure 5.33 The synthesis of the frequency characteristic of the second-order aperture corrector. (a) Impulse responses, (b) frequency characteristics

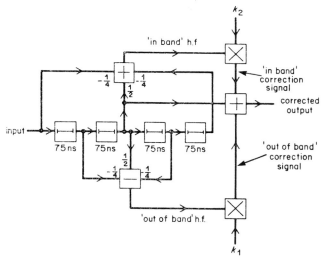

Figure 5.34 Block diagram of a second-order aperture corrector

picture lines, instead of field lines. The halving of the distance results in a doubling of the equivalent peaking frequency, resulting in a better match to the vertical loss. However, as the contributions come from a different time, such a corrector also affects the temporal behaviour of the picture. Figure 5.35 shows the two-dimensional vertical/temporal coefficient pattern, together with the vertical/temporal frequency characteristic, assuming a certain value of correction. It can be seen that the peaking also affects the region near zero vertical frequency combined with the picture frequency. Thus, picture-frequency spectral components generated by motion will be accentuated or, put another way, the filter is also a temporal aperture corrector.

Figure 5.36 shows the block diagram of such a 'picture-line' corrector which was built for experimental purposes as part of picture-enhancement equipment, incorporating noise reduction, to be described later in the chapter [115]. The notch filter was needed to ensure that the correction did not affect the chrominance band of the composite signal.

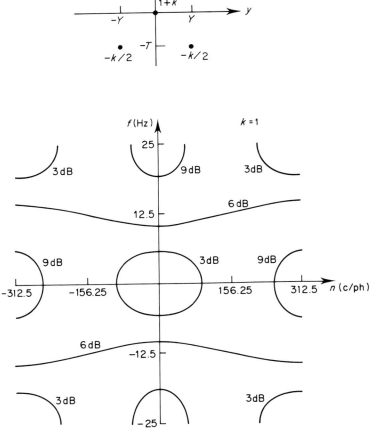

Figure 5.35 Coefficient pattern and spectral characteristic of 'picture-line' vertical aperture corrector

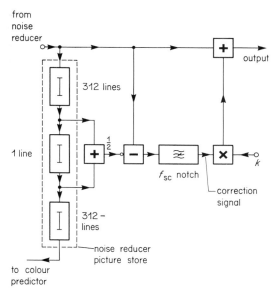

Figure 5.36 Block diagram of 'picture-line' corrector

It was found that the corrector gave a visible improvement in resolution on stationary pictures that were already relatively sharp. The minimum value of k for such pictures was found to be about 2. This value had to be increased considerably to give a noticeable improvement on unsharp pictures. As the corrector also increased the amplitude of the noise components around the peaking frequency the improvement that could be achieved before the added noise became objectionable was disappointingly low, especially with relatively unsharp input pictures.

Turning to the temporal aspects, however, the main problems with this form of corrector were caused by the resulting overcorrection of signal components around the picture frequency. In addition to motion, these can be caused by shot changes and unsteadiness, twin lens amplitude and misregistration flicker in telecine sources. Further, the judder associated with film motion was accentuated because such a corrector ignores the field pairing caused by interlaced scanning of film. The visual effect during motion could best be described as a 'halo' surrounding moving edges and increasing in visibility as the speed of motion increased. The signal could, during rapid movement, exceed its permissible range of values causing problems with sync. separators. Shot changes could give an inappropriate correction signal during the first field of the new shot, giving, in general, an output signal which, again, could exceed the permissible range of values. Inasmuch as these effects were far more objectionable than the spatial effects this shows that the spatial aperture loss of the average source is much geater than the temporal loss.

Vertical chrominance filtering

The need for this arises in colour coding systems like clean PAL [77] and MAC [19] which try to exploit the available video bandwidth in a way which is better adapted to the eye's visual acuity. Given that the horizontal bandwidth of the chrominance is usually restricted to a half of that of the luminance or less, and that the potential vertical resolution of 625- or 525-line systems is already greater than the horizontal, this results in a large disparity between the horizontal and vertical chrominance resolutions. Clean PAL attempts to trade this excess vertical resolution for elimination of luminance–chrominance cross-effects by combining the two chrominance components U and V into a single signal which is made to interact minimally with the luminance. This involves sending the signals $(U + V)/\sqrt{2}$ and $(U - V)/\sqrt{2}$ on alternate lines, which is a form of vertical subsampling. In order to optimise the use of the available bandwidth and the received noise, MAC does a similar thing by sending U and V on alternate lines [116]. At the receiver, the information on the missing lines must be interpolated from the transmitted information.

Clearly, this cannot be done without incurring a penalty. Unless the U and V signals do not contain frequencies beyond the limit that can be supported by the subsampling structure the process will result in aliasing. Thus the U and V signals must be prefiltered with a low-pass characteristic and the interpolator should have a matched frequency characteristic. This characteristic, for intra-field processing, amounts to a cut in vertical frequency of 78 c/ph (for System I), equivalent to 1.84 MHz.

Figure 5.37 shows the essential schematic of the idea. In practice, the link would contain other processes which would ideally be transparent. The two input signals are each prefiltered and then selected on alternate lines. The prefilters are transversal filters based on line delays. At the receiver the single signal is distributed on alternate lines to two interpolators which are just further filters of the same type, their inputs on the other lines being zero.

Figure 5.38 shows the spectra involved, assuming the signals are $U + V$ and $U - V$. The spectra of the input signals after prefiltering are shown in Figure 5.38a. After subsampling the spectra appear as in Figure 5.38b. Assuming the spatial origin is chosen to coincide with a line of $U + V$ the interleaved sampling of $U - V$ inverts odd-order spectra, shown dashed. The spectrum of the combined signal is the sum of these individual spectra which is equivalent to the spectrum in Figure 5.38c. It will be noted that this now appears to consist of baseband U energy and V energy centred on a carrier of 156 c/ph caused by the V-axis switch. Distribution of the signal on alternate lines, interspersed with zeros, reinstates the spectra of Figure 5.38b which are filtered by the interpolators to yield the spectra in Figure 5.38d. As the filters cannot have an infinitely sharp cut, the process creates residual spectra in the transition regions. These residual spectra can be regarded as self-aliasing of $U + V$ and $U - V$ signals or, after matrixing to U and V, as crosstalk of U into V and vice versa.

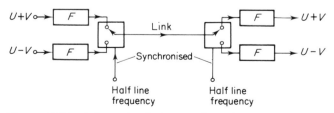

Figure 5.37 Schematic of alternate-line chrominance transmission

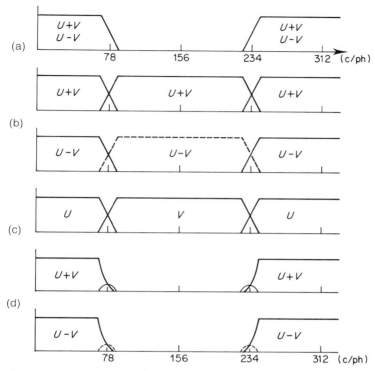

Figure 5.38 Spectra involved in alternate-line chrominance transmission: (a) prefiltered signals, (b) after subsampling, (c) combined signal, (d) after interpolation

At first sight it may appear that the ideal situation is to have filters of infinitely sharp cut which maximise the passband and minimise the residual spectra. However, such a sharp cut would produce ringing which, bearing in mind the low cut frequency, would be quite visible. Thus a compromise must be struck. As edge rise time (10% to 90%) is a succinct measure of bandwidth it proves possible to strike the compromise entirely in the time–space domain by forming a criterion function from weighted sums of the rise time, ringing overshoot and crosstalk signals. The frequency domain can be used initially to obtain a first approximation to what is required within the constraints of the number of filter coefficients. However, parameters such as transition bandwidth and maximum amplitude of crosstalk spectrum are difficult to relate to perceived effects. Thus, final adjustment is carried out using the above criterion function.

Figure 5.39 shows the frequency characteristics and time–space domain edge responses of an eight-term filter developed in this way. Eight terms was chosen as that number which would involve a reasonable amount of hardware. Such filters were built as part of an investigation into clean PAL coding. Their effect on ordinary picture material was to suppress the UV crosstalk below the perceptible limit whilst just perceptibly affecting the resolution. Even when modelled by computer simulation and used to process computer-generated colour patches, their effect was slight as shown in Figure 5.40 (see colour plates between pp 282 and 283). However, filtering by simple

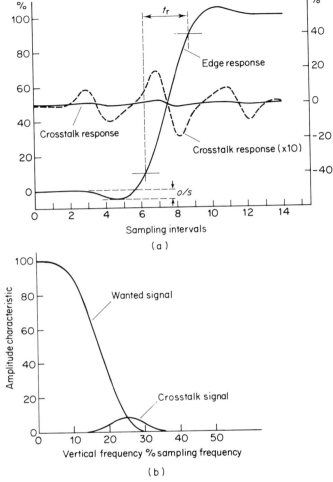

Figure 5.39 Frequency characteristics and time–space domain edge responses of the filter developed for vertical chrominance filtering. $t_r = 2.5$, $o/s = 5\%$, $c/t = 1.9\% \equiv -34$ dB

two-term averaging (first-order) was found to be inadequate for this material, as shown. It was necessary to create a graphic with periodic vertical chrominance detail before any significant effects could be revealed as shown in Figure 5.41 (see colour plates between pp 282 and 283). As well as the matched situation the figure also shows the effect of inadequate filtering at either end of the system.

Two-dimensional luminance filtering

The need for this arises in connection with both colour component and colour composite digital transmission systems where a reduction in the bit rate below the Nyquist sampling rate is sought. The earliest example of this is the colour component transmission system proposed by COMSAT for satellite transmission [117]. This was followed by a proposal for sub-Nyquist sampling of the PAL composite signal which is now in regular use on some of the BBC's links [118]. Since then, many proposals for sub-Nyquist sampling have been made in different contexts. The proposal for clean PAL coding, which has many similarities to sub-Nyquist PAL coding, but which can also be implemented in analogue form, is one such.

The underlying idea is to filter the luminance signal (possibly embedded in a composite signal) in two or more dimensions so that, when sampled at a frequency which is less than twice the highest frequency it contains (the one-dimensional limit) the repeated spectra caused by the sampling still do not overlap. The baseband spectrum, centred on the origin, can then be recovered by matched post-filtering. It is thereby assumed that certain regions of the multi-dimensional spectrum are more useful than others, which is a reasonable assumption.

The gamut of frequencies allowed by the one-dimensional Nyquist limit, regarded on a three-dimensional basis, is a cuboid. Clearly, those frequencies at the corners of the cuboid, being simultaneous combinations of high horizontal, vertical and temporal components, are less visible than those which have only one high component. In two-dimensional terms, this amounts to saying that diagonal frequencies are less visible than pure horizontal or vertical frequencies. Unfortunately, the scanning structure upsets the strict comparison between horizontal and vertical, not to mention the further complications of interlace, but this statement is qualitatively true. Thus, the earliest proposals for sub-Nyquist sampling used two-dimensional filters which removed diagonal frequencies allowing a higher order spectrum, centred on a carrier with the appropriate vertical component, to be slotted in. This carrier, being the sampling frequency, therefore corresponds to a spatial frequency having a vertical component of half the number of lines per field. In other words, the sampling frequency has a half line frequency offset.

Figure 5.42 shows the specification of a filter intended for clean PAL coding which involves sampling the luminance at a frequency of twice the colour-subcarrier frequency [79, 119]. This frequency has a half line frequency offset and so has the appropriate vertical component, as shown. The filter bound is chosen to be perpendicular to the sampling frequency vector, i.e. the line joining the origin to the sampling frequency. Assuming that the visibility of luminance spectral components is a function only of their distance from the origin, i.e. that the eye has isotropic spatial resolution, this bound then maximises the useful spectral energy. It will be noted that the highest video frequency is reduced to 5.2 MHz whilst the sampling frequency is approximately 8.8 MHz.

Such a filter, based on an orthogonal sampling grid created by a sampling frequency of four times the subcarrier frequency (neglecting the picture-frequency term), was designed using the frequency sampling technique, assuming a coefficient array of 22×4 terms. It will be appreciated that, as the filter is not variables-separable in the frequency domain, neither is the coefficient array. Therefore, such a filter cannot be realised by cascading horizontal and vertical filters and it must involve simultaneous access to each of the 88 points. The realisation of such a filter would involve an undue amount of hardware. However, its algorithm can be simulated by computer. Moreover, the subsequent processes of sub-Nyquist sampling and postfiltering with the same filter can also be simulated, together with the interpolation to and from the four-times subcarrier sampling frequency.

Figure 5.43 shows the effect on a zone plate pattern which is a two-dimensional linear

256 DIGITAL FILTERING OF TELEVISION SIGNALS

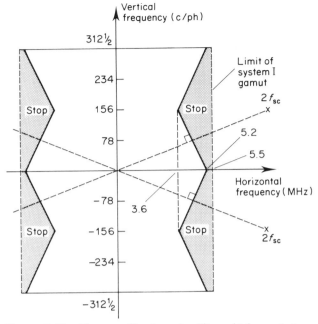

Figure 5.42 The specification of a filter which maximises the spectral energy closest to the origin consistent with a sampling frequency of twice the colour-subcarrier frequency

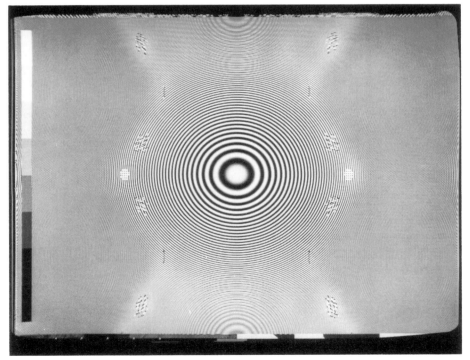

Figure 5.43 The effect on the zone plate pattern of the optimum luminance filter combined with sub-Nyquist sampling at twice the subcarrier frequency

NOISE REDUCTION 257

Figure 5.44 The effect on the zone plate pattern of a simple 'comb filter' combined with sub-Nyquist sampling at twice the subcarrier frequency

frequency sweep. Comparison with Figure 5.42 is readily apparent, the speckled regions being where there is a slight amount of overshoot. The vestiges of the first-order spectra, centred on twice the subcarrier frequency, can also be seen but these are reduced to a very low level.

Figure 5.44 shows the equivalent result for the simple first-order filter first proposed for sub-Nyquist sampling. This is variables-separable, consisting of a simple first-order vertical filter cascaded with a horizontal bandpass filter in the region of overlap. The slow rate of cut in the vertical direction allows the first-order spectra to overlap considerably with the baseband spectrum and this cannot be entirely eliminated by the postfilter. The coarse beat patterns in the transition region are caused by the simultaneous presence of the two spectra combined with the non-linearity of the display.

Figure 5.45 shows the effect on critical picture material of both the optimum filter and the first-order filter. To help the comparison, Figures 5.46 and 5.47 respectively show the filter loss and the residual aliasing in the two cases. It can be seen that the optimum filter loses slightly less picture but allows far less aliasing.

NOISE REDUCTION

Introduction

Noise is an inherent feature of any communication system and, in a broadcasting network, arises in many different ways. At the source it may be generated as photon

258   DIGITAL FILTERING OF TELEVISION SIGNALS

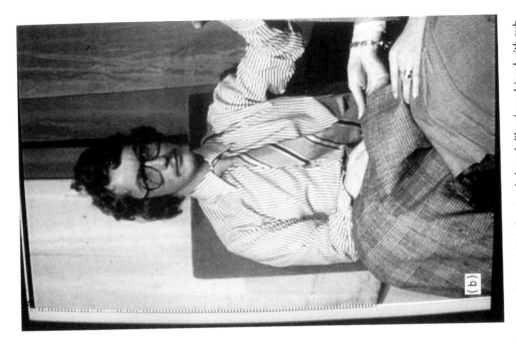

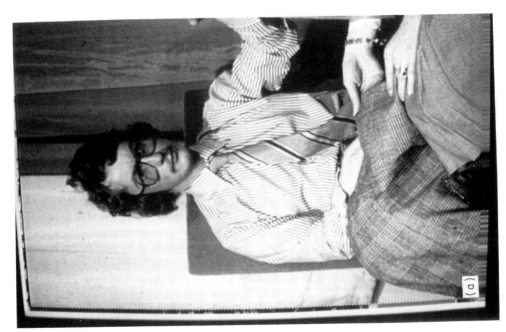

Figure 5.45 The effect on critical picture material of both (a) the optimum luminance filter and (b) the simple 'comb filter' combined with sub-Nyquist sampling

NOISE REDUCTION 259

Figure 5.46 The loss associated with both (a) the optimum luminance filter and (b) the 'comb filter'

260   DIGITAL FILTERING OF TELEVISION SIGNALS

(b)

(a)

Figure 5.47 The aliasing associated with both (a) the optimum luminance filter and (b) the 'comb filter'

noise or thermal noise in the television camera or as film grain in the cine camera. Circuits used to carry the signal then add further noise and recorders add yet more as the signal effectively recirculates round a noisy loop during post-production. Finally the signal is carried to the transmitter via the distribution network which adds more noise and radiated via a noisy transmission path. The sum total of all this is that the impairment due to noise of programmes currently broadcast varies from imperceptible to marginally acceptable.

Noise can be reduced in video signals without impairing spatial resolution by using the fact that in stationary pictures it causes the only difference between signals obtained in successive scans. Thus, averaging the signals corresponding to successive pictures leads to a noise reduction since the picture-to-picture difference is random with zero mean. Such an averaging amounts to a temporal low-pass filter, whose realisation provides a practical application for digital filters. Indeed without digital techniques it would be very difficult to obtain the accuracy and stability required for a recursive filter of the type required for noise reduction.

There are two basic ways of implementing such a low-pass filter. Firstly the signal may be passed through a series of delay elements, each of length one picture period, and the signals at each stage added together as shown in Figure 5.48. Such a filter is a transversal type and the noise reduction factor, given by the ratio of input to output noise powers, is simply the number of contributions. Thus many picture delay elements would be needed to obtain a large noise reduction. This is the arrangement used, for example, by Image Transform [120].

Alternatively, a recursive filter, as shown in Figure 5.49 can be used which requires only one picture delay element. The amount of noise reduction is controlled by the feedback factor, a, and, to avoid gain at zero frequency, the filter may be rearranged as in Figure 5.50. Here the effect of the subtractor, divider and adder is to form the output as a fraction $1/K$ of the input picture and a fraction $1 - (1/K)$ of the previous output picture. The impulse response of the filter is thus a decaying exponential sampled at the picture frequency and the effect on moving pictures is very like that of a long persistence display tube with a time constant of K picture periods. With this filter it is possible to obtain very high values of noise reduction factor simply by increasing K but the economy in having only one picture delay is partly offset by the need for an extra bit of accuracy each time K doubles, assuming the filter is realised digitally.

Figure 5.51 shows the frequency characteristic of the filter for a particular value of K. Because the delay element is one picture period the characteristic repeats at the picture frequency T_P^{-1}. The unweighted noise reduction factor is simply the ratio of the integral under the square of the characteristic to that under a flat, unity characteristic, taken over

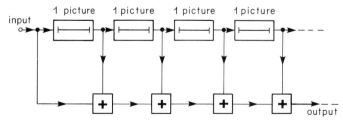

Figure 5.48 Basic transversal filter for noise reduction

262   DIGITAL FILTERING OF TELEVISION SIGNALS

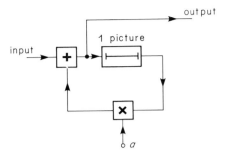

Figure 5.49 First-order recursive filter

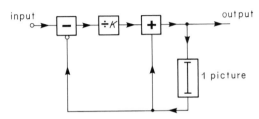

Figure 5.50 Rearrangement of the first-order recursive filter to give unity gain at zero frequency

one cycle of the characteristic. This factor can be shown to be given by

$$NF = 2K - 1$$

and is the reduction factor that would be registered by a measuring device with a time constant much longer than the picture period. However, as the reduced noise has the temporal spectral characteristic of the filter it may not appear to the eye to be reduced by the same factor.

To test this hypothesis an experiment was carried out to determine the relationship between the unweighted noise reduction factor given above and the subjective assessment. White noise, added to a suitable grey level, which had been passed through a recursive filter of the type shown in Figure 5.50, was displayed on one half of a picture monitor. Observers were asked to match this with the unfiltered signal displayed on the

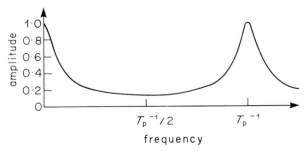

Figure 5.51 The amplitude–frequency characteristic of the recursive filter

Table 5.1
Noise reduction for temporal linear filtering

K	Unweighted noise-power reduction $10 \log_{10}(2K - 1)$ dB*	Subjective noise-power reduction†
2	4.8 dB	4.9 dB
4	8.5 dB	8.1 dB
8	11.8 dB	10.2 dB

* As would be measured by a full-band noise meter.
† Mean values over 6 observers; each observer was asked to match the two halves of a picture; one half had added white noise and was noise-reduced using the linear filter, the other half of the picture had a controllably smaller amount of added white noise and was unprocessed.

other half by varying the attenuation of the added noise. The required attenuation was taken as a measure of the subjective noise reduction and Table 5.1 shows the results. It is surprising that both the unweighted and the subjective noise reductions differ so little in view of the very disparate nature of the spectra.

In practice it is found that a K value of 4, giving an unweighted noise reduction factor of 8.5 dB, is sufficient for the majority of programme material.

Movement protection

Introduction

As the filter of Figure 5.50 behaves rather like a display with a long-persistence phosphor its effect on motion is to cause lag-like smearing of moving objects. This is normally completely unacceptable and so the filtering action must be inhibited wherever motion occurs in the picture. This implies that the noise reappears in moving areas but it is found, in practice, that this is acceptable in the majority of cases because the eye is distracted by the motion; the exception occurs where there is an isolated moving edge. The way in which the filtering action is inhibited is, however, crucial, for if it is done badly the noise-reduced picture may appear worse than the original. It will be noted that if K equals unity there is no filtering action. Thus it is only necessary to replace the simple divider in Figure 5.50 by a circuit which multiplies by a fractional factor that tends to unity as the amount of motion in the picture increases. Figure 5.52 shows such a rearrangement and Figure 5.53 shows details of a circuit, derived after considerable research, that produces such a fractional factor, varying smoothly where there is motion. Its four basic elements are a rectifier, a spatial filter, a variable attenuator and a non-linearity and these are now discussed in more detail.

Action of rectifier

The action of the rectifier in conjunction with the spatial filter H is to form the mean modulus of the picture difference signal. This is simpler than, but similar to, measuring the r.m.s. of the difference signal which is the quantity more usually found in classical

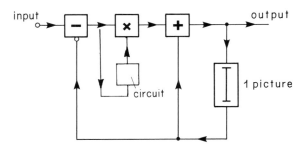

Figure 5.52 The inclusion of a motion-responsive element in the recursive filter

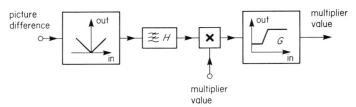

Figure 5.53 The four basic elements of the motion detector

solutions to signal processing problems. In the absence of motion this mean modulus is a good representation of the r.m.s. value of the noise which forms the only contribution to the picture difference. Because the filter, H, is finite the mean has a variation and its probability distribution function can be derived as follows.

Assuming the noise has a Gaussian probability distribution, the distribution of the picture difference signal, after rectification, is 'half-Gaussian' as shown at Figure 5.54a. It can be shown that the mean, μ, and variance, v, of this distribution are given by

$$\mu = (2/\pi)^{1/2}\sigma \quad v = \sigma'^2 = (1 - 2/\pi)\sigma^2$$
$$\approx 0.8\sigma \qquad \approx 0.36\sigma^2 \qquad (5.24)$$

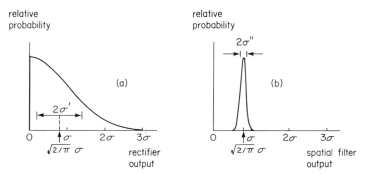

Figure 5.54 The probability distribution of the picture difference signal, assuming Gaussian statistics. (a) After rectification, (b) after rectification and spatial filtering

where σ is the standard deviation of the original Gaussian distribution, i.e. the r.m.s. noise value. Thus the standard deviation of the half-Gaussian distribution, σ', is about $\frac{3}{4}$ of the mean.

The spatial filter forms an equal-weight sum of many adjacent pixel differences and so, assuming that the differences are uncorrelated, i.e. the noise is spatially white, the probability distribution of the filter output is approximately Gaussian, according to the central limit theorem. The mean or expectation, E, of this distribution is the same as that of the input but the variance is that of the input divided by the number of terms in the filter, i.e.

$$E = \mu, \quad v_n = \sigma''^2 = v/n \qquad (5.25)$$

where there are n terms in the filter. For example, a 64-term filter reduces the standard deviation of the input by a factor of 8, so giving a probability distribution like that of Figure 5.54b; the more terms in the filter, the more peaked the distribution. The extent of this distribution determines the steepness of the non-linearity, G, for, clearly, if G rises significantly over the range of the distribution then a substantial proportion of the picture will not be noise-reduced even when there is no motion.

The spatial filter

Given that the spatial filter forms an equal-weight sum of n terms there are two opposing factors governing the choice of n. The larger n becomes, the more accurate is the measure of noise, as manifested by the lower variance of the output, and so the more readily can motion be detected by steepening the non-linearity G. On the other hand, the larger n becomes, the more slowly does the filter output respond to a change and therefore the more likely it is to miss isolated changes caused by small-area motion.

For example, consider a rectangular pulse caused by a moving edge (ignoring camera integration) propagating through the filter. In the absence of noise the filter output would be a trapezoidal-shaped pulse, being the convolution of the pulse and filter coefficient pattern. Because of noise, however, the output rises from a basal level, say E_0, with a statistical distribution about the trapezoidal shape. For a given pulse shape, the wider the filter, the earlier the output begins to rise but the less the difference between the peak output and the basal level, once the filter is wider than the pulse. On the other hand, the motion detector is more sensitive to this difference if the non-linearity is steeper. One would expect some compromise to emerge from these opposing factors and the behaviour can be analysed, to some extent, as follows.

Assume that the filter forms an equal-weight sum of $2m + 1$ input samples. Then, letting the input and output be x and y, respectively, we have

$$y = (2m + 1)^{-1} \sum_{i=0}^{2m} x_i$$

Since the x_i have a statistically varying component due to the noise, the output, y, also has a statistical variation. The expectation of the distribution of y is given by

$$E(y) = (2m + 1)^{-1} \sum_{i=0}^{2m} E(x_i) \qquad (5.26)$$

where the $E(x_i)$ are the means of the distributions of the individual input samples.

In the absence of a movement signal, all the $E(x_i)$ are equal to, say, μ_0 where μ_0 equals $(2/\pi)^{1/2}\sigma$ as previously noted.

Thus, for no movement

$$E(y) = E_0 = \mu_0 \tag{5.27}$$

However a movement pulse of amplitude V will bias the probability distribution of an individual sample by the amount V as shown in Figure 5.55. It can be shown that the mean of this distribution is given by

$$\mu = a\mu_0 + bV \tag{5.28}$$

where $a = \exp(-V^2/2\sigma^2)$ and $b = \text{erf}(V/\sigma\sqrt{2})$.

The number of samples affected in this way will depend on the position of the movement pulse in the filter. If there are j samples of the pulse in the filter then the expectation of the filter output will be

$$E(y) = (2m+1)^{-1}\left(\sum_{i=0}^{j-1}(a\mu_0 + bV) + \sum_{i=j}^{2m}\mu_0\right)$$

(using equations 5.26 and 5.28)

$$= (2m+1)^{-1}[j(a\mu_0 + bV) + (2m+1-j)\mu_0]$$
$$= (2m+1)^{-1}\{[2m+1-j(1-a)]E_0 + jbV\} \tag{5.29}$$

(using equation 5.27)

Now if the pulse is $2p + 1$ samples wide the total filter response to the pulse is $2(m + p) + 1$ samples wide. Of this, only the central $2p + 1$ elements occur during the pulse (assuming the group delay is discounted) and thus the m elements either side give time for the motion detection to build up and die away. If the pulse is to be unaffected by the recursive noise reduction process then the spatial filter output at the $(m + 1)$th

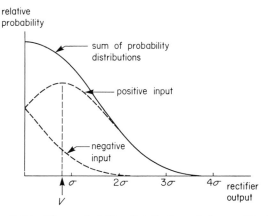

Figure 5.55 The probability distribution of the rectified picture difference signal assuming the difference is a constant, V, with added Gaussian noise of variance σ^2

sample must be such as to have completely turned off the noise reduction. That is, the non-linearity, G, must have risen to unity by the $(m + 1)$th sample.

If the pulse width is greater than $m + 1$ samples then the expected value of the $(m + 1)$th sample of the filter output, s_{m+1}, is given by equation 5.29 with j equal to $m + 1$. Otherwise it is given with j equal to $2p + 1$. Thus there are two cases

(1) $2p > m$

$$s_{m+1} = [1 - (m + 1)(1 - a)/(2m + 1)]E_0 \\ + (m + 1)bV/(2m + 1) \quad (5.30a)$$

(2) $2p < m$

$$s_{m+1} = [1 + (2p + 1)(1 - a)/(2m + 1)]E_0 \\ + (2p + 1)bV/(2m + 1) \quad (5.30b)$$

Now suppose we make the assumption that the non-linear characteristic, G, rises from its basal level, which prevails up to an input signal value of E_0, to unity at an input value which is three standard deviations (the classical 99.9% threshold) of the input distribution in the absence of motion, i.e. $3\sigma''$, above E_0. Figure 5.56 shows the relationship between these functions. The standard deviation of the input, in the absence of motion, being that of the filter output, σ'', is given by

$$\sigma'' = \sigma'/(2m + 1)^{1/2}$$

(using equation 5.25)

Then by requiring the $(m + 1)$th filter output sample to exceed the value which makes the non-linearity unity, i.e.

$$s_{m+1} \geqslant E_0 + 3\sigma'/(2m + 1)^{1/2}$$

and using equations 5.24 and 5.27 to express E_0 and σ' in terms of the r.m.s. noise, σ, we obtain an impairment threshold for the quantity V/σ, i.e. the amplitude of the movement pulse in units of r.m.s. noise, as a function of m and p, i.e. the filter and pulse widths.

For $2p > m$, as can be seen from equation 5.30a, the value of s_{m+1} is independent of p. Letting V/σ equal z, it proves convenient, in practice, to express m in terms of z for this

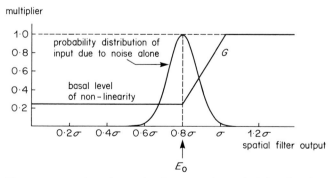

Figure 5.56 The relationship between the probability distribution of the spatial filter output and the non-linear function

range whilst to express p in terms of z and m for the other range. Thus, after manipulation, we obtain

For $2p > m$

$$(m + 1)^2/(2m + 1) = c^2$$

with solution

$$m = c^2 - 1 + [c^2(c^2 - 1)]^{1/2} \quad (5.31a)$$

For $2p < m$

$$(2p + 1)^2/(2m + 1) = c^2$$

with solution

$$p = \tfrac{1}{2}[c(2m + 1)^{1/2} - 1] \quad (5.31b)$$

where

$$c = 3(1 - 2/\pi)^{1/2}/[bz - (2/\pi)^{1/2}(1 - a)]$$

where

$$a = \exp\left(-\tfrac{1}{2}z^2\right) \qquad b = \operatorname{erf}(z/\sqrt{2}) \qquad z = V/\sigma$$

The result is shown in Figure 5.57 in which the minimum unimpaired pulse amplitude is plotted against pulse width, with filter size as parameter. For a given filter size, the amplitude of the pulse falls as its width increases until the width equals half the filter width. Thereafter, increasing the pulse width does not decrease the threshold amplitude. With large filters, small pulses are at a disadvantage compared with no filtering ($m = 0$),

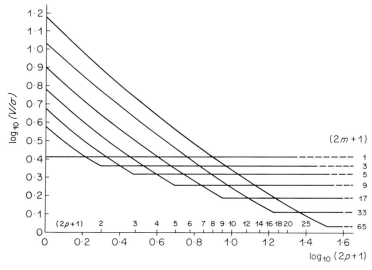

Figure 5.57 The relationship between minimum unimpaired pulse amplitude (in units or r.m.s. noise voltage) and pulse width for various widths of one-dimensional spatial filter

NOISE REDUCTION

for which the threshold amplitude is 2.6 units. For example, pulses less than 4 samples wide ($p < 1.5$) are at a disadvantage with a filter 17 elements wide ($m = 8$).

Figure 5.58 shows the effect of a two-dimensional filter of dimensions $(2m + 1)$ elements in width by $(2n + 1)$ elements in height. In this case the equations are just the same as before, assuming the pulse is one-dimensional, horizontally, except that the standard deviation of the filter output, in the absence of signal, σ'', is reduced by a factor of $(2n + 1)^{1/2}$ and hence the steepness of the non-linearity is increased by the same factor. Thus the equations become

For $2p > m$

$$(m + 1)^2/(2m + 1) = c^2/(2n + 1)$$

with solution

$$m = (c^2 - (2n + 1) + \{c^2[c^2 - (2n + 1)]\}^{1/2})/(2n + 1) \qquad (5.32a)$$

and for $2p < m$

$$(2p + 1)^2/(2m + 1) = c^2/(2n + 1)$$

with solution

$$p = \tfrac{1}{2}\{c[(2m + 1)/(2n + 1)]^{1/2} - 1\} \qquad (5.32b)$$

As can be seen, the general effect is to depress the curves of Figure 5.57 making the whole process more sensitive. But now, any square filter confers a detection advantage on any size of pulse compared with no filtering. However, shorter pulses are still harder to detect than longer ones, as might be expected. Moreover, filters whose widths are greater than their heights, e.g. the (15,5) filter in Figure 5.58 are at a disadvantage compared with square filters whereas those filters with the reciprocal aspect ratio are better off.

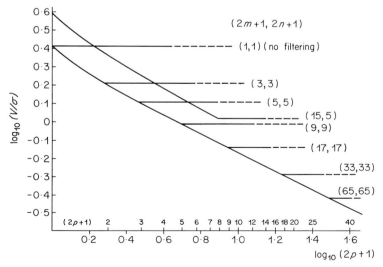

Figure 5.58 The relationship of Figure 5.57 for various areas of two-dimensional spatial filter

(This is a natural consequence of processing a horizontal pulse. The reverse would be true if the pulse were vertical.) Quite a large deviation from squareness, however, has little effect on the sensitivity for a given number of elements, once the pulse width exceeds the break point, as shown by the 81-element square filter (9,9) and the 75-element rectangular filter (15,5). As can be seen, the latter filter has a pulse width break point of eight samples ($p = 3.5$) and a break point amplitude of about 1 unit (0 on the logarithmic scale). Below the break point, the threshold amplitude rises gradually to 4 units (0.6 on the logarithmic scale) as the pulse width decreases, only pulses of width one sample being at a disadvantage, compared with no filtering. The (15,5) filter was selected as it was convenient to instrument.

The non-linearity

The foregoing analysis is based on the assumption that the non-linearity rises from the basal level to unity over the $3\sigma''$ half-range of the input distribution, the '3σ' point being that below which 99.9% of the distribution lies. If the spatial filter were of infinite extent this distribution would be infinitely sharp and so the non-linearity would rise infinitely quickly. This section will show, however, that even if the filter were of infinite extent, there is a limit to the rate of change of the non-linearity. The limit occurs because the input to the spatial filter, in the absence of motion, is not just the rectified noise *input* to the recursive filter but the rectified sum of the input and *output* noise powers. The latter quantity depends on the amount of noise reduction which, in turn, depends on the value of the non-linearity and, hence, upon the input to the non-linearity. Thus we have a feedback loop in which the input to the spatial filter depends on its output, via the form of the non-linearity. This imposes a condition on the non-linearity which can be analysed as follows.

Consider the simplified circuit of the noise reducer shown in Figure 5.59. Assuming for the moment that the spatial filter *is* of infinite extent, the input to the non-linearity, δ, has an infinitely thin probability distribution of expectation E_0, in the absence of motion, given by

$$\delta = E_0 = [(2/\pi)(\sigma_i^2 + \sigma_0^2)]^{1/2}$$

from equation 5.24 where σ_i and σ_0 are the r.m.s. input and output noise voltages.

But the output noise power is related to the input noise power through the noise factor, NF, which is a function of the recursive filter feedback factor, K, in Figure 5.50. Here, the

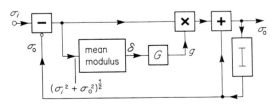

Figure 5.59 Simplified circuit of the noise reducer for analysing the dependence of its behaviour on the non-linearity

divider has been replaced by a multiplier, fed by the non-linearity, so that

$$K = g^{-1}$$

where g is the ouput of the non-linearity, G. Thus

$$(\sigma_o/\sigma_i)^2 = (2K - 1)^{-1}$$
$$= g(2 - g)^{-1} \qquad (5.33)$$

whence

$$\delta = E_O = (2/\pi)^{1/2}[2(2 - g)^{-1}]^{1/2}\sigma_i$$

or

$$g = 2[1 - (2/\pi)(\sigma_i/\delta)^2] \qquad (5.34)$$

Thus we have derived a relationship between g and δ with σ_i as a parameter. But this relationship is already governed by the shape of the non-linearity, G. Hence we could solve for δ or g in terms of σ_i. However, it is more instructive to keep the $g - \delta$ relationship explicit, to demonstrate what follows.

Figure 5.60 shows a family of curves described by equation 5.34 with σ_i as parameter, together with a possible non-linear relationship between g and δ. For any given value of σ_i the values of g and δ are given by the point of intersection of the relevant σ_i curve with the non-linearity. Now it can be seen that, in general, a curve may intersect the non-linearity any number of times, depending on the nature of the non-linearity, so giving any number of stable states. A single state is obtained only if the slope of the non-constant part of the non-linearity is less than the σ_i curve at each point.

For example, consider the discontinuous non-linearity of Figure 5.61. Any input noise level corresponding to a curve within the shaded area gives three solutions for (δ, g), one corresponding to maximum noise reduction, one to partial noise reduction and one to no noise reduction. Input noise levels lower than this are maximally reduced, those above, unaffected.

This behaviour is one of hysteresis and can be appreciated by considering what happens to the input–output noise power characteristic. Figure 5.62 shows this characteristic for the last-mentioned case. A hysteresis cycle might start by increasing

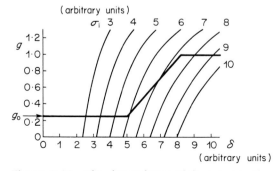

Figure 5.60 The dependence of the input to the non-linearity on its output as a function of noise input to the noise reducer together with a possible non-linearity characteristic

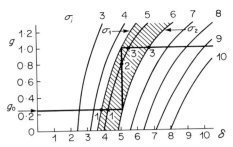

Figure 5.61 The same family of dependent curves as in Figure 5.60 together with a discontinuous non-linearity characteristic

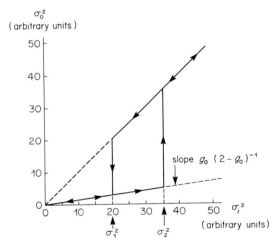

Figure 5.62 The dependence of the output noise power on the noise power input to the noise reducer, assuming the discontinuous non-linearity of Figure 5.61, showing hysteresis

the input noise from zero with the output noise following proportionally. At the input voltage σ_1 the lowest limiting curve in Figure 5.61 is reached with the intersection point at the low level of the non-linearity. As the input noise is raised beyond this point the curves in the shaded area apply and the intersection point remains at the low level, noise reduction remaining effective until the input noise reaches σ_2. Beyond this noise voltage the point must rise to the high level and noise reduction ceases so that σ_0 equals σ_i. Reducing the input noise now causes the intersection point to remain at the high level, with consequent absence of noise reduction, until the last possible moment when the input voltage σ_1 is reached, below which the intersection point drops to the low level and noise reduction becomes effective once more.

Clearly this hysteresis behaviour is undesirable and can be avoided by 'backing off' the steepness of the non-linearity until it, at least, coincides with a σ_i curve. Coincidence with the curve for, say, σ_t, shown in Figure 5.63, is the limiting case of stability and

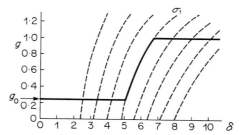

Figure 5.63 The same family of dependence curves as in Figure 5.60 together with a limiting non-linearity characteristic

corresponds to a discontinuity in the input–output noise characteristic at σ_t as shown in Figure 5.64. The arrangement is optimally efficient at detecting when σ_i exceeds σ_t or, provided σ_i remains at σ_t, at detecting motion.

Different values of σ_t can be accommodated by using the fixed relationship (equation 5.34 in another form)

$$g = 2(1 - 1/x^2) \qquad (5.35)$$

where $x = (\pi/2)^{1/2}(\delta/\sigma_t)$ and scaling the value of δ. Hence the need for the variable multiplier preceding the non-linearity. The universal plot of equation 5.35 in Figure 5.65 shows that the ratio of the abscissae where the curve cuts the values unity and zero is $\sqrt{2}$, and any practical curves must have a ratio which exceeds this value.

In practice the limiting curve of equation 5.35 is not optimal subjectively, because the spatial filter is *not* infinite. This means that, for a constant input noise, δ is statistically distributed so that the working point on the curve is not constant but has a probabilistic behaviour, as previously noted. If the centre of the probability distribution lies at the

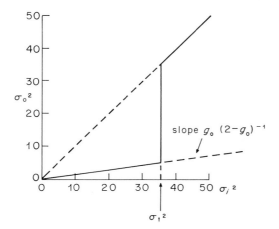

Figure 5.64 The input–output noise power characteristic of the noise reducer, assuming the limiting non-linearity characteristic

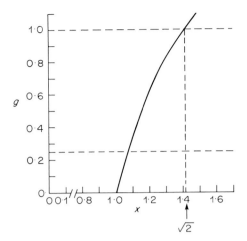

Figure 5.65 The limiting non-linearity characteristic expressed as a universal curve in terms of normalised variables

lower breakpoint of the non-linearity where the curve cuts the line $g = g_0$, then the slightest deviation causes the instantaneous working point to extend up the curve and beyond it into the unity region. This appears as lumps of input noise in a background of reduced noise, whose size depends on the extent of the spatial filter.

This behaviour can be countered by lowering the working point so that the centre of the δ distribution lies below the lower breakpoint of the non-linearity, but this only decreases the probability of occurrence of the lumps, not their objectionability.

A better solution is to 'back off' from the discontinuity in the input–output noise characteristic. Suppose a characteristic is defined which, for the sake of argument, has a linear transition of slope k between the noise-reduced and unprocessed regions, i.e.

$$\sigma_0^2 = g_0(2 - g_0)^{-1}\sigma_i^2$$

for $\sigma_i^2 < [1 - g_0 k^{-1}(2 - g_0)^{-1}]^{-1}\sigma_t^2$

$$\sigma_0^2 = k(\sigma_i^2 - \sigma_t^2)$$

for $[1 - g_0 k^{-1}(2 - g_0)^{-1}]^{-1}\sigma_t^2 < \sigma_i^2 < (1 - k^{-1})^{-1}\sigma_t^2$,

and
$$\sigma_0^2 = \sigma_i^2$$

for $[1 - k^{-1}]^{-1}\sigma_t^2 < \sigma_i^2$

as shown in Figure 5.66. Then, using equations 5.34 and 5.35 to substitute for σ_0 and then σ_i we obtain, for the transition region,

$$g = 2k(1 + k)^{-1}(1 - 1/x^2)$$

where $x = (\pi/2)^{1/2}(\delta/\sigma_t)$ as before.

This relationship is plotted in Figure 5.67. The curve for $k = \infty$ is the limiting case of stability corresponding to the discontinuous input–output noise characteristic of Figure

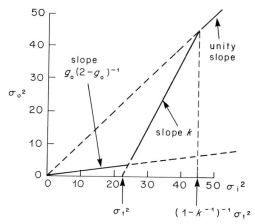

Figure 5.66 A possible input–output noise power characteristic for the noise reducer, having a linear transition region

5.64. The curve for $k = 1$ is asymptotic to unity as $x \to \infty$. For $k > 1$ the curve reaches unity when

$$x^2 = 2k/(k-1) \tag{5.36}$$

For example, the curve for $k = 2$ reaches unity when x equals 2.

The optimum value of k depends on the size of the spatial filter. We have shown that with an infinite spatial filter there is a limiting g–δ curve corresponding to $k = \infty$. On the basis of the previous theory, then, the appropriate value of k, with a finite spatial filter, is that which causes the non-linearity to rise to unity at an abscissal value of $3\sigma''$ beyond the limiting case ($\sqrt{2}$) where σ'' is the standard deviation of the filter output. For

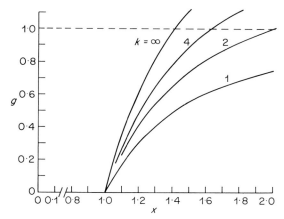

Figure 5.67 The normalised non-linearity characteristic resulting from the assumed input–output characteristic of Figure 5.66

example, a 75-term filter gives an output with a ratio of standard deviation to mean of 0.086 or 8.6%. Thus if the mean is set at the lower breakpoint of the curve the '3σ' point is 26% beyond $\sqrt{2}$, i.e. at x equal to 1.78. The corresponding value of k is, from equation 5.36, 2.8.

In practice the optimum value of k was found by patient observation of pictures to arrive at an acceptable compromise between 'hair-trigger' behaviour and smearing of low-level detail. The value finally arrived at was 1.6. However, it should be noted that the derivation of the non-linearity assumed a somewhat arbitrary linear relationship between input and output noise powers, in the transition region. Any other assumed relationship would have led to a different non-linearity which would have been equally valid. The chief value of the analysis in this subsection lies in the discovery of the limiting curve.

Colour operation

Given that the equipment must accept and deliver a composite colour signal there is a choice of two different system designs. One is to decode the incoming signal into its luminance and colour-difference components, process the three signals by three nominally identical circuits like that of Figure 5.52 and recode the three signals. The other is to use only one circuit with composite signals throughout but make due allowance for the picture-to-picture difference caused by the subcarrier. This difference is inherent in the definition of subcarrier frequency which is not an integral multiple of the picture frequency; if allowance were not made for this fact the offset would cause large difference signals in areas of high colour saturation. These would be interpreted as movement and thus saturated coloured areas would not be noise-reduced.

In such a method the composite signal in the picture store is transformed so that its colour information is appropriate to the input signal. This amounts, in the more complex case of PAL, to a 90° phase shift of the subcarrier and the reversal of the V-axis switch, the operations taking place in a single two-port network in series with the picture store, shown as a chrominance transformer in Figure 5.68. In principle the operations could be carried out by decoding and recoding the signal at that point but the transformer would then need to contain fast digital multipliers for quadrature demodulation and remodula-

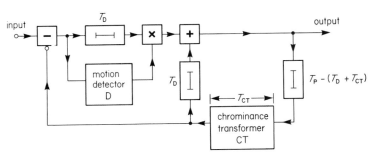

Figure 5.68 Simplified block diagram of the noise reducer showing the modification needed for composite colour signal processing

tion together with the appropriate digital filtering and means for phase locking the multipliers to the burst. Whilst this would be the most general solution which could deal with non-mathematical composite signals, the obvious complexity can be avoided by approaching the problem in a different way.

Assuming the composite signal *is* mathematical, a different approach is to consider that the transformer is trying to predict the value of the current sample, based on samples occurring nominally one picture period earlier, i.e. the transformer becomes a predictor. This approach is that used in DPCM and, indeed, the noise reducer configuration of Figure 5.68 is very similar to that of a DPCM coder. In these terms the monochrome noise reducer of Figure 5.50 predicts the value of the current sample as that occurring precisely one picture period earlier, i.e. on the assumption that there is no movement in the scene at that point.

Turning to the composite signal case, however, this prediction can no longer be used because of the aforementioned offset in the colour subcarrier. Figure 5.69 shows, in the PAL case, the phase of the subcarrier and V-axis switch polarity relative to a central reference point at various sample positions in the previous field and picture. The sampling structure assumes an instantaneous sampling frequency of nominally three times the subcarrier frequency, but line-locked. As can be seen, the phase of the subcarrier and the polarity of the V-axis switch at the sample precisely one picture period earlier are both inappropriate. In fact the V-axis switch reversal precludes the use of any samples from the line containing that sample. On the other hand, the sample exactly 624-line periods earlier has the appropriate subcarrier phase and V-axis switch polarity. (Actually the phase is not precisely the same but is in error by $0.576°$ as a consequence of the picture frequency term. This is too small to be shown in Figure 5.69.) Thus this sample can be taken as a prediction for chrominance components, provided that the positional error can be tolerated. This sample also gives a good prediction for line-locked

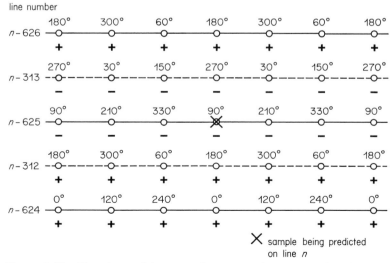

Figure 5.69 The phase of the PAL subcarrier and V-axis switch polarity at sample points in the previous field and picture relative to a reference point in the centre

luminance frequency components in the chrominance band which give rise to cross-colour of fine vertical frequency. These observations are independent of the actual subcarrier phase and V-axis switch polarity at the predicted sample which, thus, need not be known. Moreover, as samples either side of the predicted sample are not involved, the precise value of the sampling frequency is irrelevant, the only condition being that the samples are in vertical register from picture to picture.

If the chrominance band is defined by a single bandpass filter, then using the z-transform notation the required transfer function of the predictor, H_p, is

$$H_p(z) = (1 - B)z^{-625} + Bz^{-624} \tag{5.37}$$

where z^{-1} is the transfer function of a line delay and B is the transfer function of the bandpass filter. The first term in equation 5.37 represents the luminance part of the prediction for low frequencies and the second term represents the chrominance part for high frequencies, the exact form of B to be determined by experiment. This transfer function can be realised by the circuit of Figure 5.70 in conjunction with the picture store as in Figure 5.68 which must now correspond to a delay of nominally 624-line periods, less the group delay of the bandpass filter. (In fact the store must also be shortened by the group delay of the movement detector as shown in Figure 5.68 but this is a separate issue.)

Such a predictor fails where there is a sudden change of colour from one line to the next because of the positional error. To mitigate this effect the sample above the prediction point can also be taken into account to provide a mean. As shown in Figure 5.69 this has the opposite subcarrier phase and so must be inverted in taking the mean. Although this gives a better chrominance prediction the high frequency luminance prediction is worse because of the inverted contribution. Specifically, the prediction for line-locked luminance is now zero. Whether or not this predictor is preferable to the other depends, amongst other things, on the relative weight of the chrominance and high-frequency luminance contributions, in average picture material.

Using the same notation as before, the transfer function of this predictor is given by

$$H_p(z) = (1 - B)z^{-625} + \tfrac{1}{2}B(z^{-624} - z^{-626})$$

which can be realised by the circuit of Figure 5.71.

Both these predictors were realised experimentally and used in the configuration of Figure 5.68 to process composite signals in real time. The most serious drawback of the first predictor then became apparent. This was a systematic upwards movement, at the interlace strobe speed, of some of the residual noise left after the noise reduction process. This gave an impression of rising steam and was quite objectionable even though caused

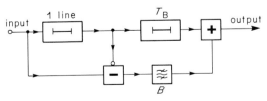

Figure 5.70 The circuit of the PAL predictor taking chrominance contributions from only one line

NOISE REDUCTION

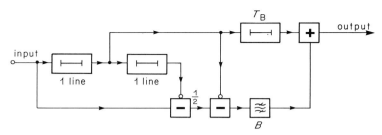

Figure 5.71 The circuit of the PAL predictor taking chrominance contributions from two lines

by only that part of the noise in the chrominance band defined by the bandpass filter. Thus the second predictor, which gave no such effect, was chosen for subsequent development. Varying the spectral characteristic of the bandpass filter, B, was found to have little effect on the efficacy of the noise reduction process, subject, of course, to the filter's having unity gain at subcarrier frequency. If anything, the movement impairment tended to be less when the characteristic was narrow. Ultimately it was decided to make the characteristic approximate to that of the luminance notch in the PAL decoder. This meant, in practice, making the narrowest characteristic possible without significant spectral overshoot with the amount of hardware available. Figure 5.72 shows the characteristic obtained.

Turning to the NTSC case, the arguments are similar except that there is no V-axis switch and the subcarrier phase shift across one picture period is $180°$. As this is also the phase shift across one line period it follows that the sample in the previous picture either above or below the predicted sample position has a suitable subcarrier phase. As far as the chrominance component is concerned, therefore, the mean of these samples is

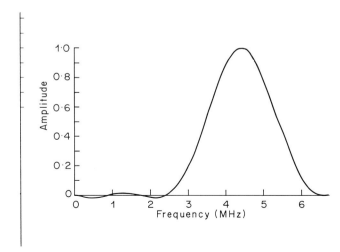

Figure 5.72 Frequency characteristic of the bandpass filter used in the PAL predictor

preferable to avoid the 'rising steam' effect and to provide some interpolation for line-to-line chrominance changes. Moreover, as both contributions are positive the mean also provides a better approximation to the high-frequency luminance component, if line-locked, than in the PAL case.

The transfer function of the predictor in this case, assuming a 525-line scanning system, is

$$H_p(z) = (1 - B)z^{-525} + \tfrac{1}{2}B(z^{-524} + z^{-526})$$

which can be realised by the circuit of Figure 5.73 in conjunction with a nominal picture delay. The form of B was, again, kept as narrow as possible and Figure 5.74 shows the characteristic used.

The practical realisation

This section discusses some of the factors affecting the design of an equipment based on the foregoing principles.

Picture store

At the outset, the picture store represented the overriding cost of the hardware. The experimental prototype equipment used a store based on 1 kbit shift registers [63] which

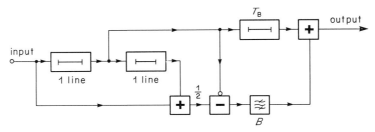

Figure 5.73 The circuit of the NTSC predictor

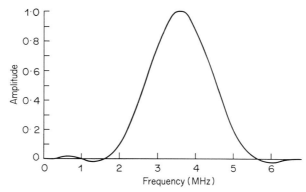

Figure 5.74 Frequency characteristic of the bandpass filter used in the NTSC predictor

was readily available and had been used for previous investigations. The store is described in Chapter 3. Unfortunately the integrity of the store was rather dubious in that the automatic test routine rarely indicated the complete absence of store errors. Fortunately, in the noise reducer, store imperfections are mistaken for movement and so are partially 'protected' by the movement detector, resulting, merely, in failure to noise reduce, the more catastrophic the fault, the greater the likelihood of protection. This, in fact, proved to be the case, and frequently the equipment could be operated for quite long periods with quite serious store errors before their effects were noticed.

Clock generation

The need for an exact picture delay in the recursive filter, at least for the low frequency luminance, implies the use of a digital clock frequency which is picture frequency locked so that it gives picture-stable samples. Moreover the theory of the colour predictor implies the use of a line-locked frequency. A frequency of 851 times the line frequency was used as clock generators working on this frequency were readily available. Such generators use the line synchronising pulses to steer an oscillator in frequency and phase and care must be taken to ensure that the stability of the pulses is adequate. Correct operation of the predictor requires that the clock frequency is stable to an accuracy of a few degrees of subcarrier phase, and this can only be achieved by using very long time constants in the oscillator control loop to average out the jitter in the separated pulses. In addition the oscillator must have a high degree of basic stability. This part of the equipment is crucial and, during the course of the project, no less than four clock pulse generators were designed. The first generator was based on an easily available *LC* oscillator and was quickly found not to be stable enough. Subsequent designs were based on crystal oscillators with means for digitally rephasing the count of the line-frequency divider, to deal with non-synchronous cuts.

Circuit block

Figure 5.75 shows a block diagram of the actual circuit of the noise reducer so far described. All the processing elements were based on TTL except for the line stores which were based on 1 kbit MOS shift registers. These had the capacity to store only the active line plus part of the back porch and so the whole recursive filter had to be bypassed by a path taking the unprocessed part of each line.

The spatial filter in the motion detector, described previously, was constructed as cascaded vertical and horizontal filters with provision for manually varying the number of terms in each up to the maximum of 5 and 15 respectively. The vertical part was constructed in straightforward transversal form so that the line delays in the filter, being substantial, could also act as the compensating delay for the main path signal. The rectification of the picture difference signal was performed by selecting the magnitude output of a sign–magnitude converter, the sign being retained for delay compensation but not passing through the ensuing filter. Selection of one, three or five terms was performed in the summing network.

The horizontal part of the spatial filter was realised using an accumulator or integrator, followed by subtraction across a delay coresponding to a number of horizontal picture elements. This technique gives a running sum of samples over an aperture

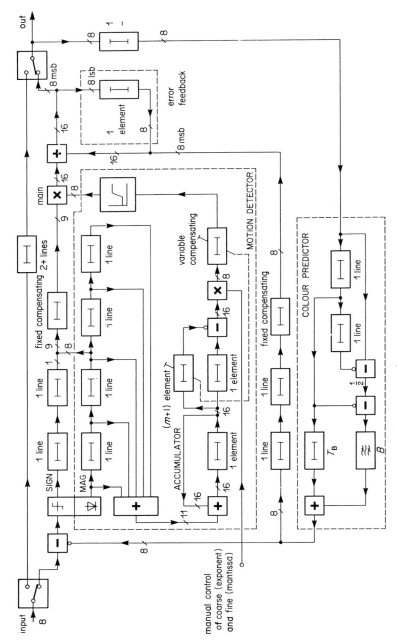

Figure 5.75 Block diagram of the complete noise reducer

Figure 1.3 Picture 'scrolling', an example of complex animated image manipulation possible with Quantel special effects equipment

Figure 1.6 Comparison of 1250-line HDTV and 625-line picture resolution. (a) conventional 625-line 4 × 3 display, full screen. (b) HDTV 1250-line display with 16 × 9 aspect ratio, full screen

Figure 1.6 (*continued*)(c) Close-up of part of conventional screen. (d) Close-up or HDTV screen

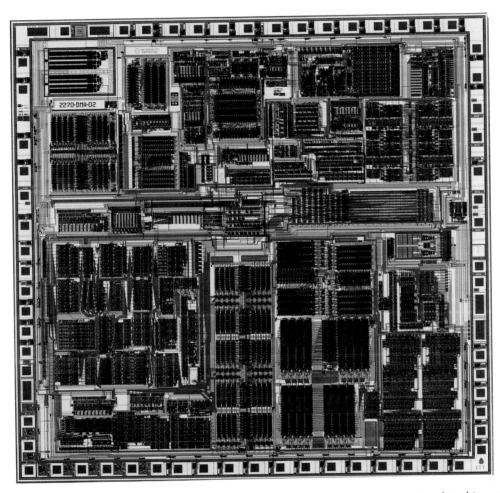

Figure 1.8 The VLSI mask pattern for one of the LSI devices in the ITT D-2 MAC-packet chip set for a digital domestic TV receiver.

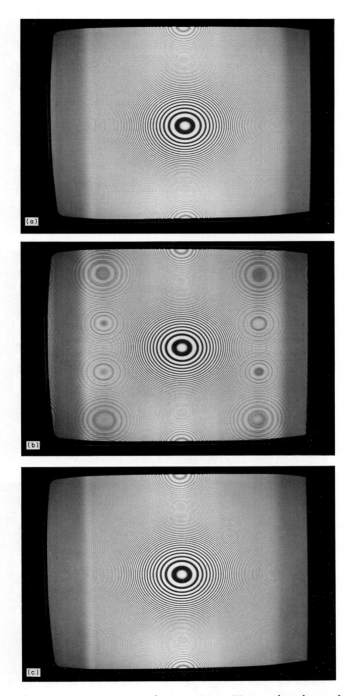

Figure 4.62 Zone plate patterns comparing the two-picture Weston decoder performance with the conventional decoded and the RGB signals. (a) RGB, (b) conventional decoder, (c) two-picture Weston decoder.

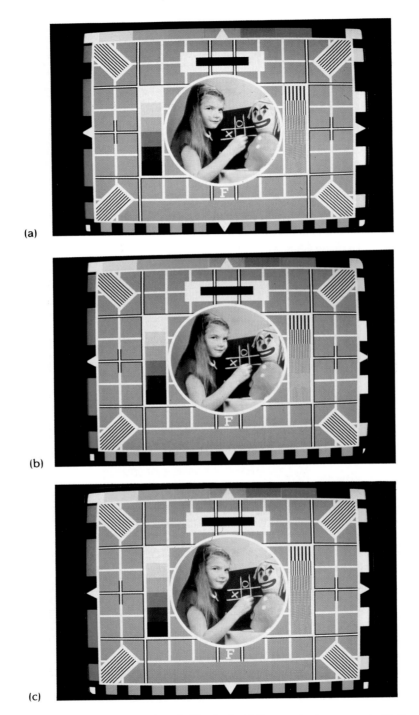

Figure 4.63 Test card F picture decoded by dual 312-line method compared with the conventional decoded and the RGB signals. (a) RGB, (b) conventional decoder, (c) dual 312-line decoder.

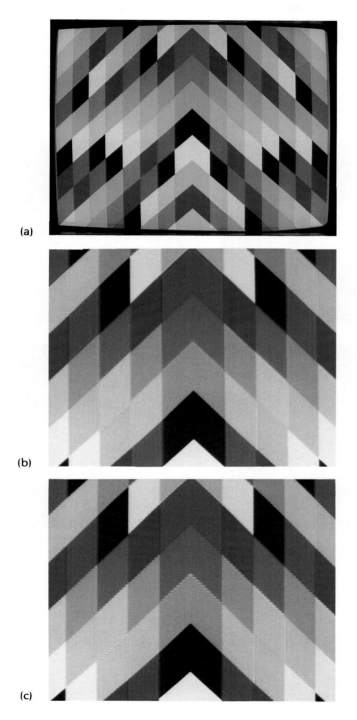

Figure 5.40 The effect of chrominance vertical filtering colour patches. (a) Full screen original, (b) Close-up of seventh-order filtered image, (c) Close-up of first order filtered image showing cross-effects at colour boundaries.

Figure 5.41 The effect of chrominance vertical filtering on a critical graphic. (a) original, (b) seventh-order matched filtering, (c) first-order prefilter, (d) first-order postfilter.

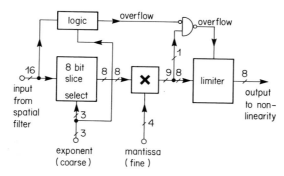

Figure 5.76 Block diagram of the motion detector multiplier

defined by the delay even though the accumulator continually overflows, provided modulo arithmetic is used. The variation in group delay as the filter delay varied was compensated by a following delay, varying in a complementary way.

As the whole spatial filter effectively adds together up to 75 terms, extra data bits are generated. The vertical filter summing network output contains 3 extra bits and the accumulator output, a further 5 bits. This poses a problem for the multiplier, following the spatial filter, which must therefore operate on a 16-bit signal. The solution was to include a limiter before the multiplier and to realise the multiplier in floating point form with independent control of binary exponent and mantissa as shown in Figure 5.76. The exponent control acts in conjunction with the limiter by selecting a 'window' of 8 bits to be passed on to the mantissa multiplier and, at the same time, testing the input to see whether or not it will be over-range after shifting. The 3-bit exponent allows multiplication in the range 2^{-3} to 2^4. A second over-range test is applied after multiplication by the 4-bit mantissa which lies in the range one to two. If the result is over-range through either of the mechanisms then it is limited to the maximum value which can be expressed using the 8-bit output. The exponent and mantissa controls were brought out to remote binary switches to form a coarse and fine control of the motion detector.

The non-linearity was contained in a programmable read-only memory (PROM) having a 256×8 word format, so arranged that the break point of the curve lay at the input value of nominally 64. (Actually this is the value where the curve cuts the horizontal axis.) In the absence of motion this is the value that should emerge from the multiplier–limiter when it is correctly adjusted. For research purposes alternative non-linearities were provided in various PROMs, selectable by a manual control.

Finally, care had to be taken to ensure that the group delay of the filter, multiplier and non-linearity path was equal to that of the main path so that the output of the non-linearity entered the main multiplier at the time appropriate to the difference signal arriving at the multiplier via the main path. This was ensured by adjusting the fixed delay in the main path to the main multiplier. Once this delay was fixed, it set the length of the fixed delay in the predictor feed to the main adder so that both signals reached the adder at the appropriate times. This delay, in turn, set the length of the nominal picture delay as the loop length round the picture store, predictor and adder must be one picture period.

284 DIGITAL FILTERING OF TELEVISION SIGNALS

Figure 5.77 Overall view of the experimental noise reducer

Figure 5.78 The Compact Noise Reduction System designed by Pye TVT Ltd

The prototype equipment

Figure 5.77 shows the prototype noise reducer as used at the output of the BBC programme feed in the initial field trials. Following extensive subjective tests further improvements were made. These included automation of the motion detector adjustment, which depends on the noise level. To allow the detector to adapt automatically a system was developed for measuring the noise level which takes into account its variation over the grey scale divided into four segments.

Field trials of the prototype machine, handling the composite PAL signal at the network output of the BBC Television Centre, confirmed the acceptability of this method of noise reduction for the vast majority of programme material containing moderate amounts of noise. Residual problems occur with material which has an excessively high level of noise or which contains extreme amounts of motion such as occurs in some sporting events but, on balance, they are outweighed by the overall improvement of subjective picture quality.

The design of the prototype was used by Pye TVT Ltd, a subsidiary of Philips as a basis for the manufacture, under licence, of the compact equipment shown in Figure 5.78. This has been used by broadcasters in several countries transmitting on both PAL and NTSC.

6 Interpolation

C. K. P. Clarke

INTRODUCTION

In the context of television processing, interpolation is used to produce intermediate values in a stream of discrete signal values. The discrete (sampled) nature of the signals makes the process particularly well suited to digital implementation.

For television signals, interpolation has two main areas of application: for changes of digital sampling rates and for the conversion of line and field rates between different scanning systems. These are particularly important for standards conversion or special effects such as changes in picture geometry. Standardisation on a single sampling frequency for digital studio signals [9] has tended to reduce the need for complex sample rate conversions. Fixed rate conversions, such as changing 4:2:2 to 4:4:4 signals are, however, still required. In addition, variable ratio interpolation is needed for picture geometry changes although, because of retiming, the sampling rate is unchanged.

The process of scanning used to create a television signal is also a sampling process. It converts the three-dimensional image falling onto the camera tube into a one-dimensional signal, that is, a signal with one degree of freedom; the brightness of the image varies with horizontal and vertical position, and with time, but the scanning process produces a single function of time to represent all three parameters. Scanning with a different number of lines per picture or of fields per second amounts to sampling at a different rate. So, standards conversion is merely a process of sampling rate conversion.

The process consists of the addition of weighted proportions from nearby lines, with the weighting factors being selected according to the relationship in time and position between the input lines and each output line. Thus, it is necessary to store the input signal in order to allow several lines, possibly from different fields, to be made available simultaneously. This combination of arithmetic and storage makes the process well suited to digital implementation. With this form of implementation, the performance characteristics predicted from theory for a particular interpolation algorithm can be accurately reproduced in a practical converter.

In subsequent sections, the theory of sampling and sample rate conversion is described, first for one-dimensional interpolation in general terms and then in the more specific applications of fixed ratio interpolation and the compression and expansion of waveforms in time. In each case, the action of the interpolator is described in both the time and frequency domains, with particular reference to frequency domain design methods. The actions of television scanning and scan conversions are also described in the frequency domain. The occurrence of various types of picture impairment is related to individual features of interpolator performance in the frequency domain and the requirements for improved performance are identified. Taking account of these factors, practical interpolation methods are derived for both 525/60 to 625/50 and 625/50 to 525/60 conversions and for separate interpolation of the colour signal components. The possible requirements of conversions associated with high definition television and with display up-conversions are also discussed.

INTERPOLATION THEORY

Signal sampling and reconstitution

Signal sampling consists of multiplying a continuous waveform, for example, Figure 6.1a, by a regular series of impulses spaced by the sampling interval T_s, as shown in Figure 6.1b. This produces the series of weighted impulses representing the sample values, Figure 6.1c.

Although signal sampling is essentially a time domain process, its effects can also be viewed in the frequency domain. Figure 6.2 shows examples of frequency spectra corresponding to the three types of signal waveform shown in Figure 6.1. Multiplication of waveforms in the time domain is equivalent to the convolution of spectra in the frequency domain; so when the baseband spectrum, Figure 6.2a, is convolved with the spectrum of the sampling impulses, Figure 6.2b, this repeats the baseband spectrum around harmonics of the sampling frequency $f_s(=1/T_s)$, as shown in Figure 6.2c.

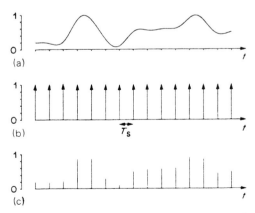

Figure 6.1 Signal sampling: (a) a continuous signal waveform, (b) a series of impulses spaced by the sampling interval T_s, and (c) the weighted impulses representing the sample values, obtained by multiplying (a) and (b)

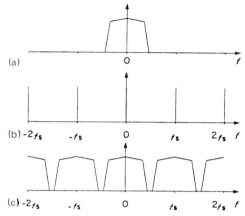

Figure 6.2 Frequency spectra corresponding to the three types of signal shown in Figure 6.1: (a) the spectrum of a continuous signal, (b) the spectrum of a series of sampling impulses spaced by $T_s = 1/f_s$, and (c) the spectrum of the sampled signal, obtained by convolving (a) and (b)

Samples can be returned to the form of a continuous signal in the time domain by low-pass filtering. This consists of convolving the series of impulses with the impulse response of the low-pass filter. In Figure 6.3 the low-pass filter impulse response is triangular and each sample is replaced by this triangular waveform scaled according to the weight of the sample. In Figure 6.3c, the individual responses, shown dashed, produce, when summed, the approximation to the original waveform shown as a full line. Although this filter does not accurately reproduce the original waveform, it is not clear how the impulse response

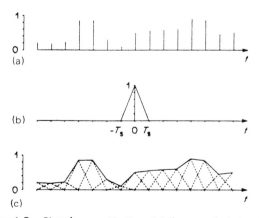

Figure 6.3 Signal reconstitution: (a) the sampled signal of Figure 6.1c, (b) an example of a low-pass filter impulse response and (c) a continuous signal reconstituted by passing the samples of (a) through the filter (b). Each sample is replaced by the filter impulse response waveform scaled according to the weight of the sample (shown dashed) and these individual responses, when summed, produce the continuous waveform (shown as a full line)

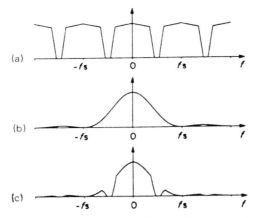

Figure 6.4 Spectra corresponding to the types of signal shown in Figure 6.3: (a) the spectrum of a sampled signal, (b) the frequency characteristic of a filter with the impulse response shown in Figure 6.3b, and (c) the spectrum of the reconstituted waveform, obtained by multiplying (a) and (b)

should be altered to make an improvement. A much clearer understanding of the action of the filter can be obtained by considering the sample reconstitution process in the frequency domain.

In the frequency domain, the equivalent process to convolution in the time domain is multiplication. Thus, the repeated spectrum of the sampled waveform is multiplied by the frequency characteristic of the filter with the intention of suppressing the harmonic components and retaining only the baseband spectrum. However, the frequency characteristic corresponding to a triangular impulse response is shown in Figure 6.4b. It is apparent from Figure 6.4c that the poor approximation to the original waveform is the result of inadequate suppression of the harmonic spectra, combined with some attenuation of wanted frequencies in the baseband spectrum.

To reproduce all frequencies in the baseband spectrum accurately up to half the sampling frequency would require a filter with the rectangular characteristic shown in Figure 6.5a. With this 'ideal' filter, the baseband spectrum would be unattenuated and

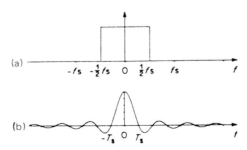

Figure 6.5 An 'ideal' low-pass filter for converting signal samples to a continuous waveform: (a) the rectangular frequency characteristic and (b) the sinc function impulse response

the harmonic spectra completely suppressed, thus returning exactly to the spectrum of Figure 6.2a. However, a rectangular filter has an impulse response defined by

$$g(t) = \frac{\sin \pi f_s t}{\pi f_s t}$$

This $\sin(\pi f_s t)/(\pi f_s t)$ or 'sinc' function, shown in Figure 6.5b, extends from $-\infty$ to $+\infty$. Therefore, in practice, it is normal to sample at a rate slightly above twice the highest frequency wanted. This leaves a margin for a practicable rate of cut in the reconstituting filter response and so ensures a finite impulse response.

This suggests that one method of improving the accuracy of the reconstituted waveform would be to use a filter with an impulse response which is a better approximation to Figure 6.5b. For example, truncating the sinc function and using a raised-cosine smoothing term so that the impulse response is defined by

$$g(t) = \tfrac{1}{2}(1 + \cos \tfrac{1}{2}\pi f_s t) \cdot \frac{\sin \pi f_s t}{\pi f_s t} \qquad \text{for } -2T_s \leqslant t \leqslant 2T_s$$

$$= 0 \qquad \text{otherwise,}$$

improves the reconstituted waveform as shown in Figure 6.6. In the frequency domain, Figure 6.7, the corresponding filter characteristic causes less attenuation of the baseband spectrum and provides better suppression of most harmonic spectra than is given by the triangular impulse response filter (Figure 6.4); however, frequencies between $\tfrac{1}{2}f_s$ and f_s are not adequately suppressed. It should be noted that although the impulse responses in Figure 6.3b, 6.5b and 6.6b all have zero values at adjacent sample points, it is not necessary that this should be so to give a good approximation to the original waveform.

If a waveform contains frequencies above the half sampling frequency, then the information content of the samples is irretrievably distorted. Such a situation is shown in

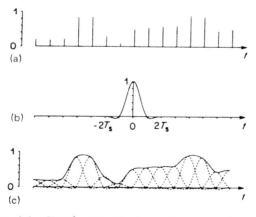

Figure 6.6 Signal reconstitution with a smoothed sinc function filter: (a) the sampled signal of Figure 6.1c, (b) the smoothed sinc function impulse response and (c) the reconstituted waveform (shown as a full line). The individual impulse responses used to make up the continuous signal are shown dashed

292   INTERPOLATION

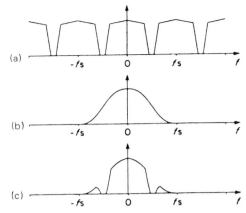

Figure 6.7 Spectra corresponding to the waveforms of Figure 6.6: (a) the spectrum of a sampled signal, (b) the frequency characteristic of a filter with the impulse response shown in Figure 6.6b and (c) the spectrum of the reconstituted waveform, obtained by multiplying (a) and (b)

Figure 6.8, in which the dashed line of Figure 6.8c represents the waveform obtained by reconstruction with an ideal rectangular filter. Clearly much of the high frequency content of Figure 6.8a has been replaced by spurious information of a lower frequency. Figure 6.9 shows the effect in the frequency domain, in which the baseband and repeated spectral components overlap. Such components, which have been shifted from their true positions in the normal band of signal frequencies, are known as alias components. In the

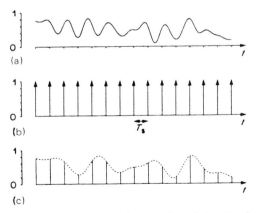

Figure 6.8 The effect of undersampling: (a) a signal waveform containing frequencies above half the sampling frequency, (b) the sampling impulses and (c) the weighted impulses representing the sample values. Returning these samples to continuous form using the ideal reconstituting filter (Figure 6.5) produces the waveform shown dashed in Figure 6.8c

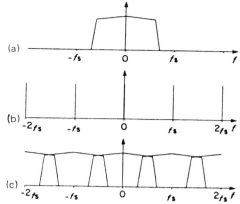

Figure 6.9 Spectra showing the effect of using too low a sampling rate: (a) the spectrum of a continuous signal with components beyond the half sampling frequency, (b) the spectrum of the sampling impulses, and (c) the spectrum of the sampled signal showing the overlap between adjacent spectra

overlap region, it is impossible to distinguish the true components of the baseband spectrum from the alias components of the repeated spectrum. Under these circumstances it is generally best to suppress all frequencies in the overlap band. This is because the loss of wanted frequencies is usually less noticeable than the presence of alias components.

Sample rate changing

Resampling

In a sampled waveform, such as Figure 6.1c, the signal value is only known at the instants the samples were taken and is undefined in between. To change the sampling rate, it is necessary to obtain the signal value at new sampling points, between the original sample positions. One method would be to return the sampled waveform to continuous (analogue) form using a low-pass filter as described in the previous section. Then the continuous signal could be resampled at the new rate in the normal way. If the reconstituted continuous signal is accurate, the resampling will not introduce any additional impairment. However, inaccuracies in the reconstituted waveform may be exaggerated by the resampling process.

For example, if the waveform of Figure 6.3c is resampled at a higher rate, interval T''_s, this gives new sample values as shown in Figure 6.10. Frequency spectra corresponding to these waveforms are shown in Figure 6.11. Here, the process of convolution causes unsuppressed remnants of the original sampling process to fall into the baseband region. The alias components, although noticeable in themselves, may interfere with the true signal to produce more noticeable beat patterns. Also, the low-pass filter used to return the original samples to a continuous waveform produces an attenuation of the baseband spectrum which affects the true components in the resampled spectrum.

294   INTERPOLATION

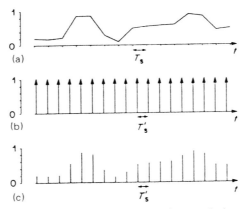

Figure 6.10 Resampling a previously sampled and reconstituted waveform: (a) the reconstituted waveform of Figure 6.3c, (b) sampling impulses spaced at T'_s, a shorter interval than in the first sampling process, and (c) the resulting sample values

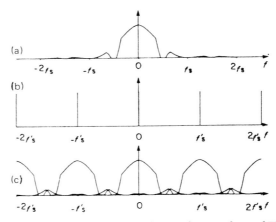

Figure 6.11 Spectra corresponding to the waveform of Figure 6.10: (a) the low-pass filtered spectrum of Figure 6.4c, (b) the spectrum of sampling impulses at a higher sampling frequency f'_s, and (c) the result of convolving (a) and (b). The residual harmonic spectra centred on f_s, $2f_s$, etc., in (a) are aliased to overlap the baseband spectrum in (c)

If the new sampling rate is lower than the original rate, then it is possible for the spectrum of the reconstituted signal to extend beyond the new half sampling frequency, even though the original waveform has been reproduced accurately. Thus, the spectrum of the new samples would include an overlap region, similar to that of Figure 6.9c. To avoid this, a further amount of low-pass filtering is required before resampling to limit the signal components to below the new half sampling frequency. This additional band-limitation can be incorporated as part of the action of the sample reconstitution low-pass filter.

The interpolation process

Although resampling of a reconstituted continuous waveform could be used for sample rate conversion, the intermediate continuous signal stage is unnecessary, as the value of each new sample can be calculated directly from the values of surrounding samples. This is known as interpolation. In practice, the interpolation method is preferable because it is inherently digital, avoiding the intervening analogue process, but, in principle, the two methods are equivalent and are capable of identical results.

As a mathematical process, interpolation consists of assuming a particular form of continuous function, usually a polynomial expression, which passes through, or near to, one or more of the original sample values. This allows new sample values to be calculated for any points between the original sample positions. For example, with linear interpolation it is assumed that the function follows a straight line between sample points as shown in Figure 6.12. The new value s depends, therefore, both on the two nearest sample values, s_1 and s_2, taken at t_1 and t_2 respectively, and on the position t of the new sample between the two; s can be calculated from the following expression

$$s = c_1 s_1 + c_2 s_2$$

where

$$c_1 = 1 - \frac{t - t_1}{t_2 - t_1}$$

and

$$c_2 = \frac{t - t_1}{t_2 - t_1}$$

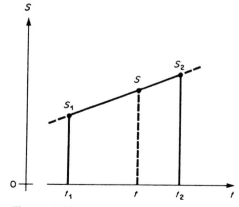

Figure 6.12 Linear interpolation: to obtain a sample value at time t, the shape of the waveform between the samples at t_1 and t_2 is assumed to follow a straight line

One form of interpolator, shown in Figure 6.13, consists of registers R to hold the input sample values, $s_1, s_2, \ldots$, multipliers to form the products, $c_1 s_1, c_2 s_2, \ldots$, and adders to sum the products, $c_1 s_1 + c_2 s_2 + \cdots$ For linear interpolation only two stages are required and the coefficients are obtained from the expressions shown above for c_1 and c_2. Additional stages (shown dashed) can be added for more complicated interpolation methods.

The way in which the interpolation coefficients vary with the relative positions of the input and output samples can be expressed in the form of an interpolation aperture function. In general, this is a continuous function because all phases of input and output samples can occur when the input and output sampling rates are unrelated. The total extent of the function is known as the aperture or aperture width and is measured in input sampling periods.

Notionally, the aperture function is applied by placing its origin at a required new sample position so that the weighting coefficient value for each input sample can be selected according to its position in relation to the aperture function. This is shown in Figure 6.14 for linear interpolation, which has a triangular interpolation aperture function as shown in Figure 6.14b. In this case, the samples produced by interpolation in Figure 6.14c are identical to the samples formed by low-pass filtering and resampling in Figure 6.3 and 6.10. This is because the interpolation aperture function, Figure 6.14b, and the low-pass filter impulse response, Figure 6.3b, have the same shape.

In the frequency domain, the interpolator performance is determined by a frequency characteristic, which can be found by evaluating the Fourier transform $G(f)$ of the interpolation aperture function $g(t)$

$$G(f) = \int_{-\infty}^{\infty} g(t) \exp(-2\pi j f t) \, dt$$

Conversely, the aperture function corresponding to any desired spectral characteristic can be obtained by the inverse Fourier transform

$$g(t) = \int_{-\infty}^{\infty} G(f) \exp(2\pi j f t) \, df$$

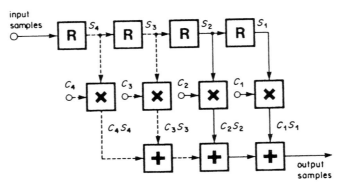

Figure 6.13 An extensible form of interpolator, consisting of registers (R), multipliers and adders

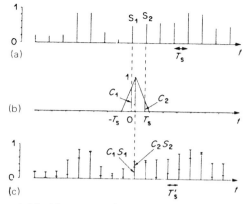

Figure 6.14 The interpolation process: (a) the input samples, (b) the interpolation aperture function, showing the coefficient values selected to multiply the input samples and (c) the new sample values, each produced by combining weighted contributions from two input samples

For linear interpolation, the aperture function can be defined, by inspection of Figure 6.14b, as

$$g(t) = t + T_s \quad -T_s \leqslant t \leqslant 0$$
$$= T_s - t \quad 0 < t \leqslant T_s$$
$$= 0 \quad |t| > T_s$$

Then

$$G(f) = \int_{-T_s}^{0} (t + T_s) \exp(-2\pi j f t) \, dt$$
$$+ \int_{0}^{T_s} (T_s - t) \exp(-2\pi j f t) \, dt$$

which after integration simplifies to

$$G(f) = \frac{\sin^2(\pi f T_s)}{(\pi f)^2}$$

This spectral characteristic for linear interpolation is shown in Figure 6.15 along with the spectra for the input and output samples; these correspond to the time domain waveforms in Figure 6.14.

As the filter characteristic of linear interpolation is the same as that shown in Figure 6.4b, the action of the interpolator is equivalent to that shown in Figure 6.4 and 6.11, although the signal never exists in an intermediate continuous form. Indeed, it is useful to visualise the interpolation process as consisting of an equivalent low-pass filtering stage, followed by resampling. Then the performance of the interpolation method is still governed by the efficacy of its filtering action in suppressing unwanted components and retaining the baseband spectrum.

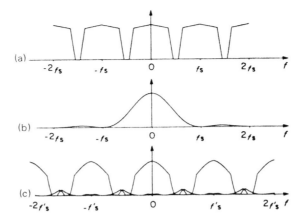

Figure 6.15 Frequency spectra for the interpolation process: (a) the spectrum of the input samples, (b) the frequency characteristic for linear interpolation and (c) the spectrum of the new samples

Synthesis of aperture functions

Although the Fourier transform provides a method of calculating the performance of any aperture function, it is preferable to design an interpolator by specifying its performance directly in the frequency domain. However, in most cases, this would produce an aperture function of unlimited width, so requiring the interpolation process to include an infinite number of samples. A truncated version could be used, but then the frequency characteristic would differ from that originally specified. What is required is a method of specifying the frequency characteristic in a manner that is constrained to produce an aperture of known width.

Such a method [121–123] consists of specifying the frequency characteristic at intervals of $(1/k)f_s$, in the knowledge that this will produce an aperture function of width k sample periods. This can be demonstrated by considering an aperture function of width kT_s as the product of a block pulse of width kT_s and a repetitive waveform of period kT_s. Figure 6.16 shows an example of such a 'fixed-point' aperture with a width of four sample periods.

The frequency spectra corresponding to these waveforms are shown in Figure 6.17: the spectrum of the repetitive waveform is a series of line spectra, Figure 6.17a, spaced at intervals of $(1/k)f_s$ ($k = 4$), with amplitudes $a_0, a_1, a_2, \ldots$ representing the Fourier series components of the waveform; the spectrum of the block pulse is a sinc function, Figure 6.17b, with its zero at $(1/k)f_s$; then the frequency characteristic corresponding to the aperture function is produced by convolution of the line spectra with the sinc function as shown in Figure 6.17c. Therefore, specifying the values of $a_0, a_1, a_2, \ldots$ in the frequency domain produces an aperture function of width k sample intervals, which can be calculated from the inverse Fourier transform of the pairs of line spectra shown in Figure 6.17a

$$\frac{2}{k}\left(a_0 + 2a_1 \cos \frac{2\pi t}{kT_s} + 2a_2 \cos \frac{4\pi t}{kT_s} + \cdots \right)$$

INTERPOLATION THEORY 299

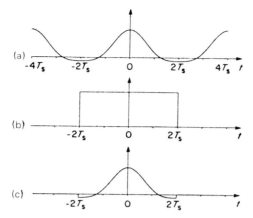

Figure 6.16 Aperture functions of set width: an aperture function of width $4T_s$, shown at (c), can be considered as the product of (a) a repetitive waveform of period $4T_s$ and (b) a block pulse of width $4T_s$.

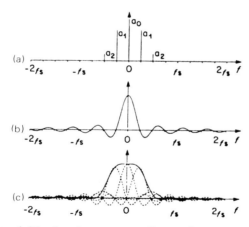

Figure 6.17 Spectra corresponding to the waveforms of Figure 6.16: (a) a series of line spectra spaced at intervals of $\frac{1}{4}f_s$, (b) a sync. function with period $\frac{1}{4}f_s$ and (c) the frequency characteristic (full line) produced by convolving (a) and (b); the individual contributions are shown dashed.

Any number of values can be specified, but usually all values beyond $\pm\frac{1}{2}f_s$ will be set to zero as indicated in Figure 6.17a. It should be noted that although asymmetrical frequency characteristics could be generated by having asymmetrical aperture functions, this would result in unwanted phase shifts in the interpolated signals.

The aperture function, Figure 6.16c, produced by specifying fixed-point values in the frequency domain has the same width, four sample periods, as the smoothed sinc function impulse used for low-pass filtering in Figure 6.6. However, in Figure 6.18, when the fixed-point aperture (Figure 6.18b) is used to reconstitute the samples from Figure

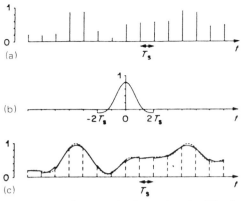

Figure 6.18 Signal reconstitution using the 'fixed-point' filter of Figure 6.17: (a) the sampled signal of Figure 6.1c, (b) the impulse response of the fixed-point filter and (c) the reconstituted waveform (full line); the original waveform and the sample impulses are shown dashed for comparison

6.1c, the resulting continuous waveform (the full line in Figure 6.18c) generally provides a closer approximation to the original waveform (the dashed line in Figure 6.18c) than that obtained in Figure 6.6. When the frequency characteristics of the two filters are compared, Figures 6.7b and 6.19b, it is apparent that the fixed point method has produced a characteristic closer to the ideal rectangular shape, Figure 6.5a. Because of this, the filter provides greater rejection of harmonic spectra in the region $\frac{1}{2}f_s$ to f_s, while introducing little in-band attenuation. The fixed-point method therefore provides a simple and effective means of synthesising interpolation aperture functions.

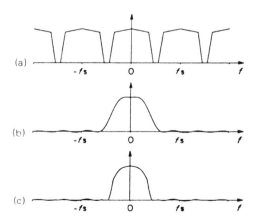

Figure 6.19 Signal reconstitution in the frequency domain using the filter of Figure 6.17: (a) the spectrum of a sampled waveform, (b) the frequency characteristic of the fixed-point filter and (c) the spectrum of the reconstituted signal

Aperture quantisation

An interpolation aperture function is continuous and, therefore, has an infinite number of values. In order to store the function in a read-only memory for digital interpolation, it is necessary to produce an approximation to the function, quantised in both the abscissa and ordinate directions. Thus, the function value is only specified at a limited number of positions and also each stored value will have only limited accuracy.

Read-only memories commonly have 2^n locations so it is convenient to store the value of the aperture function at $N = 2^n$ equi-spaced points. This even number has the additional advantage that the two end points of the function cannot both contribute at once—a situation which would require an extra multiplier.

Figure 6.20 shows the effect of representing the linear interpolation aperture function at only eight points (amounting to four values per input sample period). The eight values shown in Figure 6.20a are stored, but each value is used over a region of $\pm \frac{1}{8}T_s$, as shown in Figure 6.20b. The effective aperture function is, therefore, the piecewise constant approximation shown in Figure 6.20c which results from the convolution of Figure 6.20a and 6.20b.

In the frequency domain, sampling of the aperture function at an interval of $\frac{1}{4}T_s$ causes the frequency characteristic of the continuous aperture function, Figure 6.15b, to be repeated at harmonics of $4f_s$, as shown in Figure 6.21a. This spectrum is then multiplied by the spectrum of the block pulse Figure 6.21b (equivalent to convolution in the time domain) which attenuates the repeated spectra as shown in Figure 6.21c. When more aperture values are stored, the repeated spectra in Figure 6.21a are displaced further from the baseband spectrum and are more attenuated by the action of the block pulse spectrum, Figure 6.21b. Table 6.1 shows values of the maximum attenuation of wanted frequencies (over the band 0 to $\frac{1}{2}f_s$) and the minimum attenuation of unwanted frequencies (in the band $(N - \frac{1}{2})f_s$ to Nf_s) for $N = 2, 4, 8, 16$ and 32 values per sample

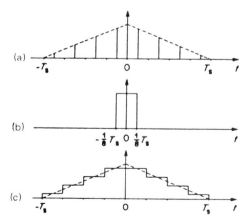

Figure 6.20 Time quantisation of aperture functions: (a) eight impulse values spaced at intervals of $\frac{1}{4}T_s$ which could be stored and used to represent the linear interpolation aperture function (shown dashed), (b) an interval of $\pm \frac{1}{8}T_s$ over which each value is used and (c) the time-quantised aperture function produced by convolving (a) and (b)

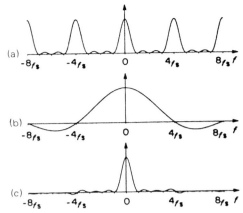

Figure 6.21 Spectra corresponding to the time functions of Figure 6.20: (a) the frequency characteristic of the sampled linear interpolation aperture, (b) the sinc function spectrum of the block pulse of Figure 6.20b, and (c) the frequency characteristic of the time-quantised aperture function, produced by multiplying (a) and (b)

Table 6.1

N	Wanted band maximum attenuation	Unwanted band minimum attenuation
2	0.637 (-4 dB)	0.300 (-10 dB)
4	0.974 (-0.2 dB)	0.139 (-17 dB)
8	0.994 (-0.05 dB)	0.066 (-24 dB)
16	0.998 (-0.02 dB)	0.032 (-30 dB)
32	1.000 (-0.01 dB)	0.016 (-36 dB)

period. The higher values of N attenuate the unwanted spectra sufficiently to reduce aliasing to a level considered acceptable in the unsampled characteristic. The attenuated sampling products centred on Nf_s are aliased back into the baseband region by the resampling effect of interpolation in the same way as shown for the signal sampling products in Figure 6.15c.

In addition to the aperture function being defined only at certain positions, the function values at those positions are also quantised. This is because the coefficients are stored and used as binary numbers so that their accuracy is limited by the number of bits available. Whereas it is straightforward to store the coefficients to high accuracy, the limiting factor is usually the capacity of the multipliers in the interpolator.

Quantisation of the values alters the effective aperture function and so changes the frequency characteristic. Direct quantisation of the individual function values, rounding each to the nearest allowed value, may cause the sets of coefficients used for each input–output sampling offset no longer to sum to unity. This would cause a variation in gain as the coefficients were changed from one set to another. In the frequency domain, the gain variation results because zeros in the frequency characteristic no longer fall at the centres of the repeated spectra.

The problem can be avoided by ensuring that the sets of coefficients used for each offset position of the aperture still sum to unity after quantisation. An effective algorithm for quantisation consists of first rounding down all the coefficients of a set and summing the results. If the sum is less than unity, sufficient coefficients are rounded up to make up the difference. The overall error is minimised by selecting coefficients in which rounding up will produce the minimum error relative to the actual value of each coefficient; even so, this may cause quantisation errors greater than half a least significant bit in individual coefficients.

The effects of quantisation can be demonstrated by taking the time quantised version of the linear function, Figure 6.20c, and quantising this in value very coarsely, so that only the values $\frac{3}{4}$ and $\frac{1}{4}$ are allowed as shown in Figure 6.22a. For any position of an output sample, one input sample is multiplied by $\frac{3}{4}$ and the other by $\frac{1}{4}$, so the requirement that each set of coefficients should sum to unity is satisfied. This modifies the frequency characteristic to the form shown in Figure 6.22b as a full line. The original characteristic for linear interpolation is shown dashed for comparison. Even with this very coarse degree of quantisation, the frequency characteristic is not greatly altered. In practice, it is convenient to use 7- or 8-bit accuracy in the coefficients, so that the characteristic for the quantised aperture function is usually very close to that of the continuous function.

Quantised aperture functions, such as those shown in Figures 6.20c and 6.22a, can be considered to be made up of a series of elements comprising symmetrical pairs of block pulses added together. Each pair has the basic form shown in Figure 6.23, with amplitude A and width T centred on $\pm t_0$. Fourier transformation of this aperture element reveals that the associated component of the frequency characteristic is given by

$$G(f) = 2AT \cos 2\pi t_0 f \frac{\sin \pi T f}{\pi T f}$$

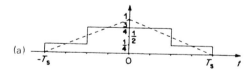

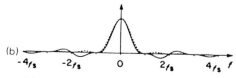

Figure 6.22 Amplitude quantisation of aperture functions: (a) the aperture function of linear interpolation (shown dashed) coarsely quantised to the values $\frac{1}{4}$ and $\frac{3}{4}$ (full line) and (b) the frequency characteristic of the quantised aperture function (full line) compared with that of the original aperture function (shown dashed)

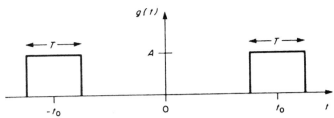

Figure 6.23 A symmetrical pair of block pulses representing a generalised element of a quantised aperture function

Therefore, the frequency characteristic of a quantised aperture function can be calculated as a summation of such terms. When the block pulses are all of width T, the sinc term is common to all the elemental components.

Fixed ratio interpolators

The concept of interpolation as a combination of low-pass filtering and resampling described in the previous section is valid for any relationship between the input and output sampling frequencies. Also the design methods described can be used for any sampling rates. However, when the sampling rates are related by a simple ratio, other design methods normally used for transversal filters can be used more conveniently. Interpolators of this type are of particular interest for conversions in both directions in the digital component signal hierarchy between, for example, 4:4:4; 4:2:2 and 3:1:1 signals.

When the sampling frequencies are related by an integer ratio, the output samples only occur at certain phases relative to the input samples, because the phases repeat in a cyclic pattern. Each phase uses a set of interpolation coefficients corresponding to discrete values of the continuous aperture function. When the discrete values are all combined, the overall effect is to sample the continuous aperture function with an interval corresponding to the lowest multiple of the input and output sampling frequencies. Since only the discrete values are used, the detail of the continuous function between these values becomes unimportant. The selection of coefficients from a linear interpolation aperture function is demonstrated in Figure 6.24 for input and output sampling intervals T_s and T'_s respectively, where

$$T'_s = \tfrac{3}{2} T_s$$

This results in two sample phases which are used alternately as shown in Figure 6.24a. When the different phases are combined, the discrete aperture function shown in Figure 6.24b is produced which has a sampling interval of $\tfrac{1}{2} T_s$.

The phasing of the sample points on the aperture function depends on the required phase of the output samples relative to the input sample positions. In many circumstances, either some of the input and output sample points will be coincident or the output samples will fall midway between the input sample points. Unless one or other of these phases is included, the discrete aperture function formed by combining the phases will not form a symmetrical impulse response, thus complicating the phase characteristics of the interpolation process.

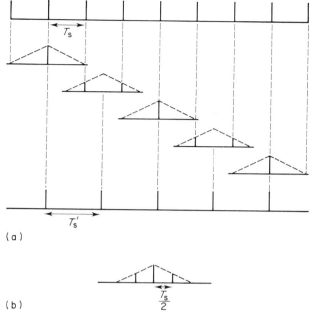

Figure 6.24 Fixed ratio interpolation. (a) Applying a linear interpolation aperture function at the required output sample positions results in two different sets of coefficients being used alternately. (b) These combine to form a discrete aperture function with a sampling interval $\frac{1}{2}T_s$.

Alternatively, the effect of related sampling frequencies can be considered in the frequency domain. Conceptually, when the sampling rates are related by a simple ratio, some of the residual harmonic components left after the low-pass filtering process are coherent with components introduced by the new sampling process. This has the effect of altering the filtering action to a frequency characteristic with a periodic form, repeating at the lowest common multiple of the sampling frequencies.

In terms of low-pass filtering and resampling, the effect of the 3:2 interpolation in the frequency domain is shown in Figure 6.25. Figure 6.25a shows the spectrum of an input signal sampled at f_s and Figure 6.25b shows the sinc-squared low-pass filter characteristic corresponding to a linear interpolation process. The spectrum produced by low-pass filtering is shown in Figure 6.25c. Resampling the signal at f'_s convolves the low-pass filtered spectrum with the series of impulses representing the new sampling frequency shown in Figure 6.25d, thus reproducing the spectrum shown in Figure 6.25c at all the impulse positions shown in Figure 6.25d. Because the two sampling frequencies are linked by the relationship

$$f'_s = \tfrac{2}{3} f_s$$

the individual harmonic components spaced at $2f_s$ in the filtered spectrum, Figure 6.25c, add together coherently to produce a periodic structure in the convolved spectrum, Figure 6.25e, which is reproduced at all harmonics of the new sampling frequency.

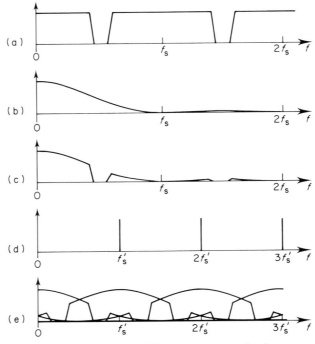

Figure 6.25 Fixed ratio 3:2 interpolation in the frequency domain. (a) The spectrum of the input signal sampled at f_s, (b) the low-pass filter response of linear interpolation, (c) the filtered input spectrum, (d) the spectrum of the output sampling frequency f'_s and (e) the output spectrum produced by convolving (c) and (d)

The periodic structure of the convolved spectrum, which results when the sampling frequencies are related, allows the interpolator action to be interpreted as a combination of resampling and filtering at the common multiple frequency followed by dropping the unwanted samples. This is shown in Figure 6.26 in the time domain, with the corresponding frequency spectra shown in Figure 6.27. As the input waveform is a series of impulses, resampling at a higher rate simply introduces extra zero-valued samples to the time waveform, as shown in Figure 6.26a. Similarly, the corresponding spectrum Figure 6.27a is unaltered except that the sampling frequency is now $2f_s$ instead of f_s. The discrete version of the linear aperture function Figure 6.26b, formed as in Figure 6.24 by combining all the phases, then corresponds to the periodic raised cosine filter characteristic shown in Figure 6.27b. Applying the filter to the resampled signal, including the null values, produces the periodic spectrum shown in Figure 6.27c which tends to suppress the odd-numbered harmonics of f_s. Dropping the unwanted samples in the time domain then resamples this periodic spectrum to produce the output spectrum in Figure 6.27e just as in Figure 6.25e.

Design of the discrete aperture function Figure 6.26b can be based on the series of processes in Figure 6.27. Primarily the filter should be designed to retain the spectra at

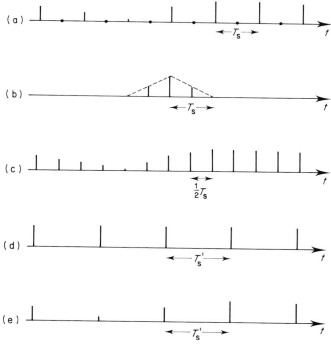

Figure 6.26 Fixed ratio interpolation with a discrete aperture function. (a) The original input samples with extra zero-valued samples added, (b) the discrete linear aperture function, (c) samples at the common multiple frequency $2f_s$, (d) the output sampling impulses and (e) the output samples produced by multiplying (c) and (d)

harmonics of the common multiple frequency and reject the intermediate spectra. Where f'_s is less than f_s, as in this case, the filter should also attenuate high frequencies in the signal which would overlap to cause aliasing in the output signal. Fourier transformation of the required spectrum then produces the sampled interpolation aperture function directly. This can be obtained for a specified aperture width using a fixed point method based on solving simultaneous equations.

The method of up-conversion shown in Figures 6.26 and 6.27 has the practical disadvantage that the digital filter has to operate at a clock rate equal to the lowest common multiple of the sampling frequencies. This can be avoided by the method [87] shown in Figure 6.28. Because some of the input samples are used directly without filtering, these are passed through a delay to the output. The alternate intermediate samples are produced in parallel by the filter operating at f_s and the results combined by switching at $2f_s$. The wanted samples at f'_s are then selected by holding their values in the register R. In the frequency domain, the effect of the filter is to invert the unwanted intermediate spectra which then tend to be cancelled when combined with those in the input spectrum. The production of samples at $2f_s$ with a filter operating at f_s is demonstrated in Figure 6.29 and 6.30.

308 INTERPOLATION

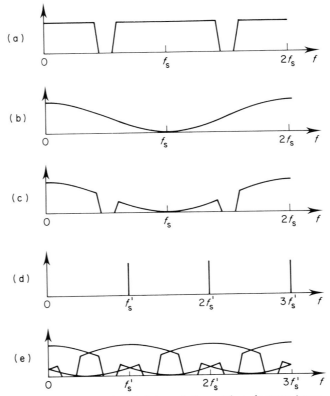

Figure 6.27 Fixed ratio interpolation with a discrete interpolation aperture function: (a) the input spectrum, (b) the raised-cosine filter characteristic, (c) the spectrum of the up-converted signal, (d) the spectrum of the output sampling frequency f_s and (e) the output spectrum produced by convolving (c) and (d)

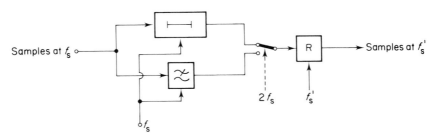

Figure 6.28 A method of fixed ratio interpolation in which the filter operates at a lower rate than the lowest common multiple of the input and output sample rates

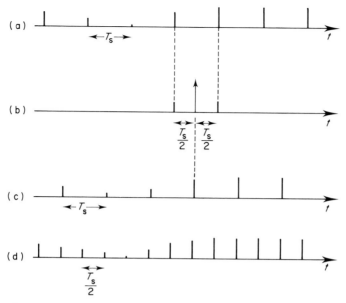

Figure 6.29 Signal waveforms for the method of Figure 6.28: (a) the delayed input samples, (b) a simple filter impulse response applied to the input samples, (c) the sampled signal produced by the filter and (d) the output samples produced by adding (a) and (c)

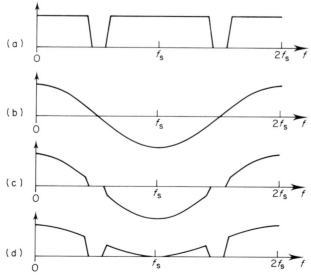

Figure 6.30 Signal spectra for the method of Figure 6.28: (a) the input spectrum, (b) the cosine response of the filter, (c) the spectrum of the filtered samples with alternate spectral centres inverted and (d) the output spectrum produced by adding (a) and (c)

310   INTERPOLATION

A particularly efficient filtering technique can be used for 1:3 conversion based on standard transversal filter modules. In this case, samples are either passed directly to the output or produced at the one-third or two-thirds positions relative to the input sampling periods. As the impulse response for samples at the one-third positions is the time reverse of that for the two-thirds positions, the adder chains of a symmetrical transversal filter module can produce both samples at once as shown in Figure 6.31. These are then switched together in the correct sequence with the unfiltered samples to produce samples at the higher rate. The effective impulse response of this filter is shown in Figure 6.32.

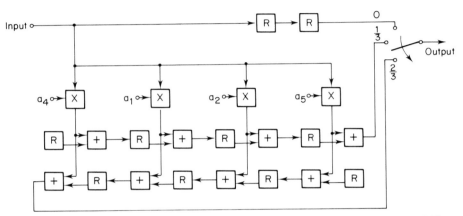

Figure 6.31 Interpolator for 1:3 up-conversion using standard digital transversal filter modules

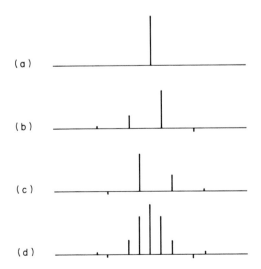

Figure 6.32 Impulse response diagrams for the interpolator of Figure 6.31: (a) the direct contribution, (b) the one-third contribution, (c) the two-thirds contribution, (d) the overall impulse response of the interpolator

Time expansion and compression

When a sampled waveform is either expanded or compressed in time, but maintains the same sample rate at the input and output, an interpolation process is required to produce samples at new positions relative to the waveform. Stretching the waveform requires more sample values to be generated, whilst for compression, fewer sample values are needed. Therefore, in each case the process consists of an interpolation stage and a retiming stage.

A circuit arrangement for waveform expansion is shown in Figure 6.33. While expansion could be performed with retiming either before or after interpolation, it is advantageous to retime the samples before interpolation [124]. This avoids the necessity of using a higher clock rate in the interpolation and allows infinite expansion without aliasing or unnecessary resolution loss.

The main circuit blocks are shown in more detail in Figure 6.34. Figure 6.34a shows a convenient arrangement of two separate stores used for retiming the samples. Whilst the incoming signal is being written into one store at the full sample rate, a selected portion of a previously stored waveform is being read at a slower rate from the second store. In systems where the overall delay is of no consequence, complete separation of the writing and reading process effectively allows the waveform to be read out before it was written, thus giving expansion in the negative time direction.

Selection of the appropriate signal samples and interpolation coefficients is controlled by a system of ratio counters [86], as shown in Figure 6.34b, which at each clock pulse provides a measure of the output sample position in terms of the number of input samples. The integer part provides the read address and the fractional part the coefficient address. Each time a carry is generated by the fractional part adder, the integer counter advances to the next read address and previous samples are handed along the string of registers in the interpolator, Figure 6.34c. Each read-only memory contains a set of coefficient values corresponding to a section of the interpolation aperture function, an example of which is shown in Figure 6.34d.

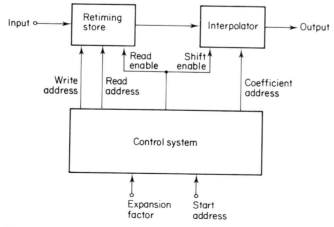

Figure 6.33 Circuit arrangement for waveform expansion in which the signal is stored for retiming before interpolation

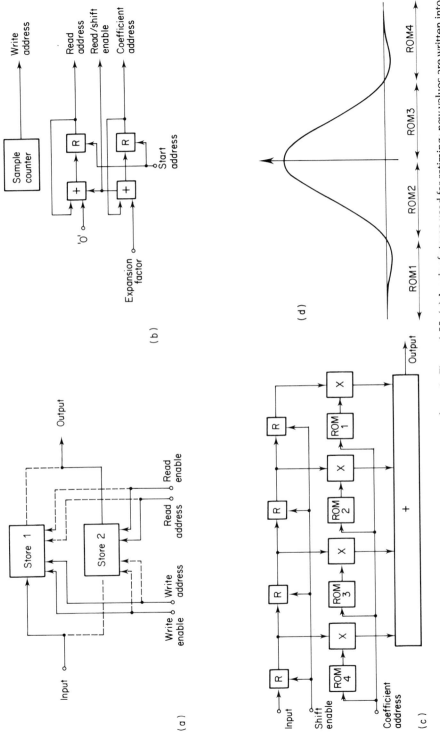

Figure 6.34 Details of the waveform expansion arrangement shown in Figure 6.33. (a) A pair of stores used for retiming; new values are written into store 1, while earlier values are read from store 2 for processing. (b) The arrangement of incrementing ratio counters used to generate control waveforms. (c) An interpolator with registers (R) before the multipliers and (d) an interpolation aperture function with the regions of the waveform stored in the ROMs of (c) marked

The operation of the expansion circuit in the time domain is demonstrated in Figure 6.35. Figure 6.35a shows a sequence of input samples which are subsequently read out of the retiming store at the slower rate shown in Figure 6.35b so that the duration of the output waveform is greater in proportion to the expansion factor. The interpolation aperture function, the shape of which is related to the input signal sampling frequency, is also effectively expanded in its action in the interpolator. Figure 6.35c shows a number of phases of the interpolation aperture function, in this case a linear function, relative to the input samples. In each case, only two input samples fall within an aperture width and are weighted to form each of the output samples shown in Figure 6.35d. The positions of the output samples in time are coincident with those of the input samples, shown in Figure 6.35a, as required.

Figure 6.36 depicts the same process in the frequency domain, with the original spectrum, shown in Figure 6.36a being scaled to form lower frequency components at Figure 6.36b. The low-pass filtering effect of the interpolation process, shown in Figure 6.36c, is also scaled to match the spectrum of the reduced rate samples in Figure 6.36b. In terms of the input frequencies, therefore, the spectrum of true signal frequencies retained by the interpolator remains the same for all expansion factors. A more ideal frequency characteristic retaining more of the baseband components and rejecting the harmonic components more effectively would require a wider interpolation aperture, thus using more input samples to produce each output sample. Synthesis of a suitable

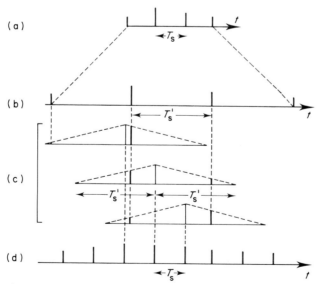

Figure 6.35 Time-domain operation of the waveform expansion circuit. (a) A sequence of input samples shown again in (b) being read out of the retiming store at a slower rate. (c) The time-expanded interpolation aperture being applied to the reduced rate samples to produce (d) the sequence of output samples for the time-stretched waveform

characteristic can be obtained by the methods previously described. Resampling of the low-pass filtered spectrum at the original sampling frequency repeats the filtered spectrum so that residual harmonic components may overlap with the baseband spectrum as shown in Figure 6.36d.

In addition to expansion, the arrangement of Figure 6.33 could be used to provide some degree of time compression, but it is preferable in this case to interpolate first and then retime the resulting samples [125], as shown in Figure 6.37. The retiming stores and address generator arrangements shown in Figure 6.34a and 6.34b can be used for compression by interchanging the read and write addresses and enables, and by interpreting the expansion factor as the reciprocal of the compression factor. The interpolator, however, needs a more substantial rearrangement to the form shown in Figure 6.37b. Here the multipliers are placed before the registers and the register chain includes a recirculating arrangement to allow the product values to accumulate. As the compression factor increases, more and more samples are combined to produce each output sample. Because of this, at some point the overall gain has to be reduced by the reciprocal of the compression factor. This can be accommodated most conveniently at the interpolator input, as shown. Storage of the interpolation coefficients is then made in a number of sections, in the same way as shown in Figure 6.34d.

The time-domain operation of the compression circuit is shown in Figure 6.38. In this case, the interpolation aperture is related to the pitch of the output samples, so more

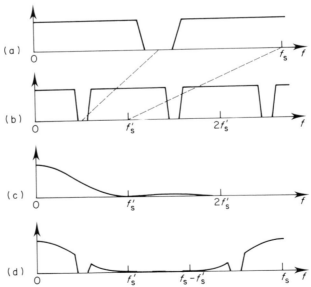

Figure 6.36 Frequency-domain operation of the waveform expansion circuit: (a) the input spectrum, (b) the reduced frequency components of the samples read more slowly from the store, (c) the low-pass characteristic of the linear interpolation aperture and (d) the spectrum of the expanded signal

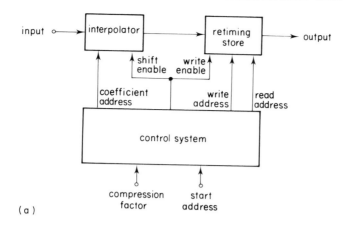

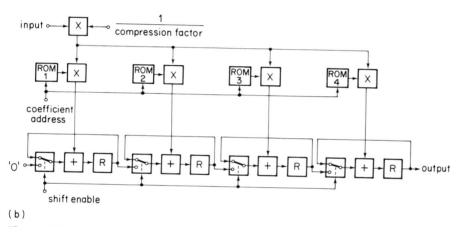

Figure 6.37 (a) A circuit arrangement for waveform compression in which the signal is stored for retiming after interpolation. While the retiming store and control circuitry have considerable similarities to those Figure 6.34, in this case the interpolator arrangement shown in (b) is advantageous

input samples from Figure 6.38a are combined at large compression factors. Ultimately, with infinite compression, all the input samples would be combined to form a single point at the output. The samples, produced in Figure 6.38c by the repeated application of the interpolation aperture in Figure 6.38b, are then read out more rapidly in the compressed form of Figure 6.38d.

The corresponding spectra for waveform compression are shown in Figure 6.39 with the input signal spectrum shown in Figure 6.39a. As the bandwidth must be reduced in compression to avoid aliasing, the low-pass filter characteristic of the interpolation aperture is matched to the maximum frequency content of the compressed samples as shown in Figure 6.39b. Resampling introduces the additional spectra of Figure 6.39c which are then scaled by the retiming operation to form Figure 6.39d. Again, a suitable frequency characteristic can be synthesised by the methods previously described.

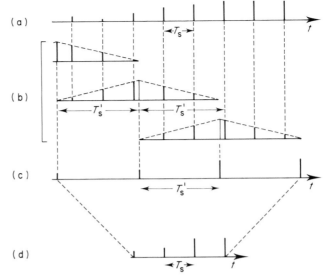

Figure 6.38 Time-domain operation of the waveform compression circuit: (a) a sequence of input samples, (b) the linear interpolation aperture used to weight the input samples, (c) the resulting output samples and (d) the output samples after being read from the retiming store at a faster rate

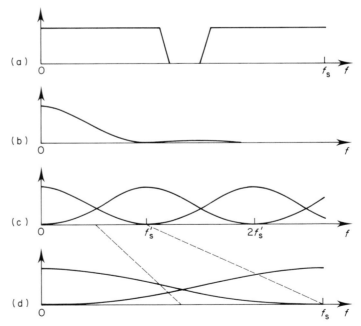

Figure 6.39 Frequency-domain operation of the waveform compression circuit: (a) the input spectrum, (b) the low-pass characteristic of the linear interpolation aperture, (c) the spectrum of the interpolated samples and (d) the stretched spectrum of the time-compressed samples read from the store

TELEVISION SCAN CONVERSIONS

Television scanning

The normal application of the one-dimensional sampling process described earlier is to convert a continuous waveform to a series of samples for PCM coding. Although this form of sampling is often applied to television signals, the process of scanning used to originate television signals from an image is also a sampling process. Conversion from one scanning standard to another is, therefore, a sampling rate conversion and follows the same basic theory as that previously described.

The brightness of an image falling onto a television camera tube is a function varying in three dimensions: horizontal position (x), vertical position (y) and time (t). The action of scanning the image with the television raster pattern converts this three-dimensional function into a one-dimensional signal.

The simplest form of scanning used to produce a television signal is sequential scanning. A 312 lines per picture, 50 fields per second sequential scan is shown diagrammatically in Figure 6.40a. Each consecutive set of lines is taken at the same vertical positions, but is offset in time from the previous set by one field period. Although not shown in the diagram, the columns of scanning points are very slightly skewed from the vertical, so that the last line of one field scan is taken just before the first line of the next. Similarly the lines themselves are not precisely horizontal (in the diagram, the horizontal dimension extends into the paper).

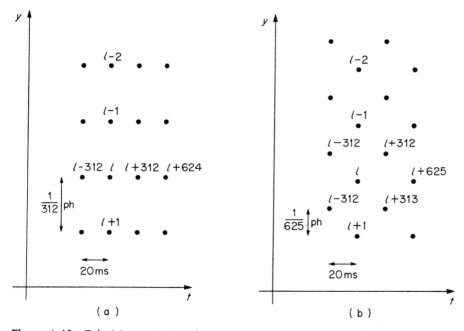

Figure 6.40 Television scanning: diagrammatical representations of (a) the pattern of lines for a 312 lines per picture, 50 fields per second sequential scan and (b) a 625 lines per picture, 50 fields per second interlaced scan. The horizontal dimension extends into the paper

318   INTERPOLATION

All broadcast television systems, however, use interlaced scanning. This is shown in Figure 6.40b for a 625 lines per picture, 50 fields per second, interlaced scan. In this case, the lines of successive fields are interleaved so that each line falls in a position vertically between those on the adjacent fields. This doubles the number of lines in the picture, while keeping the scanning rates virtually the same.

In the scanned signals the horizontal dimension is still continuous, but the image brightness is now only defined at discrete intervals in the vertical and temporal dimensions. Thus, the image brightness has been sampled in the two dimensions of vertical position and time.

The action of sampling repeats the frequency spectrum of the image at harmonics of the sampling frequency, as previously described. The image spectrum consists of horizontal frequency (m), vertical frequency (n) and temporal frequency (f) components. Horizontal frequencies in the image are not affected by scanning because the horizontal dimension of the signal remains continuous. However, in a digital standards converter, the horizontal components are also sampled; but, provided that a line-locked sampling frequency is used, this sampling is orthogonal to the vertical/temporal sampling resulting from scanning. Because of this, the two processes have no effect on one another and can be treated completely separately. Therefore, only the vertical and temporal components of the image spectrum, represented diagrammatically in Figure 6.41, are affected by scanning. Temporal frequencies are measured in hertz (Hz), whilst the units of vertical frequency are cycles per picture height (c/ph).

If the scene from which the image is formed is completely still, then the frequencies which make up the image spectrum have no temporal component and all the spectral energy lies along the vertical frequency axis. If an object in the scene appears and disappears at a regular rate, such as a light flashing, this results in temporal frequency components which are positioned away from the n axis by an amount depending on the rate of flashing. However, most temporal frequencies are produced by the movement of

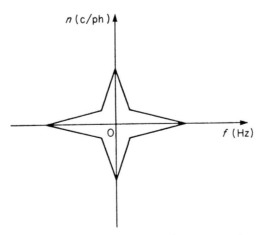

Figure 6.41 A diagrammatical representation of the vertical/temporal frequency spectrum of an image

objects in the scene. The effect of movement is more complicated because the temporal frequencies produced depend on both the rate of movement and the spatial frequencies which make up the object.

For simplicity, consider first an object consisting of a single spatial frequency component, a sine wave, Figure 6.42a. As this moves across a point in the scene, it produces a single temporal frequency, Figure 6.42b, corresponding to the time it takes to move through its own spatial period. If the rate of movement is increased or if an object of

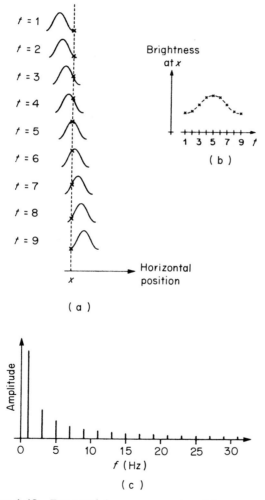

Figure 6.42 Temporal frequency produced by a moving object: (a) an object with a sinusoidal brightness profile passing a position x, shown at several equispaced time intervals, and (b) the waveform of brightness at position x, consisting of a single temporal frequency. (c) Temporal frequencies produced by a square wave object moving through its own spatial period in one second

greater spatial frequency is used with the same rate of movement, the temporal frequency resulting as the object passes a point will be increased.

More complicated shapes can be considered as a series of Fourier components. Thus, if the highest spatial frequency in the object is S times the fundamental component, as the object moves it will produce a series of temporal components. These will extend to S times the basic temporal frequency corresponding to the object moving through its own width. Figure 6.42c shows the temporal frequencies resulting from a square wave moving through its own spatial period in one second. In this case, the presence of high spatial frequencies in the square wave results in high temporal frequencies, even with a very modest rate of movement. Thus, sharp pictures produce a greater proportion of high temporal frequencies than soft pictures. Conversely, reducing the spatial frequency components in the signal, for example by including a horizontal low-pass filter, makes a corresponding reduction in those temporal frequencies produced by horizontal movement. Temporal frequencies produced by a stationary object flashing would not be affected.

For purely horizontal movement, all the spectral energy is concentrated along the f axis of the n–f spectrum. If, instead, an object is moving vertically, this produces a series of temporal components along a diagonal line passing through the origin; the angle made between this line and the n axis increases with the speed of movement of the object.

Clearly there is virtually no limit to the spatial and temporal frequencies that can occur in a scene, but some filtering is introduced by the action of the camera. The camera tube integrates the light falling on it at each point for a field period, thus attenuating the higher temporal frequencies. (It is generally assumed that interlaced line positions are discharged on each field scan.) Also, the width of the scanning spot is not infinitely small, so that the charge from adjacent areas of the tube surface contributes to the output current, thereby filtering the image vertically. This gives a probable form of camera filtering response, including aperture correction, as shown in Figure 6.43 (this model for camera response was derived by J. O. Drewery).

The effect of scanning is to repeat the spectrum of the image of the original scene, Figure 6.41, filtered by the characteristic of Figure 6.43, at harmonics of the scanning rates. The 312 lines per picture, 50 fields per second sequential scan produces the spectrum shown in Figure 6.44a, while interlaced scanning, with 625 lines per picture and 50 fields per second, produces the spectrum shown in Figure 6.44b. Although only the first quadrant is shown, the spectra extend into the other quadrants in the same pattern.

As components of the spectra can extend well beyond the outlines shown in Figure 6.44, aliasing forms an accepted part of normal television pictures. For example, with interlaced scanning, flicker arises from vertical frequency components in the region around $(0, 312\frac{1}{2})$ in the image spectrum. In the scanned spectrum these components are repeated around $(25, 0)$ arising from the spectra centred on $(25, \pm 312\frac{1}{2})$. So the vertical detail flashes with a temporal frequency of 25 Hz.

When Figures 6.43 and 6.44 are considered together, it is evident that, for the interlaced system, temporal aliasing from camera pictures is much more likely than vertical aliasing. However, electronically generated signals are, in general, completely unfiltered, temporally and vertically. Because of this, these signals often have a much greater degree of aliasing than can occur with signals from a camera. Therefore, captions and graphics are usually considerably more difficult to convert successfully from one

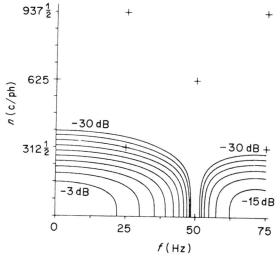

Figure 6.43 The probable form of the vertical and temporal filtering effect of a television camera, including vertical aperture correction. The characteristic shows contours of constant attenuation at 3 dB intervals down to −30 dB.

standard to another than normal camera pictures. Similarly, signals derived from scanning film in a telecine tend to produce more vertical resolution, and aliasing, than a television camera. Also, film pictures already include severe temporal aliasing resulting from the 25 (or 24) frames per second temporal sampling action of the film camera.

Spectra for other scanning standards would, in general, be similar to those of Figure 6.44 with appropriate scaling of the axes. In particular, for the 525 lines per picture, 60 fields per second, interlaced standard, the repeated spectra would be centred on (0, 525), (30, $262\frac{1}{2}$), (60, 0), (60, 525), etc. Also, the filtering effect of cameras would be a scaled version of Figure 6.43.

Ultimately the television signal is reconstituted in its three-dimensional (x, y and t) continuous form for viewing. This is achieved by scanning the brightness information onto the television cathode ray tube display. Although the display tube provides some low-pass filtering of vertical and temporal frequencies (through the profile of the scanning spot and phosphor, and the time constant of the phosphor), it has a relatively slow roll-off so that the harmonic spectra are poorly suppressed. The presence of these extra components results in the visibility of the line structure for vertical components and the presence of large-area flicker for temporal components.

With interlaced scanning, the vertical repeated spectra are separated from the baseband spectrum by a wider margin. Because of this, it is possible to display a reasonable range of vertical frequencies and still to suppress the line structure. The relatively low vertical resolution provided by cameras (Figure 6.43) suggests that the extra resolution possible with interlaced scanning is largely unused. If it were fully used, this would result in a substantial increase in the interlace flicker. It should be concluded, therefore, that interlacing is primarily a means of suppressing the line structure, rather than a method of increasing vertical resolution.

322 INTERPOLATION

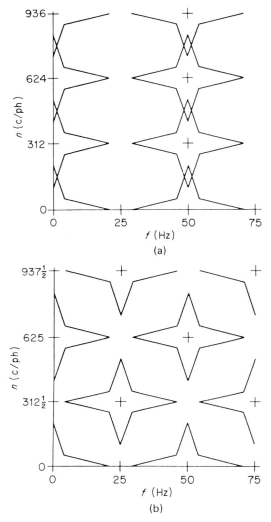

Figure 6.44 Television scanning in the frequency domain: the spectrum of the image (Figure 6.41) is repeated at harmonics of the scanning rates (a) by a 312 lines per picture, 50 fields per second sequential scan and (b) by a 625 lines per picture, 50 fields per second interlaced scan

Conversion methods

Principles of conversion

Before electronic standards converters were available, optical converters were used consisting of a display tube and a television camera. The conversion was made by displaying the input standard picture in the normal way and scanning the reconstituted image with a camera using an output standard raster. In terms of sample rate conversion, this process is equivalent to the low-pass filtering and resampling method previously described. In addition to problems of geometric distortion caused by imperfect scanning, the low-pass filtering action of an optical converter is difficult to control, since it is the result of phosphor characteristics and spot profiles. Because of this, present day electronic converters achieve better performance. Nevertheless, it is still useful, as already described, to visualise the interpolation process of an electronic converter as consisting of low-pass filtering to recover the baseband spectrum of the original image, followed by resampling on the new standard.

For interlaced scanning, the vertical/temporal low-pass filter characteristic needs to have a shape approximating that shown in Figure 6.45. This triangular shape would retain the main components of the baseband spectrum and reject the repeated spectra centred on multiples of the scanning rates.

Because television scanning is a two-dimensional sampling process, it is possible to treat vertical and temporal (line and field) interpolation either as two one-dimensional processes or as a single two-dimensional process. This section examines methods of using two separate interpolation processes and demonstrates in the frequency domain the drawbacks of this approach when applied to interlaced scanning. This is then compared with the greater flexibility and improved performance of using a single, two-dimensional interpolation process.

In a converter that uses separate vertical and temporal interpolation, the two-dimensional interpolation aperture function $g(t, y)$ is the convolution of two one-dimensional aperture functions, $g_1(t)$ and $g_2(y)$, one exclusively a function of time and the other exclusively a function of vertical position

$$g(t, y) = g_1(t) * g_2(y)$$

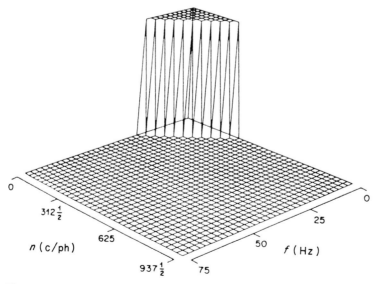

Figure 6.45 The triangular shape of the idealised vertical/temporal low-pass filter characteristic needed to retain the baseband spectrum and reject the harmonic spectra resulting from interlaced scanning

324   INTERPOLATION

Similarly, the overall two-dimensional frequency characteristic is the product of two one-dimensional characteristics:

$$G(f, n) = G_1(f)G_2(n)$$

Such two-dimensional functions are said to be variables-separable. Whenever either of the one-dimensional frequency characteristics (G_1 or G_2) has a zero value, the two-dimensional product characteristic (G) has a line of zero values, parallel to the other axis. This causes all variables-separable interpolation methods to have approximately rectangular frequency characteristics.

For sequential scanning, the rectangularly shaped characteristics of separate vertical and temporal interpolation are ideal. In this case, the interpolation processes can, in principle, be carried out in either order, as shown in Figure 6.46, with equivalent performance. However, in practice, in a digital converter it is advantageous to convert the field rate first (Figure 6.46a). This is because the results of the first interpolation need to be represented to higher accuracy to minimise rounding errors; storing these higher accuracy digital words in the smaller capacity line stores minimises the overall storage requirements.

The two separate interpolation processes are illustrated in Figure 6.47. In the first step, Figure 6.47a, lines of the intermediate standard are interpolated at a new field position; each of these new lines is produced from several lines at the same vertical position taken from neighbouring fields. The second step, Figure 6.47b, consists of interpolating between the lines of the intermediate standard to produce lines at the correct vertical positions for the output standard.

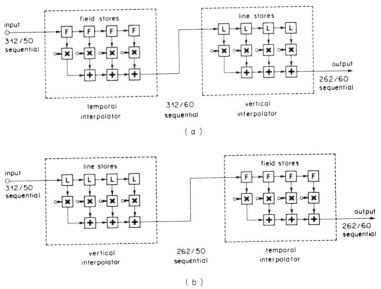

Figure 6.46 Converter arrangements for sequentially scanned signals in which vertical interpolation and temporal interpolation are carried out as separate processes: (a) temporal followed by vertical interpolation and (b) vertical followed by temporal interpolation

TELEVISION SCAN CONVERSIONS

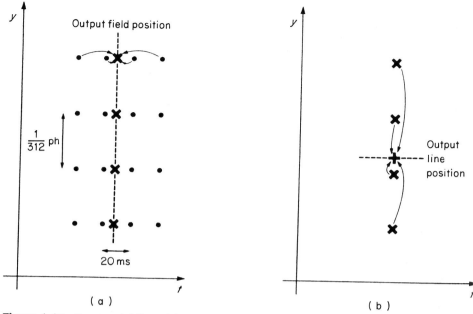

Figure 6.47 Temporal followed by vertical interpolation for sequential scanning: (a) lines of the intermediate standard are interpolated for a new temporal position, but at the same vertical positions as the input lines and (b) these intermediate standard lines are used to produce the output standard lines at new vertical positions (• input standard lines, × intermediate standard lines, + output standard lines)

In the frequency domain, the filtering effect of the temporal interpolator needs to reject all the temporal repeated spectra, leaving the vertical harmonic components unaffected. Ideally, therefore, all components in the shaded region of Figure 6.48a would be suppressed. The resampling action of interpolation then introduces new temporal harmonic components at the positions shown in Figure 6.48b. Similarly, the vertical interpolation, which follows, first suppresses all the vertical repeated spectra, Figure 6.48c, and then introduces new harmonic spectra, as shown in Figure 6.48d. In this 'down-conversion' process, the filter needs to cut at the output standard vertical half sampling frequency, 131 c/ph, to prevent the spectra overlapping on the output standard.

With interlaced scanning, applying separate vertical and temporal interpolation to the offset structure presents many problems. With separate interpolation, it is incorrect to interpolate using a mixture of lines from odd and even fields, either temporally or vertically. This is because a one-dimensional interpolation process cannot accommodate the two-dimensional offset of interlaced lines. For example, although vertical interpolation using lines from both types of fields would give good performance for still pictures, any movement would result in serious impairments because the temporal offset between the fields would be ignored. With this restriction, separate interpolation can be applied to interlaced signals in either order. Whichever order is used, the first interpolation process must produce a higher sequential intermediate standard, in order to avoid interlace components which would interfere with the second interpolation process.

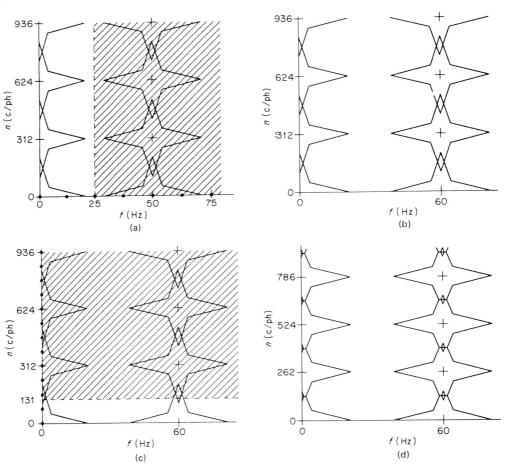

Figure 6.48 A spectral representation of temporal interpolation followed by vertical interpolation for sequential scanning: (a) first, temporal repeated spectra are rejected by the low-pass filtering action of the temporal interpolator (with a four-field aperture, the filter characteristic can be set independently at $12\frac{1}{2}$ Hz intervals as marked on the temporal frequency axis), (b) the resampling action of the interpolator substitutes repeated spectra at the new field scanning rate, (c) then the vertical interpolator rejects the vertical repeated spectra (fixed points at 78 c/ph for a four-line aperture) and (d) substitutes repeated spectra with the new line spacing

The action of an interpolator in which temporal interpolation precedes vertical interpolation for interlaced scanning is illustrated in Figure 6.49. The temporal interpolator produces new fields at each output standard field position, as shown in Figure 6.49a, using input lines at the same vertical positions. Two sets of lines are interpolated for each new field position, corresponding to the positions of the odd and even field-lines of the input standard. Therefore, in the example used in the diagram, the 625/50 interlaced input standard is converted to a 625/60 sequential intermediate standard.

The second interpolator uses lines (eight lines in this example) from one field of the intermediate standard to produce new lines at positions appropriate for the output

TELEVISION SCAN CONVERSIONS   327

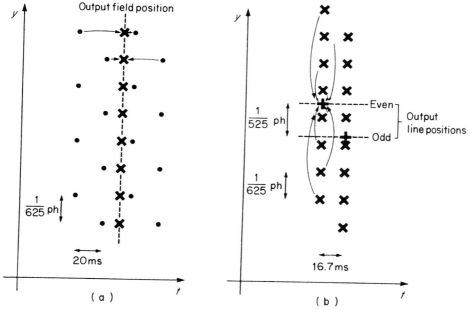

Figure 6.49 Temporal interpolation followed by vertical interpolation for interlaced scanning: (a) lines of a sequential intermediate standard are interpolated for a new temporal position, but at vertical positions corresponding to those of both even and odd input fields; (b) the intermediate standard lines are used to produce lines at new vertical positions appropriate for the even and odd positions on alternate output fields (• input standard lines, × intermediate standard lines, + output standard lines)

standard, as shown in Figure 6.49b. This includes the vertical offset from one field to the next required for interlacing. The vertical interpolator, therefore, converts the 625/60 sequential intermediate standard into the required 525/60 interlaced output standard.

Figure 6.50 shows a block diagram of an interpolator arrangement which could perform the operations shown in Figure 6.49. Note that two lines are produced by the temporal interpolator for each line of the input standard, resulting in an intermediate standard with twice the resolution of the input. For the reverse direction of conversion,

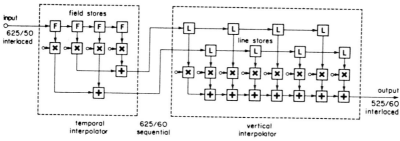

Figure 6.50 A converter arrangement for interlaced scanning in which temporal interpolation is followed by vertical interpolation

that is, 525/60 to 625/50, the intermediate standard produced by this method would be a 525/50 sequential scan.

Figure 6.51 illustrates the process of Figure 6.49 in the frequency domain. In Figure 6.51a the temporal interpolator attempts to suppress all the temporal repeated spectra; but suppression of the interlaced components centred on (25, $312\frac{1}{2}$) proves impossible without rejecting most of the wanted baseband components as well. The compromise of using a cut-off at 25 Hz retains the wanted baseband information, but leaves the interlaced spectra virtually unsuppressed. Temporal resampling at the new field rate of 60 Hz produces the 625/60 sequential spectrum of Figure 6.51b.

The vertical interpolator now requires a very abrupt cut-off at a low vertical frequency ($156\frac{1}{4}$ c/ph) in order to suppress the remnants of the interlaced components left by the temporal interpolator. This is shown in Figure 6.51c. Vertical resampling at 525 lines per field would produce only the components shown as solid lines in Figure 6.51d; however, when new lines are interpolated only at even line positions on one field and at odd line

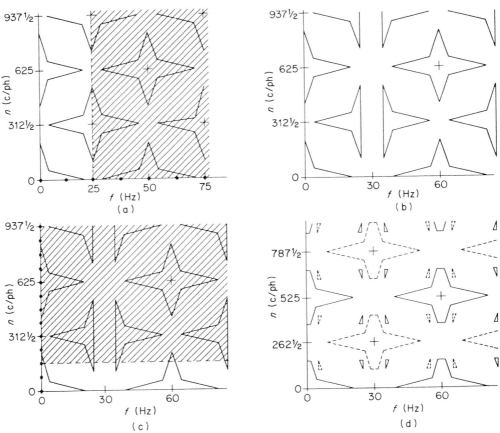

Figure 6.51 A spectral representation of temporal interpolation followed by vertical interpolation for interlaced scanning: (a) temporal low-pass filtering and (b) resampling to produce a sequential intermediate standard, followed by (c) vertical low-pass filtering and (d) resampling (the positions of spectra for interlaced output fields are shown dashed)

positions on the next, the extra components resulting from interlacing are introduced (shown dashed in Figure 6.51d).

The same performance can be obtained when vertical interpolation precedes temporal interpolation. In this case, lines from one input field are used to interpolate new lines at positions corresponding to the output standard line spacing, as shown in Figure 6.52a. The 625/50 interlaced input is therefore converted to a 525/50 sequential intermediate standard. The temporal interpolator then converts the intermediate standard into the output standard using sets of lines at the same vertical position from several fields (Figure 6.52b); for an interlaced scan, lines are produced only at the odd and then the even positions on alternate output fields.

The block diagram of a converter using vertical, followed by temporal interpolation is shown in Figure 6.53. Comparing this with Figure 6.50, the same number of multipliers and adders are used in each case, but the storage requirements differ. Figure 6.50 with its eight line stores appears the more complicated, but the field stores in Figure 6.53 need approximately twice the capacity of those in Figure 6.50 because sequentially scanned fields are being stored.

In the frequency domain, Figure 6.54, the vertical interpolator has to reject both the vertical and interlaced components to produce a 525/50 sequential intermediate standard (Figure 6.54b). This is then converted to the interlaced 525/60 standard shown in Figure 6.54d.

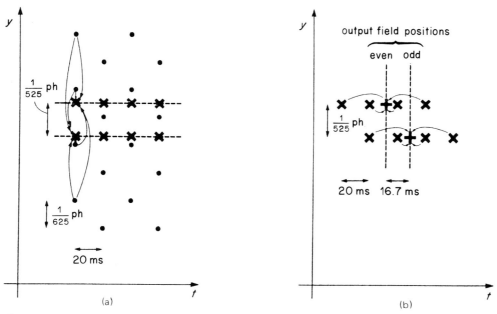

Figure 6.52 Vertical interpolation followed by temporal interpolation for interlaced scanning: (a) lines of a sequential intermediate standard are interpolated for new vertical positions, but at the same positions in time as the input fields; (b) lines from different fields of the intermediate standard are used to produce lines at new temporal positions appropriate for the even and odd positions on alternate output fields (• input standard lines, × intermediate standard lines, + output standard lines)

330 INTERPOLATION

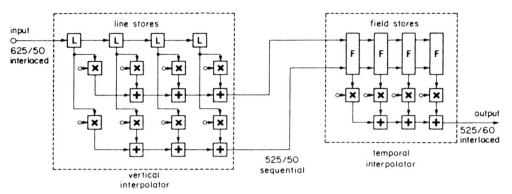

Figure 6.53 A converter arrangement for interlaced scanning in which vertical interpolation is followed by temporal interpolation. Note that the field stores each require twice the capacity of those in Figure 6.50 as sequential fields are being stored

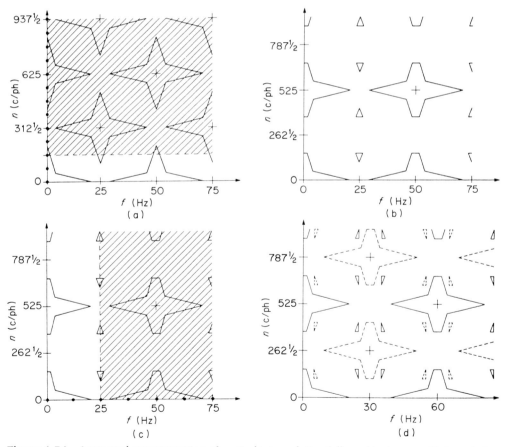

Figure 6.54 A spectral representation of vertical interpolation followed by temporal interpolation for interlaced scanning: (a) vertical low-pass filtering and (b) resampling to produce a sequential intermediate standard, followed by (c) temporal low-pass filtering and (d) resampling (the positions of the interlaced spectra are shown dashed)

TELEVISION SCAN CONVERSIONS   331

As the overall effect of the temporal and vertical interpolation processes is the same whichever order is used, the smaller storage capacity required for Figure 6.50 makes temporal followed by vertical interpolation preferable. In either case, the rectangular form of the overall frequency characteristic which, as explained earlier in this section, results with any variables-separable aperture is not a good approximation to Figure 6.45, so that vertical resolution is significantly impaired and the original interlaced components are not fully suppressed.

If an intermediate standard with a normal line rate were to be used (instead of the double line rate, sequential scan produced by the first interpolation process), this would result in the generation of extra spectral components in either Figure 6.51b or Figure 6.54b. The second one-dimensional interpolation process would be unable to suppress these extra components effectively, so that additional impairments would be introduced unnecessarily.

If vertical/temporal interpolation is treated as a single process, the two-dimensional aperture functions and frequency characteristics are not restricted to variables-separable functions and, consequently, to rectangular frequency characteristics. It is possible, therefore, to obtain a close approximation to the triangular shape of frequency characteristic required for interlaced scanning.

The process of combined interpolation is illustrated in Figure 6.55 which shows that all the input standard lines falling within the two-dimensional interpolation aperture contribute directly to one line of the output standard. Because the process is two dimensional, the weighting coefficient applied to each line can take account of both its

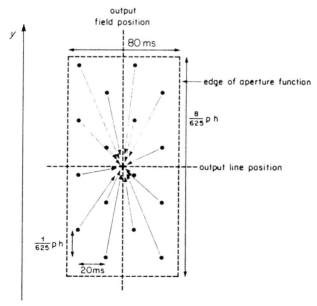

Figure 6.55 Combined vertical and temporal interpolation: each of the sixteen input lines falling within the aperture is weighted according to its individual temporal and vertical offsets from the required output line position (• input standard lines, + output standard line)

332   INTERPOLATION

temporal and vertical offsets. Conceptually, the aperture function is placed over the pattern of the input lines with its origin at the required output line position; this identifies the aperture function value (weighting coefficient) appropriate for each of the input lines.

In the combined interpolator, the extent of the aperture function determines the amount of storage required. Figure 6.56 shows an interpolator using four lines from each of four fields, which is the same as the effective aperture size given by the two one-dimensional interpolators in Figure 6.50 and 6.53. The combined two-dimensional interpolator requires rather more line stores, multipliers and adders than the separate one-dimensional interpolators. However, the line stores, although shown in Figure 6.56, can be omitted if multiple-output field stores are used [64, 126, 127]. The complexity involved in the two methods is then broadly similar.

In the frequency domain, the two-dimensional frequency characteristic rejects all the repeated spectra at once, as shown in Figure 6.57a. In this case, because it is non-variables-separable, the frequency characteristic can be specified independently at

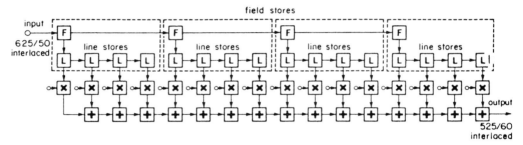

Figure 6.56 A converter arrangement for combined vertical and temporal interpolation. The line stores can be omitted if multiple output field stores are used, as indicated by the dashed outlines

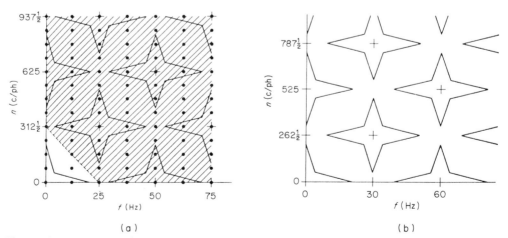

Figure 6.57 Spectra showing the effect of combined vertical and temporal interpolation for interlaced scanning: (a) all the repeated spectra are suppressed by the two-dimensional low-pass filtering action of the interpolator and (b) the resampling action produces new repeated spectra in both dimensions directly; with an aperture of four fields by four lines (Figure 6.56) the frequency characteristic can be set independently at the positions shown in (a)

$78\frac{1}{8}$ c/ph and $12\frac{1}{2}$ Hz intervals all over the vertical/temporal spectrum. Compared with variables-separable functions, this provides much more opportunity to match the frequency characteristic of the interpolator to the shape of the scanned spectrum. In particular, the vertical resolution for still pictures (along the *n* axis) can be extended, relative to that shown in Figure 6.51c and 6.54a, while still suppressing the main interlace components centred on $(25, 312\frac{1}{2})$. The repeated spectra are then reintroduced directly at the output standard positions, as shown in Figure 6.57b.

It should be noted that although the particular shape of baseband spectrum used in the frequency domain diagrams is representative of most pictures, some components can extend well beyond this outline. Therefore, in all the methods of conversion described, substantial overlapping of the spectral components can sometimes occur.

The sharp cut-off of the frequency characteristic shown in Figure 6.45, although possibly ideal in theory, is not practicable and may not even be desirable. Realisable interpolation methods do not give the abrupt cut-off filtering characteristics and the neat separation of the spectral components indicated in the frequency domain diagrams shown earlier. Next the mechanisms causing particular forms of impairment which arise from non-ideal filtering are described, first in general terms. This is followed by a consideration of the impairments introduced by two previous converters, the performance characteristics of which are well known.

A filter with a gradual cut-off may both attenuate the wanted baseband components and leave some parts of the repeated spectra unsuppressed. Much of the signal energy is concentrated along the axes of the baseband spectrum and in corresponding areas of the repeated spectra. However, the centres of the repeated spectra are easy to suppress and the higher order spectra only lead to impairments when very simple aperture functions are used. Therefore, the main impairments result from the response of the filter characteristic in the regions labelled A to F in Figure 6.58. Inadequate passband response in region A will result in blurring of movement whilst attenuation in region B will cause a loss of vertical detail. The other impairments, resulting from inadequate suppression of regions, C, D, E and F, occur when these harmonic components are aliased by the resampling process to fall within the baseband spectrum. Figure 6.59 illustrates the impairments which can be shown in a still picture. The time-varying nature of the judder (Region D) and 5 Hz flicker (Region E) impairments can, of course, not be shown in still photographs.

The aliased spectra are offset from the original by the difference between the old and new sampling frequencies. Thus, when resampled on the 525/60 standard, the components from region C, shown shaded in Figure 6.58, are centred on 100 c/ph, as shown in Figure 6.60a. In addition, the corresponding components from $(0, -625)$ are shifted to $(0, -100)$ in Figure 6.60a. The results for 525/60 to 625/50 conversion are substantially the same, although components from $(0, 525)$ and $(0, -525)$ are shifted to $(0, -100)$ and $(0, 100)$ respectively. The main components in region C are vertical frequencies from stationary pictures and these result in alias components which superimpose different vertical frequencies, offset by 100 Hz from the true frequencies. With high contrast signals, the non-linearity of the display tube produces a 100 c/ph beat frequency component between the true and aliased patterns. In pictures, this vertical aliasing appears mainly as 'knotting' on diagonal edges.

The corresponding effect for temporal frequencies is that components from region D, centred on $(50, 0)$ and $(-50, 0)$ appear as alias components centred on $(-10, 0)$ and

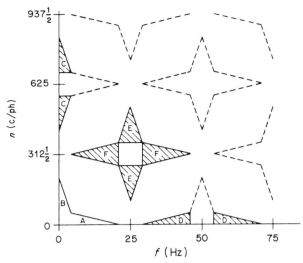

Figure 6.58 The low-pass filtered spectrum of interlaced scanned television signals: regions which produce the main impairments when converted to a new standard. Passband impairments: insufficient response in Region A causes blurring of movement and in Region B results in a loss of vertical detail. Stopband impairments: insufficient suppression of region C results in vertical aliasing, region D in movement judder, region E in flicker on vertical detail, while insufficient suppression of region F results in vertical modulation of moving detail

(10, 0) respectively in the 525/60 spectrum, as shown in Figure 6.60b. For 525/60 to 625/50 conversion, the diagram is similar although the components at (10, 0) and (−10, 0) are derived from (60, 0) and (−60, 0) respectively. This temporal aliasing appears as judder on moving objects, caused by the alias components perturbing the true rate of movement.

In Figure 6.58, frequencies along the 625 c/ph line and the 50 Hz line are easy to suppress. Consequently temporal components from the (0, 625) spectrum and vertical components from the (50, 0) spectrum usually cause no problems. However, both the vertical components (region E) and the temporal components (region F) of the interlaced spectrum centred on $(25, 312\frac{1}{2})$ can lead to individual aliasing problems.

For vertical frequencies, the shaded areas of Figure 6.60c show the positions occupied by the main alias components after resampling. In this case, the aliased spectra are offset vertically by 50 c/ph and temporally by 5 Hz so that stationary vertical frequencies result in vertical detail flashing at a 5 Hz rate. This effect is known as 5 Hz flicker and often appears as a repeated upward and downward displacement of the horizontal boundaries of objects.

For temporal frequencies, the region F components from Figure 6.58 are moved to the 50 c/ph lines of the baseband spectrum as shown in Figure 6.60d. Because of this, an object moving horizontally, which originally had no vertical detail, would have a 50 c/ph, vertical pattern superimposed on its moving vertical edges. In its most severe form, this impairment is known as 'crankshaft' distortion because of its jagged, stepped appearance.

As shown in Figure 6.60c and 6.60d, the interlaced spectra produce four separate alias components from the spectra at (± 25, $\pm 312\frac{1}{2}$) in the four quadrants. Again 525/60 to 625/50 conversion results in the same frequency offsets shown in Figure 6.60, although the alias components are reversed in their positions. All the individual alias components shown in Figure 6.60 are attenuated by the low-pass filtering effect of the interpolation process and should be superimposed, together with the original baseband spectrum, to give the overall spectrum resulting from standards conversion. This spectrum is then repeated at all multiples of the output standard scanning rates.

As already mentioned changing from a higher to a lower sampling rate can result in aliasing on the output standard. This could occur with 625 to 525 conversion in which the extra vertical detail of the 625-line system could cause 30 Hz interlace flicker on the output standard. Also, with 60 to 50 Hz field-rate conversion, the extra temporal components could result in the individual fields becoming more visible on movement. However, both these impairments are accepted as normal features of the output standard television system; so it is much less important for these to be suppressed than for the abnormal alias components shown in Figure 6.60. Even so, it is relatively easy to suppress components that would cause interlace flicker without sacrificing much useful vertical resolution. In this respect, then, the converted pictures can be of higher quality than pictures directly scanned on the output standard.

Performance of previous converters

As a further illustration of the frequency domain method, the forms of impairment produced by two well-known interpolation methods will be related to features of their frequency characteristics. The first interpolation system is that used in the analogue converters [5, 128] developed by BBC Research Department and first used in June 1968. The second method is an approximation to that used in the IBA's DICE converters, [129, 130] first used for 525/60 to 625/50 conversion in November 1972 and for the reverse direction in May 1975. Almost all of the broadcast pictures converted in the United Kingdom during the 1970s used one of these two methods.

The BBC analogue converters used quartz ultrasonic delay lines equivalent to two fields of storage, an example of which is shown in Figure 1.1, and f.m. signal averagers for interpolation. Thus, many impairments, such as spurious responses in the delay lines and slight gain variations between the lines, were the result of the analogue circuitry rather than the interpolation method. Also, the f.m. averagers were only able to produce relatively simple combinations of signals for interpolation.

However, the interpolation method is well defined and can be expressed as an interpolation aperture function one quadrant of which is shown in Figure 6.61. In terms of the time domain performance, this can be interpreted as a combination of two methods. If the required output line position is close to an input field, such as in Figure 6.62a, that is, within $\pm\frac{1}{4}$ field period, then the two closest lines of that input field are combined with weights of $\frac{3}{4}$ for the closer of the two and $\frac{1}{4}$ for the other line. When the output position is outside the $\frac{1}{4}$ field period limit, such as in Figure 6.62b, the nearest lines of the preceding and succeeding fields are averaged (one line from each). It should be

336 INTERPOLATION

Figure 6.59 Standards conversion impairments arising from inappropriate filtering in the spectral regions of Figure 6.58: (a) blurred movement arising from insufficient temporal response (Region A); (b) loss of vertical resolution arising from insufficient vertical response (Region B); (c) vertical aliasing resulting from too much vertical response (Region C); (d) vertical modulation of moving detail ('Crankshaft' effect)

caused by insufficient suppression of the interlaced spectra (Region F). The camera is moving across to follow the foreground subject. Therefore, the movement effects mainly affect the background. The low vertical resolution in (b) and the vertical aliasing in (c) are most apparent in the crowd

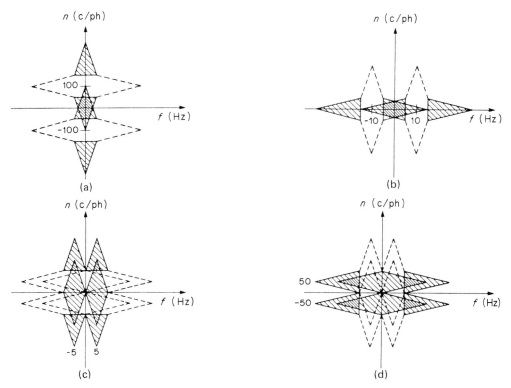

Figure 6.60 The effect of resampling the spectrum of Figure 6.58 using the 525/60 standard: (a) components from region C are shifted to be centred on (0, 100) causing 100 c/ph vertical aliasing (in addition, similar components are shifted from (0, −625) to (0, −100)), (b) region D components are shifted to (±10, 0) causing 10 Hz movement judder, (c) region E components are shifted to (±5, ±50) causing 5 Hz flicker, and (d) region F components are shifted to (±5, ±50) causing 50 c/ph modulation of moving edges

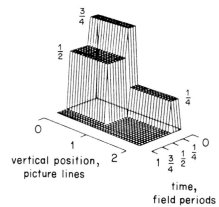

Figure 6.61 The two-dimensional aperture function corresponding to the interpolation method used in the BBC analogue converters. Only one quadrant is shown. The mode of operation of these converters was expressed in terms of an aperture function by J. O. Drewery

TELEVISION SCAN CONVERSIONS 339

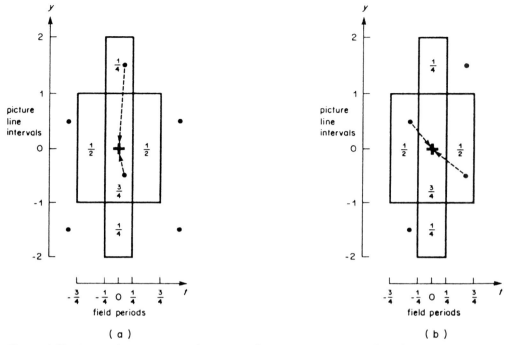

Figure 6.62 Interpolator action in the BBC analogue converters: (a) when the required output line position is within $\pm\frac{1}{4}$ field period of an input field position, two lines from that field are added with weights of $\frac{3}{4}$ for the nearest and $\frac{1}{4}$ for the next nearest; (b) when the required output line position is between two input fields, the two nearest lines are added, one from each field, with weights of $\frac{1}{2}$ (• input line, + output line position)

noted that, although each of these methods alone would be variables-separable, the combination of the two, shown in Figure 6.61, is non-variables-separable. The same interpolation method was used in both directions of conversion.

The frequency characteristic corresponding to this aperture function can be calculated by two-dimensional Fourier transformation. Figure 6.63a shows the characteristic for the 625/50 to 525/60 direction of conversion as a perspective view. However, the contour plot representation, Figure 6.63b, showing the magnitude of the characteristic, is more useful for assessing the suppression of alias components. The centres of the repeated spectra, marked with crosses in Figure 6.63b, are all well suppressed, but regions C, D and E of Figure 6.58 all contain significant components which cause vertical aliasing, movement judder and 5 Hz flicker, respectively. As there is no response along the $312\frac{1}{2}$ c/ph line, no aliasing corresponding to region F should be present. These forms of impairment agree well with the observed performance of this type of converter. The additional responses along the 50 Hz line lead to 10 Hz flicker on vertical detail. However, in pictures this would be difficult to distinguish from the higher amplitudes of 5 Hz flicker also produced on vertical detail by the region E components.

In marked contrast to this analogue conversion method, the interpolation of DICE uses digital storage and arithmetic. Because of this, the interpolation multipliers are much more sophisticated and give greater freedom for the design of aperture functions.

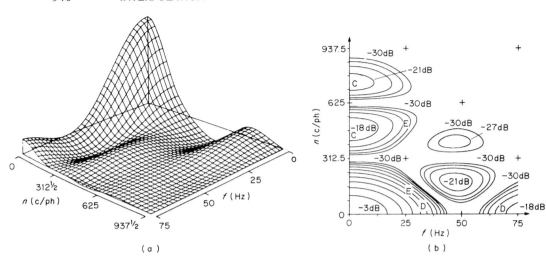

Figure 6.63 The vertical/temporal low-pass filter characteristic of the BBC analogue converters shown for 625/50 to 525/60 conversion: (a) as a perspective view and (b) as contours of equal attenuation; contours are spaced at 3 dB intervals down to −30 dB. The regions C, D, and E show the degree of attenuation provided for each of the main impairments marked in Figure 6.58

However, because it was designed at a time when digital storage was still expensive, the capacity of the main store is only sufficient for two composite interlaced fields on the 525-line NTSC colour standard. This restriction, combined with the use of separate vertical and temporal interpolation, necessitates a significantly different mode of operation in the two directions of conversion.

For 625/50 to 525/60 conversion the arrangement shown in Figure 6.64a is used. Vertical (line) interpolation using five field-lines is performed first so that the resulting 525/50 interlaced signals can be stored. Following each store output there is a second vertical interpolator using three field-lines which converts the stored 525/50 interlaced signals to the positions of 525/50 sequential scanning. The combined effect of these two interpolations is, therefore, similar to that shown in Figure 6.54a and 6.54b. This is followed by temporal (movement) interpolation using an approximation to a two-field linear aperture, resulting in spectra corresponding to those in Figure 6.54c and 6.54d.

For the reverse direction of conversion, the 525/60 interlaced signals can be stored directly in the field stores, as shown in Figure 6.64b. Again there are two vertical processes, but in this case these are placed one before and one after temporal interpolation. First, the three-line vertical interpolation produces what amounts to 625/60 sequential signals, in which the lines resulting from odd and even input fields are vertically coincident, but irregularly spaced. This allows temporal interpolation using an approximation to a two-field linear aperture function which produces 625/50 signals, although these still include the irregular line spacing. Finally, at the output, the regular, interlaced 625-line spacing is restored by the five-line vertical interpolator.

Specific details of the two vertical interpolation processes are not given [129, 130]. Even so, a reasonable indication of performance can be obtained by assuming a single sinc function approximation extending over five field-lines for the combined effect of the two processes. This leads to an optimistic view of performance because it amounts to

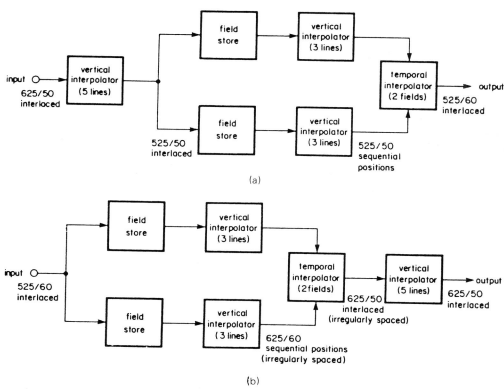

Figure 6.64 Block diagrams showing the sequence of interpolation processes in the IBAs DICE converter: (a) for 625/50 to 525/60 conversion and (b) for 525/60 to 625/50 conversion.

assuming that the three-line interpolator produces perfect results. Convolving this assumed form of vertical interpolation aperture with the quantised two-field linear aperture used for temporal interpolation (625/50 to 525/60 direction) produces the two-dimensional aperture function shown in Figure 6.65.

Fourier transformation of this aperture function produces the two-dimensional frequency characteristic shown in Figure 6.66. Particularly for vertical frequencies, this is a much more effective low-pass filter characteristic than that shown in Figure 6.63 for the BBC analogue converter. The contour plot representation, Figure 6.66b, shows that all the components from region C in Figure 6.58 are well suppressed, so that vertical aliasing is not a problem with this converter. In addition, there is greater suppression of the region D components at (75, 0), thus reducing movement judder, and less response in region E, reducing 5 Hz flicker. However, as expected for a converter with separate vertical and temporal interpolation, vertical resolution (region B) is very limited, with virtually no response beyond 156 c/ph.

Aperture synthesis

Although the preceding analysis can identify the amount and form of the impairments resulting from an interpolation method, the degree of suppression required to render each

342 INTERPOLATION

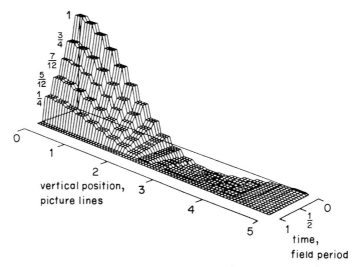

Figure 6.65 A two-dimensional interpolation aperture function approximating the interpolation methods used in the IBA's DICE converter for 625/50 to 525/60 conversion

impairment invisible can only be determined by experiment. Also when a balance has to be struck between impairments, this must take account of the visibility of each over a wide range of picture material. Because of this, a versatile experimental standards converter [127] was constructed to enable the actual performance of interpolation methods derived by frequency domain analysis to be assessed. This allowed observed impairments to be related to particular features of the vertical–temporal spectrum, so that undesirable effects could be eliminated, where possible, by appropriate alterations to the frequency characteristic.

With combined interpolation, the frequency characteristic can be set independently at fixed points spaced according to the aperture size, as shown in Figure 6.57a. However, these fixed point values correspond to a repeated version of the aperture function as previously described for the one-dimensional case. The frequency characteristic corresponding to the true aperture function is obtained by replacing each point value by a weighted two-dimensional sinc f, sinc n function and summing the individual contributions.

A frequency specification in which only one point has a non-zero value shows the form of the individual contributions. Setting (0, 0) to unity and zero values at all other specification points produces the frequency characteristic of Figure 6.67, shown both as a perspective view and a contour plot. This shows that neighbouring set points are not interdependent because the function has zero values at these positions. However, a change at one set point would affect the function value between other points at the same temporal or vertical frequency over a wide range of frequencies. The aperture function corresponding to this frequency characteristic consists of a block pulse covering an area of four field periods by four field-lines.

For interlaced scanning, it is necessary to suppress frequency components beyond a line joining (0, $312\frac{1}{2}$) and (25, 0), while minimising passband attenuation. The point

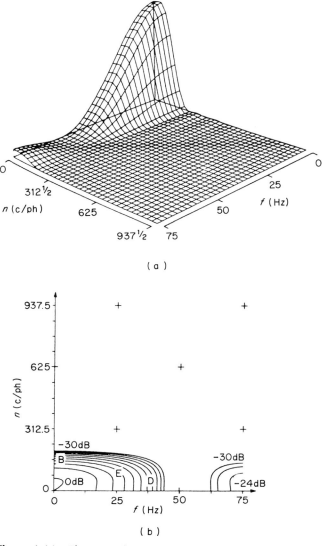

Figure 6.66 The vertical/temporal low-pass filter characteristic for the interpolation aperture function of Figure 6.65: (a) shown as a perspective view and (b) as a contour plot. The responses in regions B, D and E show the main impairments to be vertical resolution loss, movement judder and 5 Hz flicker, respectively

values specified in Table 6.2 appear consistent with this, but produce the frequency characteristic of Figure 6.68. The many overshoots in the characteristic result because the widely differing values at adjacent points in the specification make the passband to stopband transition too abrupt. The high level of stopband ripple makes the performance inadequate.

344 INTERPOLATION

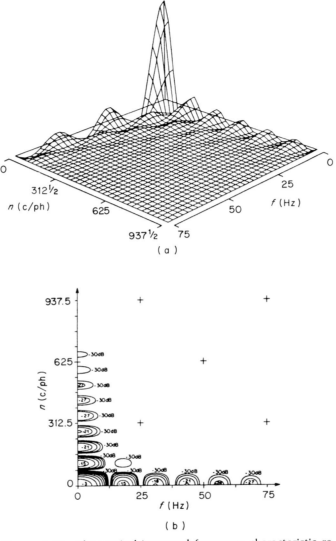

Figure 6.67 The vertical/temporal frequency characteristic resulting from a single non-zero fixed-point value for a combined two-dimensional interpolator with an aperture size of four fields by four field-lines (eight picture-lines): (a) perspective view and (b) contour plot

To obtain a smooth frequency characteristic, relatively free from overshoots, it is necessary to broaden the transition band. Even small encroachments into the stopband result in increased aliasing, so a more gradual cut-off has to be obtained primarily at the expense of passband performance. In addition, because the spacing of the temporal frequency specification points is wider, the cut-off in the temporal frequency direction will necessarily be more gradual than that for vertical frequencies. This, in turn, tends to

TELEVISION SCAN CONVERSIONS 345

Table 6.2

Vertical frequency (c/ph)	Temporal frequency (Hz)				
	0	12.5	25	37.5	5.0
$312\frac{1}{2}$	0	0	0	0	0
$234\frac{3}{8}$	1	0	0	0	0
$156\frac{1}{4}$	1	0	0	0	0
$78\frac{1}{8}$	1	1	0	0	0
0	1	1	0	0	0

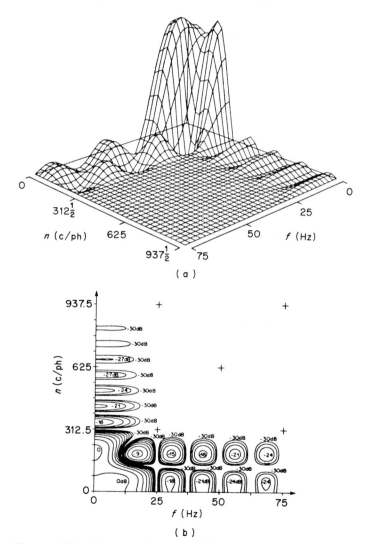

Figure 6.68 The vertical/temporal frequency characteristic resulting from the fixed-point values shown in Table 6.2: (a) perspective view and (b) contour plot

restrict the vertical resolution because, if large values were set at high vertical frequencies and the triangular shape were maintained, this would cause too abrupt a cut-off in the temporal frequency direction. Altering the frequency specification in accordance with these factors to the values of Table 6.3 produces the smooth, approximately triangular frequency characteristic shown in Figure 6.69.

Having derived an apparently satisfactory frequency characteristic, the corresponding two-dimensional aperture function can be calculated as a summation of $\cos t \cos y$ terms, weighted by the values in Table 6.3. Such a function has already been shown in its one-dimensional form. This periodic function is then truncated to four field periods and eight picture-line intervals (four field-line intervals) to produce the aperture function. The first quadrant of the aperture function resulting from the set values in Table 6.3 is shown in Figure 6.70.

For use in a digital converter, the aperture function shown in Figure 6.70 must be quantised in time, vertical position and amplitude. As the aperture function has four-quadrant symmetry, only one quadrant need be stored. In the experimental converter, the aperture function was stored using 8 values per field period and 16 values per picture line interval. For an aperture the size of Figure 6.70 therefore, a total of 1024 coefficients would be required to store one quadrant, consisting of 16 values in the temporal direction at each of 64 positions vertically.

Positional quantisation of the aperture function, in which N values are stored per sample period, repeats the baseband frequency characteristic at N times the sample rate. So, with eight values per field period and sixteen values per picture line interval, extra responses in the characteristic are centred on harmonics of 400 Hz (8 times 50 Hz) in the temporal direction and 10 000 c/ph (16 times 625 c/ph) in the vertical direction. From Table 6.1, the minimum attenuation of alias components from these regions is at least 24 dB temporally ($N = 8$) and 30 dB vertically ($N = 16$). The actual attenuation values will be greater than this because the frequency characteristic of the continuous aperture function is not flat to the half sampling frequencies. For example, the characteristic shown in Figure 6.69 has an attenuation of 14 dB at 25 Hz; so alias components due to aperture quantisation would be attenuated by a total of 38 dB in this case.

When only one quadrant is stored, only four of the sixteen coefficients (Figure 6.55) can be read directly. The remaining coefficients are found by reflecting the line positions in other quadrants temporally and vertically about the aperture centre so that they fall into the first quadrant. In practice, when reading the coefficients from a memory, this is

Table 6.3

Vertical frequency (c/ph)	Temporal frequency (Hz)				
	0	12.5	25	37.5	5.0
$312\frac{1}{2}$	0	0	0	0	0
$234\frac{3}{8}$	0.25	0.05	0	0	0
$156\frac{1}{4}$	0.75	0.35	0	0	0
$78\frac{1}{8}$	1.0	0.7	0.1	0	0
0	1.0	0.8	0.2	0	0

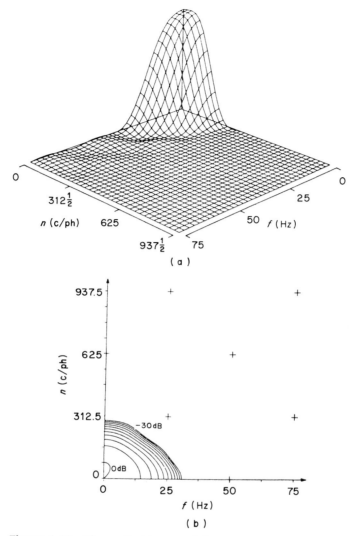

Figure 6.69 The vertical/temporal frequency characteristic resulting from the fixed-point values shown in Table 6.3, chosen to obtain a smooth response: (a) perspective view and (b) contour plot

accomplished by complementing the address bits of the temporal component or of the vertical component, or of both together.

Each of the 1024 aperture values selected for storage must also be quantised in amplitude. As aperture functions generally include some negative values, an 8-bit representation consisting of seven amplitude bits and a sign bit is convenient. For all useful aperture functions the coefficient values are less than unity, so that the amplitude of each can be expressed as a whole number of 128ths.

348   INTERPOLATION

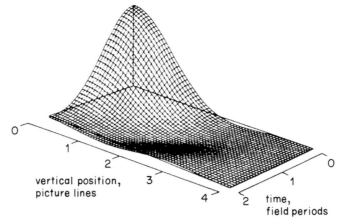

Figure 6.70 One quadrant of the two-dimensional aperture function corresponding to the fixed values of Table 6.3

As shown in Figure 6.55, the coefficients are used in sets of 16, of which there are 64 different sets. Before amplitude quantisation, each set of coefficients sums to unity. However, when the coefficients are rounded individually to the nearest level, the sum may no longer be unity. This could cause brightness variations when changing from one set of coefficients to another. To avoid this the coefficients must be quantised in sets by a method which ensures that each set maintains its unity sum, even though this may increase the quantisation error of individual coefficients.

A suitable method is illustrated in Table 6.4. The first column shows, before

Table 6.4

Unquantised values	Quantised individually	Rounded down	Quantisation error	Quantised as a set
93.990	94	93	0.990*	94
1.364	1	1	0.364	1
0.153	0	0	0.153	0
0.000	0	0	0.000	0
14.949	15	14	0.949*	15
12.708	13	12	0.708*	13
−1.761	−2	−2	0.239	−2
−1.431	−1	−2	0.569*	−1
9.356	9	9	0.356	9
7.516	8	7	0.516*	8
−1.373	−1	−2	0.627*	−1
−1.073	−1	−2	0.927*	−1
−1.498	−1	−2	0.502	−2
−2.707	−3	−3	0.293	−3
−2.198	−2	−3	0.802*	−2
0.005	0	0	0.005	0
128.000	129	120	8.000	128

quantisation, sixteen examples of aperture values to be applied as a set to sixteen input lines. The values shown have been multiplied by 128 so that after quantisation an integer value will remain; this indicates the coefficient value as a number of 128ths. The values in the first column, therefore, sum to 128. Rounding each coefficient to the nearest quantised value produces the results in column 2 which sum to 129. Using this set of coefficients would produce a d.c. gain slightly greater than unity (other sets might produce gains both greater and less than this). Therefore, as a first step in the quantisation procedure all the coefficient values are rounded down as shown in column 3. The sum of these values is less than or equal to 128, in this case, 120. The errors resulting from rounding down are shown in column 4. To obtain a sum of 128 rather than 120, eight values must be rounded up instead of down. Selecting the eight coefficients with the largest errors from rounding down (marked with asterisks in Table 6.4) minimises the errors overall and produces the quantised set of coefficients shown in column 5.

The frequency characteristic corresponding to the quantised aperture values, derived as described above, can be calculated by summing the Fourier transforms of individual block pulse elements (this was previously described for a one-dimensional aperture function). The transform of the two-dimensional elements with amplitude A shown in Figure 6.71 is given by

$$G(f, n) = 4AT \cos 2\pi t_0 f \cdot \operatorname{sinc} Tf \cdot Y \cos 2\pi y_0 n \cdot \operatorname{sinc} Yn$$

The accuracy of quantisation used in the experimental converter is such that the frequency characteristic of the quantised function is not discernibly different from that shown in Figure 6.69.

Use of the experimental converter to assess the performance of interpolation methods allowed the aperture functions to be further optimised. In particular, the main types of impairment described could be altered by adjustments to the corresponding fixed point values in the frequency specification. The results of further optimisation are described next.

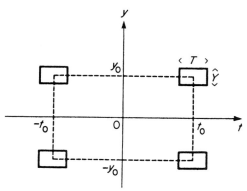

Figure 6.71 Symmetrical pairs of block pulses representing a generalised element of a quantised two-dimensional aperture function. Each block has an amplitude A

Optimised conversion apertures

Although the aperture synthesised in the previous section provides the main features required to produce an acceptable frequency characteristic, it can be improved to give better suppression of unwanted scanning products in the two directions of conversion. Also, the procedure presented there does not take account of the need to suppress luminance and chrominance cross-effects and noise. So, the latter part of this section includes these factors to derive four-field aperture functions suitable for the two directions of conversion and using separate luminance and chrominance interpolation.

However, even with two-field apertures, finer aperture quantisation and combined vertical and temporal interpolation can provide some improvement over the conversion methods already described. If this improvement were sufficient, then the extra complexity of a four-field converter would be unnecessary. Furthermore, narrower apertures are of interest to provide fall-back operation in the event of a major failure in part of a four-field interpolator. Therefore, the first part of this section assesses the performance that can be obtained with two- and three-field apertures.

With contributions from four lines on each field there is no difficulty in obtaining good suppression of vertical aliasing (region C in Figure 6.58). However, with only a two-field aperture it is necessary to compromise in temporal performance between the suppression of judder (region D) and the suppression of 5 Hz flicker (region E). The response at 25 Hz can be set to zero to suppress flicker, but this produces a characteristic of sinc f shape with large responses between 25 and 50 Hz and beyond 50 Hz. The presence of these responses results in serious movement judder. Alternatively, better suppression of judder can be obtained by using a more gradual temporal characteristic with a broad zero at 50 Hz; but then the 5 Hz flicker components remain in region E. Although flicker components, in general, occur more frequently, judder is a more damaging impairment, particularly at some rates of movement.

Therefore, it is necessary to optimise the temporal characteristic for the best suppression of judder by providing a broad zero at 50 Hz. Flicker can then be reduced by cutting the vertical response along the 25 Hz line.

With separable apertures (which allow vertical and temporal interpolation to be made separate processes) this can only be achieved by cutting all vertical frequencies, including those for still pictures. This results in a frequency characteristic such as that shown in Figure 6.72, obtained from the fixed values shown in Table 6.5.

This characteristic is broadly similar to that of the DICE converter, Figure 6.66, although it includes some small improvements. Finer quantisation of the temporal aperture values has eliminated the response at 75 Hz and reduced the response in the 25 to 50 Hz region, thus reducing judder. The reduced response at 25 Hz also reduces 5 Hz flicker. The slight reduction in vertical resolution compared to Figure 6.56 further suppresses 5 Hz flicker, but at the expense of a greater loss of vertical definition.

When vertical and temporal interpolation are combined into one process, the aperture can be made non-separable. Then, for stationary pictures, the vertical resolution obtained with a two-field aperture can be increased, while further reducing the 5 Hz flicker components at 25 Hz. Figure 6.73 shows a non-separable frequency characteristic with these features, produced from the fixed point values shown in Table 6.6. Unfortunately, the inevitable consequence of increasing the vertical resolution and cutting the temporal response more abruptly is the introduction of overshoots in the characteristic, which

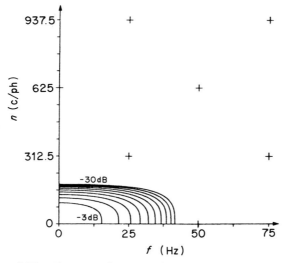

Figure 6.72 The vertical/temporal frequency characteristic of the separable two-field interpolation aperture defined by the fixed-point values of Table 6.5. Contours are shown at 3 dB intervals down to −30 dB

Table 6.5

Vertical frequency (c/ph)	Temporal frequency (Hz)		
	0	25	50
$312\tfrac{1}{2}$	0	0	0
$234\tfrac{3}{8}$	0	0	0
$156\tfrac{1}{4}$	0.18	0.0684	0
$78\tfrac{1}{8}$	0.85	0.3230	0
0	1	0.38	0

appear in Figure 6.73, centred on $(37\tfrac{1}{2}, 156\tfrac{1}{4})$ and $(62\tfrac{1}{2}, 156\tfrac{1}{4})$. Although on still pictures the vertical resolution is noticeably greater and the amount of flicker less than for Figure 6.66, the movement performance is poorer, being affected by moving high frequency detail from the interlace spectrum centred on $(25, 312\tfrac{1}{2})$. With two-field apertures, therefore, it is not possible to obtain a good approximation to the desired triangular frequency characteristic. However, the characteristic of Figure 6.72 could be used as a fall-back mode in the event of a component failure in part of a larger converter using more than two fields of storage.

Broadening the interpolation aperture to three field periods allows the frequency characteristic to be defined at $16\tfrac{2}{3}$ Hz intervals. Although a faster rate of cut can be achieved, optimising the design is more difficult because the response values at $16\tfrac{2}{3}$ and $33\tfrac{1}{3}$ Hz must be set to obtain the response required in the 25 Hz region. Also, the logic implementation of a three-field converter is generally less straightforward than that for two- or four-field converters, which benefit from being powers of two.

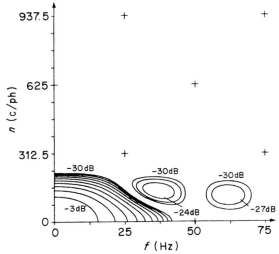

Figure 6.73 The vertical/temporal frequency characteristic of the non-separable two-field interpolation aperture defined by the fixed points of Table 6.6. The attempt to increase vertical resolution and to reduce 5 Hz flicker has caused significant overshoots in the temporal direction

Table 6.6

Vertical frequency (c/ph)	Temporal frequency (Hz)		
	0	25	50
$312\frac{1}{2}$	0	0	0
$234\frac{3}{8}$	0.01	0	0
$156\frac{1}{4}$	0.4	0.01	0
$78\frac{1}{8}$	0.9	0.22	0
0	1	0.38	0

With a three-field converter, making use of the faster rate of cut in the temporal direction produces the characteristic of Figure 6.74, calculated from the fixed values shown in Table 6.7. Whilst maintaining the same vertical resolution as Figure 6.73 for still pictures, the three-field characteristic substantially reduces judder and virtually eliminates flicker without producing overshoots in the temporal direction. However, rapid movement is blurred because of insufficient response in the $(12\frac{1}{2}, 0)$ region. Extending the characteristic along the temporal axis to reduce blurring could provide a practicable three-field aperture, but there are several clear advantages to be obtained by using four fields instead.

Extending the aperture to four field periods allows a faster rate of cut for temporal frequencies. Thus, the characteristic can be well attenuated both at and beyond 25 Hz. With a separable aperture function, this makes the temporal performance more acceptable by reducing judder and flicker with less blurring; but, because the shape of the frequency characteristic is basically rectangular, the vertical resolution cannot be

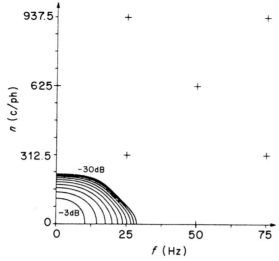

Figure 6.74 The vertical/temporal frequency characteristic of the non-separable three-field interpolation aperture defined by the fixed points of Table 6.7. With the three-field aperture it is possible to retain the improved vertical response of Figure 6.73 without introducing temporal overshoots and to reduce movement judder and 5 Hz flicker

Table 6.7

Vertical frequency (c/ph)	Temporal frequency (Hz)			
	0	$16\frac{2}{3}$	$33\frac{1}{3}$	50
$312\frac{1}{2}$	0	0	0	0
$234\frac{3}{8}$	0.01	0	0	0
$156\frac{1}{4}$	0.4	0.08	0	0
$78\frac{1}{8}$	0.9	0.3	0	0
0	1	0.4	0.003	0

increased. However, with the non-separable case, four-field apertures allow much greater freedom for setting values in the frequency characteristic, so that a closer approximation to the required triangular shape can be obtained. Also, with the finer grid of points, shown in Figure 6.57a, it is possible to develop different characteristics for luminance in the two directions of conversion and to provide a separate aperture for chrominance interpolation.

When converting from 625 to 525 lines, the output can accommodate less vertical resolution than the input standard. Therefore, in order to avoid interlace flicker on the output picture, it is necessary to suppress frequencies of $262\frac{1}{2}$ c/ph and above. Frequencies below this limit should be retained as far as possible so that reasonable picture sharpness is maintained.

354 INTERPOLATION

In contrast, with a change in field rate from 50 to 60 Hz, the output standard has the capacity for higher temporal frequencies. Even so, frequencies of 25 Hz and above must be well attenuated to avoid 5 Hz flicker and judder. These requirements combine to produce the frequency characteristic of Figure 6.75, obtained from the fixed values of Table 6.8.

Compared with the three-field characteristic, Figure 6.74, this four-field aperture retains slightly more resolution along the vertical axis and cuts more sharply to suppress frequencies above $262\frac{1}{2}$ c/ph. This is only possible because the temporal cut-off can be more abrupt, so that more vertical resolution is provided for stationary pictures, while the alias components resulting from moving high vertical frequency components are still suppressed. In addition, the on-axis temporal characteristic has a slightly sharper cut and the triangular shape is more pronounced, resulting in a characteristic much closer to the ideal than could be obtained with three fields.

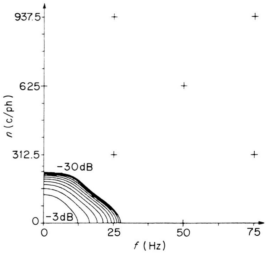

Figure 6.75 The vertical/temporal frequency characteristic of the four-field interpolation aperture for 625/50 to 525/60 conversion defined by the fixed points in Table 6.8. This four-field aperture provides a close approximation to the ideal triangular shape of Figure 6.45. Although judder is well suppressed, rapid movement is too blurred for general use

Table 6.8

Vertical frequency (c/ph)	Temporal frequency (Hz)				
	0	12.5	25	37.5	50
$312\frac{1}{2}$	0	0	0	0	0
$234\frac{3}{8}$	0.02	0	0	0	0
$156\frac{1}{4}$	0.5	0.125	0	0	0
$78\frac{1}{8}$	0.95	0.49	0.01	0	0
0	1	0.7	0.1	0	0

In practice, the performance on still and slowly moving pictures is near ideal, being free from flicker and other impairments, and providing good spatial resolution. However, rapid movement is blurred due to inadequate response in the $12\frac{1}{2}$ to 25 Hz region. The blurring can be reduced by extending the characteristic to give more response along the temporal frequency axis. Figure 6.76 shows a characteristic with the same vertical response but with an extended temporal response, obtained from the fixed point values shown in Table 6.9. This characteristic greatly reduces blurring on rapid movement, but leaves a much higher level of judder, which occurs at relatively slow rates of movement from high spatial frequencies. Components along the 25 Hz line, which cause 5 Hz flicker, remain at an acceptably low level.

It is necessary, therefore, to distinguish between two distinct types of movement. First, there are the rapid, short duration movements which occur within a scene, such as the

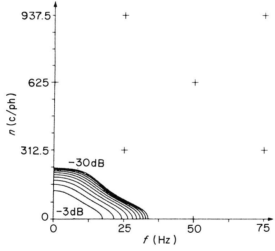

Figure 6.76 The vertical/temporal frequency characteristic of the four-field aperture for 625/50 to 525/60 conversion defined by the fixed-point values in Table 6.9. The extended temporal response avoids blurring but leaves a higher level of judder than occurs with the characteristic of Figure 6.74. This extended characteristic represents the best overall compromise for suppressing judder and avoiding blurring

Table 6.9

Vertical frequency (c/ph)	Temporal frequency (Hz)				
	0	12.5	25	37.5	50
$312\frac{1}{2}$	0	0	0	0	0
$234\frac{3}{8}$	0.02	0	0	0	0
$156\frac{1}{4}$	0.5	0.15	0	0	0
$78\frac{1}{8}$	0.95	0.6	0.07	0	0
0	1	0.9	0.32	0	0

arm and leg movements of a dancer. This type of movement predominates in studio pictures. Because each movement is of short duration, there is no opportunity for the viewer to track the movement, so that judder, if present, is not noticeable. However, blurring of these rapid movements is noticeable and can cause objectionable impairments if the temporal characteristic cuts at too low a frequency. Therefore, for studio pictures, the characteristic of Figure 6.76 is best.

The second type of movement is relatively slow, steady and of long duration, such as occurs in the background when a camera is panning to follow a moving foreground object. This camera-induced movement frequently occurs during outside broadcasts, particularly at sporting events. As the original movement is steady and of long duration, the viewer is able to recognise and follow the natural rate of movement. Then, the unnatural movement judder perturbations, due to aliasing, become more obvious. With this type of movement, although blurring of the background is noticeable, it is a much more acceptable impairment. Therefore, for outside broadcasts, the frequency characteristic of Figure 6.75 offers some advantages.

This inability to provide satisfactory performance under all conditions of movement is a direct result of the inadequacy of the pre-sampling low-pass filter (camera integration) and undersampling (using too low a field rate). Because of these factors, the wanted movement and unwanted alias components in the television signal share the same spectrum and cannot be distinguished. Therefore, blurring of true movement is a necessary consequence of suppressing judder and, if true movement is not blurred, judder will also be present. Choosing a characteristic between those of Figure 6.75 and 6.76 would result in neither type of movement being particularly well portrayed. The use of a more abrupt temporal cut-off, which would require more field stores, could provide better retention of the main wanted components and better suppression of the main unwanted components. However, as the true and alias components appear to overlap extensively, it is possible that a sharper filter might give no improvement at all. An alternative approach could be to use the most suitable aperture (corresponding to either Figure 6.25 or Figure 6.26) for each picture, under the control of a panning detection signal. While this might lead to some improvement, the abrupt change of mode itself might be found disturbing. In addition, the inclusion of such a system could result in a substantial increase in interpolator complexity.

Without this refinement, the characteristic of Figure 6.76 is generally preferred, particularly by untrained observers i.e. those used to normal picture impairments, but unfamiliar with impairments specific to standards conversion. This is because, although judder is usually a more objectionable impairment than blurring, it can go unnoticed by a viewer whose attention is following a foreground object. However, viewers whose attention is caught by excessive background judder are subsequently more likely to notice judder at lower levels. This learning process suggests that viewers will gradually find judder less acceptable.

With regard to cross-luminance, the two characteristics, especially that of Figure 6.75, give a useful degree of suppression of the main PAL subcarrier components, centred on $(6\frac{1}{4}, 234\frac{3}{8})$ and $(18\frac{3}{4}, 78\frac{1}{8})$. Nevertheless, the main consideration in selecting a luminance frequency characteristic is the suppression of scanning products and this precludes additional optimisation of cross-luminance suppression.

While many of the principles discussed above also apply for 525/60 to 625/50 conversion, the spacings of the frequency specification points change to 15 Hz for temporal frequency and $65\frac{5}{8}$ c/ph, for vertical frequency. Because of this, the temporal characteristic has to cut more rapidly to take account of the lower temporal resolution of the output standard. Also, it would be advantageous to extend the response along the vertical frequency axis to take advantage of the higher vertical resolution of the output standard. However, because of the limited rate of cut in the temporal direction, only minor changes are possible without either losing the triangular shape of the characteristic or introducing ringing in the temporal direction. Incorporating these considerations produces the characteristic of Figure 6.77 calculated from the fixed point values shown in Table 6.10.

Although the vertical resolution is rather limited, this characteristic gives output pictures free from 5 Hz flicker and vertical aliasing. As with Figure 6.75, movement is free of judder, but the more rapid movements are blurred. Again, blurring can be avoided, at the cost of reintroducing judder, by extending the temporal frequency

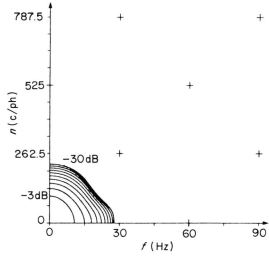

Figure 6.77 The frequency characteristic of a four-field aperture for 525/60 to 625/50 conversion defined by the fixed points in Table 6.10. As with Figure 6.75, rapid movement is too blurred for general use

Table 6.10

Vertical frequency (c/ph)	Temporal frequency (Hz)				
	0	15	30	45	60
$262\frac{1}{2}$	0	0	0	0	0
$196\frac{7}{8}$	0.1	0	0	0	0
$131\frac{1}{4}$	0.5	0.11	0	0	0
$65\frac{5}{8}$	0.9	0.35	0	0	0
0	1	0.5	0	0	0

358   INTERPOLATION

characteristic as shown in Figure 6.78, which corresponds to the fixed values of Table 6.11. As before, this extended characteristic is to be preferred for general use, although the characteristic of Figure 6.77 produces more acceptable pictures during camera panning.

The NTSC subcarrier, located at $(15, -131\frac{1}{4})$, is well attenuated, particularly with the characteristic of Figure 6.77. However, horizontal colour transitions, which produce high vertical modulating frequencies, result in components falling near the temporal axis and these receive much less attenuation. After comb filtering, much of the residual subcarrier energy falls in this region. Furthermore, for moving vertical chrominance detail, the attenuation can fall to zero and, under these circumstances, the cross-luminance pattern of the NTSC signal becomes visible on the 625/50 output standard pictures.

With chrominance signals, the retention of vertical and temporal resolution is much less important than for luminance signals. Therefore, it is advantageous to sacrifice

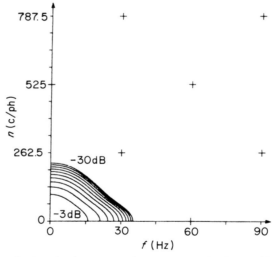

Figure 6.78 The frequency characteristic of a four-field aperture for 525/60 to 625/50 conversion defined by the fixed points in Table 6.11. As with Figure 6.76, blurring is avoided, but there is more judder than with Figure 6.77

Table 6.11

Vertical frequency (c/ph)	Temporal frequency (Hz)				
	0	15	30	45	60
$262\frac{1}{2}$	0	0	0	0	0
$196\frac{7}{8}$	0.1	0	0	0	0
$131\frac{1}{4}$	0.5	0.15	0	0	0
$65\frac{5}{8}$	0.9	0.5	0.02	0	0
0	1	0.75	0.15	0	0

resolution in order to suppress cross-colour, Hanover bars and noise. Because high temporal frequencies are virtually absent from the colour-difference signals (the result of their low spatial resolution), a very rapid temporal cut-off can be used without noticeably blurring the signals. Also, in the vertical direction, a cut-off similar to that given by a delay line decoder results in a barely perceptible loss of resolution. Placing a zero at $(12\frac{1}{2}, 156\frac{1}{4})$ ensures complete suppression of Hanover bars in areas of constant colour. This produces the characteristic of Figure 6.79, which is obtained from the fixed values shown in Table 6.12.

As well as suppressing Hanover bars, this characteristic produces pictures with substantially reduced cross-colour and chrominance noise, while introducing virtually no impairment to the wanted chrominance components. However, an alternative characteristic, Figure 6.80, calculated from the values of Table 6.13, produces greater

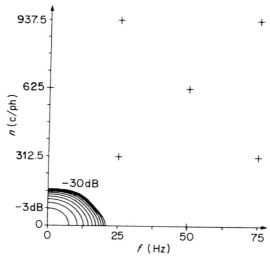

Figure 6.79 The frequency characteristic of a four-field aperture for chrominance interpolation defined by the fixed-point values in Table 6.12. The temporal response is rapidly curtailed, whilst the vertical response gives negligible softening of coloured vertical detail

Table 6.12

Vertical frequency (c/ph)	Temporal frequency (Hz)				
	0	12.5	25	37.5	50
$312\frac{1}{2}$	0	0	0	0	0
$234\frac{3}{8}$	0	0	0	0	0
$156\frac{1}{4}$	0.075	0	0	0	0
$78\frac{1}{8}$	0.7	0.2	0	0	0
0	1	0.38	0	0	0

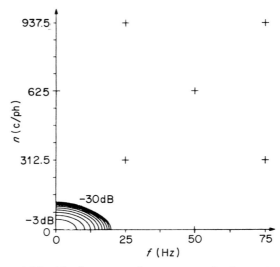

Figure 6.80 The frequency characteristic of a four-field aperture for chrominance interpolation defined by the fixed-point values in Table 6.13. Compared with Figure 6.79, the vertical response has been further reduced to provide increased suppression of chrominance noise and cross-colour

Table 6.13

Vertical frequency (c/ph)	Temporal frequency (Hz)				
	0	12.5	25	37.5	50
$312\frac{1}{2}$	0	0	0	0	0
$234\frac{3}{8}$	0	0	0	0	0
$156\frac{1}{4}$	0	0	0	0	0
$78\frac{1}{8}$	0.35	0.07	0	0	0
0	1	0.35	0	0	0

suppression of cross-colour and noise by further sacrificing vertical resolution. Although noticeable, this loss of resolution is not serious, while the additional suppression of noise has a beneficial effect on very noisy signals from poor quality sources.

The same aperture functions can also be used for 525/60 to 625/50 chrominance conversion and then scaled versions of the frequency characteristics are produced. Again, the resulting converted pictures contain substantially less cross-colour than unconverted pictures and chrominance noise components are greatly reduced.

With the experimental converter, it was possible to demonstrate the action of an interpolation method directly in the frequency domain using a test pattern generator [90]. Figure 6.81(a) shows a two-dimensional sweep pattern produced by the generator in which vertical frequencies increase in the vertical direction and temporal frequencies increase in the horizontal direction from a central origin. The centre of the screen therefore corresponds to (0, 0) and the corners to (± 25, $\pm 312\frac{1}{2}$) in the n–f spectrum.

TELEVISION SCAN CONVERSIONS 361

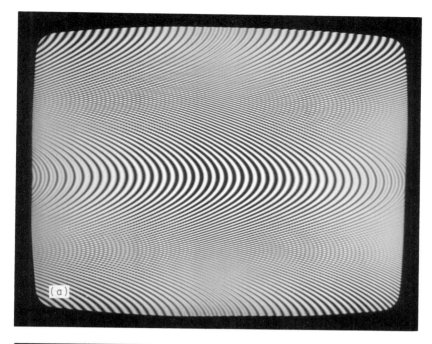

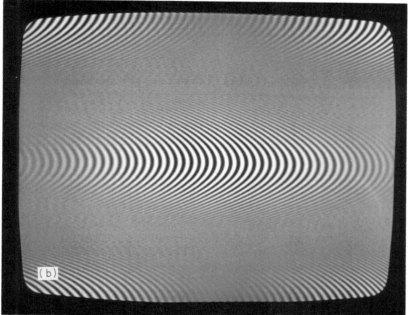

Figure 6.81 Display of the vertical/temporal frequency response of an interpolation method. (a) An electronically generated test pattern producing a sweep of vertical frequencies vertically and temporal frequencies horizontally on the 625/50 standard. The centre of the screen corresponds to (0, 0) and the corners to (± 25, $\pm 312\frac{1}{2}$). (b) The same signal converted to the 525/60 standard using an aperture function suitable for chrominance interpolation (frequency characteristic similar to that shown in Figure 6.80) in the experimental standards converter

362 INTERPOLATION

The repeated patterns at the corners of the screen are the normal result of interlaced scanning. Figure 6.81b shows the result of converting this signal using a low resolution interpolation method similar to that shown in Figure 6.80. The filtering effect of the conversion process, particularly for vertical frequencies, is readily apparent.

The results were applied to the construction of two standards converters for the BBC Television service produced by the BBC Engineering Designs Department. The equipment known as ACE—Advanced Conversion Equipment—was later made under licence by GEC–McMichael (see Figure 6.82) using the design data from the BBC prototype.

HDTV conversions

Much of what has been said in the context of conversion between normal definition broadcast standards applies equally to HDTV, but with HDTV conversions there are some additional factors to consider. Most important of these is that in terms of basic quality there may be little to choose between systems with broadly similar numbers of lines and increased field rates. In view of this, therefore, the choice of an HDTV system could reasonably be made by considering the ease and effectiveness of the conversion processes to and from existing broadcast standards.

The following paragraphs present the problems involved in conversions, primarily from possible high definition standards down to the 625/50 and 525/60 interlaced standards currently used for broadcasting; problems concerned with up-conversion are generally similar. First it is established that the number of lines used in a high definition system has little effect on the conversion process, but that the field rate is much more important. Then the use of current broadcast field rates for HDTV is considered. This is extended to the use of field rates in simple relationships to those of the broadcast standards, comparing the performance of the easy conversions to the related standards with those of the more complicated conversions to the unrelated standards. A possible compromise field rate, not simply related to either 50 or 60 Hz, is also considered. It is concluded that the adequacy of the conversion processes may determine whether a single high definition standard could be used satisfactorily in both 625/50 and 525/60 countries.

From the arguments already presented, the spectrum of an image scanned with more lines would take the form of Figure 6.83; the scanning standard assumed here is a 1251 line, 50 field per second interlaced system. The low-pass filtering required to make the conversion down to a 625/50 interlaced scan should remove the vertical repeated spectra centred on (0, 1251), (50, 1251), ..., and the interlaced spectra centred on (25, $625\frac{1}{2}$), (75, $625\frac{1}{2}$). Also, because the 625/50 output standard has lower resolution, the filter should attenuate high vertical frequency components of the (0, 0) and (50, 0) spectra. This can be approached in two ways.

The simpler method consists of using a purely vertical filter, thus needing only line delays in the interpolator. The effect of such an interpolator is shown in Figure 6.84. As the filter is only one-dimensional in its action, the filter cuts at a specific vertical frequency. The filter is, therefore, unable to suppress interlaced spectra without seriously affecting the vertical resolution of the wanted spectra as shown in Figure 6.84(a). When rescanned on the lower standard, Figure 6.84(b), the true vertical detail is largely lost and replaced by spurious high frequency detail. Such one-dimensional interpolators have

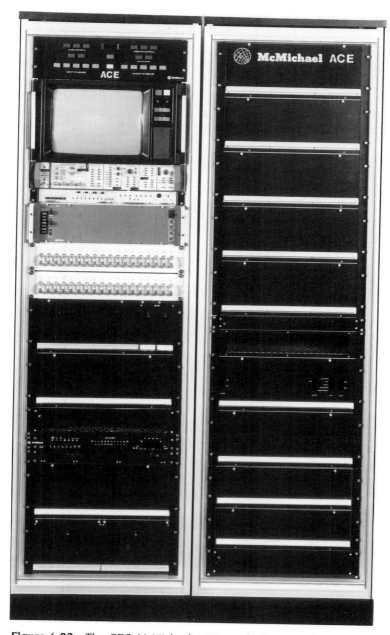

Figure 6.82 The GEC–McMichael ACE standards converter produced under licence from the BBC

always been used for line-rate conversions, for example, when converting between the 625/50 and 405/50 interlaced standards.

A preferable alternative method consists of using a two-dimensional interpolator using contributions from other lines and fields. The filtering effect can then be used to retain

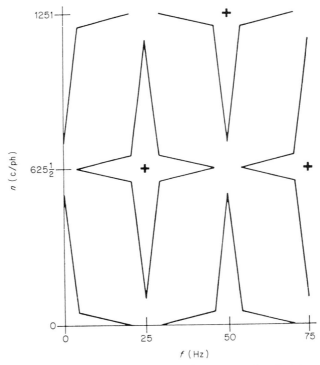

Figure 6.83 The vertical/temporal spectrum of a television signal with 1251 lines per picture, 50 fields per second, interlaced scanning

more of the vertical detail from the wanted (0, 0) and (50, 0) spectra whilst giving greater suppression of the unwanted (25, $625\frac{1}{2}$) and (75, $625\frac{1}{2}$) spectra, as shown in Figure 6.85a. When resampled as in Figure 6.85b, the spectra are then much more similar to those of Figure 6.41. Again the exact details of the response are a compromise, with the degree of attenuation in the (25, 0) region balancing the suppression of unwanted flicker on vertical detail against the avoidance of blurred movement. Similarly if too much vertical detail is left in the (0, $312\frac{1}{2}$) region, this will cause normal interlace flicker on the output standard.

It should be noted that the abrupt cut-off indicated by the dashed line in Figures 6.84 and 6.85 is purely illustrative and that a more gradual filter characteristic is to be preferred.

If the field rate is to be changed as well, as in a conversion from 1251/50 to 525/60, then the two-dimensional interpolator should also be used to suppress the (50, 0) spectrum in Figure 6.85a. However, the presence of overlap between the (0, 0) and (50, 0) spectra will cause impairments to movement. These will be substantially the same as those occurring with current 625/50 to 525/60 conversions, in which the high temporal frequency components of a moving edge move at a different rate from that of the low frequency components.

TELEVISION SCAN CONVERSIONS

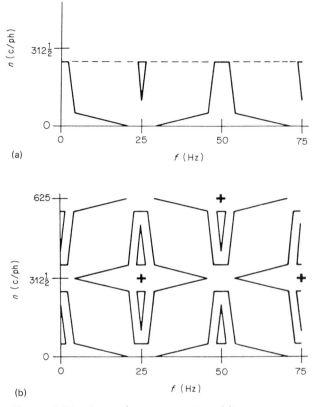

Figure 6.84 Spectral representation of line-rate conversion using a one-dimensional interpolator: (a) the 1251/50/ 2:1 spectrum of Figure 6.83 filtered to remove the main vertical and interlaced components and (b) the spectrum of the converted 625/50/2:1 output signal

A further point to be considered for HDTV systems is that there may be much more horizontal definition in the sources. Consequently, when an object moves, it will produce proportionately more high temporal frequency components, so making a low field rate less adequate for the basic quality of the standard and complicating the conversion to other standards.

Although described for an HDTV system with a 50 Hz field rate, the results would be substantially the same if a 60 Hz HDTV system were to be used; that is, improved conversion to existing 525/60 could be obtained by using a two-dimensional interpolator and conversion to 625/50 would suffer similar impairments to those incurred with current 525/60 to 625/50 conversions.

It is concluded, therefore, with conventional interpolation, that if an HDTV standard was based on either 50 or 60 Hz, then the quality of normal pictures in countries at present using the other field rate for their broadcast services would be impaired to about the same degree as that which occurs with current field standards conversions.

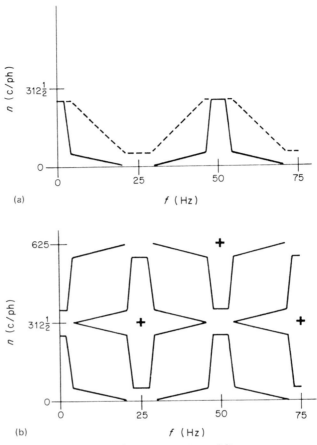

Figure 6.85 Spectral representation of line-rate conversion using a two-dimensional interpolator: (a) the 1251/50/2:1 spectrum of Figure 6.83 filtered to improve suppression of the interlace components resulting from high vertical frequencies in the image and (b) the spectrum of the converted 625/50/2:1 output signal

The problems of conversion are reduced if the HDTV system has a significantly higher field rate than that of current broadcast standards. This would also produce an HDTV system with improved basic quality, particularly in the respects of large area flicker, interlace flicker and movement portrayal. The spectrum produced by 1251/100 interlaced scanning is shown in Figure 6.86.

Conversion from this system to the 625/50 interlaced standard does not require the complicated two-dimensional interpolation needed with 50 and 60 Hz systems. Instead, the conversion can be made using two one-dimensional processes. First, the unwanted temporal frequencies can be removed, Figure 6.87a, and the signals resampled to produce a 1251/50 sequential signal, Figure 6.87b. As the field rates are simply related, a periodic filter characteristic can be used because the 100 Hz components are already

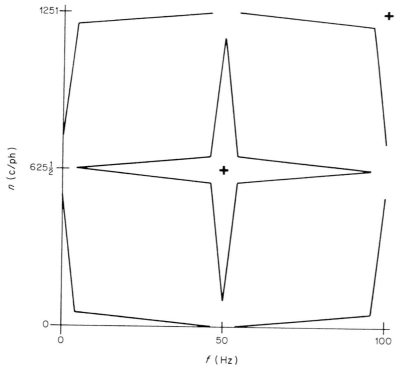

Figure 6.86 The vertical/temporal spectrum of a television signal with 1251 lines per picture, 100 fields per second, interlaced scanning

correctly positioned. Any residues of the interlace components from (50, $625\frac{1}{2}$) are converted to a much less noticeable static pattern with the 50 Hz field rate. The second interpolation process removes the unwanted vertical frequencies, Figure 6.87c, and the required interlaced structure is formed in Figure 6.87d. With such a process, the resulting 625/50 signals may well be indistinguishable from directly generated 625/50 interlaced signals.

Similar processes to those in Figure 6.87 can be used for conversion to 525/60. Although the field frequencies cannot be locked together in this case, the 100 Hz harmonics can be well suppressed without losing wanted components in the range 0 to 30 Hz. Also, the overlap of temporal components in the spectrum of 100 Hz camera signals is less because camera integration has a greater effect on alias components in the 0 to 30 Hz range. A further advantage is that the frequency offset of alias components is greater than for 50 to 60 Hz conversion (either 40 Hz between the 60 and 100 Hz spectra or 20 Hz between the 120 and 100 Hz spectra) so that the impairments are less visible.

Alternatively, an HDTV system with a 120 Hz field rate could be used with substantially equivalent results. In this case, the conversion to 525/60 would be simplified and could be expected to provide pictures indistinguishable from directly originated 525/60. On the other hand, 625/50 interlaced signals would include some movement impairments, but these would be substantially less than those incurred with current 525/60 to 625/50 conversions.

368   INTERPOLATION

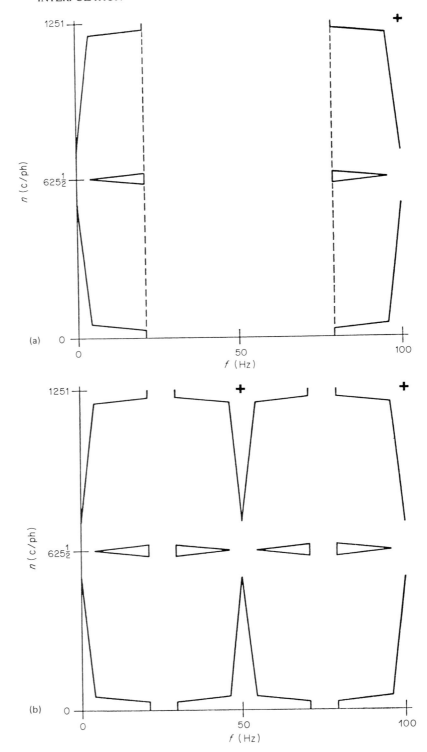

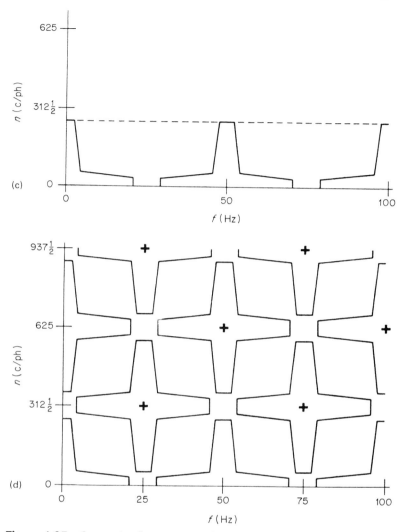

Figure 6.87 Conversion between standards with simply related field rates by consecutive one-dimensional interpolation processes: (a) the 1251/100/2:1 spectrum of Figure 6.86 filtered to remove the main temporal and interlaced components, (b) the spectrum of the intermediate 1251/50/1:1 standard, (c) the 1251/50/1:1 spectrum filtered to remove the unwanted vertical components and (d) the spectrum of the converted 625/50/2:1 output signal

It is likely, therefore, that HDTV systems using either 100 or 120 Hz field rates would allow satisfactory conversions to current broadcast standards. However, the higher field rates represent a considerable increase in data rate, so that of the two, a 100 Hz system is preferable.

If either of the 100 or 120 Hz field rates considered above were to be adopted as a worldwide standard, there would be an obvious advantage in either 50 or 60 Hz

370 INTERPOLATION

countries respectively. Also these higher field rates represent a considerable overhead in terms of bandwidth or bit rate. In view of this, there may be an advantage in choosing a field rate between 60 and 100 Hz that is not simply related to either 50 or 60 Hz.

If, for example, a field rate of 75 Hz were to be used, then the main spectral components of the scanned signal would be positioned as in Figure 6.88. As discussed earlier, high temporal frequency components from (75, 0) could overlap into the low temporal frequencies of the spectrum at the origin, but because of the higher field rate, those near the origin would be attenuated more by camera integration than if 50 or 60 Hz had been used. This would tend to improve conversions to 50 or 60 Hz field rates, although in either case the more complicated two-dimensional interpolation process would be required.

With a conversion to 50 Hz, the main interference components are 25 Hz judder from the (75, 0) spectrum and 12.5 Hz flicker from the $(37\frac{1}{2}, 625\frac{1}{2})$ spectrum, as shown in Figure 6.89a. It should be noted that the components shown in Figure 6.89a would be repeated on the alias centres of Figure 6.44b, but only the main components are shown here for clarity. In this case, the main impairment would arise from 25 Hz judder. However, this would be less noticeable than in a 60 to 50 Hz conversion, partly because it can be better attenuated and partly because it occurs at a much higher frequency. The

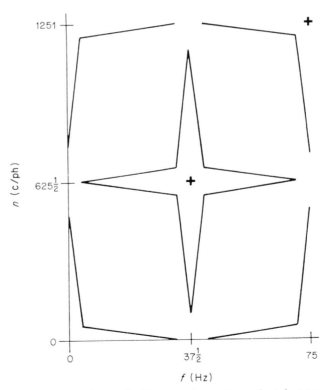

Figure 6.88 The vertical/temporal spectrum of a television signal with 1251 lines per picture, 75 fields per second, interlaced scanning

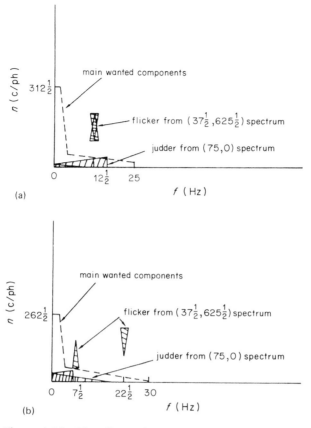

Figure 6.89 The effects of conversion from the 1251/75/2:1 spectrum of Figure 6.88 showing the positions of the main interfering components for (a) a 625/50/2:1 output standard and (b) a 525/60/2:1 output standard

amount of flicker depends on the amount of high vertical frequency information in the high definition system, but is unlikely to be noticeable.

With a conversion to 60 Hz, the interfering components would be positioned as shown in Figure 6.89b. In this case, the judder has a 15 Hz off-set from the true temporal frequencies and flicker occurs at 7.5 and 22.5 Hz. Although the judder has a lower offset than in the 50 Hz case, this would still be more attenuated and at a less noticeable frequency than occurs with a 50 to 60 Hz conversion. The 7.5 Hz flicker, although occurring at a more noticeable frequency than for 75 to 50 Hz conversion, could still be adequately suppressed unless the original 1251/75 standard had considerable high frequency energy in the $(37\frac{1}{2}, 0)$ region.

The use of a higher field rate for HDTV than is used in current broadcast standards could prove unnecessary if the development of alternative methods of standards conversion not based purely on interpolation is successful. For example, the standards converter developed by NHK for conversion of their 1125/60 HDTV system to 625/50 uses movement compensation to overcome temporal aliasing. As this constitutes the worst impairment in a converter using conventional interpolation techniques, the performance is considerably improved, although other generally less objectionable impairments can be introduced.

In broad terms, the method attempts to identify moving areas of picture and to adjust their positions in consecutive fields so that moving areas are superimposed during the interpolation process. This is particularly successful for rapid camera panning, but tends to be less successful for very slow panning or for complicated or random motion within a static scene. It remains to be seen whether these difficulties can be eliminated by further development.

The success of this technique in HDTV poses the question of whether it could be applied with equal success to conversions between normal definition broadcast standards. In general, methods using movement adaptive techniques have hitherto been relatively unsuccessful. One factor that could have a bearing on this is that accurate movement measurement may be significantly easier to achieve on the high resolution component signals used in HDTV. In normal definition composite signals, the presence of the moving colour subcarrier pattern may have a confusing effect for movement adaptive techniques.

An additional factor to be considered in HDTV conversions is that of aspect ratio conversion. While conventional broadcast standards use a 4:3 aspect ratio, HDTV is expected to use a higher ratio in the region of 5:3 or more. The solution most commonly used is to omit the extreme edges of the 5:3 picture except when significant action is occurring there. In this case steering of the displayed 4:3 area is sometimes used. Occasionally, when the edges of the picture are consistently important, the full width is displayed with black borders at the top and bottom of the picture.

Display improvement

The higher field rates and finer line structures of HDTV systems have at least in part been proposed to reduce the visibility of scanning impairments which result from the inadequate low-pass filtering action of the monitor display. Interlace is a simple and reasonably effective method of display improvement, but, with advances in technology, other methods can be considered. Indeed, as larger screen sizes are introduced, the suppression of line structure and flicker frequencies becomes more necessary. In principle, the performance of current broadcast standards could be improved in this respect to gain some of the advantages of HDTV systems without any increase in bandwidth.

The other methods consist of using higher line and field rates in the display, and possibly sequential scanning, to provide a raster in which the line structure and field flicker are invisible. An extensive programme of work has been followed [131] to overcome some of the practical problems that have been encountered. The problems arise in up-converting the conventional signals to the display scanning rates and for this there are two categories of approach.

First there are adaptive approaches which seek to detect movement and then produce output pictures using correctly positioned information, if the pictures are still, and using correctly timed information, if the pictures are moving. This has the potential for providing both better suppression of scanning defects and improved resolution, essen-

TELEVISION SCAN CONVERSIONS 373

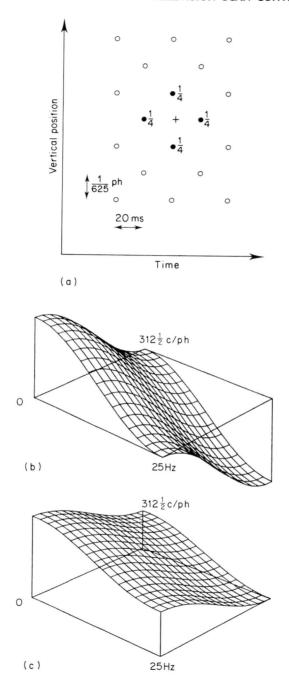

Figure 6.90 Non-adaptive interlace-to-sequential conversion: (a) simple interpolator contributions, (b) frequency characteristic of the interpolator and (c) overall characteristic with original and interpolated lines combined (• contributing lines, ○ non-contributing lines, + output line position)

tially by unscrambling the overlapping vertical detail and moving detail components. However, distinguishing real movement from vertical detail is difficult and the detection system sometimes provides incorrect indications. In this case serious impairments result from the use of an inappropriate conversion algorithm, unless DATV [23] is used to ensure that the receiver has accurate information.

The second approach is less ambitious, but fails less disastrously. Being non-adaptive it has to assume that the wanted signals lie in the area bounded by the $(0, 312\frac{1}{2})$ and $(25, 0)$ points in the vertical/temporal spectrum. The interlaced structure can then be converted to a sequential scan by filling in the gaps using an interpolator with contributions such as those shown in Figure 6.90a. This has the effect of inverting the interlaced components centred on $(25, 312\frac{1}{2})$ with the response shown in Figure 6.90b. When the original lines and the newly formed lines are combined into a single sequential scan the overall response is that of Figure 6.90c so some interlace twitter still remains but is reduced by at least 6 dB. Alternatively, a sharper cut characteristic could be obtained by using more input lines in the interpolator. The 625/50/1:1 signal is then up-converted by simple one-dimensional interpolation processes to higher line and field rates to suppress the line structure and large-area flicker.

With either of these approaches, then, the method is effective in suppressing line structure and flicker, but is less able to deal satisfactorily with the twitter components on vertical detail which result from the use of interlacing.

If the input signal is sequentially scanned, then the processes of up-conversion to suppress line structure and large-area flicker are considerably easier. Because of the orthogonal positioning of the repeat spectra, variables-separable frequency characteristics can be used without any compromise of performance. Thus, the interpolators would be one-dimensional and relatively simple. If the up-conversion processes are each 1:2, then the residual aliases have a high degree of coherence with the true components and so are less noticeable. In this way, the up-conversion processes can retain substantially the full resolution of the input signal while effectively eliminating line structure and flicker. Also, because there are no interlace components on the input, there is no need for movement detection, thus removing what has hitherto been the main problem of display improvement techniques.

7 Multipicture Storage

M. G. Croll

REQUIREMENTS FOR MULTIPICTURE STORAGE SYSTEMS

The broadcaster has a range of requirements for storing still pictures which cannot be met with video recorders, which are designed for recording and editing programme passages and cannot conveniently be used to give quick access to individual pictures. Also, the semiconductor stores described in Chapter 2 are of limited capacity and pictures stored in this way cannot be communicated between studios unless they are connected via a comprehensive data network.

The main applications for multipicture storage stem from the need to use electronic television techniques in areas where traditionally film or conventional artwork has been used but electronic picture sources are now available. This applies to where electronically drawn or computer generated graphics (see Chapter 9) are used, and to pictures obtained in the course of a television studio production or by television Electronic News Gathering (ENG) and Electronic Film Production (EFP). In the past, such pictorial illustrations and brief animated sequences were broadcast using artwork in front of a camera or a 35 mm slide prepared and broadcast using a studio caption scanner. Whilst a large number of still pictures will continue to be available originally as slides, slide scanners will continue to convert these into electronic signals for broadcasting. An electronic store can store both these electronic signals and those derived from electronic sources. Therefore, electronic multipicture stores may be universally applied regardless of the original form in which the pictorial information is first prepared. Moreover, higher degrees of reliability can be achieved using an animated logo reproduced from an electronic store than an equivalent older system using a television camera and electromechanically driven model.

To satisfy the range of possible applications of multipicture stores to broadcasting, a number of distinctly different forms of multipicture storage systems are required and are described in subsequent sections in this chapter. They vary mainly according to the number of pictures to be stored (capacity) and the rate at which new pictures must be recorded or replayed (speed of access). In a television studio, one television production seldom uses more than 30 still pictures, and a total capacity of 80 pictures would meet almost all applications for a studio multipicture store which enables still pictures to be selected for transmission in much the same way as they have traditionally been broadcast using a studio slide scanner. In this chapter it is assumed that 830 kbytes is adequate for storing one digital Y, U, V 625/50 still picture as only the active area needs to be stored. The speed of access is also required to be comparable with that of a studio slide scanner and is adequate for most purposes if new pictures can be accessed at a rate of about 1 picture each second. The capacity of a multipicture store used as a television picture library is required to be from 1000 to a few million pictures and a longer access time would be acceptable for most applications. Another range of multipicture stores is required for creating television animation. Such stores are required to record and replay television picture sequences at full television rates and the capacity should ideally be a few thousand pictures. Where hardware can be dedicated to, for instance, reproducing a logo, bit-rate reduction techniques may be applied and considerably less than 830 kbytes may be used to store each picture.

REVIEW OF SUITABLE STORAGE MEDIA AND TECHNOLOGIES

The different applications for multipicture stores mentioned above and discussed in detail later can be satisfied using a range of different storage media and technology largely based on developments initially made to satisfy the data storage requirements of computers. This section reviews media and technology currently available and being developed and discusses fundamental limitations and practical limitations imposed in storage devices designed specifically for the computer market.

Because a 625-line television picture requires at least 830 kbytes of storage, multipicture storage normally implies a requirement for at least 10 Mbytes of storage capacity. Such capacities are now provided cheaply within personal computers. The need to select pictures randomly within a second for studio still picture application, however, rules out the use of cheaper units giving slower access times. Animation requirements demand real-time replay of data at the full television rate which is an order of magnitude faster than normally provided by high performance disc drives. The need to convey picture data from one multipicture store to another is similar to computer data applications where tens of megabytes of data are transferred. This can usually be done at slower rates and the cost of the removable media may be more important than transfer speed. The fact that video data may normally be more tolerant to errors than the near 100% validity demanded for computer data can seldom be exploited within a multipicture store. The possibility of more efficient use of the media where video data are stored is forfeit in adopting, where possible unmodified computer data storage units. Here the low unit price advantages outweigh any disadvantages in terms of efficiency or facilities. A significant inconvenience in adapting standard computer units is, however, that computer data blocks are of typically 1 kbyte which is nearly three orders of magnitude smaller than the capacity required for one television picture. Thus many strings of instructions may be required before a complete picture is accessed. In some cases (see section on studio stills store) special television picture data controllers may be required to modify the operation of a disc drive to, for instance, devise alternative means of overcoming media defects. Where small blocks of data are accessed, data validity needs to be checked block by block and any repair operations initiated immediately. With television data it may be more convenient to deal with any problems of data validity between accessing television pictures. A further difficulty arises where data back-up devices are optimised to store large blocks of data corresponding to many television pictures as one continuous sequence and no means is provided to 'find' one stored picture.

The most economic media for storing still pictures is digital magnetic tape. Some different tape configurations are illustrated in Figure 7.1 ranging from open reel tape through single reel cartridge to the two reel DC 300 enclosed cartridge.

The importance of the nine track open reel tape is largely historic. Whilst the 48 picture capacity allows a useful number of pictures to be stored, the data transfer rate is

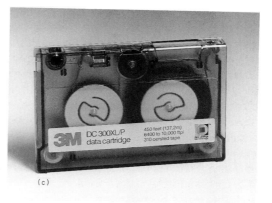

Figure 7.1 Examples of magnetic data storage: (a) $\frac{1}{2}$ inch tape reel, (b) $\frac{1}{2}$ inch tape cartridge (3480), (c) $\frac{1}{4}$ inch tape cartridge (DC 300)

only 40 kbytes per second, or 20 seconds per picture. However, the universality of this format has caused it to be used for exchanging picture data between computer based systems where no other common exchangeable media exists. Higher density and data transfer rates up to 1.2 Mbytes per second (0.8 s per picture) are now offered with this format but other formats achieve similar performances more cost effectively.

The half inch 3480 format stores 289 pictures with a transfer rate of one picture in 3.3 seconds. The quarter inch cartridge standards based on the DC 300 tape cartridges store 48 or more pictures with a transfer rate of one picture in about 12 seconds. Both of these tape cartridge systems are intended to back-up large quantities of computer data. As a result, data are stored as one block and no means are provided to skip unwanted data during a 'seek' or 'read' operation. In the case of the quarter inch cartridge a 'Serpentine' recording format is used where the tape records from one end to the other, the head steps to another position across the tape and recording progresses with the tape direction reversed. So the tape passes from end to end several times in recording a large block of data. With this format, no means of fast spooling is provided and, indeed, the mechanics of the tape cartridge is unsuitable for moving the tape significantly faster than the normal recording rate. For these reasons the larger capacity of the $\frac{1}{2}$ inch tape cartridge might not be used and only ten or less pictures may finish up being recorded on a tape cartridge. These cartridges are, however, rugged and provide a very cheap means of conveying pictures in digital form where previously boxes of slides might have been conveyed.

The digital video tape recorder described in Chapter 8 has an impressive specification for data back-up and will certainly find application with multipicture stores particularly where 'dump' and 'restore' operations may be required to be completed quickly. The capacity of the largest cartridge is about 140 000 pictures and the largest access time is likely to be less than 2 minutes as machines with fast search facilities are developed. The transfer rate is, of course, the real-time television rate. Data validity is likely to be poorer than would normally be accepted for computer data. However, this would be acceptable for multipicture store operations and the stresses in frequently accessing pictures or blocks of pictures may be little more than normally encountered in a television editing environment.

Removable disc drives provide more flexible access to television picture data, albeit at higher cost. This is an area where the optical disc drive is particularly well suited. Disc capacities of 1 Gbyte or more per side with 12 inch diameter discs allow large numbers of still pictures to be stored and accessed quickly. The formats of optical discs are well suited for this type of application as the discs allow rapid access by stepping the read/write head across tracks. Also, the tracks are in a continuous spiral so that reading or writing large blocks of data requires no special control to move the head from one track to the next. When established, this technology will certainly provide a convenient method of conveying a few tens of pictures using media of no more than 2 inch diameter. Meanwhile early media as shown in Figure 7.2a is write once and costs are high.

Removable magnetic hard disc packs, as shown in Figure 7.2b have provided rapid access to some eighty pictures where the importance of access has overweighed any cost considerations. This has occurred particularly in the support of picture design where large numbers of library pictures, backgrounds and standard texts have been required in say news or sport. It is expected that optical disc developments will displace this and provide improved data capacity with higher durability in smaller physical size at a lower

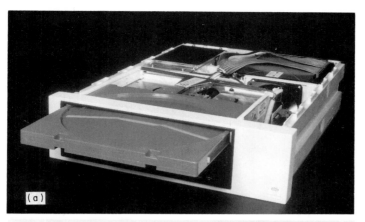

Figure 7.2 Examples of disc based media: (a) 12 inch removable write once optical disc (b) 8 inch removable magnetic disc pack (c) 8 inch Winchester sealed magnetic disc drive

Table 7.1
Tape and disc media

Media	Capacity (pictures)	Picture transfer time (s)	Longest access time	Media cost per picture	Remarks
$\frac{1}{2}$ inch open reel tape					Exchange of picture data between computer installations.
1600 bits-per-inch	48	20	16 min/48 pictures	14.5p	
6250 bits-per-inch	121	6	12 min/121 pictures	6.7p	
$\frac{1}{2}$ inch 3480 cartridge	289	3.3	16 min/289 pictures	2.6p	Economic back-up of large stores.
$\frac{1}{4}$ inch DC 300 Cartridge					Rugged means of conveying still pictures.
5 track	45	12	9 min/45 pictures	27p	
9 track	72	12	14 min/72 pictures	16p	
DVTR 16 micron tape					High capital cost. Useful particularly where picture sequences are processed and real-time recording and replay are required.
D1.S	17×10^9	0.04	1 min/17×10^9 pictures	0.1p	
D1.M	53×10^9	0.04	1.7 min/53×10^9 pictures	0.1p	
D1.L	12×10^{10}	0.04	13 min/12×10^{10} pictures	0.1p	
12 inch optical write once disc	1.2×10^3 (per side)	3.5	0.5 s/1.2×10^3 pictures	0.28p	Still picture library applications.
8 inch removable magnetic disc	96	1	0.1 s/96 pictures	300p	Where rapid access is a valuable facility.
14 inch parallel transfer	815	0.04	next picture immediately	80p	High capital cost and customised. Less than Rec. 601 resolution. Animation supported.
8 inch sealed Winchester disc	192	1	0.1s/192 pictures	non-removable	Fixed local storage for stills.
High performance parallel transfer Winchester (two drives)	2000	0.04	next picture immediately 0.1 s/2000 pictures	non-removable	High capital cost. Fixed storage for creating animation.

cost. However, some multi-headed drives using larger discs have been specially adapted for real-time television applications (see the television animation store section) and it is unlikely that optical drives will provide this high data-rate capability. The data recording format characteristics for the removable magnetic hard disc pack are similar to those of sealed non-removable Winchesters.

The sealed Winchester magnetic disc-drive has provided the main picture storage element for many picture storage systems as it is particularly economic and can hold up to 1000 pictures and provide good random access. Special Winchester drives have been equipped with parallel read/write circuitry to achieve real-time television rates with two drives working in parallel. Within a Winchester drive, data are recorded on several hard magnetic discs mounted on a common axis. Data are accessed by reading from each head in turn and then stepping the heads to the next position. That, together with the normal arrangements made to overcome media defects, usually requires a degree of customising of the data format for digital television storage purposes. Such customising is described later this chapter. Where real-time television recording or replay is required consecutive pictures must be stored in adjacent tracks as time cannot be allowed for moving the heads any greater distance. Also, large buffer stores are necessary to smooth the flow of data. This is because the disc rotation is not locked to television and that time spent moving heads or switching amplifiers, etc, interrupts the data flow. Nevertheless some 70% efficiency in transferring data is normally achieved.

The use of semiconductors for multipicture storage has increased as the level of integration increases. At 1 Mbit per chip, only eight chips are necessary to store a picture and it is feasible to store 20 pictures on one printed circuit board. With the higher levels of integration, immediate access may be constrained to blocks of bytes but, unlike other media, access to all pictures is independent of where they are stored. Thus, compared to Winchester drives, semiconductors provide very high speed access at a cost at least competitive with the cost of very high performance drives required for real-time television recording or replay. The freedom from moving parts may be seen by many potential users as adequate compensation for the higher capital cost, thus leading to their introduction in areas where the high speed access is not essential.

Table 7.1 shows a summary of different media characteristics and includes a wide range of examples of the different media discussed in this section.

DECODING TO PROVIDE A PAL INPUT TO A MULTIPLIE PICTURE STORE

In the multipicture storage applications described in this chapter, the television signal is stored in its separated luminance and colour component form and interfaces need to be provided between that and the PAL composite form of television signal used in other parts of the studio and picture contribution/distribution system.

When PAL encoded signals are fed to a stills store they must be decoded to the *YUV* component form used in the stills store. The combination of operations. where the PAL signal is first decoded and a still picture formed from one input field or picture and then finally re-encoded as PAL, which is eventually decoded a second time in the viewer's receiver, places very stringent requirements on the properties of the first decoder.

Figure 7.3 illustrates the range of cross-effects that can be generated when two PAL encoding and decoding operations are cascaded. These are described in Chapter 4, but

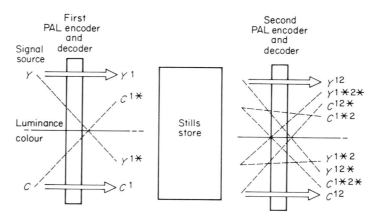

Figure 7.3 Cross-effects that may be generated when a component stills store is fed with a decoded composite picture

where a stills-store system is introduced between the two operations additional considerations must be taken into account.

The effect of forming a still picture from either one field or one picture at the output of the first decoder is to freeze the cross-products which would otherwise be periodic over an eight-field interval. Thus, after the second encoder and decoder, C^{1*2} consists of frozen cross-luminance and appears as a stationary dot pattern and could beat with C^{12*} which appears as a crawling dot pattern. In this case, the overall effect would be that a flashing dot pattern is generated. Similarly Y^{1*2} would produce a stationary coloured pattern frozen into the picture which could beat with Y^{12*} to generate a flashing coloured pattern. When the output picture is generated from one stored picture the effect is as if the relative phases of the encode/decode processes cycle through four different values on successive pictures. As a result, Y^{1*2*} becomes a luminance component which is phase modulated at a 6.25 Hz rate which, if added to Y^{12}, causes an overall flickering of high-frequency luminance components; similarly, C^{1*2*} cycles through different phases at a 6.25 Hz rate and, if added to C^{12}, causes colour flicker in the output picture.

Of the decoders that may be used, field- and picture-based decoders have been found particularly successful. In many cases, the storage that exists within a multipicture store system can be applied as part of the necessary comb filter. Where simpler forms of decoder only are available it has been found necessary to install notches in the luminance paths so that any residual subcarrier is substantially reduced. It has also been found beneficial to restrict the chrominance bandwidth to a small proportion, say one quarter, of that provided by the Recommendation 601 [9] chrominance channels.

APPLICATION OF ELECTRONIC MULTIPICTURE STORAGE TO A NEWS CENTRAL STILLS LIBRARY

Television broadcasters maintain sizeable still picture libraries. The largest of these are usually associated with news, but other libraries cover sport, current affairs and general programme production and presentation. Also, with broadcasting networks, both national and regional libraries are usually maintained. The size of these picture libraries varies from a few hundred pictures up to several hundred thousand pictures.

An example of a large picture library would be the BBC's Central Stills Library which contains a quarter of a million 35 mm colour transparencies, taken by BBC staff and commissioned photographers. There is also a collection of black-and-white agency material amounting to a further 100 000 pictures. The pictures cover news, sport and current affairs events since 1969 and include a portrait collection of personalities and location shots throughout the world. The contents of the library are constantly being updated, with some 500 new pictures added and between 100 and 200 pictures discarded each week. The main reason pictures have to be discarded is to make space for new pictures.

In the past, most pictures required to be stored in a picture library have originated in the photographic form. Photographers have been present at important functions, prepared artwork has been photographed and members of the public present at important newsworthy events have often taken a photograph and made this available to broadcasters.

News pictures have, however, been available from the main news agencies in electronic form via a facsimile system for some years. Their immediate review and use in electronic form would not only save time but, because about 98 % of all those received are immediately discarded, vast quantities of paper, currently used in printing the received pictures, would be saved. Also, there is a trend in which electronic methods are displacing photographic methods in television. Television cameras and Electronic News Gathering (ENG) are being used where in the past photographic cameras were used in covering newsworthy events. Captions and some graphic illustrations are provided at television studios using new electronic devices.

It is clear that there is an increasing need for electronic based picture libraries. Several early systems were manufactured which made use of magnetic recording on computer type disk drives. However, the storage capacity of these systems was limited. The total on-line storage capacity available with the larger such systems was about 5000 pictures. In some cases, larger capacities were achieved using removable disc packs but these systems were inconvenient to use because of the time taken to change disc packs. Moreover, the media cost for storing one still picture was as much as $1 and about 10 000 stills could be stored per cubic metre (a figure very close to that of the space required to store 35 mm slides in a filing system).

An experimental optical disc-based still picture store was demonstrated by RCA at the National Association of Broadcasters convention at Chicago in 1976. This system showed broadcasters the potential of optical discs for storing still pictures. At the present stage of development of optical disc technology, it seems feasible to store in excess of 5000 pictures in digital form on one 12 inch disc. Thus, an electronic picture library could become available with a very much higher storage density than is available using other technologies and at a media cost of a few tens of dollars per disc. With a juke box arrangement of optical discs it might be feasible to have as many as $1\frac{1}{2}$ million pictures available on-line from an optical disc-based library system.

Figure 7.4 illustrates how digital optical disc recording could be used in a television central news picture library. It is assumed that the disc recorders would be equipped with semiconductor picture store buffers so that one picture can be output while the next is being found. The picture buffers also reduce the rate at which data is required to be read

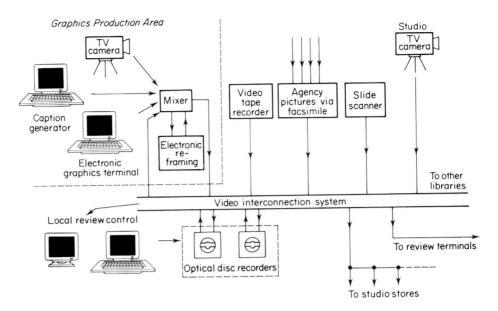

Figure 7.4 Centralised news electronic picture library system

from or written onto the disc to rates more easily accommodated than the full digital video rate. With buffering, the data transfer rate is required to be sufficiently fast so as not to unduly prolong the time to access a new still picture.

The interconnection of the disc recorders with other equipment might initially be via analogue-to-digital and digital-to-analogue converters. However, as other items of equipment become available with digital video interfaces, the number and positions of converters will change. To preserve generality, no converters have been shown in the figure and the connections are not specified as digital or analogue as in either case the schematic would be similar. The range of picture sources includes studio cameras, video tape recorders and a graphics production area. A slide scanner is also included because a large number of pictures would still be available originally in photographic form. Both local and remote view terminals would be required and the library would be interconnected with other libraries and with smaller studio stores or directly to the news studio.

To give some redundancy, it would be necessary to have at least two record/replay machines working in parallel. It would not be necessary for the recordings to be erasable so long as the control system was able to retain in memory lists of which pictures had been superseded. If some discs were reserved for pictures of a limited lifetime, these discs could be re-recorded periodically.

One of the most important difficulties to be overcome in setting up a digital optical disc based library would be in indexing the contents. To help with this, it would be beneficial to incorporate special 'browse' facilities and the ability for the operator to view several pictures at once at reduced resolution. The present slide-based system enables one set of slide pockets, containing 24 slides, to be held up to the light to enable the operator to make an initial selection.

STUDIO STILLS STORE

General

Many broadcasters now have the facilities to generate graphs, maps and other illustrations by computer [132]. There is also a range of freehand electronic drawing systems [133] and complex text generators [134] which enable the graphics designer to produce pictures using solely electronic techniques. It may not be practicable to broadcast the output of all these devices directly as such facilities are frequently shared between a number of television studios or are installed in a separate area and operated by design staff rather than members of the programme production team. Also, the devices often change pictures at too slow a rate to be used directly during a programme.

These new electronic methods of generating still pictures have, therefore, created a demand for a means of storing these pictures and selecting them for transmission in a television studio. Because of the low cost and high speed of computer data storage devices, it has become feasible to construct a digital stills store with modest storage which can fulfil this demand and can be economically installed in each television studio. Such a stills store has other important uses. It can provide a more stable and reliable method of displaying pictures derived from photographic slides than repeated scanning of these slides in an analogue slide scanner. In addition, it can be used to grab and store still pictures from any television signal supplied to it.

Although other digital stills stores are commercially available, they are in general designed to operate as large centralised library systems. The requirements of a studio stills store have been found to be different from those of a library system. Far fewer stills are required at any one time in a studio. This simplifies the access and search requirements and allows a much simpler operation by staff who need no special training.

The studio stills store to be described was developed at the BBC Research Department in conjunction with Rank Cintel who manufacture it under the product name 'Slide File'. Many of these equipments are in use at the BBC's London and Regional Studios as well as by other broadcasters. Slide File has two outputs, as shown in Figure 7.5, one of which can be used to preview stills. The other output is the main output where stills can be changed using a simple cut or cross-fade, or fade to or from black. Within the store, pictures are held on a magnetic Winchester computer type disc drive in digital component form. The two outputs are buffered using picture stores because the data record and replay rate from the disc drive is substantially slower than the full digital video rate. Pictures can be conveyed between stores using a removable streaming tape medium. Data flow between different elements is under the control of the system computer which accepts instructions from up to five remoteable control panels.

The brief specification of Slide File is:

PICTURE STORAGE
Video format: 8-bit digital *YUV* sampled at 13.5 MHz, 6.75 MHz, 6.75 MHz

Short term: Two semiconductor picture stores

Medium term: One or two 8 inch Winchester drives — capacity depends on size of drives used and is currently either 85, 170 or 300 pictures per drive. (picture transfer time — 1.0 s)

386 MULTIPICTURE STORAGE

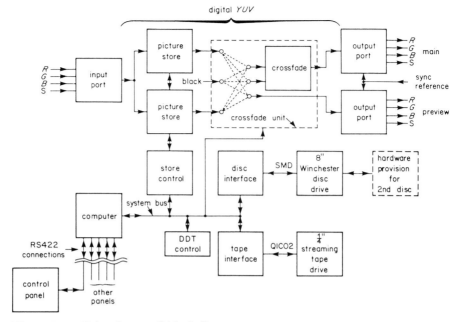

Figure 7.5 Slide File overall block diagram

Long term: $\frac{1}{4}$ inch cartridge tape drive—capacity depends on length of tape in cartridge and is currently between 45 and 72 pictures.
(picture transfer time—12 s)

PICTURE INPUT/OUTPUT
Video format: Analogue RGB + sync.

Synchronication: Output and input locked to independent sync pulses

CONTROL
1 to 5 remoteable control panels can be used.

Connections via Serial RS422 lines.

Figure 7.6 shows a standard control panel which allows control of all Slide File's facilities whereas operation is simplified in many studio installations where some facilities are only available from a centralised control panel. Thus, the normal operation for selection of pictures for transmission may be carried out with the minimum of training.

Pictures are referred to using either a 'library' number or sequences of pictures can be programmed and edited such that pictures can be referred to by their position in the sequence. Help in identifying the stored pictures is provided by a 'polyphoto' index display as shown in Figure 7.7 (see colour plates between pp 506 and 507) which can be viewed via the preview output halting the normal operation of Slide File. Each small picture has sequence and library numbers beneath it. As the store contents are browsed the polyphoto display scrolls smoothly or is completely rewritten, in a couple of seconds, if the operator jumps around the store. Achieving good quality still pictures is sometimes

Figure 7.6 Slide File control panel

difficult as movement can cause a grabbed picture to flicker and grabbing only one field may result in discarding important information from a source of still pictures. Also, if the picture source is a PAL or NTSC composite electronic picture source, grabbed residual cross-effects may give rise to additional picture impairment when the output is re-encoded and subsequently decoded at the viewer's receiver. Movement and residual cross-effects are reduced using a range of sophisticated algorithms or 'clean-up' routines, using the system computer to process pictures in non-real time after they have been grabbed.

The crucial elements in the design of Slide File are those hardware and software elements which maximise the rate at which picture information can be transferred between tape, disc and semiconductor picture stores and enable picture information to be processed quickly by the system computer. The important features of these elements are described in the following subsections.

Semiconductor picture stores

The two semiconductor picture stores use 64 kbit dynamic RAMs using the techniques described in Chapter 3. Each store has three data access ports as shown in Figure 7.8. Video data can be input asynchronously via a FIFO buffer or output, at the full digital video rate of 13.5 Megasamples per second with each sample of 16 bits being 8 bits of luminance and 8 bits of multiplexed colour difference. Data can also be applied to or from the system bus which also controls the address generators for the store's three access ports. The access times for the RAMs are multiplexed, 50/50, between the video data output and the system bus data. When video data are being input, they use the time otherwise allocated for the system bus data. This sharing causes no loss of facility as data are never required to be conveyed between a picture store and the other units of the stills store when it is being input to that store from an external video source.

The operation of the data port connected to the system bus is unusual in that the store control board contains circuitry to shift the data in the shift registers until the required word is accessible at the output. Meanwhile, the unwanted data are rotated back into the register so that it will still be available without waiting for another store RAM access. The

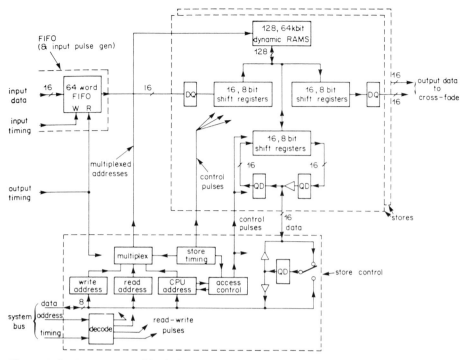

Figure 7.8 Picture store block diagram

system may read or write either or both of the two 8-bit bytes in the word. If the system modifies the block in any way, then, before accessing another block, the current block is automatically rotated back into the shift register and written back into the RAMs before reading the next block.

The system bus accesses the store by writing a 20-bit store address into a set of 'current store address' registers, and then reading from or writing to the store data ports. There is a set of eight store data locations available to the system, each accessed at a different address. One of these locations accesses the data at the 'current store address'. Other locations access the current store address plus one, plus a reset increment, minus increment, plus increment plus one, or minus the increment plus one. In each case, the current store address is left set to the address last used. So it is possible to read (or write) data from successive locations by repeatedly accessing the store via the 'address plus one' port. This greatly simplifies the transfer of pictures from disc or tape. Similarly, it is possible to access vertical columns of the picture by presetting the increment to the 'number of bytes per line' and then accessing the address plus increment (or address minus increment) port. This is when processing pictures to remove impairments caused by freezing moving interlaced or PAL decoded input pictures.

Disc

Medium term storage is provided by 'Winchester' disc drives. Winchester disc drives store data on one or more rigid/hard/non-floppy magnetic discs. To achieve a high packing density very narrow head gaps are used and the flying height is kept very low. When power is removed the heads land on the lubricated surface, usually in a special data-free landing zone. To achieve a high transfer rate and a low latency time (time waiting for the right part of the disc to come round), the discs rotate at a high speed of typically 3600 r.p.m.

All these characteristics mean that the system is very sensitive to dust and smoke particles etc. The discs and heads are, therefore, kept in sealed-for-life enclosures containing highly filtered air. Thus, unlike floppy discs, these hard discs are not normally removable and so the drives have a fixed total storage capacity.

Slide File was designed with an 84 Mbyte disc to store 85 pictures. A drive was chosen with the industry standard SMD (storage module drive/device) interface. This has allowed a 168 Mbyte device to be easily incorporated by simply altering a few constants in the control software. As disc technology improves it will be easy to use still larger drives, and a second SMD port allows for further expansion.

The 85 Mbyte Winchester disc drive used has three discs with a record/replay head associated with each of the five recording surfaces. Data are replayed from one head at a time at a rate of 1.2 Mbyte/s which is determined by the data packing density and the rotation rate of the discs (3600 r.p.m.). One picture requires 45 data recording tracks, or nine cylinders of five tracks each. The time taken to replay a picture is determined by the hardware of the disc drive and the way it is controlled. Since it takes a finite time to switch or reposition the heads, the start of each is 'skewed' with respect to the previous track. In this way, the time wasted waiting for the start of a track to pass under the replay head is minimised.

The disc drive control software is resident in the Slide File system computer. A disc interface circuit, see Figure 7.9 connects between the Slide File system bus and the SMD cables to the disc drive. This interface is 'non-intelligent' and allows the drive to respond rapidly to the control computer's instructions to, for instance, move the heads, select the correct head and enable data reading or writing as is necessary many times during the recording or replaying of one picture.

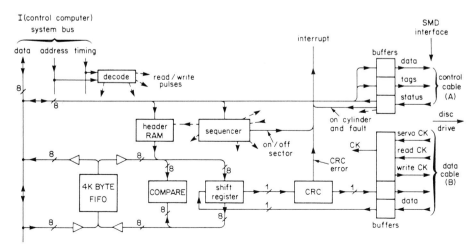

Figure 7.9 Disc interface block diagram

The interface circuit includes a sequencer and the data path through the interface includes a FIFO buffer. These allow the rate of data throughput to be maximised using the minimum of interaction with the computer which retains overall control over all data transfer operations. As data are replayed from the disc, error check codes are monitored and data transfer can be halted if errors are detected.

Each track of data written on the disc as eighteen 1 kbyte sectors is preceded by a header containing the unit number, cylinder number (head position) head number and sector number. The headers are written on the disc in an operation known as formatting which is normally carried out only once as part of system initialisation. During formatting the system computer writes the header information into a header RAM, which is installed in the disc interface, and the sequencer transfers it to the disc. The computer also loads the header RAM before any normal data read or write operation. An 8-bit comparator is used to compare the header data coming off the disc with the information in the RAM to confirm that the correct location on the disc has been found. The sequencer sends an 'on sector' interrupt as soon as data transfer starts, so that the control computer has plenty of time to alter the data in the header RAM in readiness for the start of the next sector. Successive sectors are thus transferred without interruption. When all 18 sectors have been transferred the system control computer can instruct the drive to select another head or move the heads. The sequencer provides an 'off sector' interrupt to let the computer know when it is safe to do this.

Streaming tape drive

The streaming tape media was chosen as a removable medium and for long term storage of pictures because the tapes are cheap, rugged and compact with up to 74 pictures stored in a tape cartridge measuring $150 \times 100 \times 16 \, \text{mm}^3$. The tape format is standardised as QIC-24 and is called 'Serpentine' in that nine tracks are recorded, one at a time. After a track has been transferred, the head is stepping down to the next track and the tape direction is reversed. The data transfer rate is relatively slow compared to that of the disc drive in that it takes 12 s to transfer each picture. The drives installed in Slide Files incorporate controllers which contain the necessary data buffers etc and are connected to the system bus via the QIC-02 standard interface and a simple interface circuit.

The Slide File system computer software manages tape data transfers so as to minimise the number of times that the drive data buffer empties or overfills. On such occasions, data transfer is halted while the tape is stopped and rewound a short distance, prior to streaming again.

Most of the tape is used to store the digital television pictures. At the front of the tape, a form of control header is recorded giving details of the tape contents and small versions of all pictures are also recorded. This enables the contents of a tape to be quickly reviewed without loading the entire tape contents.

The inconveniences of the tape unit, in terms of its slow rate of picture transfer, inability to access pictures randomly and need to load a tape's contents into a Slide File in order to edit a tape are accepted for the general studio applications. Here the size, storage capacity, price and ruggedness of the removable medium are particularly important. It is anticipated, however, that once optical disc recording becomes established with readily

STUDIO STILLS STORE 391

available erasable media and recording systems, they will provide a better alternative for the transfer of pictures between slide files not connected by an appropriate electronic link.

System computer hardware

The system is controlled by the system computer which is connected to up to five control panels. This microcomputer interprets the key presses and issues the necessary instructions (control bytes) to the disc, tape, store and cross-fade unit, etc. It may also read the data in the disc, tape or picture store and process it to remove impairments or to create small pictures to be used in the 'polyphoto' display. The computer is fairly heavily loaded since it is the only intelligent part of the system and its software is structured such that many different tasks appear to progress simultaneously.

The heart of the computer (Figure 7.10) is a Motorola 6809 8-bit microprocessor (CPU) connected to 64 kbytes of dynamic RAM or ROM. The five control panels are connected through RS422 drivers and receivers to five UARTs (universal asynchronous receiver transmitters). A PTM (programmable timer module) is used to ensure that the CPU performs operations such as interrogating the control panels and the disc on a regular basis. A few parts of the system are checked and controlled through a PIA (parallel interface adaptor).

The system bus connects the CPU, store control, disc interface, tape interface and cross-fade unit (see Figure 7.5). As well as handling control data given out by the computer and status information sent back to it, this bus carries the video data between disc, tape and store under the control of direct data transfer (DDT) control logic, which is in turn controlled by the computer. The bus operates at a data rate of 2 Mbyte/s which is sufficient to handle the 1.2 Mbyte/s disc data or 83 kbyte/s tape data.

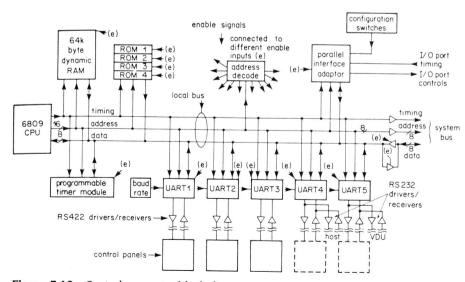

Figure 7.10 Control computer block diagram

The data and address buses on the computer board are isolated from the system bus by bidirectional buffers. These buffers allow the system bus to be used for high speed DDT controlled data transfer, while the CPU continues to operate using its own local bus to communicate with its memory, UARTs, PTM and PIA etc. The CPU may also still use the system bus at any time since the DDT controller automatically halts to give the CPU priority on the bus. This does not significantly slow down DDT operations since the CPU does not make use of a very high proportion of the bus time.

This isolation of the local bus allows the CPU to remain in full control of the system at all times. It prepares for the transfer of the next block of data during the current transfer so that there is the minimum time delay between transfers. It can control the operation of the disc drive on a sector by sector basis ensuring that the disc is transferring data to or from its FIFO buffer without any unnecessary interruptions. The CPU can thus take over the functions of an intelligent disc controller ensuring that the disc is used in the optimum way for this application.

This use of the system bus to transfer data has proved very effective. It saves having to add a separate video data routing system or separate video data bus and it allows the CPU access to all data including video data. Had the video data travelled via a different path then it would have been more difficult for the CPU to access the video data. In Slide File, the system bus is locked to the speed of the processor and is thus limited to 2 Mbyte/s, but there is, in principle, no reason why the bus could not have run at a higher speed asynchronous to the computer, had this been necessary.

System computer software

The software for the control microcomputer was mainly written in a high level programming language called 'Pascal'. The use of a high level language saves programming time, because most of the standard computing functions are supplied by the compiler. Pascal was designed to encourage the writing of well structured, easy to read software.

Program size is not a real problem in Slide File. The computer was designed with as much memory as the 6809 microprocessor can address and at present all software is automatically loaded from disc on power up.

Most of the time, speed is not a problem either. However, certain operations require the computer processing of picture data and these operations should be as fast as possible. Some low level routines were therefore rewritten in 6809 assembly language. This reduced the time taken to perform a 'clean up' operation from 5 min to 40 s whilst maintaining the high level program structure.

The main function of the computer and its software is to read the control panel(s) and carry out the required operations. It finds out the state of a control panel's buttons and faders by sending requests for data to the panel through a UART and later sending the reply from the UART. When carrying out the operations, the CPU sends LED and lamp data to the panels to inform and prompt the operator, and thus achieve a good ergonomic 'feel' to the control panels.

Most operations require the CPU to control the hardware in some way (e.g. to transfer data between store, disc and tape under DDT control or to control video flow into the stores or from the stores through the cross-fade unit to the main and preview output ports). This control is achieved by the CPU writing control bytes to hardware control registers, often after first checking that the time is right by reading status bytes from hardware status registers.

The computer memorises sequences, directory and status information, and periodically records in onto disc so that it is still available after the system has been switched off. The computer processes pictures to remove impairments and creates small versions for use in the polyphoto display. In its spare time the computer also checks the integrity of the system, especially the disc drive, and reports and, if possible, corrects any faults found.

The software is written so as to give the impression that the computer is performing all these functions simultaneously. In particular the control panel never goes dead because the computer is too busy to respond.

The full listing of the Pascal source code is some 10 000 lines long. Figure 7.11 shows the program structure to a depth of up to four levels. After initialising all the hardware and software, the program enters an indefinite loop. The loop repeatedly calls a procedure called 'Main Body' which examines the global variables describing the current state of the system. When 'Main Body' finds something that needs doing it calls another lower level procedure to do it.

Thus as the operator moves the fader or presses or releases a key, 'Main Body' will call 'Fade Respond' or one of the 'Key Respond' procedures, to initiate the appropriate action. If he holds down 'Increment' or 'Decrement' keys the 'Timed Key Respond' procedure will be invoked.

Often the action taken in response to a key press is not direct. It merely alters the contents of global variables which act as messages for other parts of the software to act on, when there is time. For example 'View Next' is called to replay a picture or move the cursor on the polyphoto display whenever the variable 'Next' changes. 'Poly Update' makes sure that the polyphoto display on the preview output matches the lists in the computer. If necessary, it plays new small pictures, numbers them and scrolls the entire display.

All the procedures called by 'Main Body' have been kept short so that the main loop cycles as fast as possible and 'Main Body' is called often enough to respond to new control panel inputs, even before the previous operations are complete. Long operations are broken up into a set of shorter procedures which are called in turn by 'Main Body'. For example, 'Poly Update' only plays one small picture at a time before returning to 'Main Body'. It is thus possible to edit a sequence while the polyphoto display is still changing as a result of a previous edit.

In deciding what to do next 'Main Body' uses information provided by the timer interrupt handler 'IRQ respond'. Every 19 ms 'IRQ respond' is called by an interrupt and records any data received from the control panels before asking for more data. (It also checks that the disc is still responding correctly and has not hung up because it cannot find the required sector. If necessary it allocates or uses a spare sector.)

Picture transfer operations are broken up into a set of short subtasks. Each of these instructs the hardware to perform some operation. When the operation is complete, the hardware interrupts the computer using the FRQ interrupt. The 'FRQ Rec' procedure identifies the source of the interrupt and calls the appropriate routine which instructs the hardware to perform the next part of the operation. Thus, every time the direct data transfer (DDT) control logic completes the transfer of a line of picture data to or from a semiconductor store, it interrupts the computer using the FRQ interrupt. 'DDT halted'

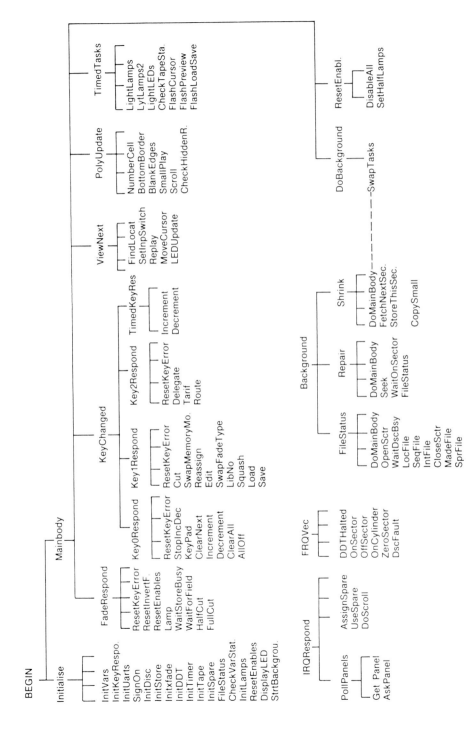

Figure 7.11 Program structure

then sets the store address registers to the start of the next TV line and restarts the DDT. Whenever the disc finds the required sector, 'On Sector' computes the address of the next sector required and sends it to the successive sectors. At the end of each sector 'Off Sector' checks that the next sector is on the current track and if necessary, instructs the disc to select the next track. If this involves moving the heads then disc data operations are suspended until an 'On Cylinder' interrupt restarts them. Disc data operations are also suspended if the disc FIFO buffer has insufficient space for replay or data for record. Operations are restarted whenever the 'Zero Sector' routine detects that it is safe to proceed.

Main Body periodically calls a procedure called 'Do Background' which transfers control to a separate task called 'Background'. The procedures in 'Background' all take a considerable time to run, so they each periodically call 'Do Main Body' to transfer control back. Thus the time-consuming background tasks are performed invisibly.

There are three background tasks:

(i) 'File Status' records the status tables on the disc whenever they are changed by, for example, recording a picture to a new location, editing the sequence, or the discovery of a bad disc sector.

(ii) 'Shrink' creates small versions of the stored pictures to be used in the polyphoto display. It reads picture data from the disc, processes it and records the resulting small picture data back on the disc. The whole process takes about 1 min.

(iii) 'Repair' tests any disc sectors which have been found to give problems during previous disc operations. If its tests confirm the problem then it attempts to repair the sector by rewriting its address header and data areas. If this fails to cure the problem, then it labels the sector as faulty so that any subsequent attempt to use it will automatically result in the use of a substituted spare sector.

Picture processing

Because each TV frame is made up of two interlaced fields scanned at slightly different times, any moving object in a grabbed picture flickers at frame rate between its two positions on the two fields. Also, decoded PAL or NTSC inputs may contain residual cross-luminance (subcarrier dots) and cross-colour (h.f. luminance misinterrupted as chrominance). These 'footprints' of the colour coding process cause serious flickering beat patterns when the picture is subsequently recorded for transmission and then decoded in a domestic receiver. Slide File software provides a set of clean-up operations, which the operator can use to remove, or at least reduce, these effects. These operations are carried out by the computer which processes the preview picture in vertical columns. They take between 10 and 50 s and when finished the cleaned up picture is automatically recorded.

The 'Adaptive' process of interpolation using the principle discussed in Chapter 6 removes movement effects from a grabbed picture by replacing one field with information derived from the other field for those areas of the picture where there is significant movement [135]. Stationary areas are unaffected so that the full picture resolution is retained in these areas. Movement is detected by comparing each sample from one field with samples from the two closest lines from the other field, i.e. those immediately above and below it. Any differences between these three samples may be due not only to movement but also to vertical detail. However, for stationary edges, we would normally expect the value of the signal from the middle line to lie somewhere between the values of the signals above and below. The signal from the middle line is left unchanged provided it lies in the range between the values of the signals above and below. If the signal lies outside values from the other two lines it is replaced by video information from those other lines. This effectively performs a field repeat in moving areas, leaving stationary areas untouched.

The PAL/NTSC clean-up is equivalent to, but faster than a frame average followed by an adaptive clean up. The frame average creates two new output fields; one of them is the average of samples 312 lines apart; the other the average of lines 313 lines apart on NTSC. Thus frame average removes subcarrier dots from one of the two fields in each system. The subsequent adaptive clean-up uses this clean field as a reference and cleans up the other field.

As well as the clean-up operations described above, picture processing using some of the concepts of standards conversion discussed in Chapter 6, is also applied within Slide File to create one-sixth size pictures which are stored on disc and assembled together in the previous store when a polyphoto index display is being used. The small pictures are created in two stages. Immediately after a new picture is input to the store, it is recorded on disc both as a full size picture and as a subsampled small picture so that it can be identified immediately. This crude small picture is later replaced by a version created by the computer as a background task from the picture data stored on disc. The picture processing operations involved in creating the second version of the small picture involve pre-filtering, subsampling and a post-filtering operation to reduce flicker. These operations are normally completed within about half a minute of a new picture being recorded.

TELEVISION ANIMATION STORE

General

This section describes an animation store system which entered service with BBC Television Graphic Design Department in 1983 to provide a means of generating animated effects using solely television techniques rather than the intermediate film medium which has been widely used for such purposes. The new television animation store uses digital video techniques and is normally operated in conjunction with a computer-controlled video rostrum camera. Figure 7.12 shows the control panels for camera and store. The camera and store together can achieve the same effects as a film rostrum camera system with the added benefits that:

(i) Effects can be built up and viewed immediately.
(ii) Changes in picture quality, as when film and television originated materials are intercut, are avoided.
(iii) The full range of television studio techniques, including electronic keying, can be applied.

Figure 7.12 Control panels for the animation picture store system and the associated video rostrum camera. The right-hand third of the panels shown in the photograph are used to control the movement of the rostrum table, the operation of the camera and the lighting; the centre third of the panels control the animation picture store system and the left-hand third are used for associated processing such as video routing and mixing, audio monitoring and control of a video tape recorder

The television rostrum camera can be preprogrammed to zoom and focus on artwork that can be moved in any direction in a horizontal plane or rotated. Also, artwork peg-bars are provided so that different elements of artwork can be moved independently.

The animation store records television pictures at rates up to the full television rate, includes all the necessary processing systems to allow sequences of pictures to be processed and recorded to combine effects and generate more sophisticated effects, and output resulting picture sequences at the full television rate.

The signal is stored and processed within the system in digital component form. Digital storage and recording enables very large numbers of generations of re-recording to be used without loss. Retaining the signal in component form, rather than PAL composite form enables signals from local sources to be stored in a more usable form. Also composite signals can be applied to the store via a sophisticated PAL decoder which makes use of picture information from adjacent television fields.

System description

The animation store provides a means of storing up to 815 television pictures (1630 fields) in digital component (YUV) form and replaying them either individually or, subject to certain restrictions, as a sequence at normal television rates. Facilities are also provided to allow the digital pictures to be combined by the mixing and colour separation overlay (CSO) described in Chapter 11. The main storage medium is a computer type disc-pack; three semiconductor picture (frame) stores are also included to provide temporary storage of pictures during internal processing operations.

A minicomputer is included in the system to perform the control functions which are necessary during the operation of the animation store. With this arrangement the operator need not be aware of the internal architecture of the system and only simple control information is necessary to define the required record and replay operations.

A block diagram of the hardware used in the animation store is shown in Figure 7.13.

The disc drive and its ancillary units are interconnected by a set of digital video buses, each of which carries a luminance (Y) signal and two colour-difference signals (U and V) sampled at 12 MHz, 4 MHz and 4 MHz. This choice had to be made in 1979, before the establishment of 13.5, 6.75 and 6.75 MHz in CCIR Recommendation 601 but was also influenced by data rate considerations.

The disc drive uses a standard computer disc-pack with 20 magnetically coated surfaces, two of which are used for control purposes; the drive unit is modified to record and replay at normal television rates with the pack rotation locked to the 50 Hz television

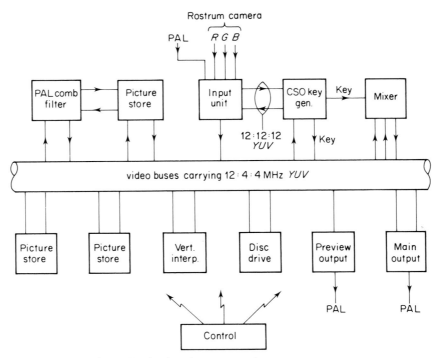

Figure 7.13 Schematic of animation store system

field frequency. It is also equipped with special heads to provide a high data-recording rate and a fast head-actuator to move the heads, from one 'cylinder' of tracks to the next adjacent cylinder, within the television field-blanking interval. One television picture (two fields) is recorded on each cylinder, and the heads are multiplexed in two groups of nine, with nine sets of record/replay electronics, so that one television field is recorded or replayed during each revolution of the pack. 815 cylinders are used for recording giving a total capacity of 815 television pictures. This corresponds to 32 s of continuous video if each picture is replayed only once. The disc drive recording and replay circuits are described in more detail in a later section.

In normal operations, the *RGB* video outputs of the rostrum camera are connected directly to the animation store. A trigger circuit is also provided between the two systems to obtain frame-accurate synchronisation.

In the input unit, the analogue *RGB* signals from the rostrum camera are digitally encoded at a 12 MHz sampling frequency and then converted to 12:4:4 MHz *YUV* signals. A 12:12:12 MHz *YUV* output signal is provided purely for wideband CSO operations.

The input unit also contains a separate A–D converter and a simple digital PAL decoder for deriving digital *YUV* signals from the analogue PAL input. One use of this input is to enable effects to be added in a post-production operation to output sequences previously recorded as PAL on a video tape recorder (VTR). The combination of operations whereby a PAL signal is first decoded to *YUV* and a still picture formed from one field or picture which is then re-encoded to PAL and eventually decoded a second time in a viewer's receiver, places very stringent conditions on the properties of the first decoder. For this reason the simple PAL decoding in the input unit is supplemented by more complex PAL decoding operations carried out in the PAL comb filter and associated picture store using techniques discussed in Chapter 4.

Additional picture stores provide buffer storage for operations which require simultaneous access to two or more stored pictures, and for editing purposes such as moving a still to a different cylinder. These are described in Chapter 3.

The purpose of the interpolator unit shown in Figure 7.13 is to generate both fields of an interlaced picture from one stored field. This facility is necessary either to generate a still picture from one field of a programme sequence or to provide slow and reverse motion effects.

The digital mixer and CSO key generator provide the means whereby pictures are combined in a similar manner to the multiple exposure and travelling matte techniques used in the film industry. The mixer, on its own, performs simple fades, or cross-fades between two pictures. The CSO circuitry discussed in Chapter 11 can generate key signals from any colour and provides smooth transitions from foreground to background pictures with no impairment resulting from the sampled nature of the signals; it can also produce transparency and shadow effects. Wideband key signals can be obtained from the *RGB* inputs via the 12:12:12 MHz *YUV* path; narrower band keys can be derived from a stored 12:4:4 MHz *YUV* signal. Either type of key signal can be stored for later use. An external key signal can be applied via the PAL input.

In each output unit, *YUV* signals are digitally encoded to PAL before digital-to-analogue conversion.

Instructions to the minicomputer controlling the store are entered from a control desk, similar to that used for the rostrum camera, with three panels and a VDU. The main panel is used to program the system and to call up information onto the VDU screen. A 'trackerball' control is included to provide electronic picture movement effects. A second panel contains preview controls which allow any stored picture to be selected and viewed either by itself or mixed with a second picture. The third panel contains the CSO controls.

A magnetic tape cartridge unit is provided for recording and replaying details of programmed picture sequences.

Operation of store

Various pages of control information can be called up onto the VDU screen. The main control pages are known as 'Run Level 1', 'Run Level 2' and 'Load'. The Level 2 page is used only if complex facilities such as CSO are required.

The first operation in the preparation of an animated sequence is to enter details of the sequence onto the 'Run' page or pages according to an animation sheet prepared by a graphics designer. A 'Run Level' page is shown in Figure 7.14. Each still to be loaded is given a unique 'still number' which is used for reference purposes in all the 'Run' and 'Load' VDU displays. The 'frame number' on each control line in Figure 7.14 indicates the point during a replay sequence when an operation is to take place. Thus, the first line indicates that still number 100 is to be repeated 200 times for frame numbers 1 to 200. A mix is programmed by defining the signal levels of the A and B inputs to the mixer and a smooth cross-fade can be produced by including the initial and final mix signal levels for the two inputs.

From the 'Run Level 1' information, the control system automatically allocates storage positions on the disc drive for each still number so that pictures are recorded in the correct order for the replay sequence.

After the run sequence has been programmed, pictures are loaded either individually or as a continuous sequence at normal television rates according to details entered on the 'Load' display. During loading, the input signals may be combined by mixing or CSO with previously stored pictures. Prior to loading, the alignment of combined pictures can be checked by means of the preview controls. Whilst a sequence of pictures is being recorded frame-by-frame a 'sequence so far' can be replayed to check animation effects and any mistakes can be rectified by re-recording.

Following completion of the load operation, some internal picture processing or 'assembling' may be required before the final replay sequence is transferred onto an external video tape recorder. This allows, for example, a cross-fade to be generated between two separate sequences of pictures on the disc store, a process which cannot be performed in real time.

By retaining the original pictures of an output sequence, a wide range of different effects can be generated without re-recording from the original artwork. For example, sequences or single stills can be repeated many times, motion can be reversed and electronic movement effects can be applied.

Consider, for example, a short animated sequence of a man walking down a road. On film a large number of cells would be required. A drawing of the background would form the bottom level, on top of which a series of cells showing the man walking would be placed in rotation. The artwork would be attached to the panning bars of the rostrum table and moved by differing amounts to give a feeling of motion and depth. Complex

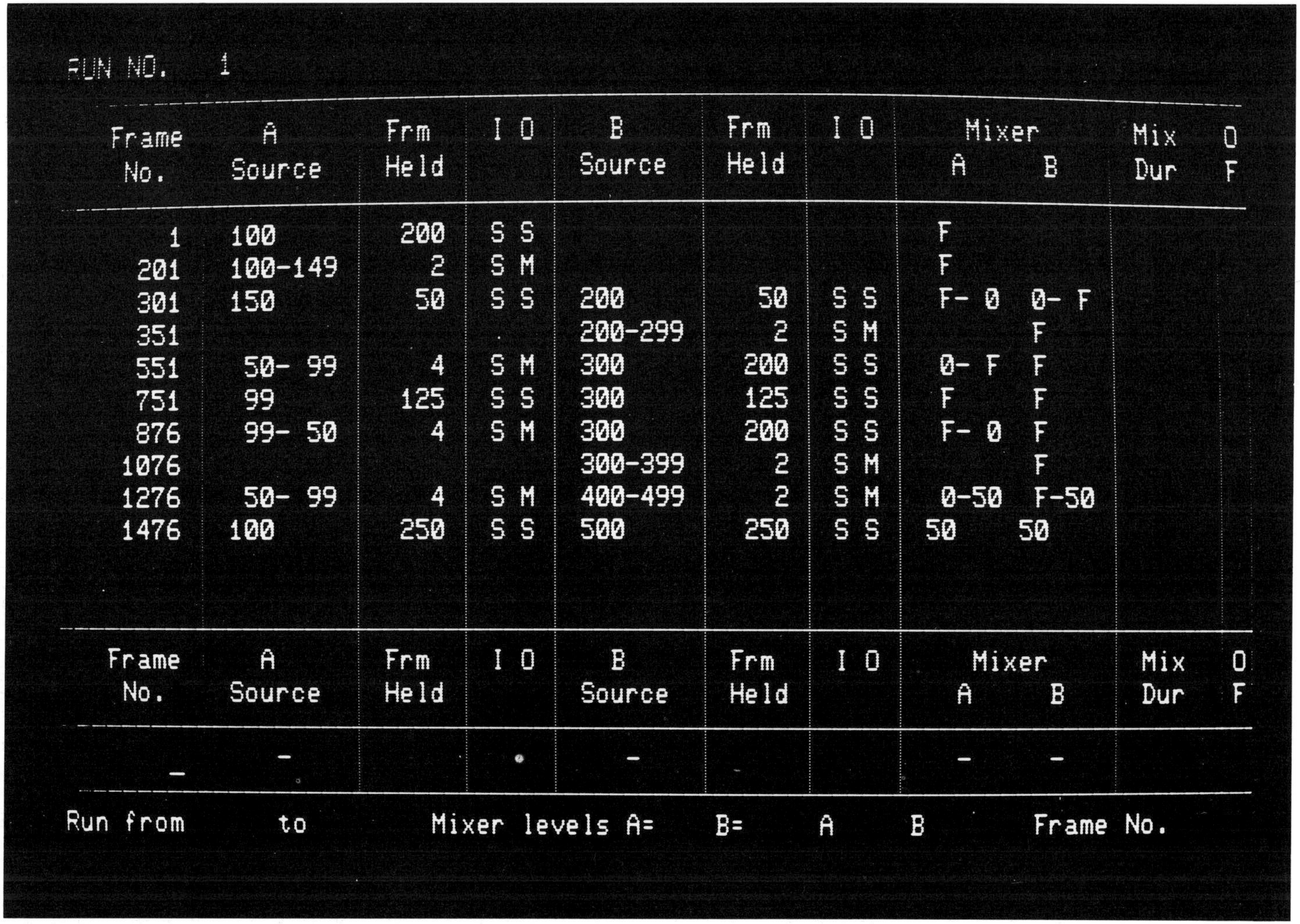

TELEVISION ANIMATION STORE 401

RUN NO. 1

Frame No.	A Source	Frm Held	I O	B Source	Frm Held	I O	Mixer A B	Mix Dur O F
1	100	200	S S				F	
201	100-149	2	S M				F	
301	150	50	S S	200	50	S S	F- 0 0- F	
351				200-299	2	S M	F	
551	50- 99	4	S M	300	200	S S	0- F	F
751	99	125	S S	300	125	S S	F	F
876	99- 50	4	S M	300	200	S S	F- 0	F
1076				300-399	2	S M	F	
1276	50- 99	4	S M	400-499	2	S M	0-50 F-50	
1476	100	250	S S	500	250	S S	50 50	

Frame No.	A Source	Frm Held	I O	B Source	Frm Held	I O	Mixer A B	Mix Dur O F
—	—	—	·	—	—	·	— —	— —

Run from   to   Mixer levels A=   B=   Frame No.

Figure 7.14 The 'Run Level 1' display used in the animation store

calculations are involved to assess how far the artwork should be moved every frame and on top of this the cells showing the man walking must be cycled through many times — a slow and painstaking process.

The same effect could be achieved in a totally different way with the video rostrum camera and animation store. A sequence of stills showing just the background movement could be recorded along with just one cycle of stills showing the man walking. The entire sequence can be created from these relatively few pictures.

The recorded cycle of the man walking is repeated as many times as necessary and moved across the screen with the picture movement facilities. This sequence is now keyed onto the moving background to produce the animation required. Additional material such as movement of clouds can be added in the same way so that a scene containing a variety of motion can be generated.

Recording pictures in real time on a parallel transfer disc drive

The storage device selected for the animation store was a modified computer drive using a removable, eleven platter IBM 3336 (Mod 11) disc pack as shown in Figure 7.15. It was modified by the manufacturers to lock its rotational rate to television synchronising pulses and by BBC Research Department staff who re-engineered each recording and replay channel to increase the data transfer rate by some 50%.

The disc-pack has 20 magnetically coated surfaces, each with its own record/replay head. Two of these surfaces are used for track position information and indexing, leaving 18 for the recording of digital video data. Nine sets of record/replay electronics are provided, with multiplexing arrangements so that data can be recorded or replayed using half of the surfaces at any time. The data disturbance generated by this head-switching arrangement is timed to coincide with the field blanking interval, so that the active picture is not corrupted.

A schematic diagram of the electronics for one of the nine channels is shown in Figures 7.16 and 7.17. Apart from those circuits involved in control and operation of the disc itself, circuits are provided for serialising, record code encoding and decoding, equalisation of the replayed data waveform, error protection, deserialising and timing correction of the replayed video.

Record channel code

A range of codebook codes [136, 137] and the Miller^2 (M^2) code [138] were tested and M^2 was selected as being better suited to the channels and their equalisers as originally supplied by the drive manufacturer.

Miller^2 is a modification to the Miller code (delay-modulation) which is intended to reduce the net d.c. content of the coded signal to zero, making it useful in digital recording channels. The minimum transition spacing is 1 input bit cell width and the maximum is 3 bit cell widths, with the result that it has a lower frequency content than the codebook codes investigated. Its disadvantage is that transitions can occur in the middle of bit cells, and, therefore, requires the output data stream to be clocked at twice the data rate. This gives a higher susceptibility to jitter on transitions of the replayed encoded signal as re-clocking of the replayed data occurs much nearer to these

TELEVISION ANIMATION STORE 403

Figure 7.15 View of the parallel transfer disc drive and its disc pack

404 MULTIPICTURE STORAGE

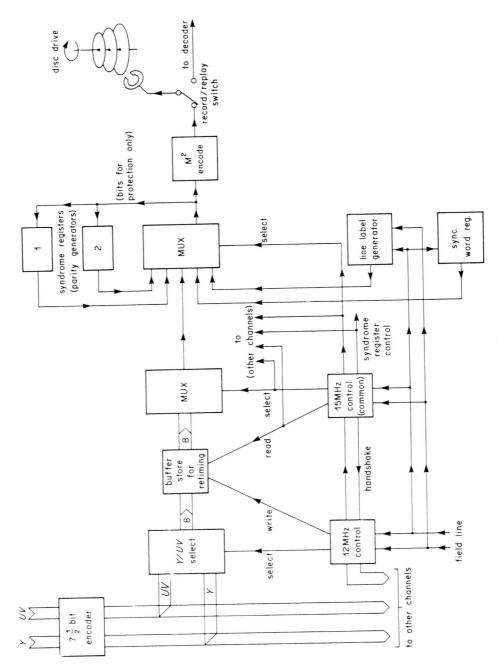

Figure 7.16 Encoding schematic for one of the nine channels

TELEVISION ANIMATION STORE 405

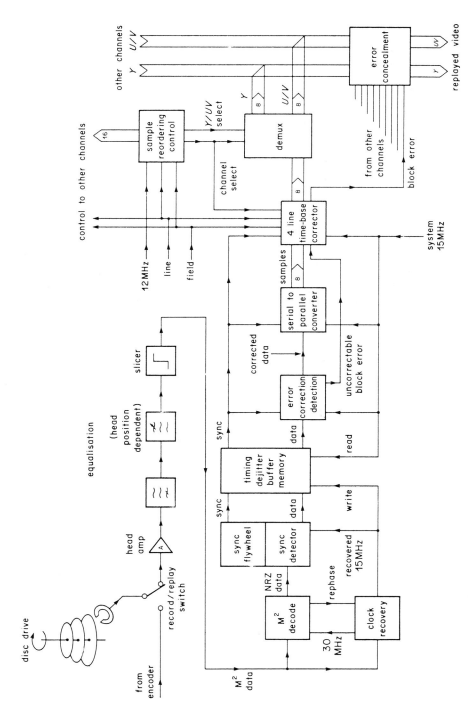

Figure 7.17 Decoding schematic for one of the nine channels

transitions. This can be a problem as pulse crowding effects can generate high levels of jitter on data transitions.

Tests involving all 18 recording surfaces with M^2 recording code and a data rate of 15 Mbit/s under a range of conditions showed that the error rate of any channel would be about 1 in 10^6 and might rise to 1 in 10^5 during a 3 month interval between services, after which the record/replay heads would require cleaning and checking. This error rate was thought to be sufficiently low to enable a modest error correction scheme to reduce it to the required output error rate, especially as this level of input error rate only occurred on the inner third of the cyclinders on the disc-pack, where recording bit density was highest.

Available redundancy and bit-rate reduction

Since the recording rate chosen for each channel is 15 Mbit/s the total data rate available is 135 Mbit/s for all nine channels. The digital television *YUV* component signals, sampled at 12 MHz, 4 MHz and 4 MHz, respectively and encoded using 8 bits per sample, gives a total rate of 160 Mbit/s and therefore some form of bit-rate reduction is required. This is achieved by two methods, and the effect of these on the available redundancy is shown in Table 7.2.

The television line blanking intervals are removed from the recorded signal. Only a total of 11 μs out of a possible 12 μs is discarded so that distortion of the blanking edges of the active video portion of each line is avoided. This reduces the mean data rate to 132.5 Mbit/s, but does not give sufficient redundancy for system overheads of error correction, etc. The field blanking interval cannot be removed since it contains the data disturbances generated when switching record heads or changing data tracks.

The chosen alternative was to reduce the sample size from 8 bits to $7\frac{1}{2}$ bits per sample. This is achieved by recording alternate 7- and 8-bit samples and allowing for the loss of each alternate least significant bit by adding it to the next 8-bit sample, as shown in Figure 7.18. This process is known as error feedback. However, only about a 3 dB increase in the quantisation noise is generated by this reduction in the data rate. This was considered to be a small sacrifice for a method which was easier to implement than

Table 7.2
Comparison of bit-rate reduction systems

Channel data rate (Mbit/s)	11 μs line blanking removal	Bits per sample	Data rate (Mbit/s)	Redundancy (%)
14	Yes	7	115.9	8.0*
15	No	6	120.0	11.1
15	No	$6\frac{1}{2}$	130.0	3.7
15	Yes	7	115.9	14.1
15	Yes	$7\frac{1}{2}$	124.2	8.0*
15	Yes	8	132.5	1.9
16	Yes	8	132.5	8.0

* Optimal

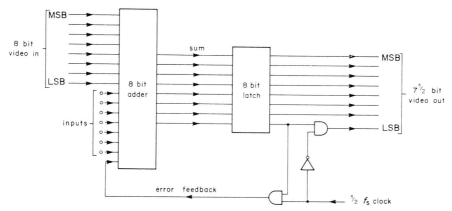

Figure 7.18 $7\frac{1}{2}$ bit error feedback encoder

other possible methods such as a reduction in sample rate or some sort of differential encoding.

The error feedback technique further reduced the mean data rate from 132.5 Mbit/s to 124.22 Mbit/s and gave an 8% redundancy with respect to the available recording rate of 135 Mbit/s. This is sufficient for system sychronisation and for the implementation of a simple error correction/detection scheme.

Error protection

An error protection scheme is required that will enable pictures of acceptable quality to be output even though the replayed data from the disc might have 1 error in 10^5 bits. The error rate after protection must be less than 1 in 10^9 if all 815 stored pictures are to appear substantially error free. Moreover, consideration must be given to the type of errors that are likely to occur and their likely distribution.

The M^2 code relies on the position of the data transitions to determine the original information. Thus, any extreme movement of these transitions or any loss of them will cause errors of decoding, which by the nature of the code could occur in bursts of several bits. The maximum burst error length has been calculated by analysing these effects for all possible transition sequences in the M^2 domain. Movement of a transition by one bit generates a maximum burst length of 3 bits in the decoded data, this being thought to be the most likely error. For the loss of a transition or insertion of an extra transition the maximum burst length is 2 bits. Other types of error were thought to be far less likely, and observations confirmed this view.

Burst errors caused by such error extention are also far more likely than those caused by dropouts, which are very rare with computer type disc packs, which are checked individually for such defects. Therefore, the error protection system should enable correction of error bursts up to 3 bits in length and provide an output error rate of 1 in 10^9 given a mean raw error rate of 1 in 10^5 affecting data replayed from the disc drive.

One helpful feature of digitally encoded television signals is that it is not necessary to protect digits of lower significance since the occasional corruption of these has a negligible effect on the quality of a displayed picture. Previous work with digitally encoded composite television signals [139] indicated that in that application the four most significant bits need protection. With a *YUV* digital component signal, the coding ranges are more fully used to describe the signal level as there is no provision for modulated subcarrier and sync. pulses. Thus, allowing for the greater visibility of errors affecting a still picture, it is suggested that protecting the four most significant digits will be adequate for the present application. By doing this, the effective redundancy is increased from 8 % (for $7\frac{1}{2}$ bits per sample) to 16.3 % giving a much wider choice of error protection and sychronisation schemes, and the possibility of selecting an overall system on the basis of ease of implementation.

As well as providing error correction over the range of error rates expected, the protection code is required to give some indication when error rates are becoming too high for all errors to be corrected. This indication could occur when a head requires maintenance and perhaps give warning of a possible 'head crash'. Such indications can also be used to enable an error concealment system to replace erroneous samples, which were not corrected, with interpolated values generated from information from other data channels. Also, the distribution of data between the nine channels is chosen to be such that adjacent samples in the reconstituted video data stream are substantially error free and can, therefore, be used in generating interpolated samples.

The code decided upon was a Hamming double error-detecting, single error-correcting code, which gives a reasonable measure of detection (for concealment) and correction [140, 141]. This family of Hamming code has a characteristic maximum code block length of n where

$$n = 2^m - 1 \quad \text{(where } m \text{ is a positive integer)}$$

and a maximum input data block length of k where

$$k = 2^m - m - 2$$

Therefore the number of parity bits added to each input data block is

$$n - k = 2^m - 1 - 2^m + m + 2 = m + 1$$

It is possible to shorten the maximum block characteristics by a constant number maintaining the number of parity bits used for a block. This is achieved by assuming that a constant number of bits at the start of the coded block are always zero and will therefore always appear so after the parity bits are added. These zeros can then be ignored because allowance can easily be made for their absence from the data blocks within the parity encoder and decoder hardware design.

Therefore, a block code (n, k) can be shortened to a block code of characteristics $(n - c, k - c)$, where c is a constant, and allow tailoring of the block size to the system requirements.

The minimum possible block length was selected to give the maximum possible error protection for the redundancy available. Blocks were then interleaved to give the required burst error protection. The simplicity of implementation of the code is shown in Figure 7.19 which outlines a typical encoder.

The system to conceal erroneous samples which cannot be corrected using the error correction system described above uses sample values on either side of an erroneous sample to generate an interpolated value which can most suitably mask the error. The greater the number of sample values used on either side becomes, the more accurate will

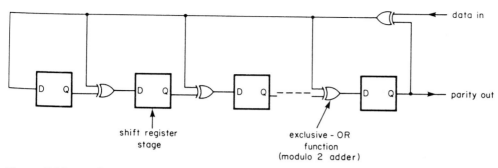

Figure 7.19 Implementation of simple block error correction codes

be the interpolated value. In the disc drive system, the samples are distributed among the nine channels so that if one channel is in error enough neighbouring samples are available from other channels for an interpolated value to be generated for it. This sample distribution is restricted because each channel requires an equal number of 7- and 8-bit samples in a particular order so that a common control system and data format can be used for all channels. The sample distribution used ensures that four samples on either side of any sample are always distributed to other channels. Thus, a five coefficient digital transversal filter with zero centre term can be used to generate the required interpolated values. The response of this filter can be tailored to give the best possible performance over the required video bandwidth and the response chosen is shown in Figure 7.20. This shows that errors occurring in picture areas containing horizontal frequencies of up to 3.6 MHz (1.2 MHz for U, V) can be accurately concealed using this interpolating filter. As large high frequency content is less common, and errors in such signals are less noticeable, it was felt that this response was well suited for error concealment.

Data format

The bit allocation in each of the nine channels for a single line of information is shown in Table 7.3. It can be seen from this table that a total of 480 bits of data per line require

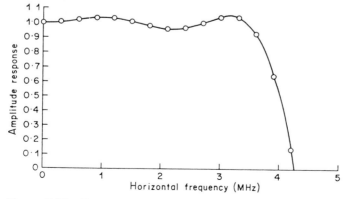

Figure 7.20 Frequency response of interpolation filter used for concealment

Table 7.3
Allocation of bits for one channel

MSBs per line = 118 × 4	472 bits (protected)
LSBs per line = 118 × 3.5	413 bits
Line label	8 bits (protected)
Sync. word	13 bits
De-jitter guard band	6 bits
	912 bits (480 bits protected)
Protection parity bits	48 bits (10% redundancy)
Total	960 bits per television line

protection and that 48 bits remain for use as parity bits. This is compatible with the Hamming code, when shortened to (88, 80) by the method described above. This is the shortest block length possible with the level of redundancy available. Pairs of these blocks are interleaved, along with the unprotected LSBs to give the required 3-bit burst error correction of the MSBs only. This interleaving will also protect against a high proportion (approximately 89%) of possible 4-bit bursts that may occur when error rates are high, as well as a small number of 5-bit bursts (23.5%). With the double error detection capabilities, all bursts of up to 7 bits in a pair of blocks will be detected.

Performance

The theoretical performance of the error correction code is illustrated in Figure 7.21. This shows the ability of the code to correct random single errors in the decoded data and its ability to correct random bursts of three errors. Since error propagation caused by the use

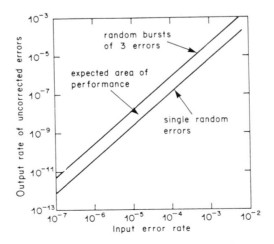

Figure 7.21 Theoretical performance of error correcting code

of the M^2 code gives bursts of errors of up to 3 bits in length, the actual output error rate will be in the area between these two curves. The concealment system will operate when random double errors occur in a block of data. Thus an output error rate after correction and concealment can be approximately calculated. This gives a final output error rate of about 1 in 10^{12} for input error rates of 1 in 10^5.

In the early period of service of the animation store, few errors were observed once a complete and careful alignment of each channel had been carried out.

Initially the most critical adjustment is that of the recovered data rate clock phase with respect to the replayed data, so that clocking of the data occurs in the middle of the data eye. Adjustment of the equalisation for different cylinders is also critical, this being optimised for an outer cylinder and inner cylinder with linear tracking being provided between these two points. Optimisation is easily achieved for the outer cylinder where the data packing density is at its minimum, with the most critical adjustment being on the inner cylinders where effects such as pulse crowding are at their worst. The record current supplied to each head is the final adjustment. This has a broad optimum range and is less critical than the other adjustments.

Severe problems were encountered in the design of the clock recovery phase lock loops. These are required to give a constant clock phase relationship with respect to the transitions in the recovered data, but should lock up very quickly after losses of data such as those that occur during head switching or movement. The use of a conventional phase locked loop in this situation requires that its loop bandwidth must be very narrow to achieve tight control on the frequency of the recovered clock. This had the disadvantages of very poor locking range and long lock-up times. To overcome these problems, information about the position of data transitions in the input signal is injected directly into the oscillator to help kick it into the correct phase and frequency relationship during the locking process. This type of phase locked loop is known as an injection locked loop and is described in greater detail in reference [142].

Lock-up times are typically less than two television line intervals after short periods of data loss; these periods being about 250 μs for a head switch and 1.1 ms for a head movement to the adjacent cylinder. The sync. word flywheel then requires two lines to lock and synchronise the rest of the replay system. As the total 'unlocked' time for head movement is about 1.4 ms, the timing of movement with respect to the video signal has to be accurate if this time interval is to fall completely within the field blanking interval. This required very careful adjustment for optimum timing, with disc control signals being sent just before the end of the active picture to allow for delays in the disc drive control unit.

The system has been operated with twenty different disc packs without further adjustment of the data circuits and only minor variations of performance have been observed. Hence, it appears that several different disc packs could be used during normal operation, as long as they are of a similar type of those in current use. Hence, a good reliable performance has been achieved, justifying the choice of error protection strategy which seems to be well suited to the type of errors generated by the disc drive. With careful alignment, the system has produced no noticeable errors even after fifty generations of copying.

ANIMATION OF LOGOS

General

In this section, two different methods of replaying animated television logos are compared. A novel method was tested in which optical discs were mastered with digital information. This method achieved the required animation effects but an alternative method using semiconductor storage and special methods to economise on the required data rate was preferred as the necessary hardware could be more cheaply replicated. Both methods are described as they illustrate several important factors which influence the choice of media for such applications.

Virtually every broadcasting organisation in the world uses channel identification logos. These usually consist of some form of animated symbol. The BBC is no exception, and for many years BBC 1 has used a rotating image of the world. Since 1969, when the channel converted to full colour transmissions, this had been a mechanical model of the world rotating in front of a mirror. The model was viewed by a monochrome video camera and the blue and yellow colours were synthesised electronically.

On the BBC Networks the same camera was used for many of the various caption sources required by the network mixer. The camera itself was remotely controlled and carried its own lights but, despite long and faithful service, had several disadvantages: it occupied a lot of space, it required quite a lot of power, it needed regular alignment and consistent results were difficult to obtain. This latter point was particularly important where the various regional centres had their own models, and required identical colours from their synthesisers.

In devising a new channel identification logo for the BBC's Network Channel 1 in 1984, means were sought for generating a new logo from dedicated equipment. Like the old logo, the new one would show a rotating world occupying about 1/6th of the picture area with static legends able to be placed in the remaining area. A high quality colour sequence of animated pictures was designed using computer graphics and a world outline database and the animation tested using the animation store described in the previous section of this chapter. The technical requirement was then to design and construct about twelve equipments which would store and replay the logo sequence which lasted about 12 s.

LaserVision optical disc with data instead of f.m. modulated video

LaserVision is the name by which Philips market an optical disc system aimed at the domestic consumer [143]. The discs contain video material derived largely from television programmes and feature films. They are pressed, rather like conventional gramophone records, from a master. The advantage of this system lies in the laser reading process which involves no physical contact with the storage medium and, therefore, no wear. Recordings may, in theory, be replayed an infinite number of times.

There are two disc formats commonly available, called Active Play and Long Play. Active Play discs provide one television picture (2 fields) per revolution of the disc. The players are capable of jumping complete tracks in the field blanking period allowing various 'stunt' modes to be achieved, such as still pictures and non-standard speed replay. The constant rotational speed used for Active Play discs results in an information density (along the direction of the track) which decreases with increasing radius. Where stunt modes are not required the total capacity may be increased by slowing down the disc when outer radii are in use so that data are recorded at a constant density at all radii. This is the Long Play format. The player recognises which format is being played from a code recorded at the beginning of the disc. For both formats, the pictures are recorded as f.m. composite video.

Experimental tests were carried out which involved mastering a LaserVision disc with segments of a short video sequence derived from the data for the new logo, and with pseudo-random data to provide means of checking error rates. Conventional pressings from this digital master were then replayed on a modified domestic player.

In these experiments the pits and spaces were used to represent binary '1's and '0's. The limiting packing density is reached where one bit cell is just less than the resolution of the system. Factors affecting the resolution are discussed in [144]. In particular the spatial cut-off frequency, f_c, is dependent on the numerical aperture, N, of the objective and the wavelength, l, of the reading laser and is given by

$$f_c = 2N/l$$

In the test $N = 0.4$ and $l = 0.63$ μm giving

$$f_c = 1270 \text{ cycles mm}^{-1}$$

The highest spatial frequency for an uncoded binary data stream would result from a sequence of 010101010. The smallest bit-cell size that may be resolved is, therefore, where $1/f_c = 0.8$ μm, equivalent to a linear packing density of 2500 bits/mm (64 000 bits/inch).

The discs were mastered at a constant rotational speed of 1500 r.p.m. This resulted in a linear packing density along the direction of the track which increased with decreasing radius. The mastering bit rate was chosen to be 34 Mbit/s so that the highest theoretical packing density would occur roughly in the middle of the recorded band. The resulting packing density varied between 100 kbit/inch at the innermost radius and 36 kbit/inch at the largest radius.

It was decided that NRZ data could be mastered directly onto the disc and recovered by a standard player without the need for any form of channel coding. To ease the demands on clock recovery circuits, the data were processed to remove strings of ones or zeros of greater than fifteen bits.

The video data to be recorded comprised a sequence of computer generated pictures of an animated, rotating world. Because of the limited data rate available from the disc, it was not possible to code the entire picture area to the CCIR digital *YUV* standard (Recommendation 601) [7] and reproduce real-time animation. Accordingly, only one sixth of the normal active picture area was recorded, comprising a square in the centre of the picture including the globe. This was the only part of the picture containing animation, the remainder being at black level.

Each line of video data contained 6 bytes of ancillary data which included a line number, a cyclic redundancy check byte, and a framing code (see Figure 7.22). The framing code was used to provide both a reference at the beginning of the line and to provide the correct phase of serial to parallel conversion on replay. The code was chosen to include a string of fifteen binary zeros, a sequence prohibited in the video data.

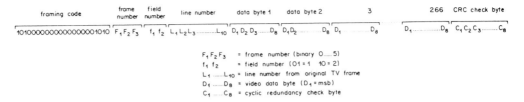

Figure 7.22 Line structure of video data

A $2^{15} - 1$ bit long sequence was chosen for pseudo-random data which was also mastered. This sequence has a wide spectral content and does not exceed the constraint on maximum string length. A commercial pseudo-random bit source (PRBS) was used.

Experiments to recover data from a digitally mastered LaserVision disc

Replay experiments used a Philips VLP830 LaserVision player modified so that control data normally derived from the replayed analogue video signal could be supplied externally. The signal paths were re-engineered to allow for the wider bandwidth of the digital data which contained frequency components both above and below those found in the normal f.m. replay signal.

The pitshapes and spectra of the pseudo-random data replayed at various radii are shown in Figures 7.23 and 7.24 respectively. Using these and other measurements, the equalisation requirements were calculated and, for a single frequency of half data-rate, are shown in Figure 7.25.

The pseudo-random data contained no d.c. component and so the extreme l.f. and d.c. components were removed by a.c. coupling the signal after the equaliser. Tracking performance for the pseudo-random data portions of the disc were considerably better than for the video data portions of the disc and this difference was traced to the lower l.f. content in the pseudo-random data. In order to measure the probability of occurrence of different error lengths, as well as overall error ratios, an error logging system was built. This comprised programmable hardware, capable of measuring error length, linked to a computer. This system was capable of logging many thousands of error events and providing a statistical analysis in a short time.

Detailed measurements on five discs were made only on the largest radius band of pseudo-random data. While the signal amplitude and noise characteristics of some inner bands was sufficient for recovery of the signal, only simple equalisers, with insufficient gain, were available for these tests. The five discs gave similar results with an overall error ratio of about 3 in 10^6 which worsened to about 1 in 10^5 when the discs became dusty. On inner tracks, where equalisation was sub-optimal, error ratios increased dramatically. The error length distribution of about 100 000 error events is plotted in Figure 7.26. It can be seen from this graph that the majority of error events (99%) are between 40 bits and 2000 bits in length.

The video data were reconstructed using hardware that included no error protection.

After careful adjustments had been made to the equalisation, the subjective error rate seemed fairly good, perhaps slightly better than the drop-out rate of a domestic VCR.

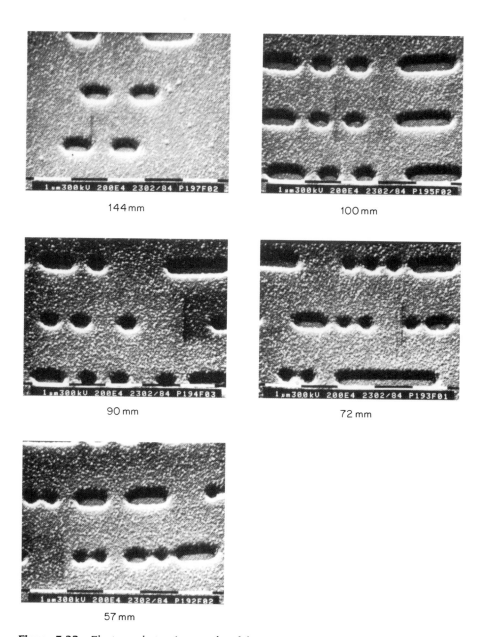

Figure 7.23 Electron photomicrographs of the surface of a digital disc at various radii

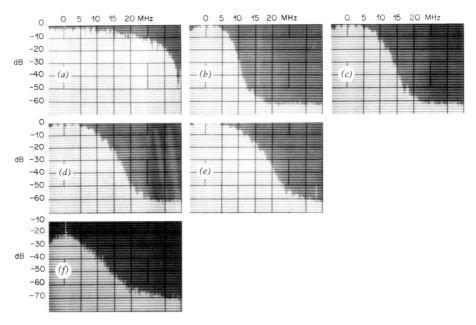

Figure 7.24 Spectra of the replayed data. (a) Spectrum of PRBS data at 34 Mbit/s direct from generator, (b) spectrum of PRBS sequence off disc at radius 60 mm, (c) spectrum at 90 mm, (d) spectrum at 110 mm, (e) spectrum at 130 mm, (f) spectrum off flat silvered disc — no data recorded (noise)

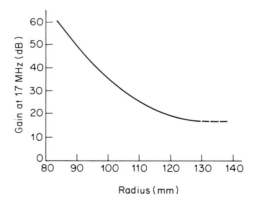

Figure 7.25 Equalisation required at 17 MHz plotted against radius

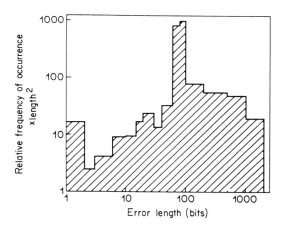

Figure 7.26 Distribution of error burst lengths

Most events appeared as short streaks of 10 or 100 samples in length, sometimes visible at the same point on the picture for several seconds. The noise-free nature of the computer-originated picture tended to increase the visibility of errors. Figure 7.27 is a photograph from a monitor showing the replayed picture.

Low-frequency components in the data (not present in analogue f.m. signals) caused a certain amount of mistracking. This became evident on the outermost radii of video data as long streaks of errors, which were correlated with picture content.

From the photomicrographs of the surface of the disc shown in Figure 7.23 it can be seen that individual pits are still capable of being physically resolved down to a radius of about 57 mm: the effective spot size of the 'cutting' laser appears to be about 0.6 μm. In these experiments, however, signals were successfully recovered only down to a radius of 115 mm. Errors at this radius were predominantly burst errors caused by disc defects and not single-bit events as would be caused by inadequate signal-to-noise ratio. A small reduction in radius from this point, caused a rapid increase in short burst errors, a sign of inadequate equalisation. At this radius the packing density was 1880 bits/mm (48 000 bits/inch).

By extrapolation from Figure 7.28, the minimum optically resolvable bit-length may be estimated as 0.4 μm. This is equivalent to a cut-off in the spatial frequency response of 1250 cycles/mm which agrees closely with the theoretical figure found in the previous section. From the noise characteristics of the disc and the response of the player, it may be estimated that with improved equalisation, single-bit 'noise' errors would begin to be significant at a radius of about 105 mm. At this radius the packing density along the track is about 2100 bits/mm (52 000 bits/inch).

By using a constant bit density along the track, the total storage capacity per side at 52 kbits/inch would be 9 Gbyte. The data rate would then vary between 18 Mbit/s and 51 Mbit/s assuming a rotational speed of the disc of 1600 r.p.m.

It was concluded from this work that the data rates at the outer radii are adequate for the real-time display of broadcast quality moving picture sequences containing highly redundant features such as the large plain areas in the 'rotating world' logo. In this

418   MULTIPICTURE STORAGE

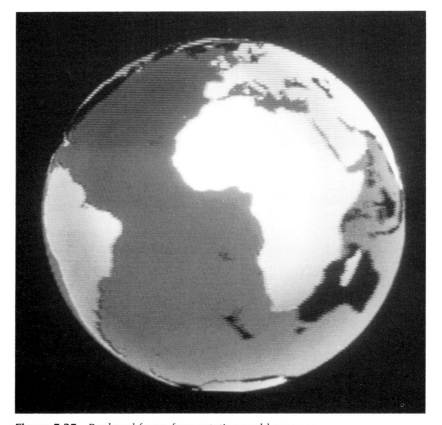

Figure 7.27 Replayed frame from rotating world sequence

application, as in others where continuous running is required, the inherent low wear of the optical replay process is an essential requirement.

A very high degree of reliability is required for the logo generator which has to be instantly ready at all times almost on a 'when all else fails' basis. Concerns about the lifetime of the first generation domestic equipment of the type used in the tests was one of the chief reasons for the BBC's choice of an entirely solid-state realisation of the rotating world logo produced by the BBC Engineering Designs Department. This version is described in the next section.

Solid state implementation of animated logo

Having opted for a solution, which involved no moving parts, to replace the electro-mechanical rotating world logo, the problem became one of devising a coding scheme so that data storage was reduced to manageable proportions.

The technique of coding data known as run-length coding is particularly effective when applied to electronic graphics. In this case it uses the principle that most areas of a

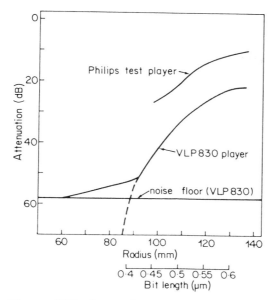

Figure 7.28 Attenuation at 17 MHz plotted against radius for a digital LaserVision disc at 1500 r.p.m

graphic are constant, and the picture does not have to be defined on every pixel. Since television is a scanning process many consecutive pixels are identical, so a number can be stored corresponding to the length of a particular 'run' of any colour.

Each run usually defines a particular colour, not in terms of red, green and blue which would use too much memory, but by a reference code. This code either looks up a colour definition table or drives an analogue colour synthesiser. There is a practical limit to the number of colours that can be displayed at any one time but not to their actual value.

This method is very efficient in terms of data storage, particularly for small or simple graphics, but does have a major deficiency: each pixel can be only 'on' or 'off'. This results in an effect known as 'aliasing' where diagonal and curved lines are stepped. To an extent, the effect depends on the pixel size (which is why home computer graphics are much coarser than those traditionally seen on television) but is in fact due to sampling a signal with only two quantisation levels. It is as much due to the horizontal sampling as to the vertical sampling (the line structure) over which there is no control.

To overcome aliasing it is necessary to define the transitions over several pixels. This is often known as multi-level or anti-aliased graphics the theory of which is discussed in detail in Chapter 9. Using this technique edges can be more accurately positioned, since the effective transition points are not restricted to pixel boundaries, and their rise-times can be controlled. This generates images that are more closer to those from a correctly aligned camera.

The BBC Engineering Designs Department has developed a method of coding digital video data which combines the advantages of conventional run-length with the greatly

420   MULTIPICTURE STORAGE

improved quality of anti-aliased transitions. There is, of course, a penalty in that the storage efficiency is reduced, but the saving is nevertheless quite substantial [145].

The data can be of four types:

Run-length This is indicated as a control byte by the most significant bit (MSB) and ensures that the current output data is held for the number of pixels defined by the seven least significant bits (LSB). The range is one to 127. For longer runs run-length can be cascaded.

Pixel mode This is indicated as a control byte by the alternative state of the MSB. The seven LSBs determine the number of pixel definitions that follow. The pixel run can be in the range zero to 127.

Pixel bytes These bytes have no restriction and determine the output level. A sequence of these is read after a pixel mode control byte. The last byte usually sets the level that will be held during the succeeding run-length.

End of line This is a special form of run-length, which appears to specify a run of 128, but causes the current data to be held for the rest of the present line. It can be used on its own to define 'blank' lines.

To give an idea of the data saving, a single monochrome field of the globe stored on a pixel basis would occupy over 200 kbytes; whereas by using the new coding method it can be stored in less than 8 kbytes; a saving of more than 25:1.

The symbol itself is a rotating image of the world with a caption displayed beneath as shown in Figure 7.29 (see colour plates between pp 506 and 507). It is larger than the previous world, and there is no reflecting mirror, but the detail and accuracy are much greater. The caption is customised for each of the various regions. The world itself is like a shell with all the sea areas etched away, leaving only the land masses. The outside of the shell is gold and appears to be lit by a spotlight from above the viewer, while the inside of the shell is black. Behind the shell is a shaded blue disc. All this shows the land on the front of the globe moving across the highlighted gold, and the land that is visible on the inside of the shell is black. Where the shell is completely transparent, that is sea on the front and back, then the shaded blue is seen.

A full rotation of the globe takes 12 s which requires 600 separate fields. It should be noted that for a smooth animation it is very important to update the image on every field, rather than on every frame.

The image is built up electronically in a similar way to the description above using hardware illustrated in Figure 7.30. Two stores hold the foreground (the highlighted gold shell) and background (the shaded blue disc with the caption). These are true frames in that they each hold a full resolution colour frame over the whole picture area and there is no restriction on their content. They are referred to as the 'fixed' memory.

The map data are stored in another memory known as the 'sequence' memory. This is a large store of 5 Mbytes, and is expandable up to 16 Mbytes. It is divided for convenience into 600 pages of 8 kbytes, each representing a single view of the globe. The data are compressed according to the method outlining above, and hold both the foreground and background information. Each page is read in sequence and decoded by the controller.

Both stores of the fixed memory are read continuously, in that the address is

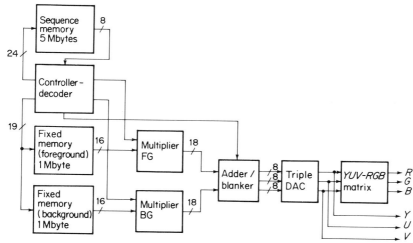

Figure 7.30 BBC 1 symbol generator, block diagram

incremented on every pixel. However the sequence memory addressing is rather more complex. The controller houses the data decoder which regulates the address incrementing according to the data; thus the address is held during run-lengths or after an end of line, but is incremented regularly during pixel data. The controller also passes data during pixel definitions and holds the current value during run-lengths.

The data from the sequence memory modulates that from the fixed memories using high speed digital multipliers. The map data from the decoder is split into its foreground and background components, and correctly scaled for the 8×8 bit multipliers. The background part is also suppressed by the foreground to eliminate any 'breakthrough'. For each video signal from the fixed memory there are two data streams: one for the luminance and one for the multiplexed chrominance. The map signal for the chrominance multiplier needs to be modified to allow for the two consecutive chrominance samples representing the same point in space.

The modulated foreground and background signals are combined in adders. To maintain quantising accuracy, 9-bit products are carried from the multipliers to the adders. There are two sets of adders, one each for the luminance and chrominance data streams. After the combining process the output signal is digitally blanked, and the chrominance signals are demultiplexed.

The background signal from the fixed memory, which is displayed in all areas around the globe, also contains the customised captions for the various regions. A facility has been included within the blanking optionally to blank an area of the screen. This is used by the network device to display or conceal a 'Ceefax' caption to indicate subtitling of the following programme.

A triple DAC is used to provide conventional component signals and an analogue matrix converts these to *RGB* for use by a PAL coder.

Since the construction of this equipment, other dedicated logo generators have been constructed using similar principles and different combinations of the hardware modules described above. Notably, a BBC Network 2 symbol is generated in this way. This symbol shows the characters TWO growing sequentially out of a plain background.

8 Digital Video Tape Recording

A. Todorovic

THE ROLE OF THE VTR

If the role that video tape recorders play in modern television production and postproduction, is considered it is obvious that the digital video tape recorder is the key element of an all digital television studio.

It would be difficult nowadays to find, or even to imagine, a television studio without video tape recorders. They are used for recording, editing and airing programmes and for storing, viewing and exchanging programmes. Their technical performance sets the upper limit for the technical quality of the final picture and sound and their operational capabilities determine the boundaries of creative possibilities. The development of video tape recorders and creative production methods is a neverending race. Sometimes the video tape recorder limits the ambitions of programme makers but can also offer more than is able to be used.

LIMITATIONS OF ANALOGUE RECORDING

Analogue video tape recorders, regardless of their precise format, have recently become extremely sophisticated. From the 1956 'Quadruplex' 2 inch format machine which, in spite of all its imperfections and limitations provoked a real revolution in the field of television production, to the present day machines with formats using 1 or $\frac{1}{2}$ inch wide tapes, we witnessed a continuous stream of improvements. However, we must accept that the analogue machines have begun to reach the limits of development. These limits are not set by the possibilities of the present day technology, or by an assessment of future development, but are the result of some basic and unavoidable constraints.

(a) All VTRs use a tape transport based on rotational video heads and a servo-controlled linear tape motion. Such systems require a very precise transport design and sophisticated servo systems. Nevertheless the residual off-tape timing error still amounts to a few microseconds and it seems unrealistic (since we deal here with a predominantly *mechanical* system) to expect much better results in the future; since such an error is intolerable, the manufacturers have developed a number of extremely efficient electronic timing correctors which reduce the residual error to less than 20 ns. Such residual error, however, is acceptable in first generation recording but, will have an adverse influence on the quality of subsequent generations.

(b) The multigeneration (re-recording) capability of analogue recorders is even more limited by the rapid degradation of a number of significant parameters such as the signal-to-noise ratio, the differential phase and gain and the Moiré or the luminance–chrominance delay. We should also mention the head band edge effects, which are not objectively measurable, but which nevertheless introduce serious impairments to the quality of the final picture in a multigeneration process. The aforementioned residual timing errors, and these degradations of different key parameters, limit the multigeneration capability of analogue video tape recorders to a maximum of between four and seven generations (depending on the stringency of the quality criteria).

(c) Magnetic video tapes, used for analogue recording are specially designed for the optimum recording and reproduction of video signals, recorded in transverse or in slant tracks. Such an approach is inevitably detrimental to audio signals recorded in longitudinal tracks on the same tape. Therefore, it is not surprising that the best available quality of an audio signal played back from a VTR is still far from the one offered by audio tape recorders.

(d) Analogue VTRs now offer a huge palette of different operational possibilities: slow motion, freeze frame, recognisable picture in 'shuttle' (fast forward or reverse) and sophisticated editing. However, the development of postproduction techniques demands some requirements which analogue machines will never be able to fulfil satisfactorily, since it is difficult to imagine the possibility of performing colour correction, colour separation overlay or reframing on the output signal of a conventional analogue composite VTR. It would admittedly always be possible to decode the reproduced colour signal into Y, U, V components, to digitise these components in accordance with the digital studio interface standard, to perform all desired manipulations of that digital signal, and then reconvert the components into their analogue form, reencode them into a PAL signal and finally record on a second analogue composite VTR. It is obvious that such a method is not a very satisfactory approach to these new operational requirements since the consecutive codings and decodings would cause a serious deterioration in the quality of the video signal.

(e) Finally the very nature of the present day analogue colour television systems, and especially of the PAL system, makes it mandatory to observe during the editing process a sequence of a given number of fields. In the PAL system, as we know, the V component of the modulating chrominance signal is reversed every other line, which results in a four field or two frame cycle. On the other hand the colour burst has a quarter cycle of subcarrier offset per line which causes a 90° phase shift of the subcarrier on successive lines. Consequently the whole cycle, or the whole sequence in a PAL system is

Frame 1: reference frame

Frame 2: Subcarrier 90° shifted and V axis reversed

Frame 3: Subcarrier 180° shifted

Frame 4: Subcarrier 270° shifted and *V* axis reversed

Therefore, if we wish to perform an ideally disturbance-free video tape edit we have to reduce our editing resolution (minimum distance between two consecutive edits) to eight fields or four television frames and to keep track of the field number one in the sequence.

It is, therefore, quite clear that analogue composite video tape recording threatens to limit the development of production and postproduction methods. At the same time it should be remembered that even the most modern analogue composite recorders require a number of delicate and time consuming alignments which are preferably performed by highly skilled operators.

The digital video tape recorder (DVTR) was, therefore, envisaged to be not only a necessary part of an all digital studio, but also a means of overcoming these limitations and opening up virgin territory in which new programme production methods could be developed.

POTENTIALITIES OF DIGITAL RECORDING

The digital approach appears to incorporate the capacity to abolish the limitations which confront analogue composite magnetic recording. The expected quasi-transparency of DVTRs should make it possible to achieve an adequate number of generations. The possibility of recording audio signals by digital means should eliminate sound quality problems. Finally, coding by components would allow the reproduced signal to be handled in all sorts of ways, while putting an end to the awkwardness caused by an eight-field PAL sequence. At the same time it seems reasonable to hope that digital machines will require neither the quality nor the accuracy of adjustment which burdens present video tape operations.

All these potential advantages of a DVTR made such a machine very attractive to broadcasters. It is not surprising, therefore, that quite early (by the mid-seventies) experiments on digital video and audio recording took place in several research laboratories of large European broadcasting organisation. One of the first achievements in that field — the BBC experimental DVTR [146] — is worth mentioning at this point.

The goal set by the BBC Research Department was not the development of a practical DVTR, but rather the creation of an experimental device which would permit some of the potentialities of digital recording of video signal to be checked and allow different problems and new aspects of that area to be studied. The equipment is shown in Figure 8.1.

In 1975, when the experimental device was developed, the concept of a component digital interface signal had not yet been established and the recorder was built for the recording and playing back of a digitized PAL composite signal. Since the work on video recording was, in a way, a continuation of successful experiments in the field of digital audio recording, it was decided to use the same approach of a multichannel longitudinal recording. Headstacks providing 42 longitudinal channels which are able to handle up to 2 MHz (in analogue terms) per track at a tape speed of 120 inches/s were already available. Testing of these heads for data recording showed that a speed of 120 inches/s

Figure 8.1 The BBC experimental digital video tape recorder

offered the possibility of recording up to 2.4 Mbit/s per track, although at such a bit rate the heads were practically at the upper limit of their possibilities and the intertrack crosstalk worsened considerably the data recovery. It was, therefore, advisable to use lower bit rates for reliable operation.

At about the time the recorder was under development, means of sampling the composite PAL signal at the sub-Nyquist rate were successfully demonstrated. It permitted the use of a sampling frequency equal to twice the subcarrier frequency with 8 bits per sample. The digital signal was fed to the recorder in eight parallel bit streams, each carrying 1 bit of each digital word of the PCM coded signal. Since 40 channels were available for video recording, it was possible to allocate five tracks to each of the bit streams, which meant recording a reasonable 1.8 Mbit/s per track.

The basic specifications of the BBC experimental DVTR were

Tape speed	120 inches/s
Tape width	1 inch
Total number of tracks	42
Number of tracks allocated to each bit stream	5
Recording time	8 min on 10 inch reels
Recording code	Delay modulation (Miller code)

The average measured uncorrected bit error rate was about 10^{-5} and both error correction and error concealment were applied. During the experiment different proportions between the relative importance of the correction and the concealment were tested.

Although this recorder used sub-Nyquist sampling, only seven bits per word (the eighth was used for parity protection), and a rather simple error protection scheme, it clearly showed the major advantages of digital recording: timing accuracy, absence of noise, moire and banding, multigeneration capability, etc.

Future development of digital video tape recording technology will cause many of the concepts applied in this machine to be abandoned. Digital components will be used as well as the digital composite signal. The technology will permit the handling of much higher bit rates such as the 216 Mbits/s required for Recommendation 601 component signals or enabling sub-Nyquist sampling to be replaced by $4f_{sc}$ for composite digital recording. The longitudinal recording will be replaced by the helical approach, but nevertheless the BBC digital recorder will remain one of the major milestones in the history of the development of digital video tape recording.

After the successful launching of their digital standards converter (DICE) the IBA engineering team led by John Baldwin began investigating the possibilities of digital video tape recording [147]. As in the case of the BBC, their goal was not to develop a production recorder, but rather to demonstrate the feasibility of such a machine and to investigate some problems related to the basic concepts of digital recording. The IBA experimental machine which was based on the rotary head concept, is shown in Figure 8.2. It used a helical scan format on one inch video tape and achieved a tape consumption identical to the analogue Format B. Again, as with the BBC machine, the recorded signal was the PAL sub-Nyquist coded composite signal. The results of this work were shown during the Montreux Symposium in 1977, and followed two years later by new results, but also by demonstrations of other experimental feasibility models coming from the laboratories of major VTR manufacturers.

European broadcaster immediately grasped the importance of the breakthrough, but they were conscious, at the same time, that it would take some time for a real usable and commercially viable recorder to be available. The basic theoretical potentialities of digital video recording were more or less known, but, at that point in time it becomes necessary to define precisely the needs and expectations of future users. The EBU (European Broadcasting Union) working parties started collecting the views of its members. The collected positions were duly processed and eventually a statement was issued by the EBU Technical Committee, entitled 'EBU Requirements Concerning Digital Video Equipment' it outlined the quality targets and the expected operational features of the future recorder. During the whole process of standardisation of a unique digital video tape recording standard, which was started at that time, the statement was used as a

Figure 8.2 The IBA helical DVTR

benchmark for all proposed concepts and specific solutions. Other broadcasting unions and engineering societies, followed the EBU and issued similar documents on 'users requirements' which expressed almost identical views [148, 149].

We shall now comment on these basic requirements as they were described in the EBU Statement. By the time this EBU document was finalized, the concept of digital components was accepted by all interested parties, and consequently the future DVTR was expected to receive at its input and to deliver at its output digital component signals conforming to the selected world digital component interface.

From the quality point of view it was expected that the future recorder would be as near as possible to the ideal of 'transparency'. It was expected that, besides the basic quality of the picture and the sound, the digital approach would permit as many as twenty almost perfect generations (it was expected that the twentieth generation would be rated 4.5 on the CCIR five-point scale).

Modern analogue VTRs have, as already mentioned, their limitations and shortcomings, but they offer a wide range of extremely useful and sometimes indispensable operational features. The state of the art was, therefore, used for the definition of the expected facilities, and the broadcasters clearly stated that the features available to them at that time were, in a way, the minimum future requirements. On the other hand, believing that some of the present day operational practices will not be changed, they expected the future recorder to offer the same facilities, like the 90 min play time in the studio configuration, 20 min minimum for portable machines and 10 min for short segment multitransport machines.

The definition of the new digital video tape recording format had, thus, to be done in such a way that the defined recording standard permits or at least does not preclude the achievement of

picture-by-picture editing capability

variable broadcastable forward and reverse slow motion, down to the still picture

jogging, inching, and variable speed shuttle with recognizable picture in either direction up to 20 times normal speed

recovery of time-code at normal replay speed, in all shuttle, slow motion and still picture modes

lock-up time from any stand-by mode should be less than 1 s (less than 2 s for portable machines)

simultaneous reproduction (confidence replay) of video and audio during recording; switchable 625/525 operation

variable forward and reverse broadcastable fast motion

four high-quality monophonic digital audio channels and a separate track dedicated to time-code signals, all independently erasable and editable; it should be possible to transfer sound signals from any audio channel to any other; pairs of audio channels may be used for stereo

it should be possible to obtain a recognisable audio output (similar to the one available in conventional analogue recorders) at speeds other than normal in both directions

the maximum allowable builded-up video to audio time shift in the post production process should not exceed 40 ms after ten generations.

Since the aforementioned document was drafted by users, special attention was paid to the maintenance aspect of the future DVTR, although it was quite clear that these requirements were not in any way connected to the definition of the recording format itself. The reason behind the inclusion of these elements in the document was the desire to take advantage of the fact that the industry was just beginning the study of a completely new machine and to put forward certain concepts which would offer definite advantages to those who would one day, operate and maintain these tape recorders.

SPECIFIC PROBLEMS OF DIGITAL VIDEO RECORDING

A DVTR has to deal with some specific problems, which are sometimes quite different from those encountered in analogue recording, or sometimes require a completely different approach. Some of these problems are inherent to digital recording and some are the result of the previously described operational requirements.

We shall mention first the fact that the DVTR has to handle much wider bandwidths than its analogue predecessors. At the same time, in order to keep the operational costs at an acceptable level, the tape consumption should be kept in the vicinity of 250 mm/s, i.e. similar to the tape speed of the present day format B and C recorders.

The gross bit rate of the source video signal is 216 Mbit/s, but it is possible to remove any non-essential part of the television waveform (such as the line and field blanking) prior to a further processing of that signal. These 'unnecessary' signals follow a strictly predetermined sequence and all their characteristics are well known, which permits their easy reconstruction whenever required. Practically, a single digital word can cope with the field and line synchronisation. Therefore, the net bit-rate calculation for recording is based on the active parts of the line and frame. Besides lowering the total bit rate, such an approach makes the bit rates for 625/50 and 525/60 virtually identical, which, in turn, leads to the realisation of recorders for both scanning standards with a high degree of commonality. A component digital recorder can thus readily be made to handle Recommendation 601 in the fullest sense, requiring only the operation of a switch when changing from the 625 line 50 Hz standard to the 525 line 60 Hz. The analogue video signal, as we know it, carries in the field interval a number of insertion signals and data signals. The inserted signals are used to correct the signals in the analogue part of the chain and they may be omitted at the moment of the A–D conversion, but the digital data signals and some other ancillary signals have to be carried through the digital chain. On the other hand the error protection of the recorded signals will require some additional bits.

As the digital video tape recording format has to offer four high quality digital audio channels, and since the easiest way to ensure the required writing speed for audio recording is to use rotary heads, it was logical to select a system where the audio channels would be recorded as separate bursts by the same heads recording the video signal. In order to permit an adequate switching between the video and audio signals, and to permit easy individual access to any individual audio channel, a guard space between audio bursts and audio and video recordings had to be provided. That guard space (usually called edit space) can be expressed in terms of a certain bit rate. Of course this factor affects only the track configuration and has no impact on the transmission capacity. If we now add the net video bit rate, the ancillary data, the error protection overhead, the audio channels (with their protection) and the edit space expressed in Mbit/s, we obtain the total bit rate of 227 Mbit/s.

Such a high bit rate can be recorded by using parallel channels, but it must be remembered that each parallel channel requires a set of read–write electronics and a separate read–write head, this leading to more complex electronics hardware. Further, in order to record this total bit rate, whilst keeping the tape consumption at an acceptable level, it is necessary to use very high packing densities.

In any form of magnetic recording the packing density can be increased by either reducing the width of the recorded track, or by reducing the minimum recorded wavelength, or, in other words, record more bits per unit of length. Both the track width and the minimum recorded wavelength have a direct influence on the off-tape signal-to-noise ratio. The signal-to-noise (S/N) ratio is related to the track width (W) by

$$S/N \propto 10 \log W \tag{8.1}$$

Therefore, if we reduce the width by a factor 2, the resulting value of S/N will be 3 dB worse. On the other hand, the relationship of the wavelength (λ) to S/N is given by

$$S/N \propto 20 \log \lambda \qquad (8.2)$$

which means that by halving the minimum recorded wavelength we lose 6 dB on the value of S/N. Figure 8.3 shows that the magnetised volume per bit, or the total number of magnetized particles, is directly proportional to the track width, and that by halving it we just reduce the volume by two. However, when we reduce the recorded wavelength by the same ratio we reduce not only the dimensions along the tape, but also the depth of the recording, i.e. we reduce the total magnetised volume by four.

In this respect we have to take two additional factors into consideration. The reduction of the minimum recorded wavelength requires the reduction of the head gap. Beside the purely mechanical problems linked with the achievement of consistent production of magnetic heads with very narrow gaps, we must also take into account the relative size of the inevitable dust particles. With narrower head gaps the same dust particles will cause larger drop-outs. For its part, the track width is closely related to the trackability, i.e. to the ability of the video head following the tracks with sufficient accuracy. The trackability is directly proportional to the track width and inversely proportional to the

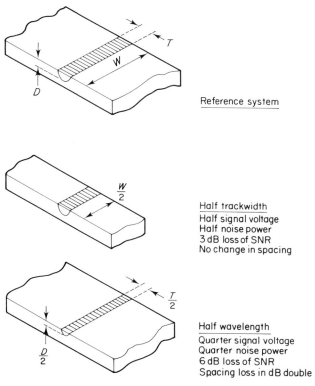

Figure 8.3 The relationship between signal-to-noise, track width and wavelength

track length. Thus correct and reliable tracking becomes more difficult to achieve as the video tracks become narrower. Therefore, in order to achieve the necessary packing density a viable compromise must be found between narrowing the video tracks and reducing the minimum recorded wavelength.

In order to design a recording system, or more precisely to define and build a VTR, it is necessary to know the sort of magnetic media on which the recording will be made.

The decision of which tape material to select had to be made between well established oxide particle tape, metal particle tape and metal evaporated tape. As we previously stated, one of the very important ways to improve the packing density is to reduce the minimum recorded wavelength. That minimum recordable wavelength depends on the characteristics of the tape (the size of the magnetic particles and the remanance). Conventional oxide particle magnetic tape, used in analogue recording tends to show regularly rather poor output characteristics at short wavelengths, especially in the region below 1 μm. Therefore, it was clear that the improvement in packing density via the reduction of the minimum wavelength required the use of a new magnetic video tape.

Since the editing and the video tape post production is in general carried out, by making a number of consecutive dubs of the recording, the final version of a television programme is unavoidably the nth generation of the original recording. As previously mentioned one of the basic limitations of the analogue recording is its limited multigeneration capability, and the primary attraction of digital recording is to overcome this limitation. It is well known that the errors in the digital domain can be either corrected or concealed. Error correction certainly offers the possibility of obtaining at the output a perfect replica of the input signal, but if the channel (and we can consider here the VTR as a transmission channel) behaviour is not fully predictable it is necessary to add a considerable overhead to the basic digital signal. Such a massive overhead will, in turn, require an even wider bandwidth and an additional improvement in packing density. On the other hand some excellent and elaborate concealment schemes would permit one to obtain an excellent picture quality in the first generation, even with a rather poor performance of the channel, but the concealed errors and their accumulation in the multigeneration process would seriously impair the picture quality of the higher order generations. Since the broadcasters requested an almost perfect picture at the tenth generation (it was thought that the quality of the picture on the tenth generation copy, when critical test material is used, should be at least equivalent to grade 4.5 on the CCIR five-point scale, and that the degradation of the quality with further generations should follow a very gentle slope), it was, therefore, necessary again to find a suitable compromise, a good balance between the correction and the concealment of errors in order to satisfy both the requirement for a manageable bit rate and for a good multigeneration capability.

Finally there is the problem of audio edits. An audio edit has to ensure the effect of a rapid change from one sound signal to another. An abrupt, or instantaneous, change will usually create an audible click. It is therefore, necessary for perfect audio editing to ensure that such a transition between two audio signals is not noticeable to the listener as an unnatural sound at the passage from one source to the other. In other words a sort of cross-fade between two signals had to occur at the edit point. The length of that 'cross-fade' is different for different frequencies, but a sort of optimum has to be chosen. The subjective tests have proven that the lower frequencies of the audio spectrum are more critical and it seemed possible to use the 70 Hz frequency as the most critical. The quality of the transition at this frequency was considered acceptable for 3.3 ms cross-fade length. By coincidence, that transition length corresponds to the transition produced with one track on a $\frac{1}{4}$ inch audio tape recorder running at 15 i.p.s when a 60° mechanical splice is performed. Since such splices are current practice in audio recording, this coincidence may be taken as a proof of the validity of the results of the subjective tests.

DIGITAL VIDEO TAPE RECORDING FORMAT

We saw that in order to be acceptable the DVTR had to achieve superior picture and sound quality, provide a number of operational features and overcome the limitations of analogue recording. Such a recorder was supposed to be able to receive at its input and to deliver at its output digital audio and video signals which would conform to CCIR Recommendations 601 and 656. To define such a recording format it was necessary to amalgamate the efforts of the European Broadcasting Union, the SMPTE and the industry. These joint efforts resulted in the definition of a single worldwide digital video tape recording format which was ultimately specified as CCIR Recommendations 657 [150, 151].

Recording formats are founded on some basic choices. In other words, the definition of all parameters relevant to a recording format presupposes the decision on

    type of recording,
    relation to different scanning standards,
    tape material,
    shortest wavelength.

We saw that the DVTR was expected to be able to record bit rates well over 200 Mbit/s, and that any sort of multichannel recording with stationary heads was obviously inadequate. Therefore, since the rotary heads were the only possible solution for the required writing speed, it was necessary to choose between the transverse and the helical scan recording. Since experience in the field of analogue video tape recording has clearly proven its inherent advantages, helical scan recording was also selected for the digital recorder [152, 153].

CCIR Recommendation 601 defines the parameters of the world digital studio interface standard. However that standard exists in two versions: one for the scanning standard of 625 lines 50 fields, and the other for 525 lines 60 fields. It is nevertheless true that the parameters were selected in a way that the differences for the two versions were reduced to the strict and unavoidable minimum [154]. The configuration of the digital video signal, therefore, makes it possible to conceive the recording format in such a way that it offers as large a commonality as possible between the 625 and 525 machines. Due to the larger production quantities and a number of common parts for the two scanning standards, it would lead to more economical machines. Furthermore, if practically no separate 525 and 625 development is required the resulting recorder will appear on the market sooner and should be less expensive. Finally such an approach will facilitate the achievement of the switchable bi-standard recorder.

We have already seen that the expected improvement in packing density implied the use of a new tape material. Three possibilities were envisaged:

evaporated metal tapes,
metal particle tapes, and
improved oxide tapes.

The possibility of using evaporated metal tape for magnetic recording was envisaged in the late sixties. That sort of magnetic media can enable very short wavelengths to be recorded. Since the thickness of the active metal layer is of the order of 0.16 μm the problem of the magnetic short-circuit, which occurs in the thicker layers of conventional materials, is avoided. However, although very intense research and development work in the field of metal tapes is under way, this technology is not yet advanced enough for mass production. Experimental tapes have shown inadequate ruggedness for professional use: a propensity to clog the heads, a tendency towards the oxidation of the metal layer and a need for much smoother base films than those currently produced. Due to the extremely thin magnetic layer every unevenness of the base film surface directly affects the performance of the tape in respect of drop-outs, spacing loss, and so on. It was clear that the evaporated metal tape had to be eliminated as a possible media for digital video tape recording.

Metal particle tape is in many ways similar to oxide tape, and, therefore, its manufacture is much easier to master. In practice the main difference lies in the nature of the magnetic particles. Metal particle tapes were currently available on the market for audio use. The drawback of metal particle tape lies in the requirement for very high magnetic fields, and therefore very high recording currents. This represents a problem for the design of recording amplifiers and was at that time out of reach for the ferrite recording heads, since the ferrite saturates at considerably lower values of the recording current. Furthermore some manufacturing difficulties and mechanical problems were observed in tests which were meant to simulate the broadcast operational environment.

All parameters relevant to the manufacturing and operational behaviour of oxide particle tape are well known. It is also known that conventional oxide tape has rather poor characteristics at very short wavelengths. However, the experience gathered in the field of consumer video products, proved to be applicable in designing and manufacturing of a new generation of oxide tape. These new tapes offer considerably improved output characteristic even at wavelengths which are shorter than 1 μm.

Bearing in mind the many other technological challenges presented by the component digital recorder it became clear that the most sensible choice for the digital recording format was to base this on the good results obtained in the field with the improved new oxide particle tape. (Indeed it was not until several years later that manufacturers again considered metal particle tapes, in this case for the analogue component and a composite digital recorder based on experience with the digital component recorder hardware, as discussed at the end of this chapter in the section describing commercial products.)

It seemed at the beginning of the development of the new digital recording format, taking into account the relationships in equations 8.1 and 8.2 concerning the track width, wavelength, trackability, drop-outs and the packing density, that the optimum solution was to select 1 μm as the shortest wavelength. Improvements in new oxide tape, and in manufacturing of video heads indicated that somewhat shorter wavelengths were also practicable. The final value of 0.9 μm, was eventually selected as a sort of compromise between the aim for higher packing densities and the concern about the overall reliability of the recording system.

THE CASSETTE

In an outline of their views on the future DVTR the broadcasters stated that an open reel machine might be accepted as the 'first generation' of digital machines, but that the ultimate goal should be a cassette configuration, since the cassette principle proved to be operationally extremely convenient. Also, the very high packing densities and the delicate magnetic material required as high a degree of protection as possible against ambient dust and handling stresses, since these could considerably increase the drop-out activity and consequently the overall bit error rate. That need for a maximum protection of the recording media, led to the conclusion that the cassette principle was not only desirable, but the only possible way for a reliable general purpose DVTR.

The first step in the definition or the design of a cassette is the choice of the tape width. Initial investigations in the field of digital video tape recording were based on the use of 1 inch video tapes and it was thought that such a width might be used later for the definition of the final format. However it soon became apparent that other widths were also practicable, and in some areas even more suitable.

It is certain, from a theoretical point of view, that any width can be selected and not only one out of the 'standard' series: 2, 1, $\frac{3}{4}$, $\frac{1}{2}$, $\frac{1}{4}$ inch; it has to be remembered, however, the world magnetic tape industry is only equipped with tools and machines designed to slit and to handle in general, tapes of the aforementioned widths. A new, for example 'metric', tape width will require a number of changes in the manufacturing process which would unavoidably affect the price of the final product. On the other hand, such a decision, it was feared at the time, might considerably delay research developments and standardisation work in the area of digital video recording.

If the bit-rates to be recorded, the expected and achievable packing density, and some operational aspects, such as the ease of handling and storing of a cassette and required playing time, are taken into consideration it is clear that the choice should be made between the values of 1 and $\frac{3}{4}$ inch. That choice had to be based on the assessment of some critical parameters:

cassette playing time,
guidability of the tape,
forces involved at different points of the tape path,
aspects of the portable DVTR,
search time.

The *cassette playing time* is set, on the one hand, by the requirement of the users, and on the other, by the maximum size and weight thought to be practical for efficient and easy handling and storage. By taking into account different requirements for playing times, the available tape materials and the mechanical parameters of the cassette itself, it is possible to compute, with a satisfactory degree of precision the dimension and the weight of the corresponding cassette. A large poll of opinions showed that the maximum acceptable size of the future cassette was about 200 × 300 × 40 mm³. Such a cassette was expected to offer 90 min playing time, which was achievable only with very thin (13 μm) tapes. A more conventional, 16 μm tape would offer about 74 min in a cassette of the same size. If a comparison is made between the cassettes for 1 inch and ¾ inch tape width for the aforementioned playing time, the rather balanced difference in size, volume and weight does not lead to a significant bias in favour of either of the two widths.

The assessment of the behaviour of 1 and ¾ inch tapes on the tape transport indicate some important differences. The mechanical analysis shows that for a given thickness of tape the *guidability* of the tape and the mechanical *forces* on the most critical points on the tape path in the scanner area depend on the tape width, and that a narrower tape offers advantages, which become more significant with very thin tapes, as those used in DVTRs.

A comparison of the theoretical designs of two possible configurations of tape transport for a *portable recorder*, based on a set of identical design assumptions, showed a slight superiority of the narrower tape. Finally, a comparison of average *search time* for two different tape widths resulted in an expected, but rather insignificant, advantage of the wider tape.

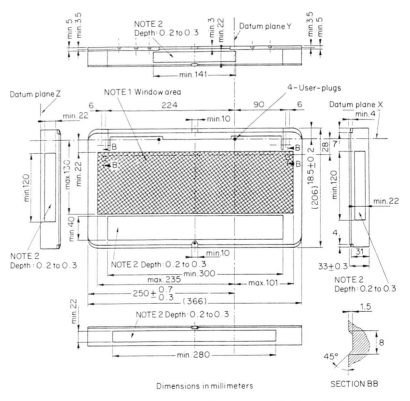

Figure 8.4 Outline dimensions in millimetres for the large D-1 cassette. Notes: 1. The cross-hatched area is available for the windows. 2. Label may be attached to this recessed area

Therefore, since the advantages of the narrower tape in the area of tape transport design are obviously more important than a somewhat shorter search time and a different 'aspect ratio' of the cassette, the tape width was set at $\frac{3}{4}$ inch (or more precisely at 19.010 ± 0.015 mm).

The design of the cassette for the DVTR, is based on the 8 mm video cassette (see Figure 8.4). The reels have two flanges, and the cassette allows top loading, front and side loading. There are four programmable holes in the base plate of the cassette and they are reserved for manufacturers' use (for the indication of the tape thickness, sort of active material, etc). Four additional holes are at the users' disposal for the 'record inhibit' and similar functions. There are three standard cassette sizes corresponding to the standard, portable and multitransport machines respectively:

Cassette type	Dimensions	Playing time	Tape thickness
Small size (S)	172 × 109 × 33 mm	11 min	16 μm
Medium size (M)	254 × 150 × 33 mm	34 min	16 μm
Large size (L)	366 × 206 × 33 mm	76 min	16 μm
		94 min	13 μm

TRACK PATTERN

The digital recording makes it feasible to record or play back the same track pattern with different transport configurations, different combinations of drum diameter and wrap angle. Theoretically the same possibility also exists with analogue recording as well; but in this type of recording there would be considerable practical difficulties if a unique, configuration of the scanner were not adhered to. Since there is flexibility in digital recording, a much larger degree of freedom is permitted for the manufacturers in the design and construction of the tape transport. The standard format is, therefore, only defined by a precise definition of the track pattern.

The optimum track pattern should be designed so that it may fulfil the following requirements

    recording of the standard component digital video signal,
    recording of four independent digital programme audio signals
    recording of a longitudinal time and control code
    recording of a control track
    the achievement of a broadcastable picture at speeds other than normal and of a recognizable picture at high shuttle speed
    provision of a 'recognisable' sound output at speeds other than normal
    ensuring a maximum 525/625 equipment commonality

The standardised track pattern for the digital video tape recording is shown in Figure 8.5 and the corresponding values in Table 8.1.

The transport for the DVTR employs a helical segmented format for video recording and, for reasons of complexity and economics, the programme audio tracks are

438 DIGITAL VIDEO TAPE RECORDING

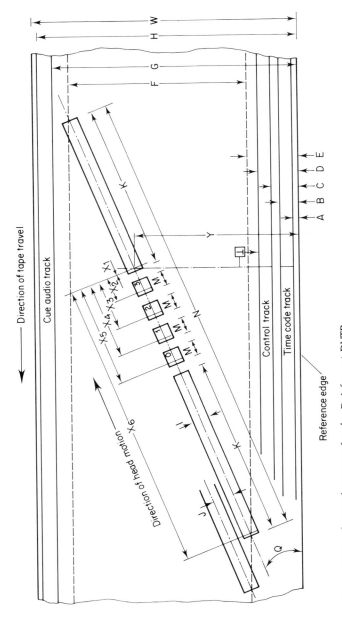

Figure 8.5 The track pattern for the D-1 format DVTR

Table 8.1
Record locations and dimensions for 525/60 and 265/50 systems

	Nominal dimensions (mm) 525/60	Nominal dimensions (mm) 625/50	Tolerance
A: Time code track lower edge	0.2	0.2	(± 0.1)
B: Time code track upper edge	0.7	0.7	(± 0.1)
C: Control track lower edge	1.0	1.0	(± 0.1)
D: Control track upper edge	1.5	1.5	(± 0.05)
E: Programme area lower edge	1.8	1.8	(Derived)
F: Programme area width	16/1.001	16.0	(Derived)
G: Audio cue track lower edge	18.1	18.1	(± 0.15)
H: Audio cue track upper edge	18.8	18.8	(± 0.2)
I: Programme track width	0.040	0.040	($+0/-0.005$)
K: Video sector length	77.71	77.79	(Derived)
M: Audio sector length	2.55	2.55	(Derived)
N: Programme track total length	170/1.001	170.0	(Derived)
P: Audio/time code head location	210.4	210.4	(± 0.3)
T: Control track record location	0	0	(± 0.10)
θ: Track angle arcsin (16/170)	(5°24'02")	(5°24'02")	(Basic)
W: Tape width	19.010	19.010	(± 0.015)
Y: Programme track reference point	10.490	10.490	(Basic)
X_1: Location of start of upper video sector	0.109	0.109	(± 0.1)
X_2: Location of start of audio sector 3	3.503	3.506	(± 0.1)
X_3: Location of start of audio sector 2	6.897	6.904	(± 0.1)
X_4: Location of start of audio sector 1	10.291	10.301	(± 0.1)
X_5: Location of start of audio sector 0	13.685	13.698	(± 0.1)
X_6: Location of start of lower video sector	92.231	92.324	(± 0.1)

All measurements shall be made at a temperature of $20 \pm 1°C$, a relative humidity of $50 \pm 2\%$, a barometric pressure of 96 ± 10 kPa and with a tape tension of 0.8 ± 0.05 N.

multiplexed with the video tracks, but in such a way that video and all audio channels can be individually recovered and edited. The channel coding, data rate and format, and the packing density, as will be discussed later in more detail, are identical for audio and video. The minimum recorded wavelength, as stated before, is 0.9 μm and the track pitch is 45 μm. There are 24 tracks per television frame in the 625/50 version and 20 tracks per frame in the 525/60 version. Gaps are provided between audio and video sectors and between audio sectors themselves, to allow independent video and audio editing. It should be noted that each audio sector contains only signals from a single audio source.

In the domain of slant tracks containing video and programme audio signals, and designated generally as 'programme tracks', two items deserve special attention

  gap azimuth, and
  position of audio sectors.

The high bit-rate to be recorded requires very high packing densities on the tape. At the same time all theoretical analysis and experiments show that good results cannot be obtained reliably if the track width is reduced below 40 μm.

Therefore, it might appear attractive to use azimuth recording which permits the recording of slant tracks without any guard band thus increasing the packing density. However, the azimuth recording method shows a clear disadvantage from the point of view of data pick-up at speeds other than normal when no track-following is used. At the same time digital recording allows for much narrower guard bands than the analogue recording. It is, therefore, possible to use very narrow guard bands of about 5 μm and conventional recording with the head gap perpendicular to the track record.

As stated before, the audio signals, for reasons of complexity and economics are located on the same slant tracks as video signals, and recorded by the same magnetic heads as separate bursts or digital audio sectors. However, there are several possible locations for these audio sectors along the programme track. They may be at the beginning or at the end of the track, or at its both ends, or finally, in its middle. The position of audio sectors has to be selected by an evaluation of the general aspect of the track pattern and the protection of video and audio signals against burst and random errors.

It is commonly accepted that there are advantages in designing a track pattern as 'balanced' (similar content in the upper and in the lower half of the tape) as possible. That leads to the conclusion that the grouping of all audio sectors at one end of the programme track should be avoided. On the other hand it should be noted that the implementation of a four head machine for the 525/60 scanning standard is rather difficult, since one field is recorded over 10 tracks—a number not divisible by four. If video tracks are not recorded as continuous ones, but with a split in the middle, it leads to 20 half-tracks or digital video sectors, and that number obviously *is* divisible by four. Finally, it should be emphasised that digital audio signals are, by definition, very vulnerable. The fact that error concealment in audio is not attractive, that audio sectors are very short, etc, leads to the conclusion that very special care has to be taken of these signals, and that every available protection should be applied.

One way of protecting audio signals is to select an optimum location of the audio sectors on the tape. Splitting audio sectors to the two ends of the programme track certainly offers better protection against horizontal scratches or grouped drop-outs, than putting all of the sectors in the same place. However, it is well known that the head-to-tape contact is the most critical at the point where the head enters into contact and the one when it leaves the tape. Mistracking errors are also more severely reflected at the ends of the tracks. Finally the grouping of all audio sectors may simplify the design of the processing circuitry. Since the splitting of the video track into two sectors is useful for the video part, there emerges an opportunity of placing the audio bursts between the two video sectors. Such a position requires more space on the tape and additional synchronisation blocks, but allows the audio sectors to be located at the safest place.

If the normal purpose of the control track is considered, it is logical to define its recording in such a way to assure the closest and the most unequivocal relationship between that signal and the corresponding slant track. Therefore, the start of the control signal corresponding to a given frame should be located (recorded) adjacent to the start of the synchronisation block containing the beginning of that frame. In mechanical terms it means that the recording of the control track should be carried out as if the control track record head were located on the drum itself. Since it is not certain that it would be possible to install a control track head on the drum with all possible scanner configurations it is necessary to define an alternate position at the exit of the scanner area. At the same time, in order to maintain the requested high degree of commonality of transports, it is necessary to define a position which will be common for both the 50 and 60 Hz scanning standards.

The data contained in the control track are discrete, periodic and closely time related by simple integer number. That means that the records on tape can be multiples of the maximum common information distance for 525 and 625 scanning standards. That greatest common distance is 4 tracks (20 track per frame in 525 and 24 tracks per frame in 625), which represents 1.9125 mm. Therefore, in order to allow for all possible scanner and tape transport configurations, the position of the control head is set at $110 \times 1.9125 = 210.375$ mm

Since two other signals, the audio cue and the time and control code, are recorded on longitudinal tracks by stationary heads, it is logical to foresee the installation of a single stack carrying all three stationary heads (or just two if the control track head is installed on the drum) and to standardize that position for the recording of the aforementioned audio and time code signals (see Figure 8.6). In practice this means that the cue audio and the time code signals are recorded 210.4 mm ahead of the corresponding video signal, or to be precise, ahead of the first sync word in the first sector allocated to the corresponding video frame.

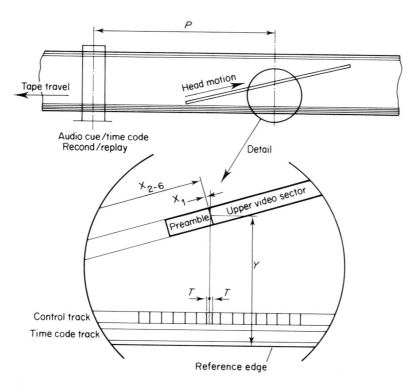

Figure 8.6 Location of the time-code and tracking control signals

RECORDING OF DIGITAL VIDEO SIGNALS

As we have already stated the digital video tape recorder has to accept at its input the digital video signals which conform to the CCIR Recommendation 601 (4:2:2 level of the family). The nature of the digital video signals, however, allows for only the selected active data to be recorded on tape. Therefore, if we omit the unnecessary redundant information and the information which can easily and harmlessly be reconstructed at the machine output, we will see that only 720 active luminance and 360 active pixels for each of the colour difference signals, and 300 lines per field (250 for 525/50 system) have to be recorded. It is important to note that the major part of these 300 lines carry digital video information, but that some of them in each field may carry ancillary data information. In precise terms, lines 11 to 310 from the first and 324 to 623 from the second field are recorded. The digital video data stream starts at line 23 in the first field and at line 336 in the second field, and since the remaining lines may carry some data, only lines 23 to 310 and 336 to 623 will be subjected to mapping and to concealment during playback.

One of the basic requirements for a DVTR is that it should be virtually 'transparent' for audio and video signals. Since the degree of 'transparency' is defined in more or less precise terms by the users, it is necessary to define the recording method and the error protection strategy in such a way as to ensure that a real recorder in a real operational environment will not depart from the 'transparency ideal'.

For the definition of the recording method and the optimum error protection strategy we do not only need a fixed quality goal (or if you prefer the expected 'level of transparency') but also a knowledge of all expected data losses and errors which may occur in the process of recording and playing back the digital video and audio signals. It means practically that we have to devise the recording method and the error protection strategy in such a way as to make them capable of facing the following three situations:

normal play
play-back at speeds other than normal
head failure.

For normal play we have to face not only the signal-to-noise related bit errors, but also random head-to-tape mechanical irregularity related drop-outs. We have also to be ready to expect some permanent defects on the tape itself, such as scratches, whose occurrence is also random and whose length can vary largely.

For slow motion, or other applications requiring play-back at speeds higher or lower than normal, the video head will not follow ideally the recorded track, and, in the extreme situation of shuttle speed, it will only cross the tracks. During that crossing the play-back head will reproduce only small parts of data from different recorded tracks, and only part of the reproduced data will be error free and, therefore, usable for the reconstruction of the 'recognisable' picture.

Since the DVTR is based on the helical scan principle, i.e. basically on the same principle as the present day format B and C machines, one cannot avoid expecting the problems witnessed in analogue recording. One of these problems is head clogging which is a purely head-to-tape interface problem. Considering that the DVTR has more than one recording head, and that data are digitised, it is realistic to expect that new recorder will offer the possibility of acceptable play-back signal quality even when one of the heads is clogged. In other words the data sequence in the recording process should be so distributed that the complete loss of one head in the recording or the play back process should leave a sufficient amount of correct data to permit a successful concealment operation.

As one might expect the best strategy for any one of the above situations is far from ideal for the others, and the final solution obviously has to be somewhat of a compromise.

We can see from Figure 8.5 that one programme track consists of two video and four audio sectors. The recording of digital video data corresponding to one television field requires 24 video sectors, and since the video data words are distributed on the tape on a four channel basis, these 24 video sectors consist of six groups of four sectors each. That set of four sectors is designated as a video segment (see Figure 8.7).

Saturation recording, used in digital video tape recording, is basically a simple process, but the signal processing required for an efficient use of that recording channel is quite complex. On the record side the processor must assemble blocks of words representing video, audio, ancilliary and internal control data, and to add check words to them which have to ensure a very high level of error detection, a good level of error correction and permit an efficient error concealment in situations when the correction fails due to overload. The record processing also includes adding necessary synchronising information and block identification to allow block recovery and orderly reassembly of the data stream. The sequence of video or audio words has also to be shuffled in order to separate adjacent samples and to spread them on the tape, which will increase the efficiency of error concealment in case of burst errors. Finally the data are coded into a recording format with appropriate spectral characteristics for the recording channel used.

Figure 8.8 shows the functional block diagram of the record path processing of a DVTR, according to the EBU–SMPTE standard.

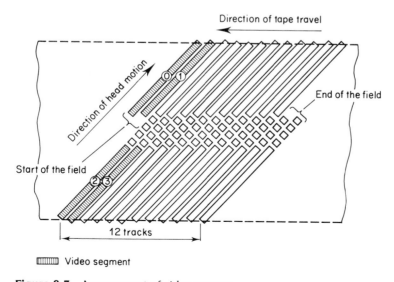

Figure 8.7 Arrangement of video sectors

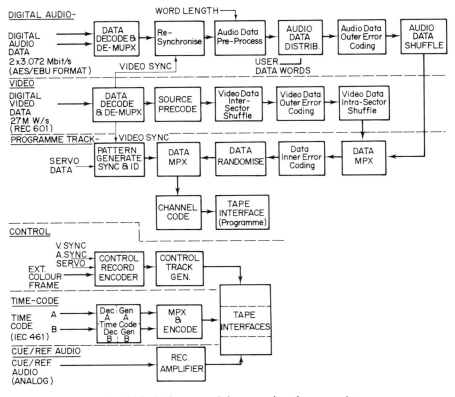

Figure 8.8 Functional block diagram of the record path processing

If we consider the part dedicated to video processing we can see that the incoming video data are first subjected to a one-to-one source video mapping or source precoding (as already mentioned, lines carrying ancillary information are not subjected to that mapping). That transcoding of video data words is done in order to reduce the peak error produced in a video sample in the case of the most probable distribution of digital errors.

The next stage in the video processing is the intersector shuffling or distribution of pixels. The pixels within each segment are evenly distributed between the four sectors belonging to that segment. This distribution results in distributing the 720 luminance words and 720 chrominance words of any line into four sectors with 180 luminance words and 180 (or 90 pairs) chrominance words per sector. The above distribution can be expressed as

$$2[(g + j) \bmod 2] + \text{int} (([j + 2(m \bmod 2)] \bmod 4)/2)$$

for the luminance component and

$$2((g + \text{int}(j/2)) \bmod 2) + \text{int} (((\text{int}(j/2) + 2(m \bmod 2)) \bmod 4)/2)$$

for the colour difference components where g is the segment where the data corresponding to a given line 'i' are recorded, m the number of that line in the segment and j the number of the pixel within the line.

After the intersector distribution the data are subjected to error protection coding. The error correction code employed consists of a block coding implemented in the form of a product code.

Although some experiments showed that uncorrected LSB did not impair noticeably the reproduced picture, it was considered that, for the sake of multigeneration capability and possible extensive postproduction work on the reproduced signal, equal weighting should be applied to all different bits of a data word. It is also important to note that all ancillary data signals provided in the interface signal are subjected to the same error protection coding as the video data words.

The product code, applied to the data words for each video sector, is arranged in a rectangular array with a row and column dimension. On the other hand two codes need to be differentiated: the *inner*, or *horizontal*, and the *outer*, or *vertical*, code. The inner code provides the basic protection against short duration random error sources, such as noise or short drop-outs, and it must be able to correct all these errors. However, it is also required that the same code should serve to reliably detect more extensive error sources, such as long drop-outs and scratches, which will be then processed by the outer code.

The video data words are arranged in a rectangular array with a row dimension of 600 bytes and a column dimension of 30 bytes. Two check words are first added to each column thus producing 600 'outer code' blocks of 32 bytes each. Each of these 32 rows, containing 600 video data bytes, is divided into blocks of 60 bytes each, to which four check bytes are then added to form 'inner code' blocks of 64 bytes. The resulting 'video product block' is shown in Figure 8.9.

The number of added check words to each inner block depends on the expected error detection and correction capability and, obviously, on the foreseeable error characteristics of the channel. Theoretically, a larger number of check words offers better protection, but at the same time increases the total overhead. If we compare, for example,

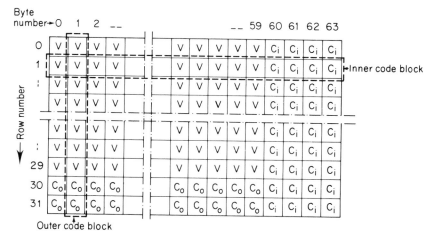

V = video data
C = check data

Figure 8.9 Layout of the video product block

two possible solutions: 60 + 6 and 60 + 4, we can see that the solution with 6 check bytes might offer a larger correction and detection capability, but that, in order to maintain the same tape consumption, it requires the shortening of the minimum recorded wavelength, or the lengthening of the video part of the programme track (at the expense of some of the other signals recorded on the tape). If we try to analyse the expected error characteristic of the channel, we can see that the actual quality of video tapes, and the expected developments in that field can easily ensure an off tape bit error rate better than 10^{-5}. On the other hand the investigation of probable error sources shows that an increased number of redundancy bytes (6 instead of 4) results in a slight reduction of uncorrected random errors, where 4 bytes in any case give good results, and that the 6 bytes scheme mainly add additional protection only for occasions of marginal performance. In the case of a slow motion play-back the results with 6 or 4 bytes would be virtually identical. Finally, an additional shortening of the minimum recorded wavelength will make the whole system more sensitive to very short drop-outs. It is therefore easy to conclude that a solution with four check bytes represents a better technical choice.

The selection of the error protection code has to be based on the requirement of a near-perfect detection of all errors, on the usefulness of a commonality of coding video and audio channels, and on the minimalisation of the required overhead.

For the outer code a shortened Reed–Solomon code over Galois Field 256 is selected.

The primitive, or field generator polynomial is given by

$$z^8 \oplus z^4 \oplus z^3 \oplus z^2 \oplus z^0$$

where the z^i are place-keeping variables in GF(2), the binary field. By convention the left-most term in such polynomials is the most significant, the 'oldest' when doing computations, and, where appropriate, the first to be written on tape.

The Reed–Solomon code generator polynomial in GF(256) is given by

$$G(x) = (x \oplus \alpha^0)(x \oplus \alpha^1)$$

in factored form. α is given by 02hex in GF(256). The Reed–Solomon check characters are given by K_0 and K_1 in

$$K_1 x^1 + K_0 x^0$$

the remainder after dividing $x^2 D(x)$, where the data polynomial $D(x)$ is

$$D(x) = \beta_{29} x^{29} + \beta_{28} x^{28} + \cdots + \beta_1 x^1 + \beta_0 \qquad (8.3)$$

by the generator polynomial $G(x)$.

The full outer code block shall then be given by

$$\beta_{29} x^{31} + \beta_{28} x^{30} + \cdots + \beta_0 x^2 + K_1 x^1 + K_0^0$$

For the inner code again a shortened Reed–Solomon code over GF(256) is used. The primitive, or field generator, polynomial is the same one as in the outer code, and the Reed–Solomon code generator polynomial is

$$G(x) = (x \oplus \alpha^0)(x \oplus \alpha^1)(x \oplus \alpha^2)(x \oplus \alpha^3) \qquad (8.4)$$

in factored form, where $\alpha = 02\text{hex}$.

The check characters are given by K_i in

$$K(x) = K_3 x^3 + K_2 x^2 + K_1 x^1 + K_0 x^0$$

the remainder after dividing $x^4 D(x)$, where the data polynomial is:

$$D(x) = \beta_{59} x^{59} + \beta_{58} x^{58} + \cdots + \beta_1 x^1 + \beta_0 x^0$$

by the generator polynomial $G(x)$.

The full inner block code is then given by

$$\beta_{59} x^{63} + \beta_{58} x^{62} + \cdots + \beta_1 x^5 + \beta_0 x^4 + K_3 x^3 + \cdots + K_1 x^1 + K_0 x^0$$

Such a coding arrangement provides for the correction of 1 byte in error and the detection of overload with a probability of failure of approximately 1 in 2^{18} overload occurrences.

In order to deal with error bursts corresponding to extended drops in level the product code uses the inner code to determine the locations(s) of the drop-out(s) by employing the error detection capability of the inner code. Once the location of the drop-out is found, then the outer code is used to correct the drop-out error(s). This outer code is, in effect, through the creation of the product code, operating on words which have been interleaved to a depth of 600 bytes.

For its part the inner code is not interleaved. Since the outer code can correct any two rows in error, then the maximum correctable drop-out length is 1200 bytes, which corresponds to 4.8 mm of track length. Further the outer code provides for double error protection, and consequently the correction of multiple short bursts, guaranteeing the correction of all double drop-outs up to 600 bytes in length. Multiple bursts, beyond two in each product block, can be corrected, but correction is not guaranteed as it depends on the drop-out lengths and locations.

In order to reduce the effect of uncorrectable drop-outs and scratches, which generally run along the length of the tape, and to improve pictures in shuttle the distribution of video words in each of the four recording channels is completed by a shuffling along each video sector (the so-called 'intra-sector' shuffle).

In the case of large errors and in shuttle mode, the error correction becomes overloaded and fails, and then the concealment of erroneous data has to be applied. To meet the requirements of an efficient concealment the shuffling has to provide a maximum number of correct samples as input to the concealment scheme. To fulfil this purpose the shuffling has to offer a reasonable spatial array, an equal distribution and adequate distance of errors and to ensure the availability of good quality pictures over all operational speeds.

Without the shuffling, a scratch or an asperity causing a drop-out would cause a simultaneous local loss of information on a given part of the reproduced picture. In the case of a longer scratch such a loss would be repeated from segment to segment and from field to field (depending on the length of the scratch). Since in such situations the correction would be overloaded, and since an uncorrected error is obviously more disturbing than a concealed error, the best policy is to conceal all the words that are reasonably suspect.

In normal play mode the concelment will be used relatively infrequently, but in shuttle mode the situation becomes completely different, and the words requiring concealment may even exceed the number of correct words. When the loss of information due to track crossing in shuttle speed is substantially equal on all segments of the reproduced picture, the quality of that picture would certainly be more than adequate for the recognition required in the editing procedure. However, at some critical speeds, the loss of information could vary significantly from segment to segment, which might further impair the quality of the resulting picture.

Therefore the shuffling scheme has to be selected in such a way as to reconcile all these situations as much as possible.

The shuffling in the digital video tape-recorder is applied after outer coding and over a range of data words corresponding to a set of 50 lines. That intrasector shuffling during the recording process can be described in terms of two successive shuffling processes:

an intraline shuffle, which shuffles video and ancillary words within a single line, prior to outer error coding, and

a sector array shufffle which shuffles data and error correction code words within the sector.

Since the previously described intersector shuffling results in 360 video data bytes per line (180 luminance and 2×90 chrominance), for the purpose of the intraline shuffle these bytes are distributed among 12 outer code blocks. If Oblk is the outer block column index $0 \leqslant \text{Oblk} \leqslant 11$, and Obyt the outer block byte number, $0 \leqslant \text{Obyt} \leqslant 31$, the intraline shuffle can be described by the following formula

$$\text{Oblk} = \text{int}(k/120) \times 4 + (k \bmod 4)$$

$$\text{Obyt} = \text{int}((k \bmod 120)/4)$$

For the purpose of sector array shuffling, the 32 rows by 600 columns sector array may be divided into 150 four-column groups ranging from 0 to 149. The four columns within a column group contain (C_B, Y, C_R, Y) pixel data bytes, respectively. Along a given row with a column group of C_B and C_R are cosited with respect to the television frame, and cosited (or nearly so) with the first Y pixel data byte, while the second Y pixel byte is horizontally offset from the first with respect to the television frame.

A column map, which is a permutation of the integers 0 to 149, is used to define the sequence in which column groups are stored in the sector array. A row map, which is a permutation of the integers 0 to 31, is used to define the sequence of rows in which data for a given column is stored in the sector array. The starting point of the row map is different for each column group, and in addition the starting point of the row map sequence for the fourth column of each column group is further offset by a constant from the starting point of the row map sequence for the first three columns of the column group.

Figure 8.10 shows a conceptual block diagram of the sector array shuffling method. The column counter is cleared at the beginning of each 50 line segment, and incremented every outer block, or twelve times per television line. The least significant 2 bits of the column counter select a column within a four-column group. The most significant 8 bits are used to address a PROM containing the column map function. The row start PROM is used to select an initial starting point of the row map sequence for each column group, except for the fourth column of the column group, which has a different initial starting point of the row map sequence. The row counter is loaded with

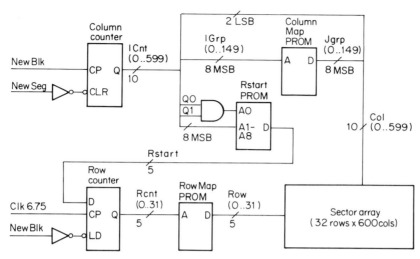

Figure 8.10 Block diagram of sector array shuffling

the row start preset data at the beginning of each outer block and increments mod 32 every data byte. The row map PROM is used to select the actual row address where the byte is stored in the sector array.

RECORDING OF DIGITAL AUDIO SIGNALS

Any method for the recording of digital programme audio signals has to fulfil the following basic requirements

  four independent digital audio signals have to be recorded in such a way that their independent erasure and/or editing are possible,
  all four audio signals have to be adequately protected, having in mind that audio signals by definition do not contain the potential redundancy we can found in the video signal,
  the means of recording should be selected in a way to permit a maximum simplification of the total processing circuitry, and
  the resulting recording method should allow a flexible handling of digital audio signals.

We saw previously that the audio sectors were placed in the middle of the programme track in order to enhance their protection.

As in the case of the video signal, the input audio signal is the serial digital data stream corresponding to the standardised EBU–AES format which subsequently led to CCIR Recommendations 646 and 647. Considering the large number of possible different applications and situations it is difficult to set a unique scheme for the organisation of the audio word length. Fortunately it is possible to satisfy almost all foreseeable applications and practices, and still to preserve the necessary compatibility, by organising the 20-bit audio words, rounded off from the original 24-bit words of the interface, into seven different modes. In these seven modes the length of the audio word varies from 16 bits (with one status, one user, one validity and one undefined but reserved bit) to 20 bits when only audio data are present (in the case, for example, when the analogue audio signals are directly encoded at the input of the DVTR). It is obvious that in the playback process the audio data are reformatted into the EBU–AES interface format, so that the output is identical to the input signal. Although there is seven modes available it is not expected that all of them will find very large application. It seems that three modes will be generally used: the already described modes 0 (with 16 bit audio words) and 7 (with 20 bits audio words) and mode 3 where 18 audio bits are accompanied with one status and one user bit.

Certain control data are added to the audio data stream at the input interface. They serve to signal informations transparently to the output interface. These control data are the *channel use*, the *preemphasis and frequency locking*, the *audio word length* and the *block sync. location*.

In principle the audio signals at the input of the video tape recorder are synchronous with respect to the video signals, i.e. the recorded audio data are sampled at a rate locked to the video sampling rate. However, in some situations, like the outside location recordings, and in some operational practices of some users, it might be difficult to ensure such a sort of synchronism. The recorder is, therefore, capable of recording non-synchronous signals, and the audio signal contains two additional bits which are used to flag such situations. Since such non-synchronous operation is considered non-standard, all tapes in the international exchange should be recorded with synchronous audio. For internal use recorders will be available on the market with special play-back circuitry for handling of asynchronous audio data. That circuitry will be triggered by the aforementioned flag and will permit an automatic resynchronisation of audio and video data.

The error protection scheme for audio signals is chosen with the following considerations in mind:

The nature of audio signals is such that it is necessary to correct almost all errors and to accept a lower level of concealment then in the case of video signals

It is desirable to implement a system having as much commonality as possible with the video error correction scheme

protection has to be ensured not only from errors due to noise, tracking imperfections, etc, but also from errors due to drop-outs, tape imperfections, scratches and even channel failure

It is required that the quality of audio signals after 20 generations should be of the order of grade 4.5 on the CCIR five-point scale.

In the recording process audio data are distributed among different sectors (16 sectors belonging to two video segments) in a way shown in Figure 8.11. We can see that different groups of samples corresponding to the four audio channels and to odd or even samples are recorded on different audio sectors, and that the same audio data are duplicated on a different programme track and on a different position inside that track.

As in the case of video protection coding, the audio data is subjected to block coding implemented in the form of a product code. In that process all different bits of a data word are equally weighted.

The data are arranged in a rectangular array with a row dimension of 24 20-bit words

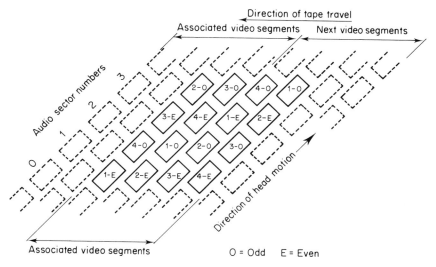

Figure 8.11 Distribution of audio samples among a set of sixteen audio sectors

and a column dimension of 7 words. Each group of 24 20-bit words is divided into 120 4-bit words. Three check words of 4 bits each are first added to each column, to produce a total of 120 'outer code' blocks of 4-bit words. Each of the new 10 rows shall be considered as a block of 60 bytes, to which 4 check bytes are added to form 'inner code' blocks of 64 bytes each. The error detection and correction capability of such protection coding is virtually the same as for the video.

The outer code will be a shortened Reed–Solomon code over GF(16). The primitive, or field generator polynomial is given by

$$z^4 \oplus z^1 \oplus z^0$$

where the z^i are place variables in GF(2), the binary field. By convention the left-most term in such polynomials is the most significant, the 'oldest' when doing computations, and where appropriate the first to be written on tape.

The Reed–Solomon code generator polynomial in GF(16) is given by

$$G(x) = (x + \alpha^0)(x + \alpha^1)(x + \alpha^2)$$

in factored form. α^1 is given by 02hex in GF(16). The Reed–Solomon check characters shall be given by K_0, K_1 and K_2 in

$$K_2 x^2 + K_1 x^1 + K_0 x^0$$

the remainder after dividing $x^3 D(x)$, where the data polynomial $D(x)$ is

$$D(x) = \beta_6 x^6 + \beta_5 x^5 + \cdots + \beta_1 x^1 + \beta_0 x^0$$

by the generator polynomial $G(x)$, as in equation 8.3. The full outer code block is given by

$$\beta_6 x^9 + \beta_5 x^8 + \cdots + \beta_0 x^3 + K_2 x^2 + K_1 x^1 + K_0 x^0$$

Similarly the inner code is identical to the one used for the video data derived from equation 8.4.

Prior to the inner coding audio data are shuffled in order to improve their protection against various types of errors encountered in digital audio recording. The shuffling is effected on the set of data words corresponding to 6.67 ms. The shuffling structure is systematic and corresponds to the lay-out shown in Figure 8.12.

During the digital processing of audio signals, processing and user words are multiplexed with audio and interface data. The processing control words have to pass control information from the record to the play-back processor. The user words are 8 bit long and their content is not specified for the moment.

Since the editing capability of a professional VTR is obviously one of the major concerns of all users, special attention is paid to that particular aspect. In order to answer the different operating practices and different situations encountered in the broadcasting environment, the editing of digital audio signals can be carried out in three different ways:

simple cut edits,

simple overlap edits, and

processed overlap edits.

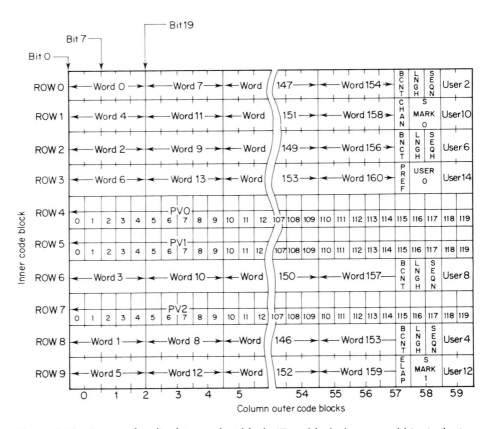

Figure 8.12 Layout of audio data product block. (Even block shown — odd is similar.)

The fact that each audio sector on tape contains data belonging only to one audio source makes it possible to individually erase or edit in any of the aforementioned ways each of the four audio channels.

For simple cut edits at the edit point, old audio sectors are simply replaced by the new ones. There is no particular processing and the audio protection data are not affected by such a procedure. It is true, however, that such a sharp transition may in some cases introduce audible transients at the edit point.

The second audio editing possibility is the simple overlap edit. In this editing mode during the overlap period (the length of the overlap is flagged by four bits being part of the aforementioned processing control words) only one out of two audio sectors is replaced by new data, and once the overlap is ended, all audio sectors belonging to that same audio channel are replaced by new ones. This method is quite simple and avoids the danger of audible transients, but the audio during the overlap is less protected due to the loss of added redundancy.

Certainly the most sophisticated and resourceful editing procedure is the processed overlap edit. This method involves the introduction of advance read heads as the process is based on the read–modify–write approach. Due to the basic nature of digital audio recording such a method offers the possibility of elaborate processing of audio edits, while maintaining unaltered the quality of the audio signal. Certainly this approach involves the rewriting of all four audio channels in all cases, regardless of the number of channels we wish to edit at that particular point. It is also clear that this editing capability involves the introduction of more complex arrangements in the head drum and playback circuitry, but it is clear that such sort of elaborate processing of the edit point is required in complex postproduction operation.

RECORDING OF SYNCHRONISATION SIGNALS

The data words on the programme track are multiplexed with synchronization and identification patterns at a constant rate along the track. This means that a synchronisation word is sent regularly after a certain number of data and protection bytes have been transmitted. The number of data and protection bytes between two successive synchronisation words defines the length of the synchronisation block. The length of this block is dependent on the possibilities for reading data at high speeds: frequent synchronisation is needed in order that the picture quality is not excessively degraded in shuttle or fast winding modes.

The length of the synchronisation block is set at 143 bytes (see Figure 8.13). The first 2 bytes are the synchronisation word, the following 4 bytes are the protected identification pattern, and the block is completed by two inner blocks of 64 bytes each.

Identification is provided for each block in order that fragments of a track recovered during cross tracking (at speeds higher than normal) might be correctly placed in a display field. This is achieved by numbering the sync. blocks along the track, the group of half tracks, the track in a group, and the field sequencing. Each identification is provided with its own error protection.

The sync. blocks have the same structure and the same content for both video and audio data.

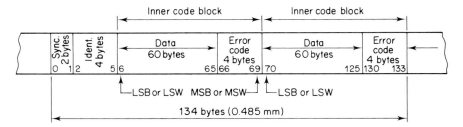

Figure 8.13 Layout of a sync block in a video or audio sector

Figure 8.14 Layout of the edit gap

All sectors, audio and video, commence with the preamble sequence. The preamble consists, in their order of recording on tape, of 20 bytes minimum of CC_{hex}, as a run-up string used for clock reference, 2 bytes of sync. words, 4 bytes of identification pattern and 4 bytes of CC_{hex} as fill data. The sync. and identification bytes are identical to those used in sync. blocks.

All sectors, audio and video, terminate with the postamble sequence consisting of 2 bytes of sync. word and 4 bytes of identification pattern. Preamble and postamble data are neither randomised nor interleaved.

As we have seen in the description of the track pattern, an edit gap is provided between each of the track sectors. The gap consists of 232 bytes of CC_{hex} fill data. During any editing process the fill data may be destroyed, but the preamble and the postamble must remain intact in order to allow correct subsequent reading of the recorded data in that particular programme track (see Figure 8.14).

CHANNEL CODING

At the output of the error protection chain a channel coding is applied to both video and audio data. The channel code (the same code is selected for both audio and video) has to ensure a maximum data throughput by optimum spectral shaping, providing at the same time an easy clock recovery over the full range of operational playback speeds.

The selection of the code is based on the analysis of the transmission channel which, in the case of the DVTR is ruled by physical laws of magnetic recording and has to take into account the specific situations of speeds other than normal in play-back. By considering for that particular case three major items in the selection of a suitable channel code (d.c. component, low frequency and high frequency potential problems), the solution for these potential problems and desirable characteristics of the code such as the simplicity of decoding and absence of overhead, a code of the NRZ (non-return to zero) type is selected.

All data and error correction check characteristics are randomised before recording. As we mentioned before the sync. identification and fill patterns are not randomised.

The randomisation is equivalent to exclusive 'or-ing' the bit serial data stream to be recorded by the bit serial data stream created by the generator polynomial

$$x^8 \oplus x^4 \oplus x^3 \oplus x^2 \oplus x^0 \quad \text{(in GF(2))}$$

The first term is the most significant and the first to enter the division computation.

Successive sync. blocks are randomised with different sequences.

RECORDING OF CUE AUDIO SIGNALS

From the concept of track pattern we can see that the format provides three longitudinal tracks. Two of them are intended for the recording of the control track and the longitudinal time-code signals. The third track arises from requirements to ensure 'recognisable' audio at speeds other than normal.

It is clear that at speeds other than normal, when the rotary heads just cross the tracks, the quantity of recovered audio data will be insufficient to ensure the sort of recognisability expected (such a recognisable audio is useful and necessary during editing operations). The simplest solution is to provide a longitudinal track which can easily be read at high speeds, and whose quality is far from critical.

Bearing in mind possible problems with spacing loss, channel equalisation, clock complexity, and so on, which would arise in the case of a digital recording of that audio track, it was obvious that the simplest and the cheapest solution was to use analogue bias recording. It is true that analogue recording is burdened with distortion and print-through problems, but considering that the quality of that channel is certainly not critical, it was decided that an a.c. bias analogue recording with an r.m.s. magnetic short circuit level of 50 ± 5 nWb m^{-1} at 1000 Hz is to be used for that cue audio track. The total cue to video timing tolerance is set at ±4 ms. This tolerance takes into account the tolerance of the mechanical positioning of the cue head (±0.3 mm) expressed in time and the possible timing error introduced in the electronic part of the recorder. Since the cue audio signal is recorded as a standard analogue audio information, that tolerance is understood as a classical 'lip sync.' tolerance.

RECORDING OF CONTROL SIGNALS

The design of the control signal waveform evolves from the need for certain information which could hardly be inserted anywhere else. The first, and the most obvious requirement is for information needed by servo systems to accomplish a fast lock-up and to control the tape speed in playback so that the heads accurately follow the slant tracks. The next information required is the five-field sequence, required only in the 525 systems, due to the existing relationship between the field rate and the selected digital audio standard. Finally, the possibility of including in this signal, as an option, the information on the four- and eight-field sequence is envisaged. The main reason behind this idea is the transitory period of combined analogue–digital operations, when the composite colour video signal might be decoded and A–D converted at the DVTR input and at the output possibly reencoded into composite PAL. It is certain that in an all-digital component environment such information is useless (see Figure 8.15).

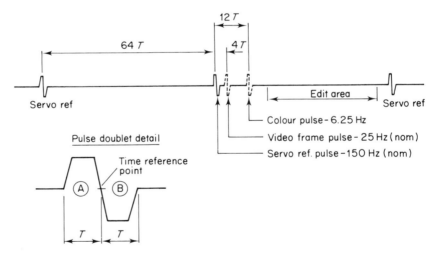

Figure 8.15 Waveform and timing of the tracking control signal. T is 1/64 the period of 4 helical tracks (i.e. 1 video segment). T = 104 μs nom. Rise–fall time of record current is 15 μs

RECORDING OF TIME-CODE SIGNALS

In editing operations it is necessary to have a longitudinal time-code track with video related time-code information. This time-code information could simply have been the existing EBU–SMPTE time and control code, which offers the advantage of using some

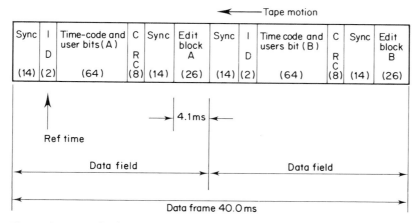

Figure 8.16 Multiplier arrangement for the time code and control signals in the data frame

existing equipment in television centres, like the time-code generators and readers, or it could also be a new time and control code which would overcome some of the limitations and shortcomings of the present one. The recorded time-code information was in any case to be recorded digitally and has to offer the possibility of recording and playing-back at speeds ranging from 0.1 to 50 times normal play speed. The recording density is set at 22 bits/mm in order to improve recovery reliability. Since the DVTR is by essence a powerful postproduction tool it is clearly desirable for it to be provided with improved time-code information. The new code consists of two time-code blocks (A and B) each with associated user bits, which can be individually edited in frame increments, without loss of data, due to the inclusion of edit gaps between sync. patterns (see Figure 8.16). Since these two codes are in fact two EBU–SMPTE time and control codes, the DVTR may be fitted with two standard code inputs and outputs, which permits the use of some of the existing time-code equipment. It is also possible to select and use only one out of these two codes (in which case it has to be code A), and, therefore, to have an operation quite similar to the one used with analogue recorders. The channel code for this time and control code is bi-phase mark and the carrier frequency is 256 times the television frame frequency. The record peak flux level is identical to the one used for the recording of the control track and is set to 185 ± 20 nWb m^{-1}.

COMMERCIAL PRODUCTS

The format corresponding to the mechanical and electrical specifications defined by the EBU–SMPTE [150] and the CCIR [11] for recording 4:2:2 component video signals and digital audio has been designated the 'D-1 Format'. The first commercial D-1 recorder (see Figure 8.17) was developed in Japan by Sony and placed on the market in 1987. This accepts both the M and L size D-1 cassettes by the use of automatic reel motor shift to adjust for different spindle spacings. The recorder uses specially designed integrated circuits some of which are mounted adjacent to the heads on the rotating part of the scanning drum [154] as shown in Figure 8.18. Each recording channel has a combined

458 DIGITAL VIDEO TAPE RECORDING

Figure 8.17 Sony D-1 digital recorder shown with large cassette

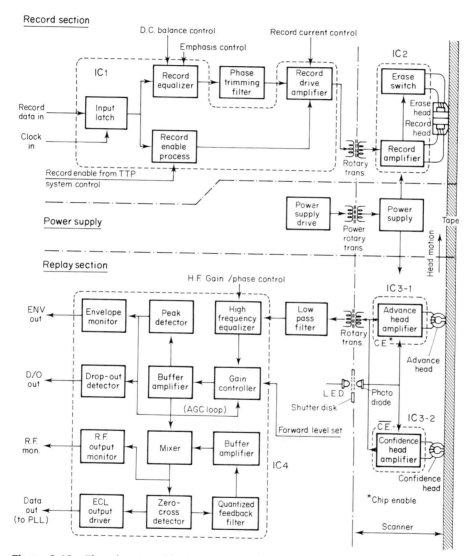

Figure 8.18 The r.f. system block diagram of the Sony recorder

460 DIGITAL VIDEO TAPE RECORDING

Figure 8.20 The scanning head assembly of the BTS D-1 recorder

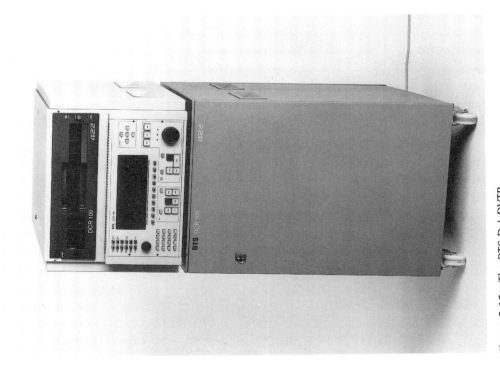

Figure 8.19 The BTS D-1 DVTR

record–erase head, an advance head for playback (and pre-reading) and a confidence playback head for read-after-write.

More recently BTS in Germany have produced a D-1 recorder (see Figure 8.19) which accepts all three (L M and S) cassette sizes. Figure 8.20 shows the construction of the scanning drum of the BTS recorder with the heads at the base and the rotary transformers along the vertical column.

Following the introduction of the digital component recorder the industry has also offered from 1989 a digital composite recorder using the same three cassettes (L, M and S) as established for the D-1 format. Figure 8.21 gives the dimensions of the L cassette. The main reason for augmenting the digital recorder product range with a composite machine is that, where the primary requirement is the multi-generation capability of the digital recorder rather than the postprocessing capability of components, the lower cost and longer playing time per cassette make it attractive for use in predominantly composite environments [155]. The format for the composite digital recorder using the 19 mm tape cassettes has been designated D-2. Figure 8.22 shows the first commercially available D-2 recorder which was produced by Ampex in the USA.

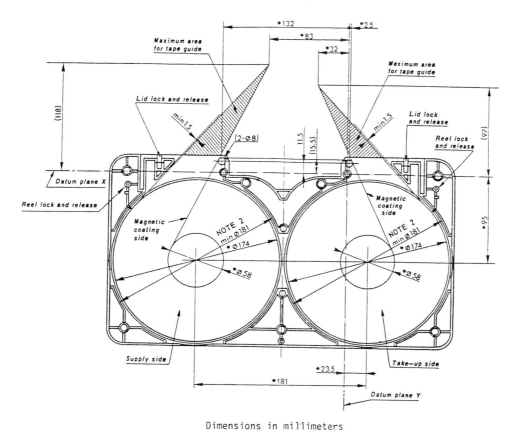

Dimensions in millimeters

Figure 8.21 The D-2L cassette structure and dimensions in millimetres (which are the same as those specified for the D-1 cassette). Notes: 1. Dimensions with an asterisk are nominal values specifying the tape path. 2. Area for the reel

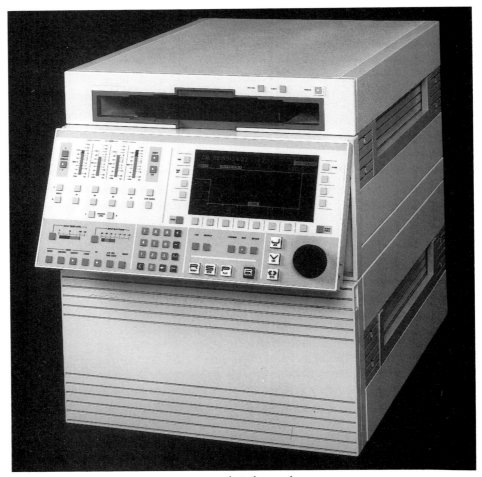

Figure 8.22 The Ampex D-2 composite digital recorder

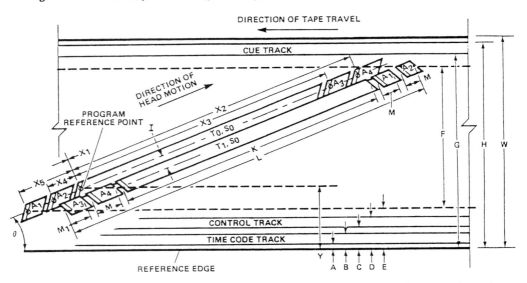

Figure 8.23 The D-2 track pattern. A_1, A_2, A_3 and A_4 are audio sectors. T_0 and T are track numbers. S_0 is a segment number side (typical). Tape viewed from magnetic coating

Table 8.2
Comparison of D-1 and D-2 parameters

	D-1 4:2:2	D-2 525 NTSC	D-2 625 PAL
Total samples per line	1420	768	948
Sampling frequency (8 bits per sample)	27 MHz	14.3 MHz	17.7 MHz
Number of recorded lines per field	250	255	304
Tape coating	850 Oe (oxide)	1500 Oe (metal)	1500 Oe (metal)
Tracks per field	10	6	8
Track length	170 mm	150 mm	150 mm
Track pitch	45 μm	39 μm	35 μm
Azimuth	0°	±15°	±15°
Channel code	Random NRZ	Miller squared	Miller squared
Playing time L cassette (13 μm)	94 min	208 min	208 min

For reasons discussed in Chapter 2 a sampling frequency of $4f_{sc}$ has been chosen for the D-2 format and the lower data rate required for this compared to D-1 is largely responsible for the lower cost and longer playing time. Additional margin has been obtained by using higher coercivity metal particle tape and alternate azimuth recording (see Figure 8.23) which allows a smaller track pitch for a given level of cross-talk since the 30° relative azimuth of adjacent tracks avoids the need for a guard-band. The sector arrangement is shown in Figure 8.24 for the PAL programme track. The 172 byte sync blocks are common to audio and video and contain two bytes each of sync and identification patterns, two inner code blocks each containing 76 data bytes and 8 inner check bytes. The edit gaps are 138 bytes nominal.

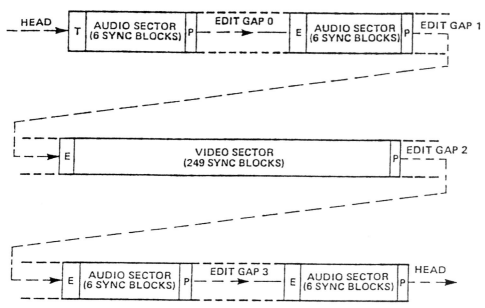

Figure 8.24 Sector arrangement for PAL programme track. T is the track preamble (62 bytes), E is the in-track preamble (28 bytes) and P is the postamble (6 bytes).

Table 8.2 compares the main parameters of the D-1 recorder with those of the two versions of D-2 which, unlike the D-1, cannot readily be switched between the 525- and 625-line modes.

In 1990 a new composite digital format, based on $\frac{1}{2}$ inch tape designated 'D-3', emerged from work by NHK and Panasonic in Japan. The signal and coding parameters are the same as for D-2. The format employs various new techniques in order to increase the packing density by a factor of about 2.5 compared with D-2. A new channel coding technique, new error correction strategy, and field shuffling of video and audio data are the main factors contributing to the achievement of the high packing density (track pitch 18 μm in the case of PAL).

Azimuth guard-bandless recording is used with metal particle tape and amorphous video heads but during editing a guard-band is generated before new data are recorded to improve the reliability of edits in the presence of tracking error.

Two cassette sixes are used: the large cassette for use in field, studio and automation recorders is $212 \times 124 \times 25 \text{ mm}^3$ and runs for 125 minutes. The small cassette, which can be used in all the machines, is $161 \times 98 \times 25 \text{ mm}^3$ and provides the 'D-3' camcorder with 64 minutes of recording time.

9 Electronic Graphics for Television

N. E. Tanton

INTRODUCTION

Electronic graphics, generated by information stored in a computer, or by an artist operating the tablet of an electronic painting system, make special demands on filter characteristics using the techniques discussed in Chapters 5 and 6. The requirements are particularly sophisticated for the reproduction of the alpha-numeric shapes of standard founts since the delicate curves and serifs are well known to the viewer who would be sensitive to any impairments introduced by the scanning process. Thus the description of the techniques for electronic generation of text illustrate rather well the general requirements for electronic graphics.

Most broadcast television programmes include sequences in which drawings or written words appear incorporated into the picture signal. Titles, explanatory captions and annotations to other graphic images within the programme, together with the closing credits, are all generated from real artwork or by means of electronic character generators, computer graphic devices or table 'painting' system such as the one shown in Figure 9.1 (see colour plates between pp 506 and 507). This equipment, built at the BBC Research Department in 1980, was one of the first to be designed specifically for TV use.

The traditional method of making text-in-vision involves preparing transparent or opaque artwork and scanning it to form an electronic picture signal. The scanned brightness of the artwork is represented in the video signal by a continuous voltage whose amplitude is a function of the transmission density or reflectivity of the artwork. Transparent artwork is usually in the form of high-contrast 35 mm photographic slides which are scanned using conventional television slide scanners. Opaque artwork, which usually comprises printed or phototypeset characters mounted on card, is scanned using a television camera.

The image prefiltering performed by slide scanners and television cameras serves to reduce the spectral components in the scanned picture signal which would otherwise give rise to aliasing distortions when the picture is displayed. Electronically generated images (including electronic captions) are sometimes unfiltered and, therefore, display noticeable aliasing distortions in the form of 25 Hz flicker, stepping and poor movement portrayal. By carefully filtering the electronically generated images and representing them in a filtered form it is possible to achieve a display image quality comparable to, or better than, that of scanned artwork. In applying these techniques to electronic graphics it is possible to make electronically generated titles and pictures with all the freedom of typographic control and with the image quality of scanned artwork.

Electronic character generators, in common with other electronic graphics devices, produce an electronic picture signal directly without direct access to any original artwork. Using simple digital techniques, they attempt to represent typographic characters by a pixellated image in which individual pages of text are built up character-by-character either under manual control from a keyboard or under the control of a remote computer. Often these images are two-level, one picture signal level representing character detail and another representing the background.

Whether from scanned artwork or electronic generator, the resulting electronic text signals are subsequently processed and presented as input to circuits which can be used to combine the text with a background picture signal as required. In the BBC Television Service, until the methods described in this chapter became established the use of electronic character generators had often been restricted to those areas of programme output (such as Sport or Current Affairs) in which operational considerations of speed or budgetary constraint had dictated their use. In general, television graphic designers had maintained strong preferences for the traditional methods of television typography in which real artwork is prepared and scanned. These preferences were, on the whole, for aesthetic reasons; scanned artwork usually produced an image which was, even to the untrained eye, considerably more pleasing in terms of typographic and image quality than that produced by many electronic character generators.

When using photoartwork, the graphic designer has a very wide choice of fount style and character size; this is very important if he is to be free to set or to suit the mood of a particular programme. Furthermore, he has complete freedom over the relative positioning of characters within the display area and one with another. Finally, if the text image is properly combined with the background picture signal, the displayed image maintains the letter forms of the original artwork and is largely free of the unpleasant image distortions which are characteristic of many synthetically generated images. These distortions include field frequency flicker on horizontal detail, stepping on curved and near-horizontal or near-vertical character strokes as well as movement judder and a modulation of character stroke width (e.g. serifs which disappear and reappear) when captions are made to roll vertically or slide horizontally.

By comparison, the repertoire of styles and sizes available on commercial character generators before the mid 1980s was small. In addition, these generators placed restrictions on the absolute and relative positions of characters in the text image. This is a direct consequence of the two-level character representation and of the physical implementation of the equipment. Furthermore, the letterforms of two-level characters (constrained by the relatively low-resolution of broadcast television) are often badly represented and bear only passing resemblance of the eponymous photosetting founts. Finally, when displayed, text images from such generators often exhibit the aforementioned distortions.

There is no fundamental reason why electronic character generators should not produce aesthetically acceptable letterforms using a digital character representation. The limitations of some commercial character generators are due to the use of only two signal levels to represent character detail. By using more sophisticated generation and filtering it is possible to simulate quite accurately the processes which an artwork image encounters on its way through the camera lens and circuits. This chapter describes methods of electronic character generation which result in an image quality equal to (or even better than) that available using traditional methods of television typography. The conclusions are equally applicable to other forms of synthetic picture generation such as computer graphics.

THE EFFECTS OF SCANNING

Sampling theory

When a continuous signal is sampled in the time domain, the spectrum of Fourier components representing the original signal in the frequency domain (Figure 9.2a) is repeated at harmonics of the sampling frequency f_s (Figure 9.2b).

The continuous signal may be reconstructed from the samples by filtering the sampled signal with a low-pass filter (Figure 9.2c) whose characteristic is such as to isolate the baseband spectrum from its harmonics (Figure 9.2d). In principle an 'ideal' filter characteristic, with the rectangular shape shown in Figure 9.2c, would pass unattenuated all frequency components up to half the sampling frequency (i.e. $f_s/2$) and suppress completely all components beyond this frequency. However, in practice, it is desirable to use filters with practicable rates-of-cut and with finite impulse reponses, unlike the 'ideal' filter of Figure 9.2c. In order to leave a margin for real filters, it is usual to sample at just above twice the frequency of the highest wanted component in the continuous signal.

If the continuous signal prior to sampling contains frequency components above half the sampling frequency, (Figure 9.2e), the repeated spectra overlap in the region around $f_s/2$ (Figure 9.2f). When reconstructed from the sampled signal using the 'ideal' low-pass filter, the resultant signal includes spurious energy from the high frequency components of the repeated spectra (Figure 9.2g). This spurious energy results in signal distortions known as 'aliasing'. The unwanted overlapping components could be removed by reconstructing the continuous signal with a filter of lower cut-off frequency (Figure 9.2h) at the expense of those wanted high frequency components which were present in the original signal in the region of potential spectral overlap. To avoid aliasing, it is necessary to pre-filter the continuous waveform prior to sampling in order to remove those spectral components (at or beyond half the sampling frequency) which would give rise to aliasing distortions.

It can be seen that a discontinuous signal such as that represented by 1-bit data cannot be perfectly represented by a sampled waveform because its spectral components extend ad infinitum. Furthermore, when the sampled waveform is reconstituted as a 'continuous' signal by the low-pass filter, the discontinuous input signal will be represented by a band-limited signal transition with finite rise-time.

Scanning as sampling

The brightness of an optical image perceived by a television camera tube or telecine photomultiplier is a three-dimensional function of horizontal position, vertical position

468 ELECTRONIC GRAPHICS FOR TELEVISION

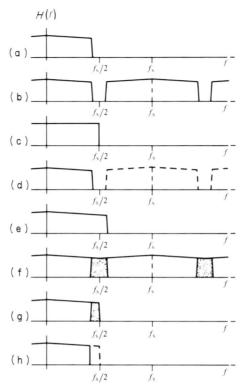

Figure 9.2 One-dimensional spectrum of a continuous signal, its sampled form and of a continuous signal reconstructed from those samples. (a) Spectrum of original continuous signal. (b) Spectrum of (a) after sampling at f_s. (c) Characteristic of the 'ideal' reconstructing low pass filter. (d) Spectrum of the continuous signal reconstructed with the filter of (c) from the sampled signal (b). (e) Spectrum of continuous signal with components beyond $f_s/2$. (f) Spectrum of (e) sampled at f_s. (g) Spectrum of continuous waveform reconstructed from (f) with ideal filter (c). (h) Spectrum of continuous waveform reconstructed with a low pass filter of lower cut-off frequency

and time. This brightness function has a corresponding three-dimensional spectrum describing the energy distribution of spectral contributions due to the picture detail. The spectrum contains horizontal and vertical spatial frequencies representing stationary detail and temporal frequencies resulting from motion within the scene. The extent and shape of this spectrum is determined by the scene detail and by the optics; it will be limited by practical factors such as aperture loss and camera integration [112], and the processes associated with these, some of which are discussed in Chapter 10.

When the analysing raster pattern of the camera or flying-spot tube scans this optical image, the three-dimensional brightness function is converted to a one-dimensional signal of voltage varying with time in which the brightness is defined only at discrete vertical positions and time instants determined by the scan. The image has, therefore, been sampled in two dimensions. As a result, the three-dimensional spectrum of the 'optical' image is repeated in the frequency domain at integral multiples of the vertical and temporal scanning frequencies (Figure 9.3a). In this chapter, vertical spatial frequency will be expessed in cycles per full picture height (c/ph) (i.e. including picture blanking). Thus the Nyquist limit of vertical spatial frequencies is expressed as 312.5 c/ph for a 625-line raster. In a 625-line, 50 fields per second scanning system, the spectrum is repeated at multiples of 625 c/ph and 50 Hz (Figure 9.3b). In addition, the 2:1 interlace structure also causes the spectrum to be repeated at odd multiples of (25 Hz, 312.5 c/ph) Figure 9.3c. This spectral repetition is common to all raster scanned images whether of natural or synthetic picture detail.

An image sampled by the scanning process is reconstructed by the television display as a continuous three-dimensional (x, y, t) brightness signal. The display tube filters the spatial and temporal components of the sampled image with a low-pass characteristic because the display spot has a finite size and because the display phosphor has a lag-like temporal response. This low-pass filtering is somewhat imperfect; there remain some direct harmonic components in the reconstructed picture of which the most noticeable is usually the field-rate flicker visible in plain areas of high brightness [113].

If there is spectral overlap in the spectrum of the scanned image, aliased components remaining in the filtered signal will appear as distortions on the displayed picture. It is the presence of these alias components in picture signals which give rise to flicker which causes sharp horizontal edges to 'twitter' at the picture repetition frequency (e.g. 25 Hz for 625/50/2:1 scanning and 30 Hz for 525/60/2:1), stepping and movement portrayal problems such as judder. Spectral overlap and therefore aliasing distortions can be reduced or removed by suitably prefiltering the image prior to forming a scanned version of it. The disparity in text image quality between scanned artwork and most present-day electronic character generator images exists because an artwork image is in effect prefiltered by all the elements of the scanning device.

Scanning artwork

The spatial definition of scanned artwork is very high, in the case of transparencies sometimes only limited by photolithographic film grain size. Nevertheless, the displayed results are considerably less sharp because the scanning, transmission and display processes impose physical limits on the picture detail which may be displayed.

There is an inherent loss of detail in the scanning process because the signal contributions of light from each point on the artwork are spread over a larger finite area (or 'aperture') of the electronic image. A simple example of such loss occurs when the camera is defocussed.

This loss of detail caused by spreading is called 'aperture loss' and results in a softening or blurring of high-frequency picture detail (including abrupt transitions). The loss is attributable to several physical causes both in the scanning and in the display processes; these causes include:

(i) lens imperfections (flare, low resolution, etc.);
(ii) image spread in the camera-tube target or film emulsion;

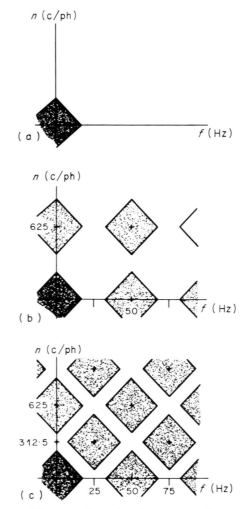

Figure 9.3 Two-dimensional spectrum of an optical image and its scanned form. (a) Spectrum of optical image (as section of three-dimensional spectrum). (b) Spectrum of image scanned with 625-line 50 fields per second sequentially scanned raster. (c) Spectrum of image scanned with 625-line 50 fields per second 2:1 interlace raster

(iii) scanning spot-size in the camera tube or slide-scanner flying-spot tube;
(iv) scanning spot-size in the display.

Partial correction for aperture loss in the scanning process is usually applied by selectively boosting the high frequency components of the picture signal; these components represent fine picture detail. Such correction is limited in scope, partly because there are still fundamental limits to the detail which may be transmitted and displayed by

THE EFFECTS OF SCANNING 471

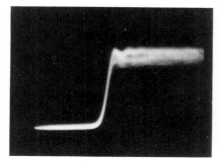 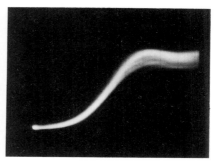

Figure 9.4 The step-response of a television slide-scanner (both traces showing signal after after-glow correction in linear signal domain). (a) 50 ns per division, (b) 20 ns per division

the television system and partly because the detail may be restored only at the expense of increased electrical noise [114].

These limits to reproducible detail mean that the abrupt changes of reflectance or transmission density which mark the boundaries of typographic characters on the artwork cannot be reproduced as perfectly abrupt changes of brightness in the displayed picture. Instead, these changes are represented by more gradual, band-limited transitions. Figure 9.4 shows the video signal from the slide scanner in which an abrupt density change in a transparency has been scanned.

Small changes to the position of detail within the scanned artwork result in changes to the timing of voltage transitions in the scanned signal and thus to the position of the displayed image detail. In this way the infinite control of the absolute and relative positions of characters which is possible within titling artwork is matched by a continuously variable adjustment of perceived position in the displayed picture.

Aperture loss in scanning is equivalent to prefiltering the image prior to sampling it by scanning. The loss is advantageous in that spectral overlap in the scanned signal is significantly reduced.

Combining scanned text with pictures

Text signals derived directly from scanned artwork are usually free from most aliasing distortions when displayed. In some studio vision mixers, however, the text signal from camera or slide-scanner is sliced to form a keying signal with which to switch colour into the background video signal.

When sliced the band-limited waveform is converted to a non-band-limited two-level signal with abrupt transitions; the advantages of prefiltering are immediately lost. The resulting characters lose much of their fine detail (especially around serifs) and display all the aliasing distortions typical of two-level graphics together with an unpleasant vacillation of character edge position both on horizontals and verticals. This is caused by electrical noise which perturbs the slicing decision level.

The keying process, in essence, involves multiplying the background video signal by the 'key'. In the frequency domain the spectrum of the background video signal is convolved with that of the 'key'. If the keying signal has unlimited spectral components (as is the case in a two-level signal) then the spectrum of the combined image signal will also have unlimited components and aliasing distortions will result.

If, however, the keying signal is filtered in the same fashion as the video signal, the combined image signal will also be band-limited. By combining scanned text and background video in a circuit such as shown in Figure 9.5 using a linear multiplication process, the prefiltering advantages provided by the scanning devices are maintained in the output image.

The principles of 'soft keying' are described in Chapter 11 and are applicable equally to analogue scanned text, synthetic images or to the multilevel digital character generator techniques described below; it is essential to use soft keying to preserve the image quality of prefiltered text signals.

Two-level electronic images

Two-level pixellated images, such as those formed by some contemporary character generators, inevitably suffer from aliasing distortions due to the spectral overlap described above. These distortions will be evident even if two-level text and video are combined in a 'soft-keyer'. Additionally, because the image is in a horizontally sampled form there is a quantisation of edge position which places limits on the accuracy of letterform of each character and on the control of inter-character and inter-line spacing which may be achieved. Some commercial character generators, which use only two signal levels to describe character detail, have made use of very high effective horizontal sampling frequencies to improve the horizontal resolution, sometimes improving the letterform without changing the spacing accuracy. As the vertical resolution is prescribed by the number of scan lines this approach is not practicable in the vertical direction.

Multilevel digital images

When an abrupt transition is low-pass filtered, the resulting response is represented by a continuum of signal levels intermediate between the two levels representing the input

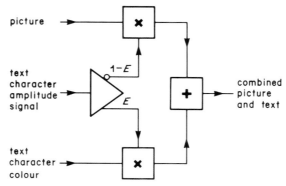

Figure 9.5 Combining pictures with text

transition. In order to describe a filtered image in digital form, it is clearly necessary, therefore, to represent the image detail with digital words of several bits or more.

By forming a suitably filtered multilevel image it is possible to generate text and any other synthetic pictures digitally with a displayed image quality comparable to, or better than, that of scanned artwork. (Because such filtering reduces aliasing distortions it is sometimes known as 'anti-aliasing'.) The final displayed image quality depends on the level of filtering and represents a balance between residual aliasing distortions and those picture impairments which are caused by the filtering itself. Such impairments include loss of detail or excess signal overshoot resulting in fringeing around edge detail.

As the nature of the filtering is important, it is appropriate to identify the spectral contributions made by various forms of picture detail and to relate particular aliasing distortions to the spectral components which cause them.

THE SPECTRAL CONTRIBUTIONS OF PICTURE DETAIL

Stationary detail

Stationary image detail results in spatial frequency contributions with no temporal component: picture detail orientated vertically as components along the horizontal frequency axis and detail orientated horizontally as components along the vertical frequency axis. Diagonal detail gives rise to spatial frequencies which may be resolved into horizontal and vertical frequency components (Figure 9.6).

Moving detail

If the image detail moves, the spatial frequency contributions due to the detail acquire temporal components whose temporal frequencies are proportional both to the magnitude of the movement and to the spatial frequency of each spatial component. Figure 9.7a shows the baseband two-dimensional spectrum of a sinusoidal grille of spatial

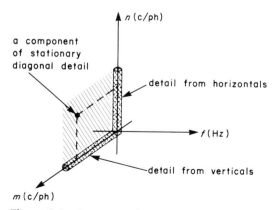

Figure 9.6 Spectrum of stationary picture detail

474 ELECTRONIC GRAPHICS FOR TELEVISION

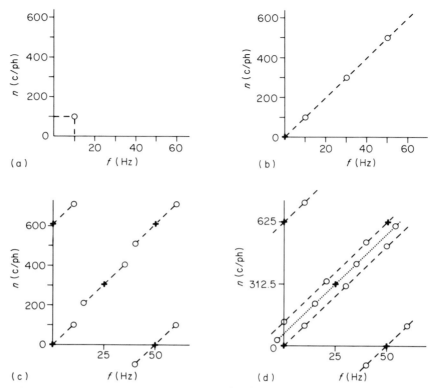

Figure 9.7 Spectrum of moving picture detail. (a) Vertical sinusoidal grille (100 c/ph moving at 0.1 ph/s). (b) Vertical square-wave grille. (c) Vertical sinusoidal grille as scanned by 625/50/2:1 system. (d) Vertical square-wave grille as scanned

frequency 100 c/ph moving vertically at a speed of 0.1 picture heights per second, a speed typical of roller captions. The corresponding temporal component is 10 Hz. As has been mentioned, abrupt steps have spatial components extending *ad infinitum*; when an abrupt transition moves, an infinite number of spatiotemporal components will result. Figure 7.9b therefore, shows part of the spectrum of a square-wave grille of the same pitch and motion as for Figure 9.7a. This is typical of electronically generated rolling captions.

When the baseband spectrum is repeated by the scanning process as discussed above, all the spatiotemporal components are repeated about the repeat sites; Figure 9.7c and 9.7d show some of the repeated components due to the moving grilles of Figure 9.7a and 9.7b.

THE CAUSES OF ALIASING

The three forms of aliasing distortion particularly associated with synthetic pictures are:

(1) picture frequency flicker on fine horizontal or near-horizontal detail (interlace twitter) (i.e. at 25 Hz for 625/50, 30 Hz for 525/60);

(2) stepping or 'jaggies' as they are commonly called because of the jagged edge seen on near-horizontal, near-vertical and curved strokes;
(3) movement judder and stroke-width modulation on moving images (e.g. roller captions).

Picture frequency flicker (interlace twitter)

If there is little high frequency vertical detail in the scanned image, the repeated spectra are well separated in the vertical and temporal frequency plane (Figure 9.8). As the high frequency vertical detail is increased (e.g. by focussing the camera) the spectra are extended further in the vertical frequency direction (Figure 9.8b) until they overlap when there are vertical frequency components at and beyond half the vertical sampling frequency (i.e. beyond 312.5 c/ph for 625/50/2:1 scanning). In particular, a purely spatial component at (0, 312.5 for 625/50/2:1 scanning) is repeated by the interlace spectrum centred on (25, 312.5) to give rise to a purely temporal component at (25, 0). In reconstructing the brightness function from the sampled waveform, the display will interpret components from this region of spectral overlay as picture frequency flicker.

When artwork is scanned by a camera or slide scanner, the aperture loss is usually such that there is relatively little spectral energy at or above 312.5 c/ph; 25 Hz flicker is then not very noticeable. The high frequency spatial components of two-level synthetic pictures often include significant energy at 312.5 c/ph: this results in visible 25 Hz flicker. Note that if a synthetic image contains some picture detail comprising an even number of television picture lines of the same brightness then that detail contributes no energy at (0, 312.5) and, therefore, generates no picture frequency flicker. For this reason founts are often readjusted or 'redesigned' so that the horizontals (including serifs if appropriate) occupy an even number of picture lines. As a consequence, the 'redesigned' fount often differs visibly from the photosetting letter forms on which it is based. This adjustment is totally unnecessary with properly filtered characters.

Stepping or 'jaggies'

When the detail is sufficient to cause the spectra to overlap in the vertical frequency direction (Figure 9.8c) vertical stepping becomes visible on near-horizontal lines and

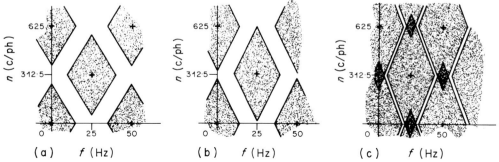

Figure 9.8 Spectrum of moving picture detail when scanned. (a) Moderate vertical detail. (b) Increased vertical detail. (c) Excessive vertical detail at and beyond 312.5 c/ph

curved strokes. This stepping (also known as 'the jaggies') (see Figure 9.9) results from coherent energy components in the baseband spectrum beyond half the vertical sampling frequency (312.5 c/ph) being folded back or aliased into the baseband region by the spectral repetition. If the amount of overlap is significant then this aliased energy is well spread over the baseband spectrum and cannot be separated out when the image is reconstructed by the display.

If the image is also sampled in the horizontal direction (as with the pixellated images of electronic character generators and most electronic or computer graphics devices) the baseband spectrum is repeated not only in the vertical spatial and temporal frequency plane but also along the horizontal spatial frequency axis. In this case the spectra are also centred at integral multiples of the horizontal sampling frequency.

The presence of horizontal spatial frequency components above half the horizontal sampling frequency will give rise to horizontal aliasing distortions. The sampled horizontal detail is reconstructed as a continuous electrical waveform by a horizontal low-pass filter before the whole image is reconstructed by the display. Horizontal aliasing is manifest as horizontal stepping on near-vertical lines and curved strokes.

Scanned artwork on the other hand has very little energy beyond 312.5 c/ph and so gives rise to very little vertical aliasing. Unless the analogue scanned signal is then inopportunely sampled horizontally (e.g. for some digital process such as picture manipulation) there will also be no horizontal stepping. Only when scanned artwork is sliced (e.g. to form a two-level keying signal) will stepping become noticeable.

Stepping is the direct consequence of representing an image in a sample system by a two-level signal. Figure 9.10a shows a typical typographic character as high resolution artwork — it is represented by a two-level unsampled signal. If this letterform is coarsely sampled vertically and horizontally a jagged image results as shown in Figure 9.10b; the artwork has been unsampled. Some electronic character generators have made use of high horizontal sampling frequencies to reduce the horizontal step-size with two-level characters. This is demonstrated by Figure 9.10c in which the horizontal sampling resolution is 2.5 times the vertical resolution. There is no opportunity of doing this vertically for a given size of character because vertical resolution is prescribed by the raster.

As the step size is reduced (by for example oversampling) stepping does become less visible, Figure 9.10d, but at the character sizes which are representative of practical television typography (i.e. between 16 and 60 picture lines height for a capital M) stepping is inevitably visible with a two-level character.

Movement portrayal

The detail in the unscanned image of a stationary scene may be described by a series of Fourier components of spatial frequency. As has been noted, when the scene moves each spatial frequency will acquire a temporal component; the greater the spatial detail or the movement speed, the greater will be the resultant temporal components. With a sharply focussed scene these temporal components may be quite large even for modest movement.

Figure 9.11a shows the two-dimensional spectrum of an arbitrary stationary object; in the absence of motion there are no temporal components. Figure 9.11b shows the

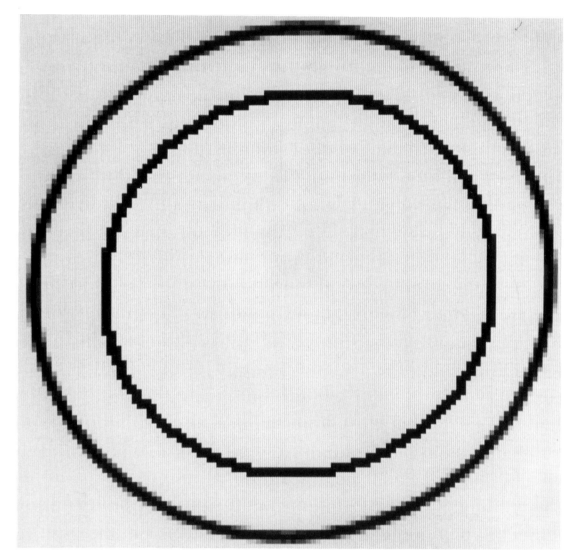

Figure 9.9 Example of stepping or 'jaggies'; inner circle without antialiasing, outer circle with antialiasing (expanded × 4)

478   ELECTRONIC GRAPHICS FOR TELEVISION

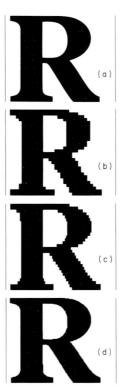

Figure 9.10 Typographic characters as represented in two-level form. (a) High-definition photolithographic quality. (b) Coarsely quantised on 30 × 30 matrix of pixels. (c) Quantised with same vertical resolution as (b) but with horizontal resolution × 2.5. (d) Quantised on 150 × 150 matrix

spectrum when the object moves vertically. The spectral repeat caused by scanning gives rise to a spectrum shown in Figure 9.11c. Due to the overlap, the display is unable to distinguish between the baseband components of fast-moving low frequency detail and the interlace spectral components of slow-moving high-frequency detail. This results in temporal aliasing, manifest as uneven movement portrayal (e.g. judder), a modulation of the position of moving edge detail (on rolling captions for example) and the well known effect of wagon wheels appearing to rotate backwards.

Although temporal aliasing is a fact of life with broadcast television the effects are mitigated by the use of electronic cameras. Aperture loss and the field integration of a camera tube serve to attenuate the higher spatial and, to a lesser extent, temporal frequencies when the optical image is scanned. Figure 9.12 shows the approximate spatial and temporal characteristic of an electronic camera based on a model by J. O.

THE CAUSES OF ALIASING

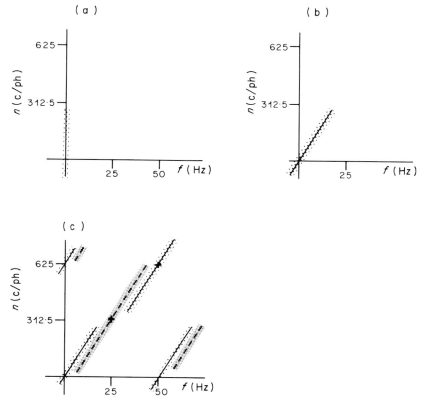

Figure 9.11 Two-dimensional spectrum of moving detail. (a) Stationary object. (b) Moving with vertical component of motion. (c) Scanned version of (b)

Drewery. The attenuation afforded to high spatial frequencies by aperture loss probably has a greater effect in reducing temporal aliasing than the temporal attenuation caused by field integration in the target of the tube.

Film cameras on the other hand provide sharp pictures and sample the scene only every 24 or 25 times a second. With less spatial frequency loss and integration over only half the frame period (180° shutter angle) temporal aliasing is more severe, as exemplified by the wagon wheels effect and judder on rapid camera movements, such as pans. Experienced cinematographers usually go to great lengths to control camera movements and motion within the scene in order to minimise film judder.

Electronically generated pictures are often totally unfiltered spatially or temporally. High frequency components in the image thus give rise to high temporal components when the image is made to move, which results in severe temporal aliasing. The movement judder and stroke-width modulation evident on digitally generated roller captions provide obvious examples of these effects which are considerably less noticeable on scanned artwork captions.

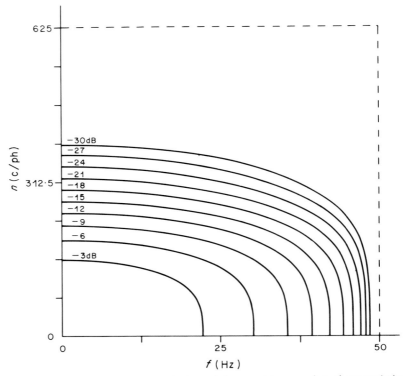

Figure 9.12 Approximate vertical spatial, n, and temporal, t, characteristic of an electronic camera

FILTERED TEXT IMAGES

From the preceding discussion it is clear that in order to minimise aliasing, electronic images should be generated in filtered form and that spatial frequency components at or beyond half the vertical and horizontal sampling frequencies should be attenuated to mitigate interlace twitter, stepping and, to a lesser extent, movement distortions.

If the image is to be seen to move then some temporal filtering may also be desirable to further reduce temporal aliasing effects. One possible two-dimensional spatio-temporal filter template suitable amongst other purposes for electronic graphics is shown in Figure 9.13.

The filtering described below applies to the two spatial dimensions.

Source of data

Master fount data are available in digital form from a large number of type foundries. The data are derived from reference artwork, which has been carefully prepared by typographers, and is digitised to two levels at very high resolution. The data are edited to make minor modifications and to eliminate obvious digitising errors and then stored

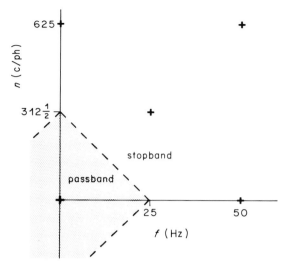

Figure 9.13 A possible two-dimensional spatio-temporal filter template for electronic graphics

character-by-character on a suitable magnetic storage medium. The exact format of the stored data varies from foundry to foundry (e.g. run-length, vector, feature description, etc.)

As supplied, these master data are unsuitable for direct use in television being in two-level form, not necessarily in scan line format and of far higher resolution (typically greater than the equivalent of 1000 lines character height for a capital M) than is required for normal television typography, where character heights of 16 to 60 lines are more appropriate. It is thus essential to undersample the character images and, if necessary, to convert them to scan-line format for display.

Using software, the master data are low-pass filtered in the horizontal and vertical directions and the resultant multilevel signal resampled at the requisite output resolution. The filter characteristics are scaled to suit the ratio between the input and output sampling resolutions and are chosen according to the criteria described above. In this way well-shaped characters may be generated from the master data for any character size practicable for broadcast television. A particular advantage of multilevel text prepared in this way is that all features of the master data make a contribution to the final image — within the limitations of the television display system, the displayed character should, therefore, display the correct proportions of strokes, serifs and counters that are embodied in the master data. In this context the author would like particularly to acknowledge the loan of master fount data by The Monotype Corporation and the advice of the BBC Graphic Design Department, in particular Charles McGhie and John Cook.

Choice of filter function

The complexity of the low-pass filter used to process the master data is less important than the image quality because the filtering is performed by software and need only be performed once per character for a given fount style and output character size. The criteria for the choice of filter function are:

(i) pass-band: too soft a filter will give indistinct characters with degraded edges, indistinct serifs and filled-in counters.
(ii) stopband: insufficient filtering will result in aliasing distortions.
(iii) pulse/step response: too sharp a low-pass filter will result in fringeing around and within character strokes, due to the unbounded step response.

Although signal overshoot will give rise to a distinct image, the fringeing caused by sharp-cut low-pass filters may be typographically unsatisfactory.

PRACTICAL INVESTIGATION

Master fount data provided by The Monotype Corporation were filtered and displayed using the electronic graphics system known as 'Flair', a commercial product from Logica based on the equipment shown in Figure 9.1. This equipment comprises a microcomputer-controlled pixel-addressable frame store in which a digital image is created and stored and from which it can be displayed. With each pixel is associated 8 bits of storage, the displayed image being 702 pixels wide by 575 pixels (lines) high. Monochrome images could therefore be displayed with 256 grey levels and a horizontal pixel resolution of 74 ns.

Purpose-written software running on the microcomputer enabled the effects of different filtering functions on image quality and letterform to be investigated. The program allowed vertical and horizontal filter functions to be independently specified in analytic terms by defining either the frequency characteristic or the impulse response of the filter function. The one-dimensional frequency characteristic and the impulse response of each filter was plotted on the display for reference and a filtering aperture calculated. After specifying the master fount-data file and the character to be taken from that file the filtering apertures were convolved with the master character data and the resulting multilevel letterform displayed.

The output from the frame-store was fed to digital-to-analogue converters via a look-up table which enabled the picture signal to be gamma-corrected to suit the display. Incorrect gamma-correction can greatly reduce the effect of filtering — in the limit a high gamma effectively results in a two-level text image and the filtering advantages are nullified. On the other hand, if the text signal is being used in the circuit of Figure 9.5 as a soft-key control then it should be totally linear (i.e. gamma = 1).

The filtering functions investigated can be simply divided into two categories, those with monotonic and those with non-monotonic step responses. Filters with monotonic step responses do not give rise to pre- and post-transition signal overshoot — there is no ringing on a step. Although inherently less sharp-cut in their frequency characteristic, these filters are of particular interest for two reasons.

First there is no fringeing around character detail due to the lack of overshoot on edges — this makes letterforms more acceptable typographically. Second, present practical soft-key circuits (such as shown schematically in Figure 9.5) usually operate on analogue input signals for which a 'negative' background signal has no meaning. For such circuits monotonic step response filter functions are appropriate for processing the key signal whereas the sharper cut-off filter functions with non-monotonic step responses may be practicable in future digital keying circuits designed to accommodate signed keying signals.

Filters with monotonic step responses

(1) Area averaging. This involves summing equal contributions from all the input data samples within a prescribed area centred on the output sample site. It is often used for filtering because it is easy to envisage and implement. The appropriate impulse response and its frequency characteristic is shown in one dimension in Figures 9.14a and 9.14b. The resultant letterform is rather soft. Flicker may be eliminated by choosing the height of the averaging aperture to encompass exactly two output picture lines—Figure 9.14c—this causes a zero in the vertical frequency characteristic at 312.5 c/ph. Stepping is just visible due to the incomplete attenuation beyond the first zero in the characteristic.

A character image is shown expanded in Figure 9.14f. Each pixel in the image has been repeated a number of times horizontally and vertically in order to demonstrate the use of intermediate grey levels.

(2) Linear interpolation. This is related to area averaging by convolution. The contribution made by an input data sample to the output sample value can be weighted linearly from being zero at a prescribed distance to being a maximum value at the output sample site. The impulse response and filter characteristic of such a linear-interpolation filter is shown in Figure 9.15. As with area-averaging, judicious choice of vertical aperture width will result in a zero at 312.5 c/ph and no flicker. The corresponding frequency characteristic slopes more quickly and has more attenuation beyond the first zero than for area-averaging; the resulting image is thus softer but with less stepping.

(3) Filters with an impulse response having a 'raised cosine' transition. If the filter function has an impulse response of raised-cosine shape as shown in Figure 9.16, the image quality can be made to be quite sharp but only at the expense of flicker and stepping.

(4) Gaussian impulse response filters. A filter function with an impulse response of the form $h(t) \propto \exp(-kt^2)$ has a Gaussian impulse response and Gaussian frequency characteristic (Figure 9.17). The scanning spot of a telecine flying-spot tube has a radial spread-function of approximately Gaussian shape. A Gaussian filter may thus be used approximately to model (in a variables-separable form) the non-aperture-corrected image of scanned artwork. If the argument of the Gaussian is chosen approximately to model the vertical frequency characteristic of a television camera tube (i.e. approximately -4 dB at 156.25 c/ph), the filtered character is slightly soft and largely free from aliasing distortion. Decreasing the width of the impulse response results in a sharper image at the expense of flicker and a small amount of stepping (this is true of the Gaussian depicted in Figure 9.17).

Such a filter function probably represents a practical balance between image sharpness and aliasing for monotonic filter functions.

The generation of two-level characters by subsampling without filtering can be represented by convolution with a function whose impulse response is unity at the centre of the output sample site and zero elsewhere (Figure 9.18).

484 ELECTRONIC GRAPHICS FOR TELEVISION

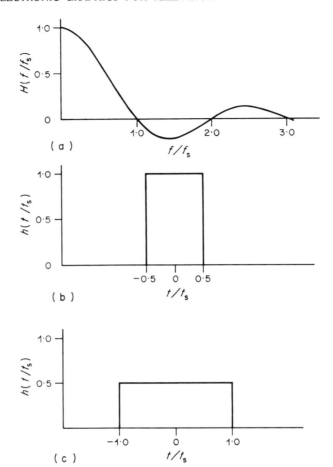

(a)

(b)

(c)

(d)

(e)

(f)

Figure 9.14 Area-averaging filter. (a) Frequency characteristic of filter a, (b) Impulse response of filter a (spatial aperture width one sample or line) (c) Impulse response of filter b (spatial aperture twice width of filter a). (d) Characters filtered with area-averaging filter a. (e) Characters filtered with filter b. (f) Capital A filtered with area-averaging filter a and expanded to demonstrate the use of intermediate grey levels to represent the signal transitions

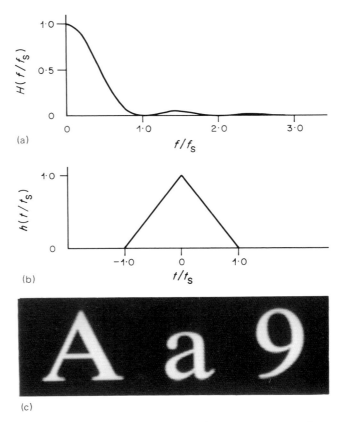

Figure 9.15 Linear interpolation filter. (a) Frequency characteristic of filter. (b) Impulse response of filter. (c) Characters filtered using this filter

This was implemented in order to test the software and to allow side-by-side comparisons of two-level characters with filtered ones.

Filters with non-monotonic step responses

Filters with non-monotonic step-responses have pre- and post-transition signal overshoots, producing a sharper image at the expense of fringeing around, and within, the letterform.

(1) Aperture-corrected Gaussian characteristic. The filtering effect of an aperture-corrected slide scanner scanning real artwork can be synthesised very approximately by using a filter function which combines the Gaussian filtering of the scanning spot with the first-order vertical aperture-correction usually applied to the scanner signal. This correction [113] can be arranged to provide a maximally flat characteristic fully correcting the aperture loss at 156.25 c/ph vertical frequency. Figure 9.19 shows the impulse response and frequency characteristic; the dashed line represents the characteristic of the uncorrected Gaussian filter. The resulting text image is quite sharp, aliasing

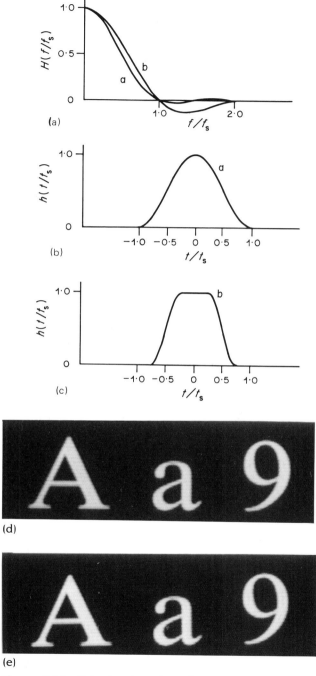

Figure 9.16 Raised-cosine time domain transition filters. (a) Frequency characteristics a and b. (b) Impulse response of filter a. (c) Impulse reponse of filter b. (d) Characters filtered using filter a. (e) Characters filtered using filter b

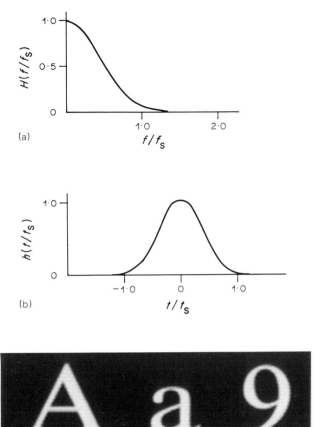

Figure 9.17 Gaussian filter. (a) Frequency characteristic of filter. (b) Impulse response of filter. (c) Characters filtered using this filter

distortions are hardly visible and any fringeing is restricted in extent to the immediate vicinity of the character edges.

Although the scanner-like image is not a perfect simulation, this filtering provides a useful comparison with which to judge other functions.

(2) 'Ideal' low-pass filter. The 'Ideal' low-pass filter with its sharp-cut characteristic may successfully be implemented in software. The resulting images are sharp and free from aliasing when stationary but there is very noticeable fringeing (known as 'ringing') around and within the character strokes (Figure 9.20).

This fringeing may be reduced in amplitude (at the expense of image sharpness) by choosing filter functions with less sharp-cut characteristics or it may be reduced in amplitude and spatial extent by multiplying the impulse response of the 'ideal' sharp-cut filter with a monotonic function which suitably attenuates all but the fringe closest to

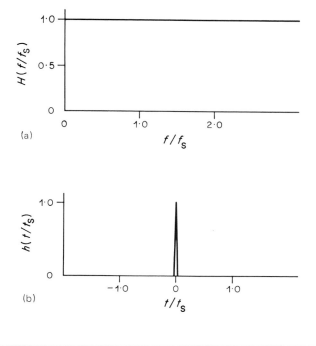

Figure 9.18 Undersampling. (a) Frequency characteristic of undersampling. (b) Effective 'impulse response' of undersampling. (c) Characters undersampled showing characteristic stepping

detail transitions. This process is called 'windowing' and necessarily produces a less sharp-cut filter characteristic.

(3) Less sharp-cut low-pass filters. One suitable low-pass filter based on the 'ideal' characteristic has a transition band described by one half-cycle of a raised-cosine waveform (Figure 9.21) As the width of the transition region widens so the fringeing amplitude is reduced but both the amplitude and the spatial extent of the ringing remain unacceptable until the image has become quite unsharp.

(4) Windowed impulse response filters. The impulse response of a fairly sharp-cut filter can also be modified by multiplying the impulse response by a weighting function which progressively attenuates the ripples in the impulse response. This modification affects the final filter characteristic which becomes the convolution of the original characteristic with the Fourier transform of the windowing function.

490 ELECTRONIC GRAPHICS FOR TELEVISION

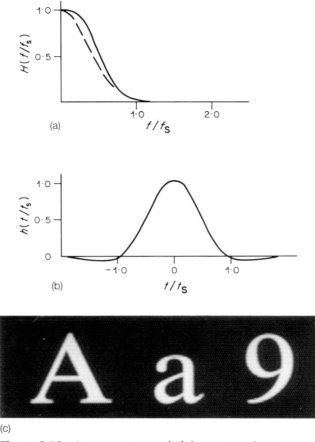

Figure 9.19 Aperture-corrected slide-scanner characteristic. (a) Frequency characteristic — full line represents characteristic of corrected filter, dashed line that of uncorrected Gaussian. (b) Impulse response of 'corrected Gaussian' filter. (c) Characters filtered using 'corrected Gaussian' filter

Figure 9.22 shows the impulse response and frequency characteristic of the combination of a rectangular low-pass filter with raised-cosine transition band (as in (3) above and Figure 9.21) and a Gaussian window function applied to the impulse response. The result is a pleasingly sharp letterform free of flicker and stepping but with fringeing restricted to the immediate vicinity of each edge. The image is sharper than that available from real scanned artwork. Similar results are obtained with 'raised-cosine' window functions.

Combined filtering

In the investigations described above the use of separate horizontal and vertical filtering functions (a variables-separable approach) was prescribed by the size and architecture of

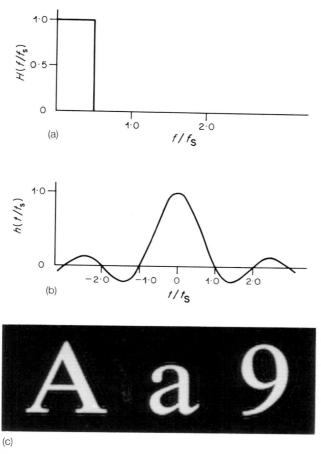

Figure 9.20 'Ideal' low-pass filter. (a) Frequency characteristic of filter. (b) Impulse response of filter. (c) Characters filtered with 'ideal' impulse response showing extent of ringing around signal transitions

the computer used and the size of the input character images (>1.2 Mbits as bit-map images). It is doubtful whether greatly improved results would be obtained from variables-non-separable filtering (combined horizonal and vertical filtering) except if the sampling structure is non-orthogonal (i.e. the horizontal sampling frequency is not line-locked to an integral multiple of the matrix line frequency).

It is likely that improved movement portrayal would be achievable for moving images (e.g. rolling captions) if some combined spatio-temporal filtering were to be performed (Figure 9.13). This was beyond the scope of the equipment used and the spatial filtering techniques described here will certainly produce better motion portrayal than is available with unfiltered synthetic images. More recent experience of spatio-temporal filtering has demonstrated that a purely vertical low-pass filter characteristic with loss at 312.5 c/ph is sufficient to significantly reduce temporal aliasing on rolling captions. Further work needs to be done on this aspect of filtering.

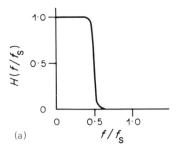
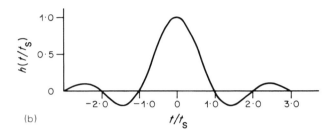

Figure 9.21 Less sharp-cut low-pass filter. (a) Filter characteristic with 'raised-cosine' transition band. (b) Impulse response of filter. (c) Characters filtered using this filter

Under certain circumstances, by reducing the number of bits representing the display image signal transitions a respectable image can be displayed using 16 grey levels (4 bits). These conditions are when the filter function is monotonic (i.e. there are no overshoots to be represented) and when the image is linearly described as for soft-keying. Aliasing does become evident as the bits are further reduced and although no work has yet been done to confirm it, it is likely that movement portrayal will be degraded on rolling captions.

If the master data has been properly filtered the perceived shape of the output letterform will be a perfect representation of the master artwork. Unlike two-level character images, every input sample in the master data makes a contribution to the final displayed image. There should be no need therefore to edit pixels or to adjust the sampling phase of the filtering function in order to achieve a better typographic representation of the master artwork. Residual aliasing, insufficient display grey levels or

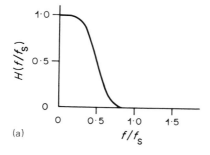

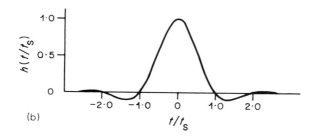

Figure 9.22 Windowed impulse response filter. (a) Filter characteristic of a suitable filter. (b) Impulse response of filter. (c) Characters filtered using this filter

incorrect display gamma may, however, slightly corrupt the letterform and its proportions. With the practical filters described in this chapter the levels of residual aliasing are unlikely to affect the letterforms or their proportions.

One of the complaints often levelled at contemporary electronic character generators concerns the coarseness of control of inter-character spacing which is usually provided. Conventional artwork may be spaced with miniscule precision but character generators are usually limited to the horizontal pixel dimension or wider. By adjusting the exact phase of the output sample lattice with respect to the input sample lattice, filtered text may be adjusted in position and displayed to an accuracy determined by the input sample size.

For example, if the two-level master data has a horizontal resolution of eight times the final output image then that output image may be positioned to within one eighth of a horizontal pixel dimension. Figure 9.23 shows the reconstructed filtered edges corresponding to a step in master artwork of eight times the requisite output resolution. The edge has been filtered with a perfect low-pass filter (quite feasible using software) and the continuous waveform reconstructed using a practical (finite rate-of-cut) low-pass filter. The artwork has then been moved by the input pixel width and a new output waveform constructed.

From this it can be seen that inter-character spacing can be adjusted with filtered images to a finer accuracy than the output pixel dimensions. In the case of Figure 9.23 if the eight edges were all stored as separate character strokes and the horizontal output pixel resolution was 74 ns a spacing resolution of 9 ns would be possible. It is, therefore, not necessary to use high horizontal sampling frequencies to achieve good positional accuracy or control.

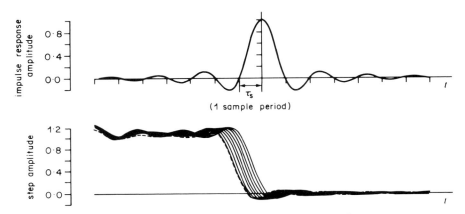

Figure 9.23 Sub-pixel adjustment of intercharacter spacing. The continuous waveforms of the lower trace have been reconstructed from data sample values using a reconstructing low-pass filter whose impulse response is shown in the upper trace. The signal transition has been sampled at eight different sampling phases corresponding to successive shifts of 1/8th sample period increments even though it is represented by data values at sample period increments

USE OF FILTERED TEXT FOR TELETEXT AND VDUs

Another advantage of filtered text is that it is possible to make characters which are more legible than their elementary two-level or alphamosaic counterparts, especially at the small character heights typical of teletext displays or of VDUs. Figure 9.24 shows a comparison between a teletext page displayed using a conventional alpha-mosaic teletext character generator and the same page displayed using filtered (and proportionally spaced) characters. The experimental decoder used for this latter display used a frame store with a 74 ns pixel width in accordance with CCIR Recommendation 601 [9]. The filtered characters are between 13 and 14 picture lines capital M height.

During these investigations, some filtered text in a sans-serif fount was generated at approximately ten picture lines capital M height and was found to be perfectly legible.

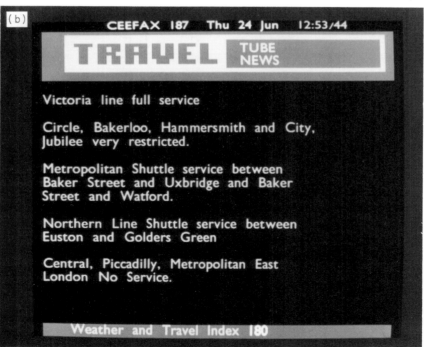

Figure 9.24 Filtered text used for improved teletext displays. (a) Conventional alphamosaic characters. (b) Filtered pixel images

APPLICATIONS OF ELECTRONIC GRAPHICS

Some of the applications of the graphics techniques described in this chapter have been mentioned in chapter 7 in connection with animation, etc. Perhaps the most demanding is the type of News and Current Affairs programme in which camera pictures, drawings and up-to-the-minute text are combined in a live programme such as the coverage of a general election.

The last time that 'traditional' methods were used by the BBC to present the statistics during an election broadcast was in 1979. The 'swingometer' seen in Figure 9.25 (see colour plates between pp 506 and 507), a pointer operated by hand from behind a chart in the studio, seems a very crude device compared to the methods based on the graphics techniques described in this chapter which were first used for the General Election broadcast of 1983. An example from this programme is shown in Figure 9.26 (see colour plates between pp 506 and 507) in which electronic painting systems were computer controlled to relate the graphics image to the statistics of the moment.

A much more sophisticated arrangement was used during the 1987 General Election broadcast, but in presentational terms was closer to the 'swingometer' than the 1983 broadcast. In 1983 the programme would essentially switch between the graphics and the presenters. In 1987 the electronic graphics again became very much a part of the studio set but with all the flexibility offered by the technology. The block diagram is reproduced in Figure 9.27.

The studio set included a 12 foot by 8 foot display constructed from a 3 × 2 matrix of images rear projected using six carefully aligned GE light-value projectors. This was called the 'Battleground'. Each projector was driven from one of the outputs of an *RGB* switching matrix. The various inputs to this matrix included pictures from Rank Cintel 'Slide-File' still stores, decoded PAL signals from outside broadcast vehicles and the six independent *RGB* outputs from a specially commissioned array of six MEIKO 'Computing Surface' graphics boards based on the Inmos Transputer. These graphics boards were used to compose maps and text of constituency information for display on the 'Battleground' either as individual screens or together as a kind of composite HDTV display capable of 2160 active pixels horizontally by 1150 active lines vertically. Thus maps could be combined with stills (Figure 9.28a) or with a composite of outside broadcast pictures (Figure 9.28b) (see colour plates between pp 506 and 507).

Synoptic and individual constituency results were represented by animated graphics generated in microVAX-controlled Quantel 'Paintboxes'. These are 'soft' Paintboxes in which the standard Quantel painting software is supplanted by BBC written software which enables them to double as real-time graphics generators. The structure and style of these graphics were designed before transmission; the detailed content of each image was added by the automatic typesetting of antialiassed characters as each constituency result was declared (and verified). More than 400 individual separate images were designed and characterised for use during the programme and each was updated as appropriate using data from the database management computers (2 DEC VAX 11-750s). Some of these graphics included real-time animations (growing histograms showing proportion of votes cast, etc.) and some were prefixed by 'stings', short pre-prepared computer animations which had been stored in digital form on an Abekas A64 digital disc recorder. These were controlled by another microVAX connected by a LAN to the database computers and a microVAX controlling the Transputer boards.

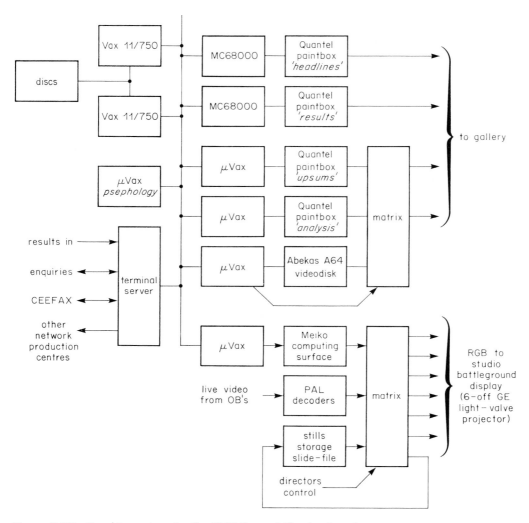

Figure 9.27 Graphics system for the 1987 General Election broadcast

If a Paintbox graphic was remotely requested to be prefixed by a sting, the microVAX synchronously switched the local *RGB*s switching matrix to output the pre-recorded animation from the A64, ran the animation and awaited its end, switched the matrix to show the appropriate Paintbox output and finally signalled to the remote computer controlling the whole graphic that the Paintbox was now 'on air' and could commence its real-time animation. This part of the system is also shown in Figure 9.27. All graphics generation was remotely controlled or triggered from the studio floor via the LAN. Figures 9.29a to 9.29d (see colour plates between pp 506 and 507) show stills from one of the pre-recorded 'stings' replayed from the A64 and from the real-time Paintbox animation which immediately followed it.

Some of the techniques described in this chapter have now been applied commercially for digital text generation in versions of the Quantel 'Paintbox' and in the Quantel 'Cypher' picture and text manipulator. They are also being used in the Aston IV and Chyron 'Scribe' character generators, examples of which are shown in Figures 9.30a and 9.30b, respectively (see colour plates between pp 506 and 507). Figure 9.31 (see colour plates between pp 506 and 507) shows the Aston equipment in use at the BBC Television Centre.

10 Telecine and Cameras

I. Childs

INTRODUCTION

This chapter will consider some of the opportunities for, and some of the problems of, the applications of digital techniques to television picture origination equipment. With the exception of electronic generation of graphics, which has been discussed in the previous chapter, there are two principal sources of television signals. The first of these is the television camera itself; this is the main workhorse of the broadcasting organisation. It is responsible for an enormous variety of programme material from studio-based drama, through light entertainment and news to the most testing of live outside broadcasts. The second major source of television programmes is from motion-picture film (which still currently provides a significant percentage of the broadcast television output); in this case it is the film scanner, or telecine, that translates the film images into television signals. A variation of the telecine, the slide scanner, is used, as its name suggests, to produce still pictures from photographic slides; these are typically used as background visual material in news programmes or for continuity announcements made during breaks between programmes.

Unlike electronic graphics equipment, cameras and telecines use essentially analogue signal processing, even with solid state sources. However, as digital techniques improve, their advantages will outweigh the cost of change and this is happening in the case of telecine. Although these two picture sources, the camera and the telecine, both perform the same basic function — to convert an optical image into a scanned electrical signal — they have very different technical and ergonomic requirements. The major difference can be stated very simply — a camera must be mobile (and often highly so) but a telecine need not be. Thus major efforts have been, and will continue to be, made in order to reduce the size, weight and power requirements of television cameras; currently the programme makers have a choice between small, hand-held cameras able to go almost anywhere and the larger, pedestal mounted studio cameras which, although not so versatile, have a wider range of facilities coupled with a higher picture quality. In contrast the major design objectives of telecine manufacturers are to produce a machine which is as reliable, and as easily maintained, as possible and which produces the best quality television pictures; as a consequence of this, telecines are usually fairly large (indeed a minimum size is usually set by the requirement to accommodate large reels of film) and frequently incorporate more sophisticated signal processing than is used in a camera.

Types of camera

At the present time, all television cameras are based on the use of either vacuum tubes or charge coupled solid state devices as the basic opto-electronic sensor device. Figure 10.1a shows some of the major circuit elements in a typical studio camera of the first type which would use plumbicon or saticon electron beam tubes. An optical image of the scene is formed on the targets of the three camera tubes, after being split into red, green and blue components by the colour splitter block; an electron beam scanning the rear surface of these targets senses the illumination at each point of the image and converts it to a corresponding electrical signal (for a more detailed description of the mechanisms involved, the reader is referred to) [156]. Each camera tube output appears as a variable *current* source in parallel with a small capacitance; this variable current is then converted to a variable voltage by a head amplifier, which usually operates on the transresistance principle [157]. At this stage, therefore, the video signal appears as a

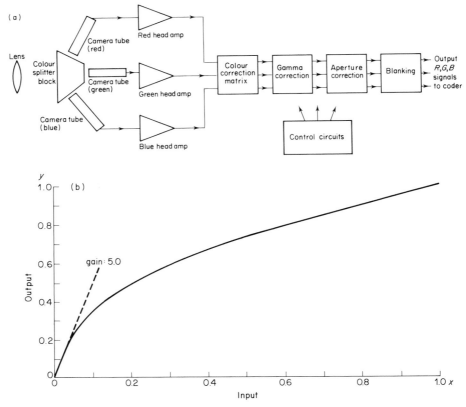

Figure 10.1 (a) Block diagram of a typical colour camera. (b) Transfer characteristics of a typical camera gamma corrector

voltage waveform whose instantaneous value is directly proportional to the illumination falling on the corresponding area of the camera tube target.

Such a signal is not directly suitable for display on a television receiver. One reason for this is that display cathode-ray tubes are not linear; the light output intensity, I, is related to the voltage applied to the input, V, by an equation of the form

$$I \propto V^\gamma \qquad (10.1)$$

where γ, the display gamma, is in the range 2.7 to 3.0 depending on display tube type and driving mode chosen (cathode drive or grid drive). Correction must be made for this display gamma and it is preferable to do this in the camera, rather than in the receiver, in order to minimise the costs of television receivers (there are also technical benefits in that the visibility of noise and transmission defects is reduced in dark areas of the picture where they would otherwise be annoying). Such a gamma corrector is shown in Figure 10.1a; its effect is to produce the law shown in Figure 10.1b where the darker areas of the picture have a higher gain applied to them than the lighter areas. Following gamma correction the signals are aperture corrected, to restore any loss of high spatial frequencies encountered in the camera tube, blanked, to remove any disturbances occurring in the flyback intervals, and are then usually coded into a composite form as has been described in Chapter 4. The coder may be remote from the camera, in which case the individual red, green and blue signals must be transmitted down the camera cable, or alternatively it may be included in the camera head. Also shown in Figure 10.1a is a colour-correcting matrix; this corrects the outputs from the head amplifiers for colour analysis errors in order to optimise the camera colorimetry [158]. Finally there is usually a fairly complex control network in order to control factors such as the registration of the scan positions of the individual camera tubes and the uniformity of the black and white signal levels.

The second type of camera is based on the use of a solid-state image sensor such as the charge coupled device (CCD) [159] to replace the electron beam tubes. These solid state sensors are integrated circuits containing an array of individual photosensitive elements, each of which senses the brightness of one part of the image. Scanning electronics, incorporated on the chip, serves to interrogate each of the elements in sequence and so build up a television waveform at the output of the sensor. Figure 10.2a shows a frame transfer solid state area-array sensor produced by Philips in Holland. In the close-up of the CCD structure on the silicon surface, the individual photosensitive elements can just be distinguished. Cameras using such solid state arrays are different in many respects from cameras using plumbicon or saticon tubes. One difference is due to the fact that the sensing is carried out by many thousands of discrete detectors; this causes a spatial sampling of the incident image, the effects of which will be considered more fully later. A second difference is that the signal-to-noise ratio can be quite different for solid state sensors than it is for plumbicon tubes; again the reasons for this, and some of the consequences, will be considered later.

Types of telecine

In contrast to the use of solid state sensors in cameras, which have only gained limited acceptance for broadcast applications, the use of such sensors in telecine applications is

502 TELECINE AND CAMERAS

growing rapidly. This is because the motion of the film itself can be used to give one of the scans; the solid state sensors need only then consist of a single line of photosensitive elements (a line array). The sensor is then considerably simplified (to only around 1000 elements compared with half a million as required for a camera sensor) and both performance and manufacturing yield are thus increased. Figure 10.2b shows a Fairchild CCD line-array sensor produced in the USA; comparison with the area-array sensor of

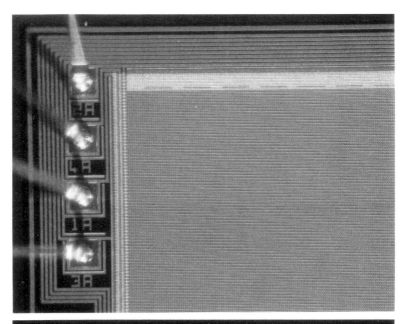

(a)

Figure 10.2 (a) A typical solid state area-array image sensor, the Philips frame transfer CCD (shown in close-up at the top)

INTRODUCTION 503

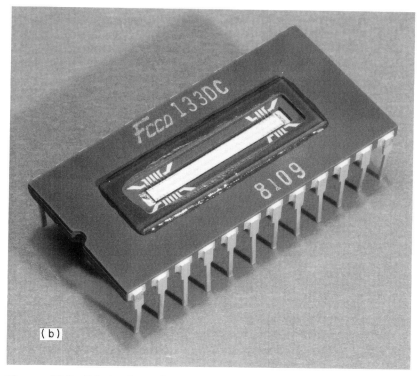

Figure 10.2 (b) A view of a typical solid state line-array sensor, the Fairchild CCD 133

Figure 10.2a clearly demonstrates the smaller chip area, and hence higher yield and lower cost, of this type of device. A block diagram of a telecine using such line-array sensors is shown in Figure 10.3. An image of the film is thrown onto the three line-array sensors, one for each of the red, green and blue channels. The outputs of the three sensors must first be corrected for variations in sensitivity between the various elements of which they are composed [160]; this correction can also be made to eliminate any vignetting or shading due to non-uniform illumination of the film. The signals are then gamma corrected; in contrast to the gamma correctors used in television cameras, which are usually based on simple diode shaping networks, the gamma correction in telecines is of higher performance and is based on amplifiers having a logarithmic and an exponential response, with a variable gain between the two. The circuit thus accurately achieves the correct characteristic and the variable gain allows the exact value of the gamma being corrected for to be adjusted to compensate for any non-linearities in the film itself. Again, as in television cameras, it is usual to incorporate a colour-correcting matrix. This may either be located in front of the logarithmic amplifier, before gamma correction, or in between the logarithmic and exponential amplifiers; matrixing the logarithmic signals has some advantage as it corrects more accurately for colour errors produced by the film itself rather than merely correcting the telecine analysis [161]. Following the gamma correction, the signals are aperture corrected and blanked as for a camera; minor differences may exist in the characteristics of the aperture correction but these will be

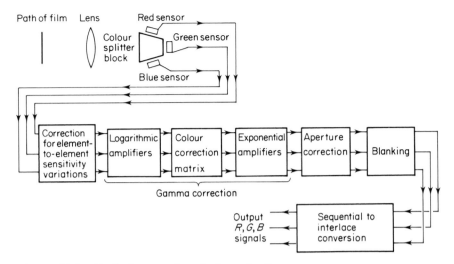

Figure 10.3 Block diagram of a telecine using line-array sensors

dealt with in a later section. The signals are still not suitable for display, however; because the line-array sensors were used to scan only across the width of the film, the vertical scan being achieved by the motion of the film through the telecine, the outputs from the sensors emerge in a non-standard sequential order. This must be converted to standard interlaced form; while such conversion could be achieved at the sensor outputs before gamma correction, carrying it out after gamma and aperture correction will be shown in a later section to yield considerable simplifications.

Other types of telecine in current use are based on flying-spot cathode ray tubes or television cameras. The camera, or photoconductive, telecine uses a normal television camera on which the projected film is imaged. It has the advantage of being able to show films at rates unrelated to the television field rate by using the storage characteristics of the sensing tubes (which are usually vidicons rather than plumbicons); it also has the advantage of being able to show very dense films if the light level in the film projector is increased. Nevertheless telecines of this type combine many of the disadvantages of both the film medium and television cameras; for this reason, and also because their characteristics are very similar to those of basic television cameras, photoconductive telecines will not be covered in this chapter.

Figure 10.4 shows the essential components of a flying-spot telecine. The scanning of the film image is accomplished optically by imaging a scanning spot of light, produced on the screen of a special-purpose CRT, onto the film via a lens. The light that passes through the film is sensed by three photomultipliers, one for each of the red, green and blue channels. Scanning the film in this fashion gives several advantages over a photoconductive approach. Firstly there are inherently no registration errors between the three colours, except for those caused by chromatic aberration in the main copying lens (which can be kept very low by good optical design). Secondly there is a reduced amount of maintenance, when compared to camera telecines. Finally the use of photomultipliers to sense the light transmission of the film results in a noise level that decreases in dark areas of the picture; this is a good situation to achieve, since subsequent

INTRODUCTION 505

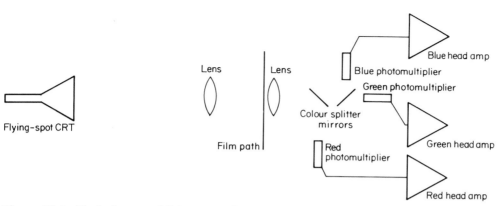

Figure 10.4 Block diagram of flying-spot telecine (processing channel omitted)

gamma correction will boost any noise occurring in such dark areas, and is historically the reason for the improved accuracy of gamma correctors in telecine machines as compared to cameras.

The signals from the photomultipliers are processed electronically according to the block diagram in Figure 10.5. This is shown in greater detail than in the relevant section of Figure 10.3; nevertheless, it is clear that there is a considerable similarity between the two figures. Indeed the only difference in practice is that the correction for element-to-element sensitivity variations in Figure 10.3 is replaced by a corrector to compensate for the effects of afterglow in the flying-spot CRT.

Flying-spot telecines can be made to produce interlaced signals directly. Early machines achieved this by using a dual optical path; odd or even fields were then selected by a mechanical shutter masking off the undesired optical path [162]. Later generations of machines used a 'jump scan' approach to switch the position of the raster on the flying-spot CRT; this had the advantage that the five raster positions required for 525/60 operation could be more easily achieved [163]. More recently still, however, it has become accepted practice to scan the film in a sequential manner and use electronic processing to convert the signals to interlaced form, in a similar manner to a line-array telecine [164].

Areas where digital processing may bring benefits

Table 10.1 shows some of the possible applications of digital processing in conventional television cameras (some of the special requirements of cameras using solid state area arrays are discussed in a later section). Currently the most important application lies in the control system; the use of complex digital elements, including microprocessors, in this area allows a more complex, and at the same time more flexible, control strategy to be employed. The benefits of this will be outlined in a later section.

Digital signal processing may also be used in the gamma correction and masking stages of the camera; the implications of this are discussed later. At the time of writing this is not such an attractive application since analogue processing is more compact and of adequate performance (the stability of most cameras is limited by the characteristics of

506 TELECINE AND CAMERAS

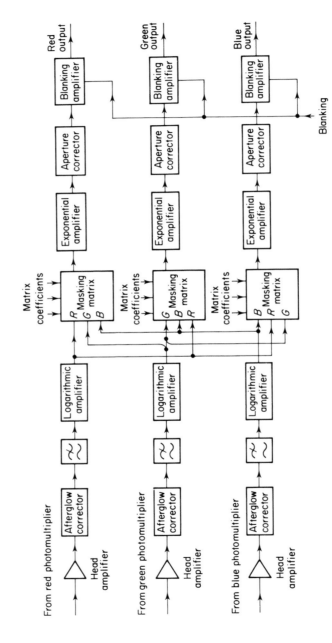

Figure 10.5 Block diagram of an analogue processing channel for a flying spot telecine

Figure 7.7 'Polyphoto' display showing contents of electronic 'slide' store.

Figure 7.29 The BBC 1 animated logo

Figure 9.1 Electronic painting system using a graphics tablet.

Figure 9.25 1979 General Election broadcast showing the 'Swingometer'

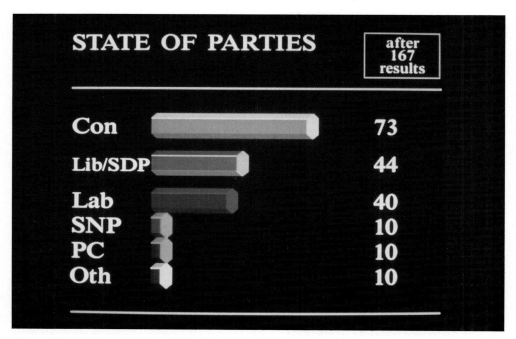

Figure 9.26 Computer controlled electronic painting system output as used in 1983 General Election programme.

Figure 9.28 (a) Composited text and map on 'Battleground' (b) Computed map and four live outside broadcast contributions on 'Battleground' as used in 1987 General Election programme.

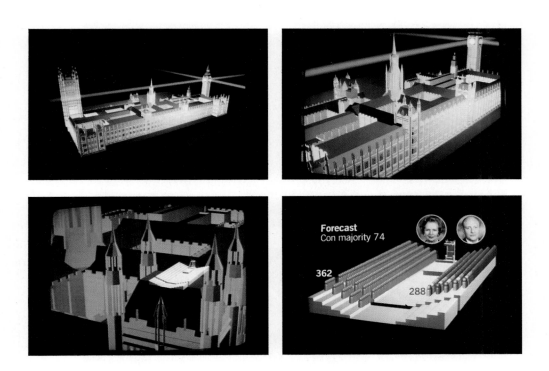

Figure 9.29 Four stills from the 'sting' animation, the last of which contains images and text under computer control to show the latest results.

Figure 11.1 Stages in the production of a chroma-keyed picture. (a) Original foreground scene. (b) Foreground scene with colour supressed.

Figure 11.1 (*cont.*) (c) Background scene multiplied by key. (d) Composite scene given by (b) plus (c).

Figure 9.30 Filtered text from recent commercial character generators.

Figure 9.31 The Aston IV Generator in use at the Television Centre.

Table 10.1
Applications of digital techniques to cameras

(a) Video signals	gamma correction
	colour matrixing
	aperture correction
(b) Control system	convergence control
	shading control
(c) Audio signals	transmission of remote sound

the photoconductive tubes themselves and not by the processing circuits) but forthcoming developments in digital circuit performance may change this situation. In any case, the emergence of the YC_RC_B digital standard outlined in Chapter 1 will mean the signals will leave the camera in digital YC_RC_B (or RGB) form rather than in analogue PAL and this will increase the attraction of digital signal handling throughout the processing chain.

A third application for digital signals is in the transmission of audio information. Frequently, and particularly on outside broadcasts, cameras are used in locations remote from the main control area. It is convenient to combine any audio signals, picked up at the camera point, with the main video output of the camera thus reducing the number of cables required. The use of digital coding for the audio signals reduces the effect of any distortion and crosstalk occurring in the camera cable and gives a higher quality.

Table 10.2 shows the applications of digital techniques to telecines. As with cameras, there are considerable advantages to the use of digital circuits for control purposes and several examples will be described in a later section. In addition, the increased cost and weight of gamma correctors in digital form are not so much of a disadvantage to telecine machines; because of this, advantage can be taken of the greater accuracy and stability of digital systems and the parameters required from a digital processing channel for a telecine are considered. The sequential-to-interlace converter required by line-array, and some flying spot, telecines is another major area where digital circuits bring benefit since it contains a large amount of storage; the performance given by this part of the telecine is outlined in a later section. As with cameras, there is also an application for digital techniques in the audio stages of the telecine; a digital audio delay may be used to compensate for any time delays occurring in the video signals (due, perhaps, to the storage in the sequential-to-interlace converter) or to allow the sound pickup heads to be in a physically more convenient location inside the machine transport.

Table 10.2
Applications of digital techniques to telecines

(a) Video signals	gamma correction
	colour masking
	aperture correction
	sequential-to-interlace conversion
(b) Control systems	variable film speed scanning
	automatic colour correction
(c) Audio signals	audio delay to match film speed changes or film gauge differences

508   TELECINE AND CAMERAS

These different applications (with the exception of those concerned with the audio signal, which are outside the scope of this book) will be considered in more detail in the following sections. Because of the higher signal quality required from telecine systems, and because of the lower constraints placed on size, weight and power consumption, there is a greater range of applications in telecine machines than in television cameras — indeed in many instances those applications in cameras are simplified versions of the corresponding telecine applications. For this reason, it is convenient to consider telecine first.

DIGITAL TECHNIQUES IN TELECINE MACHINES

A digital telecine processing channel

The need for a processing channel to accomplish gamma correction, aperture correction and colour matrixing has already been described. Using digital techniques in this processing channel would bring about several advantages; among them would be increased precision, reduced variation from telecine to telecine, the reduced need for maintenance and adjustment and an improvement in stability. Some second-order effects at present occurring in analogue systems (for example a reduction in signal bandwidth at the higher-gain portions of the gamma characteristic) may also be eliminated.

Reference [114] describes some basic experiments to investigate the parameters required from a digital channel. A flying-spot slide scanner was used as the picture source, so that the highest possible picture quality would be achievable, and the system of Figure 10.6 assembled. The afterglow correction was accomplished in analogue form prior to analogue-to-digital conversion; the use of a digital afterglow corrector was not

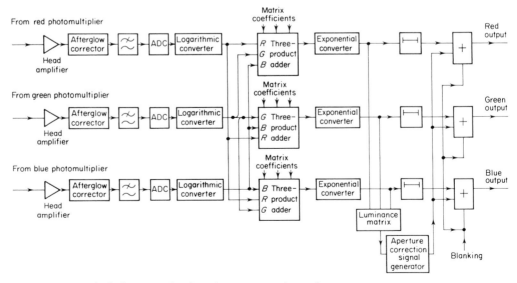

Figure 10.6 Block diagram of a digital processing channel

considered beneficial, since it would have involved an extremely complicated digital filter. In contrast, analogue afterglow correctors are extremely simple and stable.

The outputs of the afterglow correctors were then applied via suitable lowpass filters to three analogue-to-digital converters (ADCs), one for each of the red, green and blue signals. When signals are digitally coded before gamma correction, the visibility of quantisation is increased in lowlights; this is because gamma correction increases the gain for signals near black. Low level signals, therefore, need to be coded to more than the usual 8 bits resolution. Previous work [165] suggested that 10 bits per sample might be required, at least in the green channel. As quantising errors were not as visible in the red and blue channels, a reduced number of bits per sample, together with the addition of dither signals to break up contours caused by quantisation, sufficed; however, if a large degree of masking or considerable aperture correction had been required, quantisation effects in the red and blue channels might have been transferred into the green channel. Thus all three ADCs ideally need at least 10-bit resolution.

This problem was overcome in the system of [114] by realising that the extra resolution was only required for signals near black. A 'range-changing' ADC was developed by modifying an ordinary 8-bit coder so as to insert analogue preamplification wherever the input signal fell below a predetermined threshold. The preamplification factor was an exact power of 2; thus the 8-bit word from the ADC could be located within a longer word by simply displacing it by the appropriate number of binary places.

Figure 10.7 shows the visibility of quantising effects on signals that have been 8-bit linearly encoded and subsequently gamma corrected (adapted from [165]). The visibility is shown as a fractional change in perceived luminance, $\Delta Y/Y$. The smallest change in luminance that can be seen is about 2% (the Fechner fraction). It can be seen from Figure 10.7 that $\Delta Y/Y$ is greater than 2% when the input signal falls below 53% of the peak level. However, it does not rise rapidly until the signal level falls below 15%. Thus a factor of 8 was chosen for the pre-amplification factor giving three extra bits of resolution

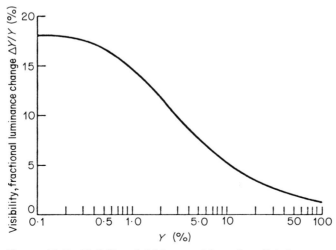

Figure 10.7 Visibility of 8-bit quantising after digital gamma correction

for signal levels below 12.5%. The maximum value of $\Delta Y/Y$ was then 2% for signals in the pre-amplified range and 4% elsewhere.

A block diagram of the ADC is shown in Figure 10.8; the pre-amplification and switching was accomplished in a dual sample-and-hold circuit preceding the 8-bit ADC. The ratio of the gains of the two paths was adjusted to better than ± 1 part in 64 (approximately $\pm 1\%$); it was also found important to match the delays throughout the ADC and to control the d.c. offsets of the two video paths accurately. These factors were ensured by the use of the matching delays and clamp circuits shown in Figure 10.8.

After the ADCs the digital signals passed to a logarithmic converter which used read-only-memories (ROMs) to generate a 12-bit number (4-bit characteristic plus 8-bit mantissa); the block diagram for this is shown in Figure 10.9. A small offset was added to the signal at the input of the circuit to prevent the logarithm of a true input signal of zero being generated ($\log 0 = -\infty$). Gamma-correction and logarithmic matrixing (masking) was then carried out in a set of three-product adders. Each three-product adder produced the sum

$$\text{output} = k_1(\text{input}_1) + k_2(\text{input}_2) + k_3(\text{input}_3) \tag{10.2}$$

where k_1, k_2 and k_3 are constants, set by external control, and specified to six bits. The constants could be programmed to lie in the ranges

$$0 \leqslant k_1 \leqslant 1\tfrac{31}{32}$$
$$-1 \leqslant k_2 \leqslant \tfrac{31}{32}$$
$$-1 \leqslant k_3 \leqslant \tfrac{31}{32}$$

The video input and output signals were specified to 12 bits, but, because of internal rounding errors, the output was only accurate to 10 bits.

From the three-product adders, which together produced the matrix multiplication

$$\begin{bmatrix} \log R_{\text{out}} \\ \log G_{\text{out}} \\ \log B_{\text{out}} \end{bmatrix} = \begin{bmatrix} k_{11} & k_{12} & k_{12} \\ k_{21} & k_{22} & k_{23} \\ k_{31} & k_{32} & k_{33} \end{bmatrix} \times \begin{bmatrix} \log R_{\text{in}} \\ \log G_{\text{in}} \\ \log B_{\text{in}} \end{bmatrix} \tag{10.3}$$

The video signals were sent to three exponential converters. As with the logarithmic converters, these were based on the use of ROMs and each generated a 12-bit output signal. Thus the gamma correction and masking circuitry was an exact equivalent of the processing carried out in the analogue channels of Figure 10.3 and 10.5.

Gamma correction is a non-linear process and, as such, generates harmonics of the input signal. If these harmonics occur at frequencies above $f_s/2$ (where f_s is the sampling frequency of the digital system) then they may cause moiré patterns due to aliasing. Figure 10.10 illustrates (in the time domain) how these alias signals are produced in a digital gamma corrector. Figure 10.10a shows a 100% sine wave signal of 1/3 sampling frequency sampled in two different phases; Figure 10.10b shows that the mean level of the gamma corrected output samples depends on the sampling phase. If the signal frequency is not exactly $f_s/3$ the sampling phase will vary periodically. As the phase changes, the mean level will change producing a low frequency beat added to the high frequency signal. Similar beats are, in principle, produced by inputs having frequencies near all submultiples of the sampling frequency.

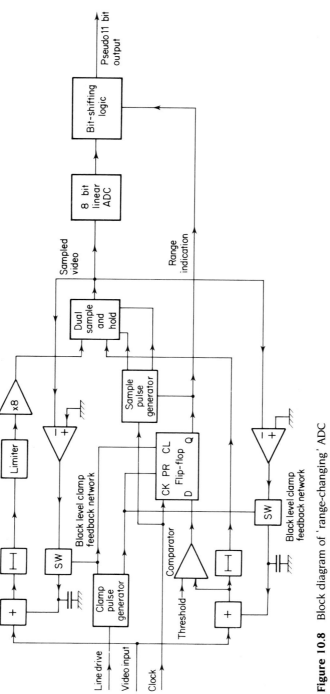

Figure 10.8 Block diagram of 'range-changing' ADC

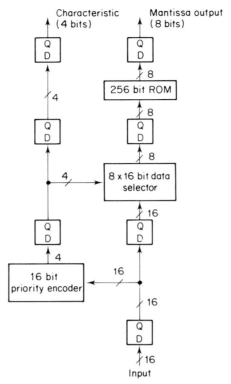

Figure 10.9 Block diagram of log converter

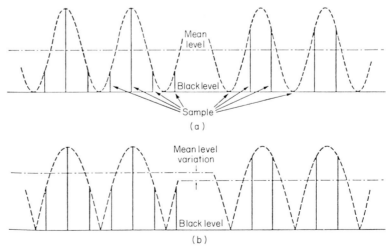

Figure 10.10 (a) 100% sine wave inputs, sampled at different phases. (b) Gamma-corrected output showing variation in mean level of samples

The amplitude of the beat depends on the input signal frequency, its amplitude, and the sit (added d.c. component). Table 10.3 shows how the amplitude of the beat varies with input frequency for a given value of f_s for a 100% amplitude input sine wave. The amplitude of the beat also decreases very rapidly as the lowest part of the sine wave departs from black level — that is, as the sit of the signal is increased.

Another non-linear process which generates harmonics of the input signal is clipping such as would occur if the ADCs are overdriven. The amplitude of the harmonics generated in this way will be greater than that generated by gamma correction, which is a smooth non-linearity generating sufficient harmonics only with large input signals. It is interesting to note that moiré patterns arising from this cause can be seen in analogue channels in the presence of vertical detail, arising from the sampling action of the 625-line scanning system; this moiré from vertical detail is not usually considered to be a serious impairment.

Reference [114] indicates that, at sampling frequencies of around 13.3 MHz (i.e. very close to those of the CCIR standard) aliasing effects were not a major problem. Moiré patterning due to gamma correction was just visible with electronically generated sine wave signals, but very rarely visible with real pictures; moiré from clipping was easily visible with electronically generated signals, but could be seen with real pictures only if the signals were grossly distorted by severe clipping. Thus the generation of beats in a non-linear digital processing channel is unlikely to be a problem at the sampling frequency used in the CCIR standard.

The final step in the digital processing channel of Figure 10.6, was to aperture correct and blank the gamma-corrected *RGB* signals. Aperture correction is required to compensate for high frequency losses due to lens imperfections, image spreading in the film emulsion (or the camera tube target, for television cameras), the scanning spot size in the flying-spot tube and, to some extent, the display spot size [112]. These effects all spread the image of each picture element over a finite area or aperture, thus softening the displayed picture. Aperture correctors attempt to reduce the effects of this spread by subtracting from the signal a proportion of the signals from neighbouring picture elements as in the block diagram in Figure 10.11. Line delays can also be used to give access to picture points above and below the element of interest on adjacent lines.

In analogue aperture correctors, the Gaussian aperture losses usually encountered are most accurately corrected by using small values of delay and large values of *k* (the amount of correction). This puts the peak correction at very high frequency so that the correction rises at an ever-increasing rate from d.c. to beyond the edge of the video band.

Digital aperture correctors, in contrast, must operate on sampled signals and, as a result, the minimum delay which can be used in aperture correction is restricted to one

Table 10.3
Amplitude of alias signal

Input frequency	Alias signal level (dB)
$f_s/2$	-13
$f_s/3$	-21
$f_s/4$	-26
$f_s/5$	-30
$f_s/6$	-33

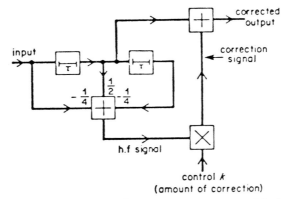

Figure 10.11 Simple aperture corrector—block diagram

sampling interval (approximately 75 ns for the CCIR standard). However, an improved performance can be achieved by taking contributions from other delays spaced at multiples of the sampling interval. This is shown diagramatically in Figure 10.12; this second-order corrector subtracts 'in band' correction (set by k_2) to prevent overcorrection of low frequencies and allows a relatively large amount of 'out of band' correction to be applied while maintaining a maximally flat response. In principle such an approach may be extended to higher and higher orders giving greater and greater flexibility along the lines illustrated by Chapter 5; in practice the second-order correction shown in Figure 10.12 is a reasonable compromise.

Figure 10.13 shows a block diagram of the digital aperture corrector constructed for the experiments of [114]; it is very similar to the system of Figure 10.12 but with the

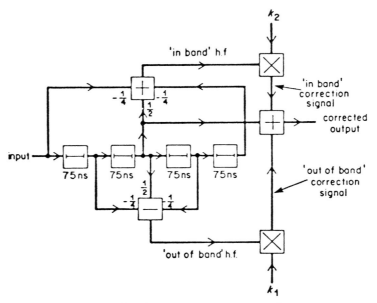

Figure 10.12 Second-order aperture corrector—block diagram

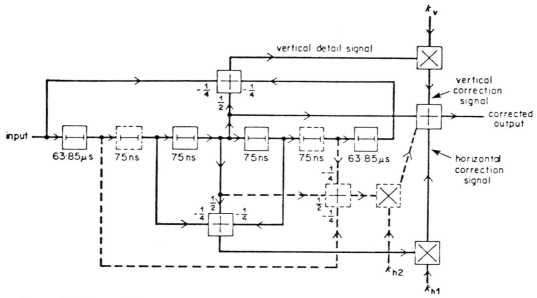

Figure 10.13 Parallel two-dimensional aperture corrector—block diagram

addition of vertical aperture correction using line delays. The contributions used to give this vertical correction were all from the same field and were, therefore, spaced by 2/625 of a picture height (equivalent to 135 ns horizontally). This gives the vertical detail signal a maximum response at $156\frac{1}{4}$ cycles per picture height (equivalent to 3.7 MHz horizontally) which was considerably 'in band' (i.e. it was unable to correct the highest possible vertical frequencies in the region of $31\frac{1}{2}$ cycles per picture height). This was in practice an advantage since correction of these frequencies would have accentuated the 25 Hz flicker components caused by interlaced scanning; however, if desired, it is possible to extend the correction to these higher frequencies either by using a field store to give access to the lines of the other field, which are closer, or by carrying out the aperture correction before sequential-to-interlace conversion (in picture sources where this is relevant).

Second-order vertical correction would have only affected relatively low spatial frequencies and could not have overcome the basic limitations of 'in band' correction mentioned above. It was not, therefore, used in the digital corrector. Also, while the sampling structure of the CCIR standard is exactly locked to the line frequency, it was established in [112] that adequate operation of the aperture corrector could be maintained even for sampling structures which are not vertically aligned but instead slope at angles of up to $\pm 16°$ from the vertical.

In order to achieve colour operation, the corrector of Figure 10.13 could have been used in each of the red, green and blue channels. In practice, however, aperture correction is only necessary for the luminance components and so it was possible to avoid triplicating the circuits by using the arrangement of Figure 10.14. As the corrector delayed the luminance signal by more than one television line it was necessary to delay the chrominance components to match. Some colour cameras achieve this by mis-

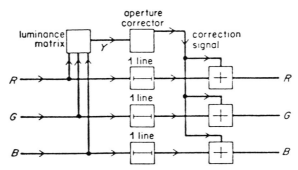

Figure 10.14 Aperture correction of colour signals—block diagram

registering the colouring tubes but this is not possible in flying spot scanners and the approach usually used for analogue correctors is to form R–Y and B–Y signals, delay them to compensate for the delayed luminance (Y) signal and recombine them with the Y signal to reform corrected R, G and B. This would have involved a large number of multiplications in a digital version and instead a different approach was used in Figure 10.14 which reduced the number of multipliers required at the expense of an increase in delay circuits. Consequently, all three R, G and B signals were delayed and a correction signal added to each of them to give corrected R, G and B directly. The correction signal was obtained from a luminance signal matrixed from the input R, G and B. The matrix coefficients did not need to be exact and for simplicity $Y = \frac{1}{4}R + \frac{5}{8}G + \frac{1}{8}B$ was used.

The performance of the complete digital processing channel was assessed by performing subjective tests to indentify any quantisation effects. Two colour slides were used that had been found to be rather critical in earlier tests [165]. The first of these, 'Boy', had a dark, plain background to show up any contouring due to quantisation effects in low lights; the second slide 'Fish Lure', contained low level, high frequency detail which would be lost if too few bits were used. This use of stationary pictures made the tests easily repeatable and gave observers the chance to find and examine the most critical parts of the picture. However, quantising effects were known to be more visible in the presence of slow brightness changes, so a 1 Hz sinusoidally varying sit of amplitude 0.1 % of peak white was added to the green channel. Preliminary tests had established that this sit variation made no difference to the assessment of pictures derived by analogue processing, but that quantising effects were increased in visibility when too few bits were used in the digital coding.

For the experiments, the range-changing ADC was used in the green channel only; the red and blue signals were each coded to 8 bits with added dither being used to reduce the visibility of quantising. All ADCs were used to within 1 or 2 dB of their full conversion range and switching logic could be used to discard bits from the 11-bit output of the range-changing ADC. The digital channel was set to a gamma of 0.4 with a medium-saturation mask and the output was displayed, without PAL coding, on a high quality shadowmask monitor whose peak brightness was set to 70 cd m^2. The results of the tests are shown in Figure 10.15 where the degree of picture impairment is plotted against the number of bits used in the green channel for both slides. Dither, as used in the red and blue channels, could also be added to the green channel and curves are shown in Figure

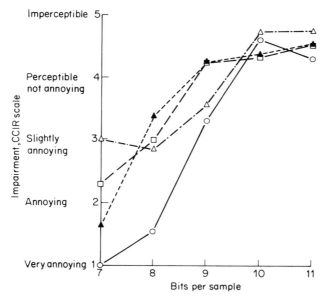

Figure 10.15 Subjective test results (○—○ Boy, no dither; ▲---▲ Boy, with dither; △–·–△ Fish Lure, no dither; □--□ Fish Lure, with dither)

10.15 for both conditions. The impairment was classified according to the CCIR 5-point impairment scale in Table 10.4. Results showed that there was little difference between pictures produced using 10 or 11 bits in the green channel. Dither made no difference under these circumstances; 9 bits with dither was considered almost as good as 10 or 11 bits. Figure 10.15 also shows that the effects of quantisation are more severe on the 'Boy' slide (where it produced contouring in dark areas) than on the 'Fish Lure' slide (where it produced a loss of low level detail); this implies that the accuracy of quantisation need not be so high in areas of detail as it is in relatively plain areas and has implications in the field of bit-rate reduction (see Chapter 13).

The probable reason why dither made no difference to the 10- and 11-bit signals is that the noise generated from the head amplifiers of the slide scanner was sufficiently high in level to conceal the differences between 10- and 11-bit quantisation effects. Previous work [69] has shown that quantising will be invisible if the video signal-to-noise ratio is less than $(6n + 5)$ dB, where n is the number of bits used. The r.m.s. noise level produced

Table 10.4
CCIR five-point impairment scale

Grade	Impairment
5	Imperceptible
4	Perceptible, but not annoying
3	Slightly annoying
2	Annoying
1	Very annoying

518   TELECINE AND CAMERAS

by the head amplifiers of the slide scanner was about -65 dB with respect to the peak signal level; thus it would be expected that quantisation of 10 or 11 bits, with or without the addition of dither, would be largely invisible. However, if the noise level were to be decreased it might be necessary to increase the quantisation accuracy, or to re-introduce dither, in order to keep quantisation effects invisible. Figure 10.16 shows the Marconi telecine which uses three line-array CCD sensors and has been designed around a digital processing channel similar to that of [114]. The digital electronics are located in racks behind the doors underneath the main control panel of the equipment.

Sequential-to-interlace conversion

Telecines using solid state line-array sensors, and also later generations of flying-spot telecine, scan the film in a non-standard sequential fashion, as has already been outlined.

Figure 10.16 A broadcast CCD line-array telecine which uses all-digital signal processing, produced by Marconi

Consequently they require a sequential-to-interlace converter in order to transform the signals to the standard interlaced format; this converter uses a large amount of storage and is therefore a major area of application for digital techniques.

The first use of such a sequential-to-interlace converter was demonstrated in 1977 [166], the converter had a storage capacity of one television field which was sufficient for the replay of 16 mm film running at 25 frames per second synchronised to the 50 Hz field rate. Table 10.5 shows the sequence of line numbers being fed to, and retrieved from the store. For ease of explanation, the lines are numbered sequentially from the top of the picture; following this convention, the lines of a given television field will, therefore, follow the sequence 1.3.5.7.9, etc., or 2.4.6.8.10, etc., rather than the sequence defined in Chapter 1. During the first 20 ms lines 1 to 312 are written into the store, the output from the store consisting of the previous film frame. After this time, line 1 is read from the store and line 313 written in its place. Line 3 is read out next and line 314 written in its place. The read sequence then continues with lines 5, 7, 9, etc., lines 315, 316, 317, etc., being written into the store. Eventually line 625 is required to be read from the store at the time that it is generated at the input; the store is, therefore, bypassed for this one line. At the end of this period, lines 2, 4, 6, 8, 10, etc., remain in the store and are read out while lines 1 to 312 of the following film frame are being written in. Thus the store capacity must be at least 312 lines, or approximately one television field.

Table 10.5
Sequence of line numbers in sequential to interlace store

Line no. being written	Line no. being read
1	
2	
3	lines 2–624
4	of previous
⋮	frame
311	
312	
313	1
314	3
315	5
⋮	⋮
623	621
624	623
625	625
1'	2
2'	4
3'	6
4'	8
⋮	⋮
311'	622
312'	624
313'	1'
314'	3'
315'	5'
⋮	⋮

520 TELECINE AND CAMERAS

Operation of line-array telecines in this fashion is only possible because of a coincidence, however, Figure 10.17a shows a number of film frames. The area between successive film frames is not to be reproduced and is arranged to be eliminated by the television blanking. Thus the number of *active* television lines (575 for the system I standard) must all be generated from the active frame height x. Because the vertical scan is produced by the motion of the film, if the ratio of this frame height to the inter-frame pitch, y, does not bear the same relationship as that of the number of active television lines (575) to the total number of television lines, including blanking (625), the scan rate emerging from the line-array sensors will not be synchronised to the television line rate. Such lack of synchronisation is also encountered with sequentially scanned flying-spot telecines, where, in order to avoid scanning a single line on the CRT with the attendant risk of damaging the tube, the line scan rate is raised above its normal value [164]. Fortunately, 16 mm film does have the correct relationship; 35 mm film does not. Thus if 35 mm film is to be televised the sequential-to-interlace converter must also be used as a buffer to synchronise the incoming line rate; in order to do this, its storage capacity must be increased. Similar considerations arise, for both line-array and sequentially scanned flying-spot telecines, if the film is required to run at speeds other than 25 frames per second.

Since the amount of storage provided in the sequential-to-interlace converter is fixed at the time the telecine is designed, it is clearly of crucial interest to study what the worst-case requirements are. Such a study has been carried out [167], for line-array telecines; similar results may be derived for flying-spot machines.

Figure 10.17b is a diagrammatic representation of the operation of the sequential-to-interlace store. Information is continuously being written into the store from the film being scanned. A particular film frame can only be used for a given television field if it will have been completely scanned by the end of the field period; thus the second line of Figure 10.17b shows the sequence of film frames used for each field.

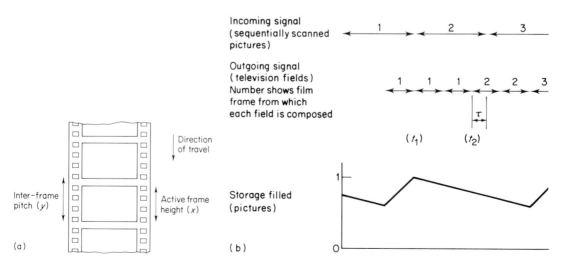

Figure 10.17 (a) Frame pitch and height (for 35 mm motion-picture film). (b) Operation of sequential-to-interlace store

A given film frame can start to be erased from the store at the beginning of the last television picture to use information from that frame (a television picture being composed of two interlaced fields). In Figure 10.17b it is assumed that the television picture rate (half the field frequency) is higher than the film frame rate; thus the instant when the film information starts to be erased ((t_1) in Figure 10.17b) is the time at which the store is most full. Likewise instant (t_2) is the time at which the store is most empty. A similar method can be used to analyse the situation when the television picture rate is lower than the film frame rate; this is carried out in [167] and shown to lead to similar results to those derived below.

If the time between successive film pictures if T_f, the time between successive television pictures T_t and the offset time (defined in Figure 10.17) is τ then the storage at (t_1) is

$$\text{Picture 1} + \frac{(T_f - T_t - \tau)}{T_f} \times \text{Picture 2} \qquad (10.4)$$

or

$$2 - \frac{(T_t + \tau)}{T_t} \text{ pictures}$$

Likewise at (t_2) the storage is

$$\frac{(T_f - \tau)}{T_f} \times \text{Picture 2} \qquad (10.5)$$

or

$$1 - \tau \text{ pictures}$$

The storage used throughout the cycle varies linearly-between these values as shown in the bottom line of Figure 10.17b. In practice only the active lines need to be stored; this introduces an additional complication into the analysis. Reference [167] considers the implications of this and concludes that, subject to certain restrictions on the ratio of film frame height to inter-frame pitch (which are, in practice, easily met), equations 10.4 and 10.5 remain valid.

It is clear, therefore, that the offset time, τ, is a significant factor in determining the storage required to achieve a given speed of operation. In general, if the film is not locked to some simple submultiple of the television field rate, τ is free to take any value between $T_f/2$ and just above zero (the actual value of zero corresponds exactly to the case $\tau = T_f/2$). Sooner or later, however, τ will assume a value that is very close to, but not equal to, zero. The storage then required will be approximately $(2 - T_t/T_f)$ pictures.

If the film speed is locked to the television field rate, the situation is considerably simplified. Only a certain number of values are now allowed for τ, one of which is usually made to be zero (giving a storage requirement equal to that required by $\tau = T_t/2$). Thus, of the other values allowed for τ, there is a minimum value, τ_{min}, which gives rise to the greatest storage requirement. For example, if the film rate is $16\frac{2}{3}$ frames per second (i.e. $\frac{1}{3}$ of the field rate), τ_{min} becomes $0.33T_f$ and T_t is $0.67T_f$. The storage requirement is then one picture; if the film speed had not been locked the storage requirement would have been $1\frac{1}{3}$ pictures.

The sequential-to-interlace conversion can, as has been suggested, be accomplished at several stages in the signal processing path (before or after gamma correction for example). If the converter is placed before the gamma corrector, however, the ADCs necessary to turn the signals into digital form must have 10- or 11-bit accuracy, as has been discussed in the previous section. In contrast, if the conversion is carried out after gamma correction, 8-bit ADCs will suffice; the store itself need then only handle 8 bits and, in view of the size of store required, this represents a worthwhile simplification. Similarly, if the sequential-to-interlace converter is placed after the colour-correcting matrix it is no longer necessary to preserve the individual red, green and blue signal components; matrixing to Y, U, V form may be accomplished with the U and V signals being stored at lower bandwidths. This second simplification has the further advantage that the signals can be taken directly from the store in the correct CCIR digital format thus enabling an easy interface to other digital equipment.

Movement interpolation

In the previous section it has been shown that one of the benefits of including digital storage at the output of the telecine is the ability to run film at any speed (a suitable control system to enable this to be achieved will be described in the next section). However, doing this can result in problems when there is movement in the scene. The store will replay m sequentially scanned film frames to occupy n interlaced television fields, where m and n are unrelated numbers. This is achieved by replaying each film picture n/m times. Where n/m is not a whole number the repetition rate may cycle. For example $n/m = 2\frac{1}{2}$ can be achieved by repeating the first frame twice, the second three times, the third twice and and so (corresponding to the case shown in Figure 10.17). Unfortunately this leads to unacceptable movement judder of the type discussed in Chapter 6. This is a minor annoyance in 625/50 countries and limits the film replay speed to 25 frames per second at which speed $n/m = 2$; the penalty is a 4% error in sound pitch and running time on material shot for cinema release at 24 frames per second. In 525/60 countries, however, the movement judder is a more serious problem since replay at 24 frames per second yields $n/m = 2\frac{1}{2}$ as above; the alternative of replaying the film at 30 frames per second would give an unacceptable speed error.

To combat movement judder, movement interpolation can be used (see [168]). Figure 10.18a illustrates the situation. The film pictures are samples of the initial scene at discrete moments in time, shown by the circles; each circle shows the time at which a given line was sampled. For correct movement portrayal it is necessary to generate an approximation of what the samples would have been if the scene had been televised by a television camera. The crosses in Figure 10.18a correspond to the time at which each line would have been scanned with a television camera. The slope of the television fields with respect to time is a complication which can be neglected with very little penalty; Figure 10.18b shows the situation if this is done. A simple movement interpolator can then generate approximations to the signal that would have been produced by a television camera by adding together a proportion, S_1, of film frame 1 to a different proportion, S_2, of film frame 2. In order to avoid flicker in the final picture

$$S_2 = 1 - S_1 \tag{10.6}$$

Several possible laws relating S_1 (and hence S_2) to τ (the relative time relationship between television fields and film frames, as in Figure 10.18b) are shown in Figure 10.19. Law A corresponds to the case where there is no movement interpolation. In

DIGITAL TECHNIQUES IN TELECINE MACHINES

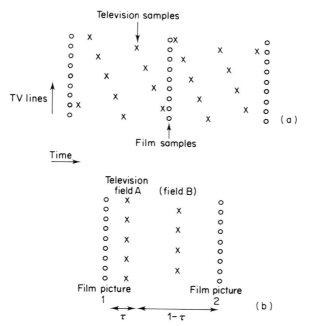

Figure 10.18 Temporal relationship of film frames and television fields

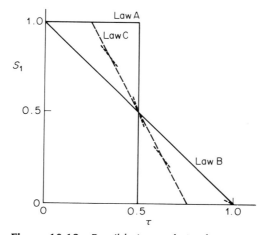

Figure 10.19 Possible interpolation laws

contrast, Law B corresponds to the situation where there is an excessive amount of movement interpolation; no judder is visible on movement, using this law, but instead there is a pronounced loss of resolution in moving areas. There is an infinite number of intermediate laws (such as Law C in Figure 10.19) which bring back some of the lost resolution at the expense of re-introducing a small amount of movement judder; in general, for most movement speeds, the optimum interpolation characteristic was found, in the experiments of [168], to be close to law C as shown, and this was judged to be a significant improvement over law A or B. Slight departures from this characteristic were beneficial for some film speeds, however.

In order to allow movement interpolation to be carried out, the amount of storage used in the sequential-to-interlace converter must be increased over that calculated in the previous section. The amount of this increase depends on the law chosen for the movement interpolation; law B in Figure 10.19 would require an extra 1 picture of storage. As the duration of the cross-fade (the sloping portion of the law) reduces, so the extra storage reduces; law C, having a cross-fade duration of approx. 50% as drawn, requires only $\frac{1}{2}$ picture extra storage.

Telecine control systems

Scanning control and variable speed operation

Telecines using storage, as described in the previous sections, are capable of producing high quality pictures from film running at non-standard, unlocked speeds. In order to do this, however, they must incorporate suitable control systems to arrange that the film is scanned in the correct fashion. Reference [168] describes a control system; it used a high speed microprocessor system, together with peripheral digital interfaces, to lock the scanning rate of the telecine to the film speed being used. Although the system was developed for use with a line-array telecine, very little of the processing was specific to either a line-array or a flying-spot approach; the requirements of each, are, in fact, very similar. As the line-array approach introduces one or two additional problems, however, the description of the processing will be given in terms of a line-array telecine.

It is apparent, from the results of the previous section, that, because the scanning speed of the line-array sensors must both track the film speed and also be adjusted to obtain the correct displayed height, the sensors are only rarely scanning at the same rate as the television line standard. Low sensor scan rates can be accommodated relatively easily; one convenient approach is to use a fixed clock frequency and insert 'gaps' of variable numbers of clock pulses between successive scans. The maximum scan rate of which a sensor is capable is fixed, however, and so there is a maximum speed at which a film can be run if the sensors are to scan every line. For example, as we have seen, the scan rate required by normal aspect ratio 35 mm film running at 25 frames per second and producing an output intended for the 625/50 standard is 18.075 kHz (1 line every 55.3 μs). If a telecine were to be designed with this as a limit then 16 mm film could be run at speeds up to 28.9 frames per second; any increase over these speeds, or any increase in the displayed picture height, would then mean that the sensor could scan only every other line.

Figure 10.20 shows the control system used in the experiments of [168]. The heart of the system was a microprocessor which linked the servo-circuits of the telecine transport to the scan-generating circuit for the sensor. This microprocessor measured the speed at which the film was running by timing the interval between two successive pulses produced by a tachometer fitted to the capstan motor which drove the film. Suitable tachometers could also sense in which direction the film was being driven. Provided that the film shrinkage was known, the scan rate required to produce television pictures of the desired height from the particular film gauge could be calculated relatively easily. This

DIGITAL TECHNIQUES IN TELECINE MACHINES

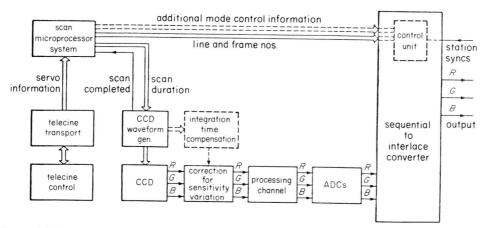

Figure 10.20 Control system for line-array telecine

shrinkage was measured by examining at regular intervals the output of an idling sprocket wheel driven by the film perforations. Any change in the film framing from one frame to the next can be interpreted as an error in the shrinkage being assumed in the calculation of scan rates. The idling sprocket also enabled the calculations to be adjusted so that the film was brought into a correct framing relationship.

When the microprocessor had calculated the desired scan rate, the information was sent to a waveform generator which produced the signals required to drive the sensors. The outputs from the sensors (one for each of the red, green and blue channels) were corrected for variations in element-to-element sensitivity, processed in the normal manner, and then written into the sequential-to-interlace store. A 'label' was sent to the control unit of the sequential-to-interlace converter (itself another microprocessor) to identify the particular line being scanned. This label was generated by the scan microprocessor which, in order to do this, required an additional signal from the waveform generator indicating the start of every scan. The sequential-to-interlace converter control microprocessor used the label to decide where in the store the particular scan signals should be put so that they could be retrieved correctly. It was also responsible for retrieving the scan information in the correct sequence as required by the ouput television signal.

In this way all the critical elements of the control system were digital and there was a rigid relationship between the film speed and the scan rate throughout the machine. Thus, once the correct framing relationship and film shrinkage information had been acquired, it could be maintained whatever the film speed fluctuations. Broadcast quality signals could thus be generated, in perfect framing, at all speeds from still frame up to the maximum rate at which the sensors were capable of operating.

In order to achieve this flexibility, however, it was necessary to solve the problem of integration time in the line-array sensors. Each sensor integrated the total amount of light falling on it between successive scans. If the interval between scans were to become very long, as it would have for low film speeds, then there would have been a danger that the sensors could overload. In order to overcome this, more scans were generated than were actually required with the additional 'dummy' scans being discarded before the sequential-to-interlace converter. The generation of dummy scans commenced when the film speed fell below that at which the scan rate was half the maximum possible. Under those conditions the sensors could generate two scans in the time the film took to move one line. As the film slowed down still further, the rate of these two scans also reduced until more dummy scans could be generated, and so on.

Thus the integration time of the sensors varied only over a 2:1 range as the speed was changed. This caused a 2:1 change in the amplitude which was compensated by adjusting the gain subsequently applied to the signal; such compensation was most easily achieved by applying an additional gain-controlling input, generated from the scan rate signal fed to the waveform generator, to the variable-gain unit already required to correct the element-to-element sensitivity variations of the sensor. This connection is shown dotted in Figure 10.20. Nevertheless, it was considered good practice not to expose any errors in the compensation more than was necessary. Thus, for the lower film speeds, the valid scans were all kept at a constant duration and the number and duration of the dummy scans were varied to track film speed variations. By this means the integration times of all the valid scans were identical.

These techniques allowed the reproduction, with full broadcast quality, of film running at slower than normal speeds. However, film running at faster than normal speeds led to the problem mentioned above (that the sensor could not scan fast enough). It was, therefore, not possible to scan every line of every film frame. Two solutions were possible; either the television signal could have been assembled over several film frames, or one film frame only could have been scanned and the missing lines generated by interpolation. Following the first approach, if the film were to move at four times normal speed (i.e. approximately 100 frames per second) lines, 1, 5, 9, 13, etc., would have been scanned from one film frame, lines, 2, 6, 10, 14, etc., from the next frame and so on. Using the second method, lines 1, 5, 9, 13, etc., of the first frame only would have been stored and the missing lines generated by repetition or interpolation of these stored lines. Under most circumstances this latter approach was found to generate by far the more acceptable results for review purposes. Figure 10.21 shows a telecine produced by Rank Cintel, which is also based on the use of CCD line-array sensors, which uses a scanning control and variable speed operating systems similar to that described here. The telecine uses a control bay and up to three film transport bays (only two of which are shown in the photograph). The microprocessors for variable-speed scanning are located in racks within the control unit—behind the door in the lower half of the control bay.

Automatic colour correction

Despite the care taken in shooting, printing and processing film, small variations in colour and brightness usually occur. While these are not often noticeable in a darkened cinema, as the eye adapts readily, they can frequently be noticed in the critical conditions under which television is viewed. A television receiver is usually watched in a moderately lit room with familiar objects around and with the screen occupying only a small part of the field of view; in these circumstances the eye does not adapt to anything approaching the same extent. It is thus common practice to make corrections to the telecine output to adjust the colour and brightness of each scene [169]. These corrections may be made manually when the film is televised. Alternatively, the corrections may be established during a rehearsal run and stored in a suitable form [172]; they are then applied auto-

DIGITAL TECHNIQUES IN TELECINE MACHINES 527

Figure 10.21 The Rank Cintel telecine having a similar variable-speed control system to that described in the text

matically during transmission [170]. The second approach eliminates the transient colour errors caused at scene changes while the controls are readjusted but requires extended rehearsal time. This is not always available, particularly in the case of news and current affairs programmes. Unfortunately, the type of film and the filming conditions used for such programmes frequently lead to large changes in colour balance and exposure.

It is, therefore, desirable to have an automatic colour corrector capable of calculating and applying corrections very quickly (in one or two TV fields). Such a corrector is described in [171]; while its performance was not as consistent as would be achieved by a trained operator, it could apply corrections very quickly and an operator could then make any small adjustments that might be necessary.

The corrector of [171] in fact used analogue techniques throughout. Later work, however, used a digital microprocessor as the central control unit carrying out the calculation of the corrections required. The corrections themselves were still applied to the video signals using analogue circuitry although, in principle, they could be combined with the digital approach used previously with little difficulty.

528   TELECINE AND CAMERAS

Figure 10.22 shows a block diagram of the automatic colour corrector. It operated on the red, green and blue signals at the output of the telecine processing channel, varying the black level (sit) and the gain of each according to certain rules. These were

(1) Adjust the black levels until the darkest parts of each red, green or blue signal reaches a pre-determined level—nominally 5% of white level.

(2) Adjust the differential gains so that the means of the three signals are all equal.

(3) Adjust the overall gain so that the most positive of the peaks of the three waveforms just reaches white level.

In order to achieve this operation the incoming video signals were clamped and analogue circuits used to measure their maximum, minimum and mean levels. Calibration pulses appropriate to the desired values of these levels were added to the signals

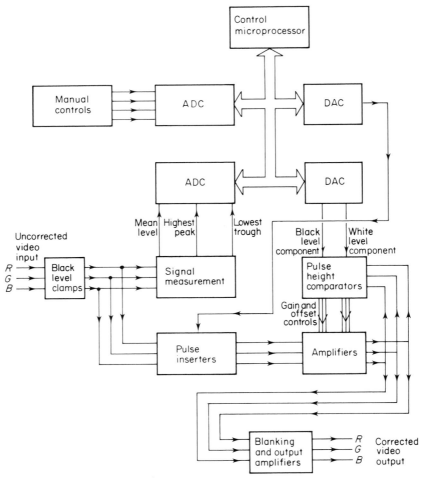

Figure 10.22 Automatic colour corrector

during the field blanking interval and the offsets and gains of the (analogue) video amplifiers adjusted until the amplitudes of the pulses representing the mean levels were equal and the pulses representing minima and maxima reached black-and-white level respectively; the pulses were then removed by blanking in the output stages. To prevent momentary drops in gain when scenes contained flashing lights or intermittent specular reflections, a high-pass-filtered version of the detected input peak was added to the white reference. The output peak level, therefore, followed the input peak level through sudden changes, returning to white level when the input was substantially constant.

Alternative modes of operation were possible. The black levels could be equated, not to the reference, but to whichever input black level was lowest. Similarly, the white reference could be replaced by the highest detected maximum level causing the output peak level to follow the highest input peak level. Information on the mode required, the signal maxima, minima and means, and the desired manual trim adjustment (set by two joysticks) were entered into the microprocessor control system via 8-bit ADCs. The microprocessor then calculated the necessary pulse amplitudes and reference levels and output them, via DACs, to the correction circuitry. The system operated satisfactorily and the additional flexibility of microprocessor control would allow much more sophisticated control routines to be developed; for example, incorporating a shot-change detector would allow the settings to be 'frozen' shortly after a scene change to prevent movement in the scene causing fluctuations in the colour correction settings. The stability of the digital storage of the correction settings would then be used to great advantage.

DIGITAL TECHNIQUES IN CAMERAS

A digital camera processing channel

At the time of writing, there has not been a pressing requirement to produce a digital processing channel for use in television cameras since analogue techniques have a perfectly adequate performance. However, as has been pointed out, the emergence of the *YUV* digital standard is bound to increase the attraction of handling the signals in digital form throughout the camera rather than merely digitising the output of conventional cameras. It is, therefore, of interest to examine the likely form of such a digital processing channel.

The telecine processing channel is not immediately suitable for inclusion in a television camera; it is too complicated, and hence too bulky, and its performance is unnecessarily high. Considerable simplification is not only possible, therefore, but also desirable.

The first area where simplification is possible is the ADC itself. The ADC was required to have 11-bit resolution for signals near black because of the low noise level produced by flying-spot slide scanners. The predominant noise source in plumbicon-based television cameras is the tube head amplifier; this typically produces a noise level of about -50 to -55 dB with respect to the peak white level. The quantisation noise produced by 8-bit digital coding is -53 dB with respect to full range; thus 8-bit coding would appear to be sufficient. In practice, 9-bit coding would be preferable because of the tendency of 8-bit quantising to 'sharpen up' the appearance of the head amplifier noise, making it more objectionable. Nevertheless, the reduction of accuracy has a significantly simplifying effect on the ADC circuitry.

Television cameras using solid state sensors have the potential to achieve higher signal-to-noise ratios than those quoted for photoconductive tubes. This has long-term implications for the accuracy required from camera ADCs; nevertheless the use of added dither and noise, as already mentioned above, would be an adequate interim solution should the physical size of high-accuracy ADCs be such as to prohibit their use in small, handheld cameras in the near term.

In further contrast to telecines, which are required to cope with many different film stocks, cameras are usually designed to incorporate only one individual type of image pickup tube. Consequently, the colorimetry is much simplified and a fixed matrix, operating directly on the incoming digital signals, can be used. This matrix does not need operational adjustment and can, therefore, rely heavily on the use of PROMs to simplify the circuitry. Finally, because of the higher noise level produced by camera head amplifiers, the gamma correction circuitry usually has a more restricted maximum gain. This, combined with the facts that the gamma law does not need operational adjustment and that colour correction is not required on the logarithmic signal, means that the gamma correctors can also use PROM look-up tables.

The aperture correction circuitry cannot be simplified to the same extent, however, and extra circuitry will be required to allow compensation for tube shading (see the following section) and to allow the black and white levels of the individual colour channels to be adjusted; this latter requirement is to allow 'painting' to be carried out, either to produce special effects or, more commonly, to cancel coloured shadows produced under difficult lighting conditions. Nevertheless, it is clear that a camera processing channel is considerably simpler than the corresponding telecine channel.

Camera control systems

As with telecine machines, the use of digital techniques (and, more specifically, the use of microprocessor devices) in camera control systems can bring many benefits. Such control circuits are usually required to perform many functions; potentially the most important of these are the compensation for registration and shading errors between the three colour channels.

Registration errors can arise if the instantaneous positions of the scanning electron beams in the red, green and blue camera tubes are not exactly correct. In previous generations of cameras, matched sets of tubes and deflection yokes were often employed and simple correction waveforms generated to cancel out the most noticeable error components; the proportions of these correction signals were adjusted manually for best registration performance, registration errors being visible as coloured fringes around displayed objects. The tendency to use smaller and smaller camera tubes has led to an increasing need to control registration errors, however, and an alternative technique is rapidly becoming popular. In this method the horizontal and vertical correction signals required to register the scan tubes are stored for a grid of points; there may be more than a hundred such points in this grid [173]. High speed interpolation circuitry then smooths these signals to generate the correction waveforms for the intermediate points; in this way extremely tight control may be exercised over the registration performance. Initial setting up can be achieved by addressing each point in the grid separately and manually cancelling out any registration error.

Alternatively, the microprocessor system can access all the points simultaneously to generate a simpler control interface; for example a 'differential width' control could be produced, with the microprocessor translating the desired setting of the control into individual signals for each grid point. Alternatively, again, an automatic measurement system can be added enabling the microprocessor system to calculate the corrections required. The camera is then merely pointed at a suitable test object and the registration achieved automatically [174]; a second stage of slight manual adjustment may then be required to achieve optimum results.

Shading errors can be produced by illumination levels and/or target sensitivities that vary over the face of the tubes. Such variations are most objectionable when they occur differentially between colour channels; the colour of a highlight then appears to change with its position in the raster. Variations in the black level can also occur due to variations in target dark current and/or any bias illumination (used to combat tube lag). Both types of shading may be compensated for using a similar technique to that described for registration correction. A grid of black and white corrections is stored and the necessary interpolations carried out for intermediate scan positions. The corrections for black level shading are then subtracted from each camera tube output; following this, the gain of each channel is adjusted under the control of the while level shading corrections. Again, as for the registration correction, the shading corrections can either be derived manually or else measured automatically.

When such sophisticated control systems are incorporated in a camera, the final performance is potentially only limited by the variations in the lens errors as the zoom angle is altered. Some systems offer the facility of also taking account of these errors, however, by storing several different sets of correction for different settings of the zoom angle. By interpolating between these to suit the zoom angle actually in use it is possible to reduce registration and shading/vignetting errors still further. Thus the digital control systems will enable smaller and simpler camera/lens combinations to be used (giving advantages of further savings in size and weight) without making sacrifices in terms of picture quality. Figure 10.23 shows a full-facilities plumbicon television camera with a control system of the type considered in this section. Such a camera is able to achieve an excellent level of picture quality from a wide range of available lenses without penalties in increased weight or restricted manoeuverability.

Use of solid state area arrays in television cameras

In contrast to the use of solid state sensors in telecine applications, solid state area arrays have been much slower to be accepted for use in studio quality television cameras. The initial applications for such devices have been in small, handheld cameras; an example of such a camera is shown in Figure 10.24. The technical problems of the use of area arrays are considerably greater than those occurring in a line-array telecine; this is summarised in Table 10.6. Firstly there is a greatly increased number of elements in an area array sensor; this can lead to problems of low manufacturing yield which in turn implies high device costs. In the less expensive types of camera, therefore, sensors having a small number of blemishes are often used, particularly in the less critical red and blue channels; techniques for concealing these blemishes must then be included in the camera signal processing. Such techniques have been investigated [175] and found to be

Figure 10.23 A view of a Thomson studio camera which uses a digital registration control system to correct for aberrations in the optical system. The correction waveforms are generated in the printed circuit shown in the foreground of the photograph

Figure 10.24 The BTS lightweight camera employing three frame-transfer CCD area-array sensor image pickup devices of the type shown in Figure 10.2(a)

Table 10.6
A comparison between the performance required of solid state sensors in cameras and telecines.

Parameter	Requirement for telecine applications	Requirement for camera applications
Number of elements	1024	414 000
Integration time	64 μsec	20 ms
Dark signal (corrected for different integration times)	0.15% of peak white per ms	0.002% of peak white per ms
Overload protection	Desirable, but not essential	Essential

adequate. A second result of the increased number of elements is that any circuits used to compensate for element-to-element sensitivity variations must have a correspondingly increased store requirement (one picture instead of one line). Fortunately, the visibility of these variations is considerably less in an area array. In a line-array telecine they are correlated from line to line and appear as a set of vertical stripes; in an area array they appear more random from one line to the next and having a more grainy structure. To date, therefore, it has been possible to manage without correction of the sensitivity variations; an alternative approach might be to correct only for the low frequency components of the variations, giving some reduction in the storage required.

Thirdly, the integration time of television cameras using area arrays is considerably longer than that of the line-array sensors used in telecines (20 or 40 ms rather than 50 to 100 μs). Thus, whereas the dark current of the sensor is not a problem in line-array telecines, it is potentially a problem in cameras. The dark current varies from one element to another and it is these variations which produce the visible disturbance, resulting in a background of frozen noise. The dark current performance of modern sensors is, in general, just adequate for use in handheld cameras at moderate lighting levels. Further improvement could be achieved by removal of element-to-element variations in a similar way to that suggested for removal of sensitivity variations. This would again require a store of one picture capacity, however, and a better method, commercially available in the higher performance cameras of this type, is to cool the sensors (or at least to cool the sensor in the blue channel, which operates at the lowest signal level) using a Peltier-effect device. Leakage currents in silicon follow an exponential characteristic as the temperature is varied, with each 10 °C rise approximately doubling the leakage. Cooling the sensors by 50 to 60 °C (relatively simple to achieve) would therefore reduce the dark current problems by a factor of 16 to 32; in view of the reduced gamma correction range used in cameras this improvement may be sufficient. Care must, however, be taken to avoid the formation of dew or frost on the sensor or any nearby glass surfaces, and an alternative solution is to employ the Peltier devices merely to prevent excessive temperature rise in conditions of high ambient temperature, e.g. when operating the camera in direct sunlight.

A fourth problem associated with the use of solid state arrays in television cameras is that of optical overload. Some early sensors, based on a CCD structure, showed very severe blooming when illuminated above a saturation threshold. This produced similar results to those illustrated in Figure 10.25, which was produced by using a line sensor in a telecine; the excess charge bleeds into adjacent areas. While such overloading can be prevented in a telecine, by careful control of the film illumination level, it is not realistic to do so for a camera since specular highlights can occur unpredictably. Fortunately, it is possible, by incorporating additional structures during the fabrication of the sensor, to control the majority of the overload charge [176]. There are, however, some residual effects which vary in severity between one type of sensor and another. Broadly speaking, of the range of sensor technologies that have been investigated, currently the most popular is that based on charge-coupled device, or CCD, architecture. Such CCD sensors have principally been made using one of two alternative structures — the frame transfer structure or the interline transfer structure. The internal organisation of frame transfer sensors is such as to make it susceptible to light falling on it during the field synchronisation period (for an explanation of the reasons for this, the reader is referred to some of the excellent articles dedicated to CCD devices, e.g. [177]); such light produces vertical streaks on highlights, similar to those caused by very slight overload. In order to overcome this streaking, broadcast-quality cameras that use frame transfer sensors also usually employ a synchronised shutter to cut off the incident light during the field blanking period.

Interline transfer sensors do not require such a shutter. They do, however, also suffer from some residual vertical streaking on scene highlights, due to a proportion of the charge generated in these areas migrating through the sensor structure (again for a fuller explanation, the reader is referred to the CCD literature).

Figure 10.25 Illustration of the problem of blooming caused by optical overload

More recently a new type of CCD structure has been developed. This structure, known as the frame-interline transfer device [178], combines some of the attributes of both frame transfer and interline transfer devices. Although it uses a more complex electrode design, and is in consequence more expensive to produce, it does have a much lower level of highlight streaking. This issue, therefore, together with other factors such as spectral response, sensitivity and electrical noise level, is clearly very much associated with the technology of solid state sensors, which is improving rapidly and will continue to do so. There is one feature that is fundamental to the use of solid state sensors, however, and that is the spatial sampling that they impose on the input scene [179]. The sampling structure may be locked to that of the CCIR digital standard, or, alternatively, the sensor can be made to look like an unsampled, analogue source feeding its output into a suitable lowpass filter. There are then two aspects of the situation which need to be considered:

(a) The effect of optical pre-filtering (or the absence of it) on the signal produced by the sensors, and the alias components thereby produced.

(b) The efficiency of removal of frequency components consisting of sidebands of the sensor sampling frequency. This is important because if any such components are left they will be converted into alias components in the second sampling process.

Figure 10.26 illustrates the situation. In Figure 10.26a the spectrum of the optical image is shown. This incorporates any effects due to the camera lens and is likely to be approximately Gaussian; as an example it is assumed that the required bandwidth is 5.5 MHz, the signal energy has become negligible at about 11 MHz and the sampling frequency of the sensor is 12 MHz.

Figure 10.26b shows the modification of the spectrum due to the aperture of the sensing elements. This aperture will be a function of the geometric layout of the light

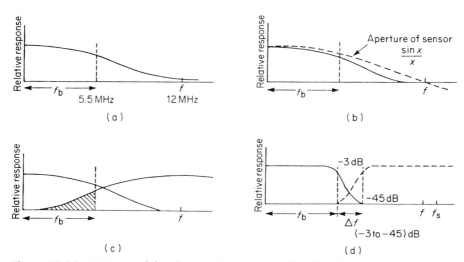

Figure 10.26 (a) Spectral distribution of scene as modified by the optics of the camera. (b) Further modification caused by the aperture of the sensor, assumed (sin x/x). (c) Formation of alias components by sampling at a frequency f; aliased components shown hatched. (d) Characteristics of the filter necessary to avoid further alias components caused by sampling at a second frequency, f_s

sensing elements and the shape of the potential wells beneath them, but the figure illustrates the effect of a simple rectangular aperture of width equal to the sample pitch; this gives a sin x/x spectrum with the first zero at the sensor sampling frequency. The overall spectral response is given by this sin x/x function multiplied by the curve in Figure 10.26a.

Figure 10.26c shows the result of the sampling of Figure 10.26b by the sensor structure; it represents the spectral characteristics of the signal emerging from the sensor. It can be seen that there is an overlap between adjacent orders of spectrum (alias components) which extends well into the wanted region of the band, i.e. 0 to 5.5 MHz. If this output is to be resampled at a different frequency (that of the subsequent digital processing) it must be low-pass filtered to remove components above the video bandwidth. Assuming that this low-pass filter defines the bandwidth of the wanted information by its -3 dB point, a criterion can be deduced to define the necessary characteristics of such a filter. With reference to Figure 10.26d we see that

$$f_{s2} = 2f_b + \Delta f_{(-3 \text{ to } -45)\text{dB}} \tag{10.7}$$

where f_{s2} is the sampling frequency of the second sampling process (i.e. the digital processing), f_b is the required bandwidth and $\Delta f_{(-3 \text{ to } -45)\text{dB}}$ is the frequency difference between the -3 dB and -45 dB points of the filter (-45 dB being assumed as a reasonable target for the degree of alias suppression required).

Thus, only the resampling frequency, f_{s2}, is involved and the sampling frequency of the sensor is no longer relevant. However, this relationship only gives the characteristic of a filter designed to remove unwanted frequency components arising out of the energy distribution in the sampled scene. In particular, it does not take into account spurious frequencies caused by clock pulse breakthrough occurring in the sensor. Such a breakthrough can occur at the sampling frequency, f_{s1}, and at multiples of it; in some arrangements it can even occur at $f_{s1}/2$. The level of breakthrough can be substantial (0.1 V peak–peak for example) and, if it occurred at the sampling frequency, f_{s1}, it would give rise to a beat frequency of $f_{s2} - f_{s1}$ in the final digital output. Under these circumstances it might well be necessary to supplement the low-pass filter response by an additional notch filter at f_{s1}. Nevertheless, it is clear that there are no theoretical obstacles to resampling the output at a frequency different to that of the sensor; the problem is purely one of filter design optimised for the particular circumstances involved.

A more difficult question to resolve is the tolerable level of alias components produced by the sensor itself as a result of trying to resolve an optical image containing higher resolution than the sensor can reproduce (the alias components illustrated in Figure 10.26c). With a larger number of elements in the sensor, i.e. with a higher value of f_{s1}, there will be less likelihood of serious interference from such components but this is a parameter over which there is very little control, since it depends on the quality of the camera optics. Although the camera lens itself will reduce the level of high spatial frequency components in the image, it is normal to include an additional optical element, which is designed to act as an optical low-pass filter; typically this will use either an array of miniature prisms or a series of birefringent plates [180]. While this certainly reduces the level of the most serious alias components—those which appear coarsest in the reproduced image—alias components in the vicinity of the Nyquist limit are relatively unaffected, as it is virtually impossible to attain an idealised sharp cut spatial filter in this way.

For this reason, the broadcaster is tempted to ask for very high resolution, well in excess of the Nyquist minimum, purely in order to minimise the danger of generating alias components. While this can be achieved relatively easily with line-array sensors in telecine applications, the chip area required to achieve it for camera arrays means that the sensor manufacturer has to strike a balance between chip complexity and acceptable device yield particularly for HDTV applications. Thus a compromise between alias components and resolution will have to be found for television cameras using area arrays. This compromise may well change as advances are made in sensor technology. Initially sensors have had to be used having fewer elements than the Nyquist ideal. Eventually, sensors with higher numbers of elements may become available; reference [179] suggests that an adequate number of elements may be in the region 600 to 1000. While there may be advantages in ultimately achieving 1000 horizontal elements, the use of sensors having in the region of (but not exactly) 720 horizontal elements may create practical problems due to the very low frequency beat between f_{s1} and f_{s2} (f_{s2} is 720 samples per active line for the CCIR standard); such a low frequency beat will demand a very high degree of rejection in the low-pass filter following the sensor. Under these circumstances it will probably be preferable to lock the sensor sampling frequency to the digital standard.

Thus three distinct stages of evolution can be predicted:

(a) Sensors with significantly less than 720 horizontal elements are used with processing designed to minimise the visibility of aliasing (see below).

(b) Sensors with 720 horizontal elements are used, the output sampling frequency being locked to that of the digital standard. A few extra elements above 720 may be used to give a degree of overscan.

(c) Sensors with significantly more than 720 horizontal elements are used in order to reduce optical aliasing if solution (b) is inadequate for the most demanding circumstances. The outputs are low-pass filtered and then resampled at the CCIR standard.

Cameras in the first stage of evolution can show significant aliasing on certain types of material. Some improvements can be made by offsetting the sampling lattices of the sensors used for the red, green and blue channels so that the aliasing is in different phases in the three colour channels [181]. Alternatively, if the colour separation is achieved by a colour filter matrix deposited directly onto the solid state sensors (as is becoming an accepted technique in cameras intended for the consumer product market where only one CCD chip is used) reduction of the aliasing can be achieved by suitable design of the colour filter pattern [182] or by using two such sensors, spatially offset, in conjunction with a neutral beamsplitter [181]. Additional improvements can be obtained by taking advantage of signal processing techniques developed for single-tube vidicon cameras (e.g. [183] which also employ colour-stripe filters and which, therefore, share some of the problems of colour aliasing. Such sophisticated processing systems are a further application for digital techniques in order to take the utmost advantage from the extremely stable sensor sampling lattice.

11 Digital Chroma-key and Mixer Units

V. G. Devereux

INTRODUCTION

The process of colour separation overlay, or chroma-key as it is now known, has long been an essential production tool, particularly for News and Current Affairs programmes. In its basic mode of operation, a chroma-key unit and an associated video mixer are used to replace areas of a chosen colour in a 'foreground' picture with corresponding areas taken from a 'background' picture. In this way, foreground objects can be made to appear as if they are situated in front of the background scene. Alternatively, the chroma-key signal derived from the foreground picture can be used to switch between two unrelated pictures.

The principles of chroma-keying have been evolving over many years [184–188], having their basis in techniques originally used in monochrome television. Two of the later developments, namely the use of linear key processing and an additive rather than a multiplicative technique for removing the key-colour areas in the foreground picture, have been found to be particularly beneficial in equipment operating on digital video signals [190]; similar techniques for chroma-keying from *RGB* analogue video signals are used in equipment described by Hughes [188]. Digital chroma-key techniques allow very precise keying at positions which are numerically defined in the raster.

The digital chroma-key and mixer units discussed in this chapter form part of a digital video storage system developed by the BBC which, as described in Chapter 7, is used in the preparation of animated picture sequences obtained from an electronic rostrum camera. Within the animation store, the video signals are stored in YC_RC_B digital component form with the Y (luminance) component sampled at 12 MHz and the C_R and C_B (colour) components sampled at 4 MHz. (Here the terminology of CCIR Recommendation 601 is used to represent $R-Y$ and $B-Y$ signals which have been scaled to fit the digital coding range from -112 to $+112$ 8-bit quantum levels.) Although the chroma-key unit can operate on these stored signals, it normally operates on wider bandwidth C_R and C_B signals sampled at 12 MHz, these signals being available before storage takes place. The equipment also includes the means for deriving a key signal from the luminance component of the foreground pictures.

The sampling rates given above had to be fixed before the standard rates of 13.5 and 6.75 MHz for luminance and colour components given in CCIR Recommendation 601 were established. The techniques which are described would not be affected by a change to these standard sampling rates but there would be a significant improvement in the quality of 'downstream' chroma-key operations [189, 190] i.e. operations in which the key signal has to be obtained from the narrow-band, 4 or 6.75 MHz sampled, colour components.

Different stages in the production of a chroma-keyed picture are shown in Figure 11.1 (see colour plates between pp 506 and 507). The original foreground scene shown in Figure 11.1(a) contains a number of testing features. Critical parts include the transparent objects, the fine detail in the stalks of the flowers and there are difficult soft edges in the fur of the toy lion. Figure 11.1(b) shows this picture after the blue key colour has been suppressed to black. This is achieved by subtracting a key signal derived from the foreground C_B and C_R colour components. Figure 11.1(c) shows the background picture after it has been multiplied by the key signal. One point to notice is that the background scene has been allowed through at reduced gain where the transparent objects and shadows reduced the intensity of the blue key-colour in the foreground picture. This indicates that a linear key process which can provide a gradual cross-fade between the foreground and background pictures is being employed. The final composite picture shown in Figure 11.1(d) is obtained by adding the processed foreground and background pictures. 12 MHz sampled C_R and C_B signals were used for the key processing illustrated in this figure.

PRINCIPLES OF OPERATION

General

The principles of operation of the chroma-key unit will be explained first for the case where a wideband key signal is derived from C_R and C_B signals sampled at 12 MHz. A block diagram of the circuitry used in the animation store is shown in Figure 11.2. The changes required when these signals are sampled at 4 MHz are explained later and only involve changes to filtering and rate-changing processes. For both wide and narrow-band keying, the mixer connected to the chroma-key unit operates on 12:4:4 MHz YC_RC_B signals. Inputs marked c.bus in Figure 11.2 are connected to the computer control system of the animation store. The individual blocks of Figure 11.2 are described below.

Key generator

In this block, the C_R and C_B components of the foreground video signal are processed to provide a key signal which differentiates between colours close to a chosen key colour and the remaining colours in the foreground scene. This process will be explained with reference to a plot of C_R against C_B values as illustrated in Figure 11.3. On this figure, the coordinates of saturated colours have been plotted for reference purposes.

The C_B and C_R colour components are first transformed to obtain coordinates measured relative to the X and Z axes shown in Figure 11.3. The key signal is then derived from these X and Z colour components. A manual control is provided which can continuously rotate the 'hue' angle θ between the X and C_B axes in steps of about 1° to

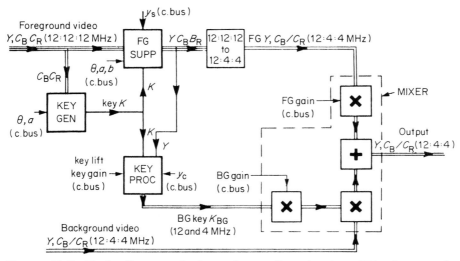

Figure 11.2 Block diagram of chroma-key unit and mixer (FG = foreground, BG = background, c.bus = control bus)

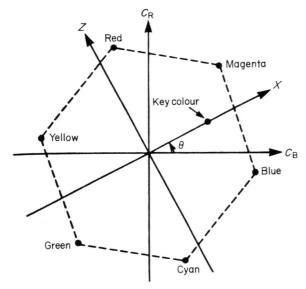

Figure 11.3 X and Z axes obtained by rotating C_B and C_R axes through hue angle θ

give keying off any colour. Optimum keying occurs when the X axis passes through the key-colour coordinates.

The X and Z components of any foreground colour are related to its C_B and C_R components and the hue angle θ by the relationships

$$X = C_B \cos \theta + C_R \sin \theta \tag{11.1}$$

$$Z = C_R \cos \theta - C_B \sin \theta \tag{11.2}$$

The key signal K is derived from the X and Z colour components using the expression

$$K = X - a|Z| \quad K = 0 \text{ if } X < a|Z| \tag{11.3}$$

where a is an adjustable constant.

The resulting values of K obtained for X and Z values based on normalised values of C_B and C_R lying within the range -1 to $+1$ are illustrated in Figure 11.4. Any given non-zero value of K corresponds to a pair of straight lines as shown for $K = 0.5$. The 'acceptance' angle α of the sector where K is non-zero depends on the value of a in equation 11.3, the relationship being given by

$$\alpha = 2\tan^{-1}(1/a) \tag{11.4}$$

In the equipment which has been constructed, values of $a = \frac{1}{2}$, 1, 2 and 4 are available giving values of $\alpha = 126°$, $90°$, $53°$ and $28°$.

It can be seen that the magnitude of K obtained from any colour in the acceptance sector increases in a linear manner as its hue approaches the main key colour and as its saturation increases.

The key signal is filtered at the output of the key generator as discussed in a later section.

Foreground suppressor

Coloured areas in the foreground picture having colour-difference components lying within the acceptance angle have their Y, C_B and C_R components reduced in the 'FG

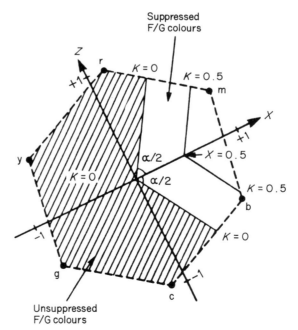

Figure 11.4 Values of key signal K used for suppression of foreground signal. α = acceptance angle

PRINCIPLES OF OPERATION 543

Suppressor' of Figure 11.2 by subtracting suitable proportions of the key signal K, the key colour being completely suppressed to black level. In the past, removal of the key colour has more commonly been achieved by multiplying the foreground signal components by a clipped version of the key signal giving unity gain in wanted foreground areas and zero gain in key-colour areas. The main advantages of the subtraction technique are that it gives softer edges at the key colour boundaries and it avoids, in digital chroma-key, the generation of in-band alias components caused by multiplication and clipping processes.

For the simplest form of suppression, the foreground suppressor effectively reduces the foreground chrominance by subtracting the key signal from the chrominance signal X, i.e. the suppressed $X = X - K$; Z is not affected. In practice the circuitry performs the equivalent operations on the C_B and C_R signals, namely

$$C_{\text{BSUP}} = C_B - K \cos \theta \tag{11.5}$$

$$C_{\text{RSUP}} = C_R - K \sin \theta \tag{11.6}$$

where C_{BSUP} and C_{RSUP} are the suppressed values of C_B and C_R. These operations on C_B and C_R are preferred because no processing of the C_B and C_R signals is required when $K = 0$.

The resulting effect on the colour-difference signals obtained along a line in the foreground picture in which the colour changes from red, in a wanted foreground subject, to a blue key colour is illustrated in Figure 11.5a.

The suppressed luminance component Y_{SUP} is given by

$$Y_{\text{SUP}} = Y - y_s K \quad Y_{\text{SUP}} = 0 \quad \text{if} \quad y_s K > Y \tag{11.7}$$

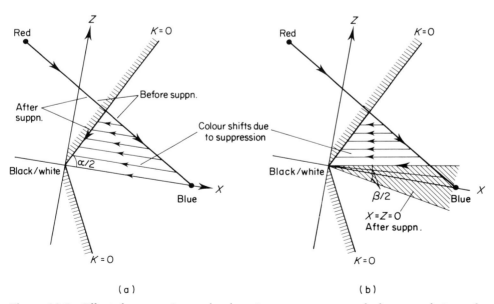

Figure 11.5 Effect of suppression on the chrominance components of a foreground picture for a gradual change from a wanted red subject to a blue key colour. (a) Simple suppression. (b) Improved suppression. β = suppression angle

where y_s is an adjustable constant. The luminance suppression parameter y_s is normally adjusted so that Y_{SUP} is just clipped at zero, i.e. at black level, in the main key colour areas, as discussed later.

One disadvantage of this simple subtraction technique is that the C_B and C_R components of the foreground signal are set to zero only if they are lying exactly on the X axis. As a result, any hue variations in key colour areas caused by noise or irregularities in the colour of the original pictures are not entirely suppressed. To overcome this problem, circuitry has been included which sets the C_B and C_R components to zero for all colours which lie within a 'suppression' sector of the C_B, C_R plane. Like the acceptance sector, this suppression sector is symmetrically disposed on either side of the X axis. Its included angle β is adjustable from zero, as in simple subtraction, to a maximum of about one third the acceptance angle. For $\alpha = 90°$, the available values of $\beta = 0°$, $12°$, $22°$ and $36°$. The suppression of colours using a non-zero value of the suppression angle β is illustrated in Figure 11.5b.

In addition to removing the key colour from the unwanted areas of the foreground picture, the suppressor also removes key-colour fringes on the wanted foreground subject caused by overspill of the key-colour; such fringe suppression has been used for some years together with multiplicative removal of the unwanted key-colour areas [184].

Key processor

This unit processes the key signal K obtained from the key generator to provide a background key signal K_{BG} which is used for removing the areas of the background picture where the foreground picture is to appear in the final chroma-keyed picture. This removal of foreground areas from the background picture is achieved by multiplying the background video signal by the background key signal K_{BG}. K_{BG} must therefore be equal to zero in wanted foreground picture areas and equal to unity where the background picture is required with no attenuation. K_{BG} is obtained from the key signal K by applying lift and gain adjustments followed by clipping at zero and unity values of K_{BG}. For the range between $K_{BG} = 0$ and $K_{BG} = 1$, K_{BG} is related to K by the expression

$$K_{BG} = (K - k_L)k_G \qquad (11.8)$$

where the key lift (k_L) and key gain (k_G) parameters are applied to the key processor via the control system of the animation store and are continuously variable within their adjustable range.

Typical values of K_{BG} obtained from different colours in a foreground picture containing a blue backdrop for keying purposes are shown in Figure 11.6.

The transition band between $K_{BG} = 0$ and $K_{BG} = 1$ is made as wide as possible in order to minimise any discontinuities in the transitions between foreground and background areas in the combined picture. Optimum adjustments of key lift and key gain are discussed in more detail later.

When shadows are cast by foreground objects onto key colour areas of the foreground picture, the resulting foreground chrominance values will typically lie along the X-axis in the linear range of K_{BG} between 0 and 1. As a result the background signal is partially attenuated giving rise to shadow effects on the background picture. In a similar manner, attenuated background pictures can appear through transparent foreground objects

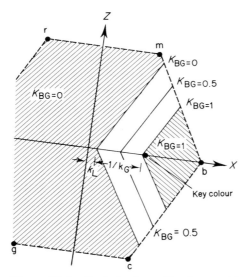

Figure 11.6 Typical values of background key signal K_{BG} for different foreground chrominance components. $K_{BG} = (K - k_L)k_G$ for $0 < K_{BG} < 1$

situated in front of key-colour areas. If the shadows are not required, they can be removed by adjustment of the key gain control so that the $K_{BG} = 1$ region is moved closer to the $K_{BG} = 0$ region.

A further feature of the key processor enables the magnitude of the background key signal to be reduced as the luminance of the foreground signal increases. Thus, for example, it is possible to arrange that the background signal is not turned on by a wanted light blue object in the foreground scene although this object has the same $C_B C_R$ components as the lower luminance blue key colour.

The key signal K' obtained after compensation for increasing luminance is given by

$$K' = K - y_c Y_{SUP} \quad K' = 0 \quad \text{if} \quad y_c Y_{SUP} > K \qquad (11.9)$$

where y_c is an adjustable constant.

In the main key-colour areas of the foreground picture, the luminance suppression is normally adjusted to give $Y_{SUP} = 0$, so that in these areas $K' = K$. In addition, $K' = K = 0$ for wanted objects in the foreground picture whose colour components lie outside the acceptance angle. Thus the only areas in the foreground picture for which K' is less than K are those whose colour lies within the acceptance sector and whose luminance is greater than that which is suppressed to zero.

The final background key signal is transmitted to the mixer as two 8-bit signals, one sampled at 12 MHz for applying to the luminance component of the background video signal, the other being sampled at 4 MHz for applying to the C_R and C_B components. The 4 MHz-sampled signal is derived from the 12 MHz sampled signal via an interpolating filter as discussed in a later section. No filtering is applied to the 12 MHz signal in the key processor.

Mixer

The final chroma-keyed picture is obtained by adding the suppressed foreground signal to the background signal multiplied by its key signal K_{BR}. This addition takes place in a mixer which also allows the gain of each video signal to be separately adjusted by 8-bit control signals giving a maximum gain of unity. A minimum of 8 bits for these gain control signals was found to be desirable in order to give the appearance of completely smooth fading for the most critical video signals e.g. a peak white signal over the whole picture area. Precise unity gain was achieved by allowing the video signal to by-pass the eight-bit multipliers used in the mixer when the 8-bit code for the gain control signal was equal to 11111111. This by-pass arrangement avoided the need for the extra bit required for multiplication by the 9-bit binary code 100000000. Using the larger and hence more expensive 12×12-bit multipliers which are now available, there would be no need for the by-pass circuits.

It should be noted that the simple multiplication of a video signal of magnitude V by a gain factor A as shown in Figure 11.2 causes the output signal $A \times V$ to fade to zero when $A = 0$ rather than to the video level corresponding to zero luminance or chrominance, i.e. the 8-bit quantum levels 16 and 128 respectively. One method of overcoming this problem is to subtract the zero luminance or chrominance level V_0 from V before multiplying by A and to add V_0 again afterwards giving the required result

$$\text{Output from fader} = (V - V_0)A + V_0 \tag{11.10}$$

Although it is applicable with larger multipliers, this process was not used in the animation store because the subtraction of two 8-bit numbers V and V_0 give a 9-bit number causing problems when only 8×8 bit multipliers were available. Instead, the problem of allowing for V_0 was overcome by adding a correction factor $(1 - A)V_0$ to the product $A \times V$. This can be seen to give the same result since equation 11.10 can be rewritten as

$$\text{Output from fader} = AV + (1 - A)V_0 \tag{11.11}$$

The circuitry required to form the correction factor $(1 - A)V_0$ is relatively simple because (a) with V_0 equal to quantum levels 16 or 128, the multiplication by V_0 simply involves a binary shifting operating and (b) $1 - A$ can be obtained by simply complementing A with a correction for the resulting LSB (least significant bit) error being applied in the same adder as used for adding $(1 - A)V_0$ to AV.

Circuits corresponding to equations 11.10 and 11.11 for mixing two video signals V_1 and V_2 multiplied by gain factors A_1 and A_2 respectively are shown in Figure 11.7. If the two signals V_1 and V_2 are to be cross-faded with $A_1 + A_2 = 1$, a simpler arrangement as shown in Figure 11.8 can be used. Note that no input of V_0 is required for this condition.

In the mixer employed in the animation store, the signals following the multipliers were handled with 12-bit accuracy. The final output of the mixer was reduced from 12 bits to 8 bits using error feedback circuitry as discussed later.

Filtering and sample rate changing

Three separate digital filters are used in the process of obtaining chroma-key pictures from 12:12:12 MHz YC_BC_R foreground video signals.

PRINCIPLES OF OPERATION 547

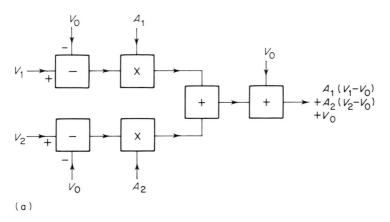

(a)

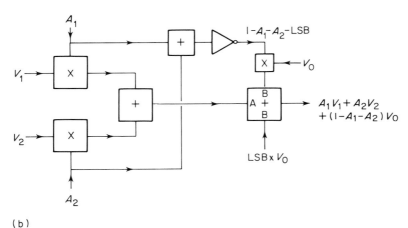

(b)

Figure 11.7 Alternative circuits for mixing two video signals V_1 and V_2 multiplied by gain factors A_1 and A_2, respectively; V_0 is the zero video level. LSB $\times$ V_0 is obtained by setting any unused LSBs of input B and carry-in to 1. (a) Arrangement based on principles of equation 11.10. (b) Arrangement based on principles of equation 11.11

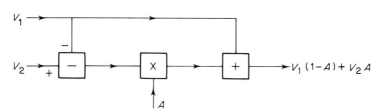

Figure 11.8 Circuit for cross-fading between two video signals V_1 and V_2 with $A_1 + A_2 = 1$

Firstly, the 12 MHz-sampled C_B and C_R signals obtained from the foreground suppressor are passed through a low-pass filter in the 12:12:12 MHz to 12:4:4 MHz converter shown in Figure 11.2. These filters are flat to about 1.4 MHz and -15dB at 2 MHz. This filtering occurs prior to subsampling of the C_B and C_R signals at 4 MHz and is necessary to eliminate the alias components which would otherwise be generated by the subsampling process.

Secondly, low-pass filtering of the key signal at the output of the key generator was found to be desirable but not essential. The purpose of this filter is partly to reduce high-frequency noise in the key signal but is mainly to prevent unwanted keying from high-frequency picture detail in the wanted foreground areas. This detail can generate key signals, even though no key colour is present in the original foreground picture, as a result of camera misregistration or unequal high-frequency gains in the *RGB* camera channels. Further work is required to determine the optimum low-pass filter for these purposes, but the existing slow-cut filter which is flat to about 3 MHz, -6dB at 4 MHz and -40dB at 5 MHz appears to be satisfactory.

Thirdly, the background key signal K_{BG} sampled at 12 MHz is passed through a 2 MHz low-pass filter and then subsampled at 4 MHz in order to obtain a suitable version of K_{BG} for applying to the 4 MHz sampled C_B and C_R components of the background video signals. A simple slow-cut filter was found to be adequate for this purpose, its amplitude characteristic being -6dB at 2 MHz and -40dB at 4 MHz.

OTHER KEY FACILITIES

General

Other facilities provided by the chroma-key and mixer units in the animation store will be explained with reference to the more detailed block diagram shown in Figure 11.9. The video routing system (VRS) shown in this figure carried 12:4:4 MHz YC_BC_R video signals with C_B and C_R multiplexed into one data stream and interconnects all the units in the animation store.

With the switches in the positions shown in Figure 11.9, the processing provided is the same as that described in the previous section, apart from the inclusion of a non-additive mixer where an external key signal and/or garbage matte may be added to the chroma-key signal. In particular, the magnitude of the foreground key signal K_{FG} applied to the mixer is equal to unity for the given positions of SW1, SW2 and SW3. Thus no keying is applied in the mixer to the foreground video signal as required by the additive key technique.

With SW1 in its other position, the foreground video signal is multiplied in the mixer by $(1 - K_{BG})$ so that the foreground component of the mixer output is then turned off when the background component is turned on and vice versa. This multiplicative keying of the foreground video signal is used for luminance keying and also when the video signal applied to the key generator is not related to the foreground video signal applied to the mixer; under these conditions, suppression of the foreground signal by subtracting the key signal as already described (sometimes referred to as an additive key process) is not applicable.

OTHER KEY FACILITIES

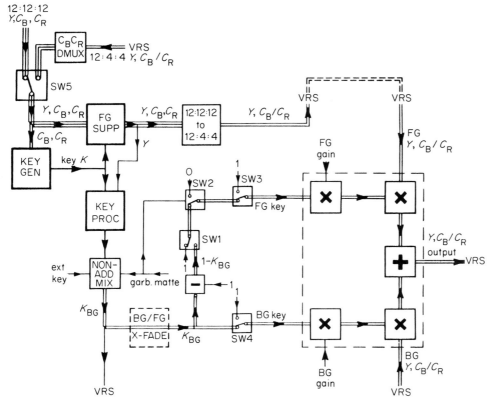

Figure 11.9 More detailed block diagram of chroma-key and mixer units

The alternative positions of SW3 and SW4 making $K_{FG} = 1$ and $K_{BG} = 1$ are used when simple mixing with no keying is required.

Further details of the facilities provided by the arrangement shown in Figure 11.9 are discussed below.

Keying from 12:4:4 $YC_B C_R$ signals

Key signals limited in bandwidth to slightly under 2 MHz can be obtained from the 12:4:4 MHz $YC_B C_R$ signals stored on the disc drive. These stored video signals are supplied via the video routing system to the C_B/C_R demultiplexer selected by SW5.

In the C_B/C_R demultiplexer, two separate 4 MHz sampled C_B and C_R data streams are obtained from the multiplexed C_B and C_R data on the video routing system. These separate data streams are then re-clocked at 12 MHz to make them compatible with C_B and C_R data in the 12:12:12 MHz $YC_B C_R$ input obtained directly from the rostrum camera signals.

There are two main further alterations in the processing compared to that used for keying from 12:12:12 MHz $YC_B C_R$ signals as previously described.

Firstly, the cut-off frequency of the low-pass filter at the output of the key generator is reduced to 2 MHz to remove high frequency alias components. This filtering would not have been required if 2 MHz low-pass interpolating filters had been included in the C_B and C_R sample rate changing performed prior to key generation; however, the equivalent processing would have required two filters instead of one. The characteristic which is employed is flat to about 1.5 MHz and $-$ 6dB at 2 MHz; this characteristic includes the effect of re-clocking the 4 MHz C_B and C_R data at 12 MHz. By making this characteristic skew-symmetrical about 2 MHz, every third sample of the key signal is identical to that obtained with no filtering. Thus an unfiltered 4 MHz sampled key signal for applying to the C_B and C_R components can be obtained by subsampling the 12 MHz key in the correct phase.

The second alteration involves a replacement of the complex 12:12:12 MHz to 12:4:4 MHz YC_BC_R converter including a 2 MHz low-pass filter which follows the foreground suppressor by a simple converter in which the 12 MHz C_B and C_R data is subsampled at 4 MHz without filtering. No filtering is required in this sample-rate conversion process or for the key signals used to suppress the C_B and C_R data because the suppression of the C_B and C_R signals is effectively carried out at a 4 MHz sample rate with each sample being temporarily clocked at 12 MHz. This change freed the complex sample rate changer for processing the signals from the rostrum camera.

Luminance keying

A key signal can be derived from the luminance component of the foreground signal as an alternative to keying off its C_B and C_R components. Circuitry is available for insetting the background picture into either 'white' or 'black' parts of the foreground picture. For keying off 'white' areas in the foreground picture, the keyed picture consists of entirely background for foreground luminance levels above an adjustable value Y_U and entirely foreground below for foreground luminance levels below a second value Y_L $(Y_L < Y_U)$.

For foreground levels between Y_L and Y_U, there is a linear cross-fade between the foreground and background pictures. The key gain control adjusts the difference $Y_U - Y_L$ and the key lift increases or decreases Y_U and Y_L equally.

In the luminance key process, the colour key signal K is set to zero so that no foreground suppression is applied and the key signal K_{BG} is generated in the key processor unit. The parts of the foreground picture which are to be replaced by the background picture are removed by multiplying the foreground signal by $(1 - K_{BG})$. Removal of the foreground signal by means of the suppression technique used for chrominance keying is not used for luminance keying, because the magnitude of C_B and C_R components are not normally related to the magnitudes of the luminance key signal.

Garbage matte

The garbage matte is the colloquial name for a device generating a key signal which forces the mixer to produce the background signal alone either on the outside or on the inside of a rectangular area in the picture with normal keying taking place elsewhere. The position of the edges of this rectangle are adjustable. A typical use of this garbage matte would be for keying in the background scene at the edges of a picture when a backdrop providing the key colour in the foreground scene is not large enough to completely fill the entire picture area. As shown in Figure 11.9, this garbage matte is added to the background key signal in a non-additive mixer (i.e. a mixer whose output is equal to its greatest input) and also it switches the foreground key signal to zero at SW2.

Stored key signals

The animation store is capable of recording background key signals for later use. Stored key signals are distributed as a 12 MHz Y signal along the 12:4:4 YC_BC_R video routing system. This facility allows wideband keying from recorded data, thus largely overcoming the problem that only narrow-band colour keying is possible from the recorded 12:4:4 MHz YC_BC_R video signals.

Foreground–background cross-fade

This facility allows the foreground to be gradually faded out and replaced by the background picture. The associated processing is illustrated in Figure 11.10. This figure shows that the required effect is obtained by further processing of the background key signal obtained from the non-additive mixer of Figure 11.9 and by a further connection to switch SW1.

As indicated in Figure 11.10, the new background key signal K'_{BG} and foreground key signal K'_{FG} are given by

$$K'_{BG} = nK_{BG} + 1 - n \qquad (11.12)$$

$$K'_{FG} = n \qquad \text{additive key} \qquad (11.13)$$

$$\phantom{K'_{FG}} = n(1 - K_{BG}) \qquad \text{multiplicative key} \qquad (11.14)$$

where n is an adjustable cross-fade parameter.

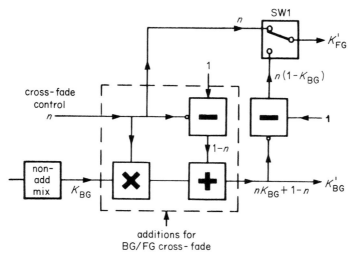

Figure 11.10 Arrangement for producing a cross-fade between a keyed picture and a background picture alone. $n = 1$ gives keyed picture; $n = 0$ gives background picture alone

It can be seen that when $n = 1$, then $K'_{BG} = K_{BG}$ and $K'_{FG} = 1$ or $(1 - K_{BG})$ as in normal keying. When $n = 0$, K'_{BG} is set to unity and K'_{FG} is set to zero thus resulting in a background only picture. Intermediate values of n give varying proportions of the foreground and background pictures in the normally wanted foreground areas where $K_{BG} = 0$. However, for areas of a keyed picture containing only background detail i.e. where $K_{BG} = 1$, variations in n have no effect because $K'_{BG} = 1$ (see equation 11.12) thus leaving the background picture fully turned on and the foreground picture is turned off because it is suppressed with additive keying or it is multiplied by zero, i.e. $K'_{FG} = 0$, for multiplicative keying (see equation 11.14).

PROBLEMS ASSOCIATED WITH DIGITAL PROCESSING

General

While digital processing provides improved stability and control of chroma-key compared to analogue processing, it can also introduce two forms of picture impairment, namely aliasing and quantising distortion, which do not occur with analogue processing and which can be more troublesome in chroma-keying than in most other digital video processing operations.

Aliasing distortion

As discussed in previous chapters, the normal method of ensuring that there are no significant alias components in the output signal from digital equipment is to employ the appropriate low-pass filters at the input and output which remove components above one half of the sampling frequency. However, non-linear clipping or multiplication of digital signals, as employed in chroma-key equipment, can introduce alias components into the wanted frequency band even though the input and output filtering is satisfactory for normal linear processing. Multiplication is more likely to cause trouble if the digital signals being multiplied are correlated, which is the case for the foreground video and key signals but not for the background video and key signals.

In the equipment being described, aliasing effects have been kept to a minimum by using an additive rather than a multiplicative technique for removing the key-colour areas from the foreground picture and by making the key signal processing as linear as possible.

Alias components in the wanted frequency band of the key signal can also be generated as a result of a change in the sample rate of the key or C_B and C_R signals from 12 MHz to 4 MHz or vice versa. In these processes, alias components are removed by the use of low-pass filters with a cut-off frequency of about one half the 4 MHz sample rate i.e. 2 MHz.

If necessary, alias effects caused by non-linear processing can be reduced by decreasing the bandwidths of the key signals to be somewhat less than half their sampling frequency but this will lower the resolution of the chroma-key effects. For keying from 12:12:12 YC_BC_R signals, no difficulty was experienced in obtaining sufficient resolution without introducing visible aliasing effects. However, for keying from 12:4:4 YC_BC_R signals, although aliasing effects were not obvious, there was a noticeable lack of resolution in the chroma-key process on critical picture material such as the stalks of the flowers shown in Figure 11.1. The most relevant filters used for the 12:4:4 chroma-key were those in C_B and C_R signal paths prior to the key generator which were flat to about 1.4 MHz and -15 dB at 2 MHz, and the interpolating filter used in changing the sample rate of the key signal from 4 MHz to 12 MHz which was flat to about 1.5 MHz and -6 dB at 2 MHz. The results obtained with keying from 12:4:4 MHz YC_BC_R signals indicate that an increase in the bandwidth of the C_B and C_R filters, and possibly that of the key filter, should give a better balance between resolution and freedom from aliasing.

Quantising distortion and use of error feedback

The quantised nature of digital signals can be more noticeable in chroma-key processing than in most other digital video processing operations because of the limited range of C_B and C_R values available for keying purposes. With pictures from television cameras, the change in the C_B and/or C_R signals between the fully on and off values of the background key signal is typically about one quarter of the maximum possible changes in these signals. The resulting coarseness in the quantisation of the key signal is sufficient to cause visible quantising noise effects with 8 bits per sample C_B and C_R signals when there is a gradual change between foreground and background pictures or when shadows are being transmitted via the key signal. It is difficult to see how this problem can be overcome except by employing more bits per sample for the C_B and C_R signals. However, for the vast majority of chroma-keyed pictures, quantising effects are not noticeable, because transitions between foreground and background pictures are normally quite sharp.

It should be noted that quantising artefacts can be significantly affected by the techniques used to handle additional fractional bits generated by processes such as mixing and filtering. Where possible, the information in these fractional bits should be maintained to at least 12-bit accuracy. However, where a restriction to 8-bit accuracy is imposed as, for example by the video routing system between the chroma-key, mixer and other units in the animation store described in Chapter 7, a reduction to 8 bits is necessary. Previous work had shown that simple truncation of fractional bits at 8-bit interfaces can produce visible picture impairment in plain areas of a picture, the effects being particularly noticeable towards the end of a fade to black. This problem has been overcome by means of an error feedback rounding process which requires only a few additional components [191, 192].

Figure 11.11 shows the implementation of an error feedback circuit where 12 bits are reduced to 8 bits. Its principle of operation is that fractional bits (bits 8 to 11) are accumulated sample by sample until a carry bit is generated. This carry bit is then added to the least significant bit being output.

This processing has two important features. Firstly, the mean level of the discarded bits is transferred to the output signal so that the resulting quantising errors have a predominantly high frequency spectrum with no d.c. component. Secondly, contouring effects which would have been produced by simple truncation are broken up and replaced by much less visible high-frequency variations between two adjacent output quantum levels.

These advantages of error feedback are illustrated in Figure 11.12. Only 4 bits were used for Figure 11.12b and 11.12c in order to make quantising effects more visible. It

554 DIGITAL CHROMA-KEY AND MIXER UNITS

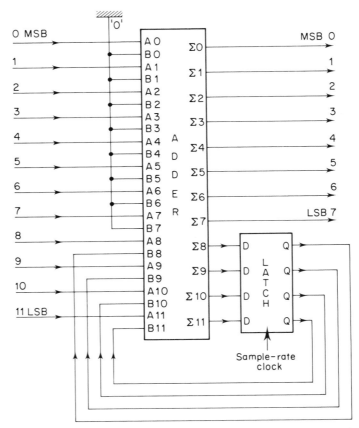

Figure 11.11 Error feedback applied where 12 bits are reduced to 8 bits.

(a)

Figure 11.12 Comparison of 'error feedback' and 'simple truncation' methods of rounding down number of bits per sample. Processing applied to black-to-white ramp signal with 700 samples per active line (13.5 MHz sampling). (a) Original: 8 bits per sample. (b) Rounding to 4 bits by simple truncation. (c) Rounding to 4 bits by error feedback

can be seen that some fine patterning is just visible after the error feedback process but this patterning is much less noticeable than the errors caused by 4-bit truncation. Furthermore, the line-to-line correlation in the patterning caused by error feedback on a line-repetitive picture as shown in Figure 11.12c is not obtained with a normal non-line-repetitive picture, thus further reducing its visibility.

Even when no contouring effects are involved, error feedback rounding has the advantage that the weighted r.m.s. magnitude of the predominantly high-frequency quantising errors that it introduces is lower than the weighted r.m.s. magnitude of the flat spectrum errors introduced by simple truncation.

This error feedback rounding technique is to be recommended for any digital video processing equipment, including electronic picture generation, where fractional bits are generated and the output is to be conveyed to other equipment via an 8-bit interface as specified in CCIR Recommendation 656.

ADJUSTMENT OF CONTROLS

The chroma-key parameters are set via the computer which controls the entire operation of the animation store. Adjustments can be made using dedicated manual controls or by entering details on to a VDU display from a numeric keypad. Correct settings are most easily determined by examining a display of the output of the mixer using the procedure described below.

With the background picture turned off, the hue, θ, control is adjusted so that the chrominance in the key-colour areas of the foreground picture is suppressed as far as possible. The acceptance angle is made as wide as possible without causing any significant suppression of wanted foreground colours and the suppression angle is set as narrow as possible consistent with good suppression of the foreground chrominance in the key colour areas. Narrow acceptance and wide suppression angles are undesirable because they tend to cause harsh and noisy transitions between foreground and background pictures. In practice, it has been found that values of the acceptance angle $\alpha = 90°$ and suppression angle $\beta = 0°$ give close to optimum results for the vast majority of pictures. Thus it is questionable whether the benefits to be gained by allowing these angles to be varied is worth the extra circuit complexity, particularly that required for non-zero values of the suppression angle. The processing of the foreground picture is completed by adjusting the luminance suppression y_s control so that the luminance in the main key colour areas is just clipped at black level.

Adjustment of the keying of the background picture is carried out in the following manner with the foreground picture turned off.

Firstly, with the luminance compensation switched off ($y_c = 0$), and with no key lift ($k_L = 0$), the key gain k_G control is adjusted so that the contrast of the background picture just reaches its maximum value in all the main key colour areas. This occurs when the background key K_{BG} is just clipped at $K_{BG} = 1$.

Secondly, the key lift control is adjusted, if necessary, to prevent unwanted keying of very dark objects in the foreground picture such as might be caused by reflections of the key colour from a shiny black telephone.

Thirdly, the luminance compensation parameter, y_c, can be increased to improve keying to black of the background picture for wanted foreground colours close to the key colour but of higher luminance. However, increasing y_c tends to worsen the appearance of some foreground to background transitions and this should be checked on the combined foreground and background pictures. For most pictures, $y_c = 0$ gives the best overall results.

Finally, slight adjustments to the controls can be made on the combined foreground and background pictures. For example, small adjustments of the key gain k_G and luminance suppression y_s controls may slightly improve the overall appearance of the chroma-key process.

CONCLUSIONS

A 12:12:12/12:4:4 MHz YC_BC_R digital chroma-key system has been designed and implemented in the electronic rostrum camera animation system described in Chapter 7. Excellent results are achieved at 12:12:12 MHz. Results at 12:4:4 MHz are good but they support the view that the CCIR move to 13.5:6.75:6.75 MHz is worthwhile because of the benefit to chroma-key of wide bandwidth C_B and C_R signals, irrespective of other benefits.

12 Digital Video Interfaces

D. J. Bradshaw

INTRODUCTION

CCIR Recommendation 601 [9] provides a framework for the generation of high-quality digital video component signals. In order to preserve the quality that Recommendation 601 permits, the signals should remain in digital form when they are transferred from one equipment or area to another. To permit such equipments to be interconnected, a standardised interface is required and internationally agreed specifications have been produced setting out the manner in which digital video studio equipments should be interfaced to each other.

OPTIONS

The 8 bit per sample Recommendation 601 signal can either be sent in a bit-parallel format along 10 bearers (i.e. 8 plus one clock and one earth wire) or multiplexed into a single high bit rate serial stream which would normally be carried by a coaxial or optical cable [6]. Much of the circuitry within the equipment in a digital television studio will operate on digital video signals in a bit-parallel format for the foreseeable future. Conversion from parallel format to a bit-serial format for the interconnection adds complexity and cost and an interface is clearly required that is capable of carrying the signal in bit-parallel form over the relatively short distances found within a television studio. For inter-studio connections, a serial-format link is required, to take advantage of the lower cost of coaxial cable over multi-way cable.

It is worth noting that the basic data rate corresponding to the three component signals amounts to 27 Mword/s, that is 216 Mbit/s.

There are at least six ways of carrying the data, based on bit-parallel and bit-serial versions of the following:

three separate bearers for the three component signals;

two bearers, each operating at 13.5 Mword/s, one for the luminance signal, Y, and the other for a multiplex of the two colour-difference signals, C_R and C_B;

a single bearer operating at 27 Mword/s, and carrying a multiplex of the luminance and colour-difference signals.

The use of several bearers for the individual component signals is operationally inconvenient, raises problems of maintaining the correct timing relationship between the signals and is a less cost-effective solution than the single bearer (on account of the increased cable, connector and termination costs).

CCIR Specifications Recommendation 656 have been produced for interfaces in both bit-parallel and bit-serial formats [10], based on the digital component signals conforming to Recommendation 601. As discussed in Chapter 11, there are circumstances where more than 8 bits per sample are required for certain processes and if subsystems in different equipment need to be interconnected at this level an interface with more than 12 bits per sample may in future be useful. However, for most studio purposes, remaining at, or rounding down to, the Recommendation 601 standard would be the most attractive practice so that the parallel or serial interface described in this chapter can be used.

INTERFACE SIGNAL FORMAT

CCIR Recommendation 601 states that there will be 720 luminance samples and 360 samples of each colour-difference signal on each active line, i.e. a total of 1440 samples on each active line, and this is the same for both 525- and 625-line systems. They are co-sited with the first, third, etc, luminance samples, as discussed in Chapter 1 and colour-difference samples shown in Figure 1.4. In addition, for 625-line signals, there is a total of 288 luminance and chrominance samples on each line outside the active line, i.e. in digital blanking (for 525-line signals there are 272 luminance and chrominance samples in the blanking period for each line).

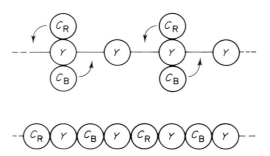

Figure 12.1 Rearrangement of orthogonal samples in sequence for transmission in the multiplex

The sampling structure is orthogonal, that is, corresponding samples on adjacent scanning lines are aligned vertically, so that the positions of corresponding samples on all lines are the same.

In order to create the multiplex of samples from the luminance and chrominance samples, data words are interleaved as shown in Figure 12.1. The orthogonal sampling structure means that the sequence is repeated on each line. Thus there is a repeated sequence of four words on each line

$$\langle (C_R, Y, C_B), Y \rangle$$

where (C_R, Y, C_B) represent the cosited samples and the final luminance sample of the four-word sequence is not cosited with chrominance samples.

Synchronisation

Having multiplexed the component signals into a single data-stream, means must be provided to permit the accurate demultiplexing of the data into the individual component signals. This is achieved through the use of synchronising signals embedded in the multiplex signal.

These synchronising signals are necessary also to permit digital video signals from different areas to be synchronised and to enable the active portion of each digital line to be separated from the blanked portion. They also have a role to play in the serial interface, as described below. In addition, they facilitate the regeneration of an analogue synchronising waveform at the point where the digital component signals are converted to analogue form.

In order to achieve a high degree of commonality between the 625- and 525-line specifications, the synchronising signals—known as timing reference signals—are placed at the beginning and end of the active line data, that is, there are 1440 samples of video data between them in both 525- and 625-line systems (the systems differ only in the number of samples in blanking).

A number of criteria can be established for the synchronising signals. They must be unique so that video or other data are not mistaken for them, they must be easily detected and they must be robust—ideally, capable of reliable detection in the presence of errors.

The multiplex signal contains a basic four-word sequence as described above, and the timing reference signals are also based on a sequence of four words:

FF 00 00 XY   (hexadecimal notation)

The first three words form a unique fixed preamble through the use of the data words FF and 00 which are reserved for 'housekeeping' purposes and are excluded from the video data coding range. The final word, XY, is made up as shown in Table 12.1: it contains three bits, F, V and H, the polarity of which indicates odd–even field, field blanking on–off and line blanking on–off. The point at which F, V and H are all set to 0 is the beginning of the active picture on the first field of the two-field sequence. The four least significant bits of the final word are protection bits to enable the F, V and H bits to be decoded even in the presence of a single-bit error, so providing robustness. The protection is based on Hamming coding, and the way in which the F, V and H bits are decoded for all the 256

Table 12.1
Timing reference signal format

Data-bit number (MSB)	Word 1 (FF)*	Word 2 (00)*	Word 3 (00)*	Word 4 (XY)
7	1	0	0	1
6	1	0	0	F
5	1	0	0	V
4	1	0	0	H
3	1	0	0	P3
2	1	0	0	P2
1	1	0	0	P1
0	1	0	0	P0

*In hexadecimal notation.

possible states of the final word is shown in Table 12.2. The most significant bit is always set to 1.

The way in which the F, V and H bits change state during field blanking is illustrated in Figure 12.2 and the timing relationship between the timing reference signals and analogue synchronising signals is shown in Figure 12.3.

It should be noted that, unlike the case for an analogue video signal, the signal at the digital video interface contains no half-lines and, consequently, nothing corresponding to

Table 12.2.
Error correction and detection in the video timing reference signal. The following look-up table enables single-bit errors in the fourth byte of the video timing reference signal to be corrected. Double- and some multiple-bit errors are detected but not corrected

Received P3-P0	Received FVH (bits 4–6)							
	000	001	010	011	100	101	110	111
0000	000	000	000	*	000	*	*	111
0001	000	*	*	111	*	111	111	111
0010	000	*	*	011	*	101	*	*
0011	*	*	010	*	100	*	*	111
0100	000	*	*	011	*	*	110	*
0101	*	001	*	*	100	*	*	111
0110	*	011	011	011	100	*	*	011
0111	100	*	*	011	100	100	100	*
1000	000	*	*	*	*	101	110	*
1001	*	001	010	*	*	*	*	111
1010	*	101	010	*	101	101	*	101
1011	010	*	010	010	*	110	010	*
1100	*	001	110	*	110	*	110	110
1101	001	001	*	001	*	001	110	*
1110	*	*	*	011	*	101	110	*
1111	*	001	010	*	100	*	*	*

The table gives corrected values for FVH where possible. Multiple errors are denoted by an asterisk.

INTERFACE SIGNAL FORMAT

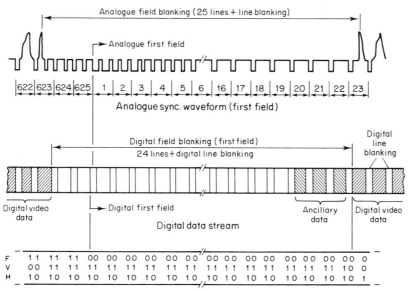

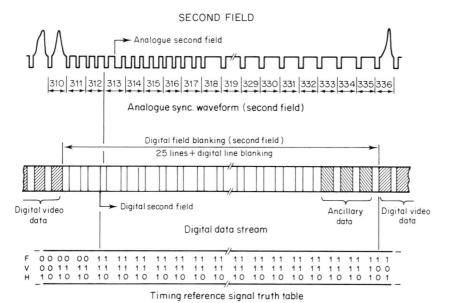

Figure 12.2 Relationship between the digital and analogue fields, showing also the position of the digital field-blanking interval in the 625-line system

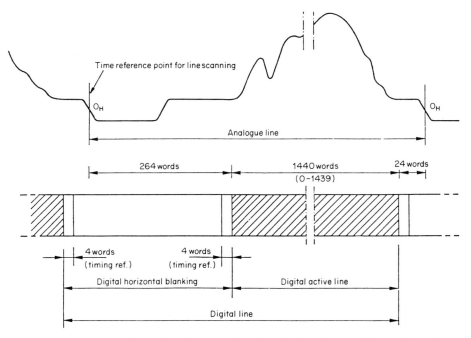

Figure 12.3 Timing relationship between the digital video data and the analogue line synchronisation (reference for line scanning in the 625-line system)

equalising pulses. All lines are of the full length, to facilitate signal processing and to maintain commonality between the 625- and 525-line systems.

The timing reference signal which occurs immediately prior to the start of the digital active line is also known as the SAV (start of active video) signal and the one which follows the active video data as the EAV (end of active video) signal, particularly in North America.

THE BIT-PARALLEL INTERFACE

Electrical format

Desirable features of the line sending and receiving arrangements include

the use of low-cost, general-purpose integrated circuits;

simple circuitry in senders and receivers;

capability to operate with long lengths of cable;

freedom from adjustment for length of cable;

freedom from crosstalk.

The last requirement can be largely met by the use of balanced transmission over terminated cables, while the first and second requirements favour the use of the standard ECL 10000 series devices as these operate reliably at a 27 MHz clock frequency and have

THE BIT-PARALLEL INTERFACE

balanced inputs and outputs. Over short distances, say up to 30 m, simple senders and receivers can be made from TTL-to-ECL and ECL-to-TTL translators respectively.

Since the signal is in digital form, the accuracy of equalisation required to extend the range beyond a few tens of metres is much less critical than is the case for analogue transmission and the same equaliser will operate on links of very different lengths without adjustment. An equaliser characteristic has been established which has been found to permit operation on special cables up to 200 m in length and this is shown in Figure 12.4. An implementation of the equaliser is shown in Figure 12.5: note that this circuit incorporates the 110 Ω termination for the cable.

(In early work the equaliser was fitted permanently and was found to allow operation from zero to 200 m. However, some users found that the equaliser worsens the effect of

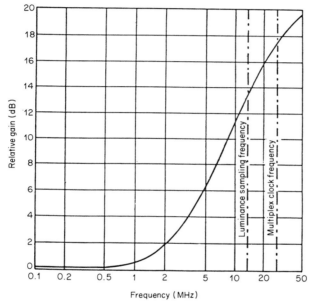

Figure 12.4 Receiver equalisation characteristic (small signals)

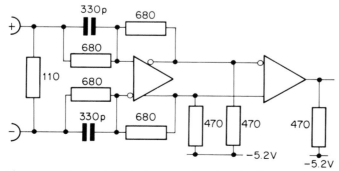

Figure 12.5 Parallel interface equalisation circuit

crosstalk in an environment where there are many relatively short links in use. Consequently, it is now preferred for the equaliser to be made optional, being fitted when the link exceeds about 30 m in length.)

In the interests of simple sending and receiving circuitry, the data multiplex is transmitted in NRZ form, i.e. it is not inherently self-clocking. Therefore, a clock signal is transmitted along with the data, on a ninth signal pair.

The frequency of the clock signal is 27 MHz. Whilst the use of 27 MHz represents the most straightforward approach, it requires the link to carry a signal that is much higher in frequency than the spectrum of the data signals. Experiments showed that while it was possible to transmit a clock signal over greater distances by using a lower frequency clock (e.g. 6.75 MHz or 13.5 MHz), this advantage was negated by the need for additional clock regeneration circuitry such as a phase-locked loop oscillator. Since the great majority of links in a studio are less than thirty metres in length, well within the capability of simple receivers, the additional complexity could not be justified.

The clock transitions are specified as occurring mid-way between data transitions. This was chosen to reduce the sensitivity of the clock edge to jitter produced by crosstalk from the video data and to differences in path length between the clock and the data signals. The timing relationship between the clock and data signals is illustrated in Figure 12.6.

Mechanical implementation

Cables

The quality of cable used for a digital video link can be selected according to the length of the link. The most important feature of the cable is that all the signal pairs should have identical path lengths and this has to be taken into account in the design of the cable if unacceptable timing differences are not to occur.

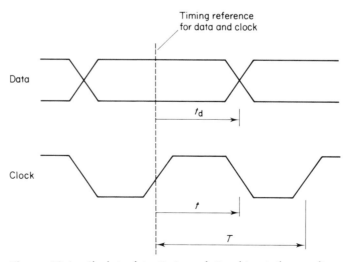

Figure 12.6 Clock-to-data timing relationship at the sending end. Clock period $T = 1/1728$ $f_L = 37$ ns, clock pulse width $t = 18.5 \pm 3$ ns, data timing at sending end $t_d = 18.5 \pm 3$ ns

THE BIT-PARALLEL INTERFACE

Since digital apparatus will generate a significant amount of energy at the clock frequency of 13.5 MHz and its harmonics (including 121.5 MHz, an international aeronautical distress frequency), cables should be screened and the screen continued through the connector to the apparatus.

Suitable cables have been produced for links up to 200 m. These tend to have an outside diameter of about 11 mm, slightly larger than conventional studio-quanlity coaxial cable. For short links of up to about 10 m, relatively inexpensive general purpose screened multi-pair cable can be used. Whilst suitable for very short links, the use of flat ribbon cable is limited to within screened enclosures as it is not itself screened.

Connectors

The choice of connector for the interface is a compromise, attempting to meet the following, conflicting requirements

- ease of assembly and repair;
- small size;
- rugged when mated;
- high minimum rated insertions and withdrawals;
- multiple sourcing and worldwide availability;
- low cost.

The connector specified, the 25-way subminiature type D, has the advantages of being widely used for other applications, well accepted and available in a variety of contact forms and versions. The contacts themselves, when gold-plated, offer adequate reliability. The assignment of signals to contacts in the connector is shown in Table 12.3. Contacts 11,24 and 12,25 may be used only for data bits of lower significance than those on contacts 10,23.

A locking mechanism was considered essential and the slide-lock was selected on the basis of operational suitability: it occupies little panel space and requires no tools for its operation.

Table 12.3
Parallel interface connector contact assignment

Contact	Assignment	Contact	Assignment
1	Clock	14	Clock return
2	System ground	15	System ground
3	Data 7 (MSB)	16	Data 7 return
4	Data 6	17	Data 6 return
5	Data 5	18	Data 5 return
6	Data 4	19	Data 4 return
7	Data 3	20	Data 3 return
8	Data 2	21	Data 2 return
9	Data 1	22	Data 1 return
10	Data 0	23	Data 0 return
11	—	24	—
12	—	25	—
13	Chassis ground		

Whilst the type-D connector is suitable as an installation connector, it is not considered suitable for jackfield-type applications, for which a different style of connector is required. No connector has, however, been standardised for this application in Recommendation 656 [10].

Experience with the parallel interface

The interface has been taken up rapidly by industry and proved operationally satisfactory based on the experience of a number of organisation which have installed digital video equipment interconnected by means of the parallel interface. The major problem experienced has been due to the size of the type-D connector, which limits the number of connectors that can be mounted on a panel, and the large number of signals that have to pass between the interfaces and electronics modules. This is a particular problem in the case of studio routing matrices [193, 197].

Clock regeneration has not proved necessary, except at the very limit of long links, where signal regenerators have been fitted to re-establish the correct clock-to-data timing.

THE BIT-SERIAL INTERFACE

Whilst the majority of signal processing operations within a studio require access to the signal in the bit-parallel form, presented by the bit-parallel interface, there is a requirement for a bit serial interface for use on long inter-area links. This would permit transmission over a single bearer such as coaxial or optical-fibre cable and would avoid problems caused by clock-to-data timing skew that can occur on long multi-pair cables.

Since the bit-parallel and bit-serial interfaces have to coexist within a studio complex, one of the basic requirements was that the serial format should be derived directly from the parallel format, to allow easy transformation from one format to the other.

Coding strategy

Probably the most important aspect of the serial interface is the coding strategy to be employed as it is not possible to transmit a directly-serialised version of the parallel-format signal over any significant distance. It is necessary to code the signal in some way for transmission. A number of criteria for the serial signal need to be met, including:

The serial format should be derived from the parallel format. (As a consequence the presence of timing reference signals in the data stream can be assumed.)

The code must be suitable for transmission on both coaxial and optical fibre cables. This implies that multi-level codes would be unacceptable.

The coded signal should have a very small low-frequency content (to allow a.c. coupling and to simplify cable equalisation).

The code should require limited overhead in order to restrict the signal bandwidth and clocking rate of the high-speed circuitry, to simplify equalisation and maximize the transmission distance.

The coding system should be simple to implement.

The latter requirement, simplicity of implementation is particularly important in the context of television production centres, where a large number of links are employed.

In addition, there are a number of practical requirements that the coding strategy must meet, for example:

The coding system must be transparent; all combinations of data words must be transmitted faithfully.

Adequate timing information must be contained in the serial signal to permit reliable and straightforward extraction of a clock signal in the receiver.

Error detection, although not considered essential, would provide a means of detecting link or equipment malfunction.

Signal-bit errors in the serial signal should not result in extension of the error to more than one word in the receiver.

Assessed against these criteria, the coding methods of scrambling and block encoding appear suitable. Scrambling would result in a relatively simple implementation, suited to large-scale integration, but the technique carries the risk of error extension in the descrambler and poorer control over the low-frequency spectrum than block encoding. In view of the requirements for use on optical-fibre cables and simple equalisation, scrambling was, initially, rejected in favour of the block-encoding technique described below.

The 8-bit–9-bit block code

The coding system approved by the CCIR for the digital video serial interface is a bit-mapping block-encoding system in which each of the 256 8-bit data words is 'mapped' to one of the 512 possible 9-bit words for transmission (8B–9B) [194]. (A set of transmission words with more than nine bits could have been used, but this would have resulted in a higher clock frequency and wider transmission bandwidth.) Of the 512 9-bit words, those used are selected for their transmission properties and there is no algorithmic relationship between the data words and the corresponding transmission words.

The desirable features of the nine-bit transmission words are:

Minimum d.c. component; this implies selection of words containing the smallest number of identical bits, i.e. five 1s or four 1s (5:4 or 4:5 words). There are 252 words of this type.

Maximum number of transitions; some of the 5:4 words contain an inadequate number of transitions for reliable clock regeneration and a further selection is made of those words which contain a maximum of four consecutive 1s or 0s and which begin or end with not more than three consecutive 1s or 0s.

Only 226 words from the set of 512 meet these criteria. For the additional 30 words required, 29 are selected from the set having six 1s or 0s (6:3 words) and the maximum number of transitions. One word of type 7:2 is used for synchronisation purposes, as described below.

Control of the d.c. component

Clearly, a result of using transmission words with an odd number of bits is that individual transmission words will have a d.c. component, albeit small. In the case of 5:4 words, a random sequence of such words will have a near-zero d.c. component.

For 6:3 words there is a significant d.c. component, but if a 6:3 word is combined with a 3:6 word the d.c. component is restored to zero. Therefore it is arranged that when it is desired to transmit a 6:3 word, the word is sent 'true' or 'complemented' depending on the sense of the previous 6:3 word transmitted. In other words, successive 6:3 words are sent true, complement, true, and so on [217]. In this way the d.c. component is maintained very close to zero.

The 8-bit–9-bit map

The data to be transmitted relates to video signals and is not, therefore, random, certainly not in the short term. Significant correlation exists between samples that are adjacent horizontally, vertically and temporally—particularly in the case of electronically generated video signals such as computer graphics. It is worthwhile examining the properties of the digital video signal to see if these can be exploited in any way.

An analysis of the occupancy of coding levels for several digitised test slides showed the coding levels in a narrow range centred on level 128 (corresponding to zero chrominance) were the most frequently occupied, see Figure 12.7. This is the area to which the most desirable transmission words—those with the largest number of transitions—are assigned. The assignment of transmission words of the various types is shown in Figure 12.8, from which it can be seen that the transmission words with the largest number of transitions are concentrated around level 128 while the words containing fewer transitions are assigned to the infrequently-occupied coding levels.

As a result of this assignment, the received serial signal should contain an abundance of transitions to aid clock regeneration and a well controlled low-frequency component.

The complete map of 9-bit transmission words to 8-bit data words is shown in Table 12.4. A functional block diagram of a serialiser is shown in Figure 12.9.

Link synchronisation

In the receiver the incoming signal has to be deserialised and for this to be performed correctly the start of each 9-bit word has to be identified. This operation, word synchronisation, can make use of information either contained within the serial signal itself or derived from the deserialised data which confirms that correct synchronisation has been achieved.

THE BIT-SERIAL INTERFACE 571

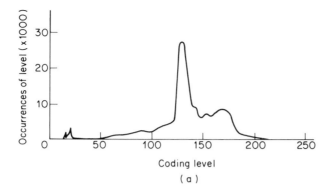

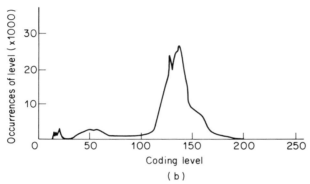

Figure 12.7 Analysis of the occupancy of coding levels for two EBU test slides: (a) 'toys' (b) 'boats and lighthouse'

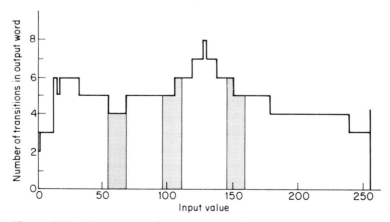

Figure 12.8 Occurrence of transitions in the transmission word as a function of the input data value, showing also the areas where 5:4 (□) and 6:3 (▨) transmission words are used

Table 12.4
8B–9B coding table

8B	9B	$\overline{9B}$	8B	9B	$\overline{9B}$	8B	9B	$\overline{9B}$	8B	9B	$\overline{9B}$	8B	9B	$\overline{9B}$	8B	9B	$\overline{9B}$
00	0FE	101	2C	1AC		58	099		84	0D5		B0	11A		DC	131	
01	027		2D	057		59	166		85	12A		B1	0E9		DD	0DC	
02	1D8		2E	09B		5A	09B		86	095		B2	116		DE	127	
03	033		2F	059		5B	164		87	16A		B3	02E		DF	0E2	
04	1CC		30	1A6		5C	09D		88	0B5		B4	1D1		E0	123	
05	037		31	05B		5D	162		89	14A		B5	036		E1	0E4	
06	1C8		32	05D		5E	0A3		8A	09A		B6	1C9		E2	11D	
07	039		33	1A4		5F	15C		8B	165		B7	03A		E3	0E6	
08	1C6		34	065		60	0A7		8C	0A6		B8	1C5		E4	11B	
09	03B		35	19A		61	158		8D	159		B9	04E		E5	0E8	
0A	1C4		36	069		62	025	1DA	8E	0AC		BA	1B1		E6	119	
0B	03D		37	196		63	0A1	15E	8F	153		BB	05C		E7	0EC	
0C	1C2		38	026	1D9	64	029	1D6	90	0AE		BC	1A3		E8	117	
0D	14D		39	08C	173	65	091	16E	91	151		BD	05E		E9	0F2	
0E	0B4		3A	02C	1D3	66	045	1BA	92	02A	1D5	BE	1A1		EA	113	
0F	14B		3B	098	167	67	089	176	93	092	16D	BF	066		EB	0F4	
10	1A2		3C	032	1CD	68	049	1B6	94	04A	1B5	C0	199		EC	10D	
11	0B6		3D	0BE	141	69	085	17A	95	094	16B	C1	06C		ED	076	
12	149		3E	034	1CB	6A	051	1AE	96	0A8	157	C2	193		EE	10B	
13	0BA		3F	0C2	13D	6B	08A	175	97	0B7	148	C3	06E		EF	0C7	

14	145	40	046		6C	0A4	98	0F5	C4	191	F0	13C
15	0CA	41	0C4		6D	054	99	0BB	C5	072	F1	047
16	135	42	04C		6E	0A2	9A	0ED	C6	18D	F2	1B8
17	0D2	43	0C8		6F	052	9B	0BD	C7	074	F3	067
18	12D	44	058		70	056	9C	0EB	C8	18B	F4	19C
19	0D4	45	0B1		71	1A9	9D	0D7	C9	07A	F5	071
1A	129	46	14E		72	05A	9E	0DD	CA	189	F6	198
1B	0D6	47	0B3		73	1A5	9F	0DB	CB	08E	F7	073
1C	125	48	14C		74	06A	A0	146	CC	185	F8	18E
1D	0DA	49	0B9		75	195	A1	0C5	CD	09C	F9	079
1E	115	4A	06B		76	096	A2	13A	CE	171	FA	18C
1F	0EA	4B	194	1B9	77	169	A3	0C9	CF	09E	FB	087
20	0B2	4C	06D	13B	78	0A9	A4	136	D0	163	FC	186
21	02B	4D	192	1B3	79	156	A5	0CB	D1	0B8	FD	0C3
22	1D4	4E	075	137	7A	0AB	A6	134	D2	161	FE	178
23	02D	4F	18A	1A7	7B	154	A7	0CD	D3	0BC	FF	062
24	1D2	50	08B		7C	0A5	A8	132	D4	147		
25	035	51	174		7D	15A	A9	0D1	D5	0C6		
26	1CA	52	08D		7E	0AD	AA	12E	D6	143		
27	04B	53	172		7F	152	AB	0D3	D7	0CC		
28	1B4	54	093		80	155	AC	12C	D8	139		
29	04D	55	16C		81	0AA	AD	0D9	D9	0CE		
2A	1B2	56	097		82	055	AE	126	DA	133		
2B	053	57	168		83	1AA	AF	0E5	DB	0D8		

15B 1AB 15D 1AD (at rows near 4B-4E)

10A 144 112 142 114 129 122 124 (C4-CB)

19D (near top right)

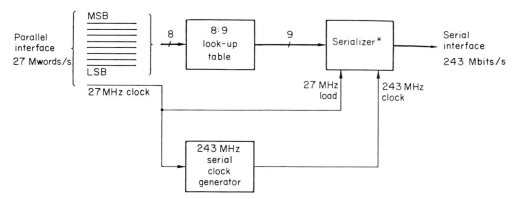

Figure 12.9 Functional block diagram of a serialiser. *parallel-in–serial-out shift register

Since there is a one-to-one relationship between the data words and the transmission words, a given input sequence will always result in the same transmission sequence. One such input sequence is the digital video timing reference signal (TRS) with its unique FF 00 00 preamble and, as the presence of the TRS in the input video data stream can be assumed at all times, it can be used as a framing signal. The sequence of transmission words corresponding to the TRS preamble could be detected as a bit sequence in the serial bit stream or as a word sequence in the parallel format after deserialisation.

Figure 12.10 shows the functional block diagram of a deserialiser based on detection of the TRS in the parallel domain. When the deserialiser phase is correct, the TRS preamble will be detected twice per line, i.e. twice per 1728 words. If the preamble is not detected within this period, the start point of the 9-bit words can be slipped along one bit by altering the division ratio of the divide-by-nine stage to divide-by-eight for one count: the detection process then begins again, the deserialiser phase being slipped by one bit each line until the TRS preamble is detected.

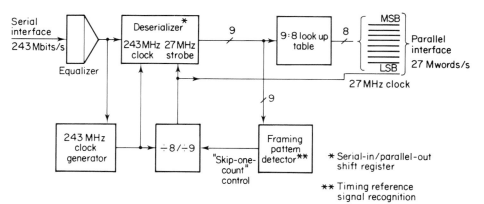

Figure 12.10 Functional block diagram of a deserialiser using the technique of skipping one count in the divide-by-nine counter for achieving correct phasing. *serial-in–parallel-out shift register, **timing reference signal recognition

A faster method of establishing the correct deserialiser phase would be to examine all the nine possible phases simultaneously. This can be achieved by the use of custom LSI devices but is unwieldy to implement with conventional integrated circuits.

Scrambled 10-bit interface

Where the transmission of 10-bit video data words is essential, there is a 10-bit serial interface based on the alternative coding technique, scrambling. Whilst not yet an internationally-agreed standard, this 270 Mbit/s interface is supported by several equipment manufacturers.

The sending interface accepts data words of eight- or ten-bit resolution, serialises them and scrambles the data-stream for transmission. Scrambling breaks up the long runs of consecutive 1s and 0s that occur when the digital video data signal is merely serialised. These would cause problems in clock extraction and cable equalisation.

The encoding rule employed in the proposed system is based on the use of a generator polynomial $G(X)$ which is the product of two polynomials, $G1(X).G2(X)$:

$$G1(X) = x^9 + x^4 + 1 \quad \text{and} \quad G2(X) = X + 1.$$

The scrambling action can be considered as passing the serialised signal through a shift register with feedback applied according to the scrambling algorithm used, as shown in Figure 12.11 for this application.

The first polynomial $G1(X)$ carries out the scrambling action, while the second, $G2(X)$, has the effect of breaking up runs of 1s appearing at the output of $G1(X)$. Each logic 1 appearing at the output of the first polynomial is converted to a transition from either state at the output of $G2(X)$, while no transition occurs for a logic 0 output from $G1(X)$. Thus the scrambled data is carried as changes of states rather than 1s or 0s, so making the system independent of polarity.

The effect of scrambling is to produce a signal power spectrum approximating to random noise. However, one of the potential problems with a scrambler is that it is possible to devise an input which will result in a long run of 0s at the output. For the particular algorithm used for this interface, the maximum theoretical run of 0s is 38. Fortunately, because of the limitations on the use of levels 00 and FF in the video data coding range, the practical limit on consecutive 0s becomes 25. Nevertheless, this represents a substantial low-frequency component and it remains to be seen whether it

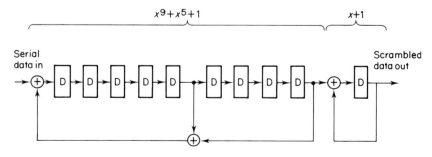

Figure 12.11 Shift register scrambler

will introduce a limitation to the distance over which the scrambling interface can operate.

Descrambling is the reverse of scrambling and a diagram of the descrambling operation is shown in Figure 12.12. The feedback extends over ten latches, so any single-bit error in the incoming data will be extended over ten bits. This means that two ten-bit words (but no more) will be affected by any signal bit-error.

Just as in the case of the block-encoding technique, the correct deserialiser phase has to be established and, again, use is made of the presence of the unique preamble of the TRS in the data-stream.

Electrical format

To permit the use of the serial signal directly for driving an optical link, a two-level code is necessary. The serial sender is specified as having an output signal amplitude of 700 mV peak-to-peak, compatible with the output of emitter-coupled logic devices. It is intended that a 75 Ω coaxial cable shall be the normal signal bearer, so a 75 Ω output impedance is specified.

Mechanical implementation

The widely used BNC connector is specified for the serial interface. Because of its greater ruggedness, the 50 Ω version of the connector is used.

Experience with the bit-serial interface

At the time of writing, several versions of the bit-serial interface have been implemented and successfully interconnected. Transmission over distances of 350 m using studio-quality coaxial cable and 500 m with CATV cable has been demonstrated. Equipment incorporating the serial interface is available commercially, including all-serial routing matrices.

Operation over optical-fibre links has been demonstrated also, though it is anticipated that it will be several years before these become economically attractive for studio use when compared with coaxial cable. A particular problem associated with the use of optical cables in studios is the difficulty of switching in the optical regime. Until this is

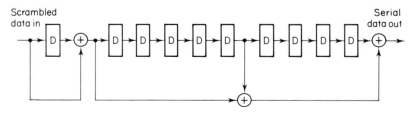

Figure 12.12 Shift register descrambler

solved the need to convert to and from optical to electrical signals at the matrix makes it more attractive to remain electrical throughout.

ANCILLARY DATA CHANNEL

It is common practice in analogue television studios to make use of the lines which occur during field blanking for carrying ancillary signals. These signals include test signals, such as insertion test signals used for checking the performance of circuits, and time code signals which enable each television frame of a recorded signal to be identified for editing purposes. In addition, the synchronising pulse itself is used to carry a digitised sound signal over the links between studios and transmitters.

Digital video interfaces provide a very large capacity for carrying ancillary signals in both the digital field- and line-blanking periods (1720 bytes per line). At the time of writing, work is still in progress to define a standardised way in which ancillary signals are to be handled.

APPLICATION OF DIGITAL INTERFACES IN STUDIOS

The completion of work on the specification of interface standards means that it is now possible to interconnect a wide range of digital video equipment operating to the CCIR Recommendation 601. With the realisation of the DVTR, the way is clear for all-digital production facilities.

At the time of writing, digital production studios are still at the experimental stage and are limited in the facilities available. Many of the applications within the BBC are described in other chapters and this equipment is increasingly being interconnected by one of the interfaces discussed in this chapter. It is also appropriate to refer to the activities of some other colleagues in broadcasting who are pioneering the introduction of digital equipment in studios. The most significant installations are at the facilities of Telediffusion de France (TDF) at Rennes, France, the Independent Television Association (ITVA) laboratory at Thames Television, Teddington, England, and the Canadian Broadcasting (CBC), Montreal (and Toronto), Canada.

The objective in assembling these facilities has been to evaluate the possibilities offered by digital production and postproduction, to familiarise production staff in any new working methods that the new techniques may require and to identify any difficulties that may arise when the use of digital techniques is extended to full-scale studios.

A full range of studio equipment has been installed in these facilities, including switching matrices, transparency scanners, still stores, caption generators, vision mixers, colour correctors and effects systems. They are, therefore, able to be used for the production of programmes and material produced in the first two of the installations has been used in transmitted programmes. In practice, however, they are used mainly for post production work where the transparency of the digital signal path permits multi-generation pictures to be generated without perceptible degradation.

A number of organisations which specialise in post production work are equipping themselves with digital post production facilities, to take advantage of the multi-

578   DIGITAL VIDEO INTERFACES

generation capability of the DVTR. Indeed, it is likely that this is the area in which digital operations will have the greatest impact, at least in the immediate future. Quantel have provided one of the first commercially available packages for digital editing and post production.

In television studio centres, this will result in the creation of digital islands around the digital video-tape facilities, with conversion between a composite video standard and the digital standard at the interface to the island. The interface requires high-quality decoding (using the techniques described in Chapter 4) and encoding if the signal quality is not to be degraded unnecessarily.

In all these installations, the studio equipment operates in bit-parallel form internally. The installations are, however, different. For example, in the BBC facility for off-line electronic graphics generation the simple digital mixer shown in Figure 12.13 is most conveniently connected to the other equipment shown in Figure 12.14 by the bit-parallel interface [196]. On the other hand, in the TDF–FRS–TUE installation shown in Figures 12.15 and 12.16, bit-serial interconnections are used.

Operation with the bit-serial interface has the advantage of allowing standard coaxial cable to be used in the studio cabling and permits a more compact routing matrix, there being only one wire per cross-point as opposed to the nine pairs required for the bit-

Figure 12.13 The BBC designed digital mixer in use at the Television Centre

parallel interface. However, circuitry to operate at the 243 Mbit/s speed of the serial interface is relatively expensive and adds to the cost of every equipment. Widespread use of the serial interface will no doubt take place when a low-cost LSI implementation is available. Meanwhile most early installations will continue to use the parallel interface

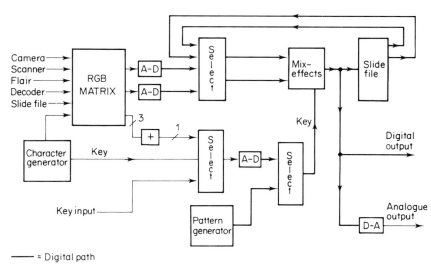

Figure 12.14 Schematic of BBC electronic caption preparation system, 1986

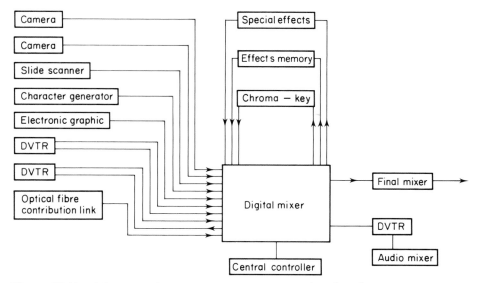

Figure 12.15 Schematic of TDF–FRS–TVE joint venture digital studio in Rennes

580   DIGITAL VIDEO INTERFACES

Figure 12.16   The control room at the digital studio in Rennes

within the studio, but the bit-serial interface will be used for the longer interconnections between studios via some form of central routing system. Figures 12.17 and 12.18 show the scope of the ITVA installation which used the parallel interface (designated EBU 27, a reference to the 27 Mword/s of the EBU parallel interface which formed the basis of Recommendation 656 [10]).

Experience gained from the experimental studios has confirmed the basic robustness of digital signals and the high quality of signal handling that accompanies the use of digital component equipment, provided that care is taken in the design and installation of the studio [198].

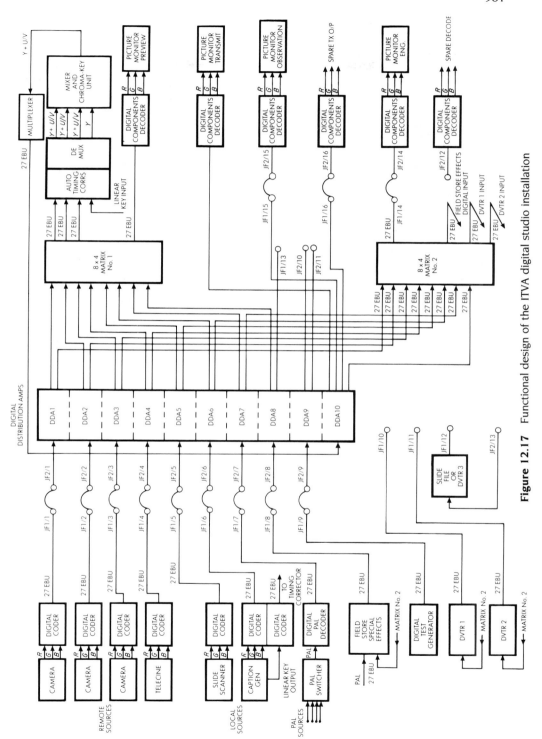

Figure 12.17 Functional design of the ITVA digital studio installation

Figure 12.18 The operations area of the ITVA digital studio

13 The Bit-rate Reduction of Television Signals For Long Distance Transmission

N. D. Wells

INTRODUCTION

The advantages of using digital techniques for long-distance transmission are well known. Signals coded in digital form are resistant in the main to the effects of distortion and noise introduced by a transmission link and can be regenerated at intervals without loss of fidelity. Accordingly, the quality of a television signal (video and audio) which has been transmitted in digital form is determined primarily by the coding method used rather than by the link itself. Digital techniques also offer greater flexibility in multiplexing and switching operations.

For these reasons among others, telecommunications administrations around the world are changing to digital networks which can carry telephony, television and data services. There are standard hierarchical bit rates for access to the digital network. In Europe at the present time, these hierarchical levels are at the following approximate levels; 64 kbit/s, 2048 kbit/s, 8448 kbit/s, 34.368 Mbit/s and 139.264 Mbit/s. (These last two bit rates are often referred to as 34 Mbit/s and 140 Mbit/s respectively.)

At present, and for many years in the UK, the television signal will both originate and be radiated in a composite PAL form. Thus there will be a continuing requirement for the distribution of composite signals possibly in digital form. Straightforward PCM coding of PAL generates a digital signal with a bit-rate of 100 Mbit/s which if transmitted without bit-rate reduction would be wasteful of channel capacity.

As all-digital studios come into service there will be a need to interconnect them using a signal format based on the internationally agreed, 4:2:2, format for digital *YUV* components used within the studio. This standard requires a sampling frequency of 13.5 MHz for the *Y* luminance component and 6.75 MHz for each of the *U* and *V* chrominance components. This standard generates a source signal with a total bit rate of 216 Mbit/s. Bit-rate reduction is obviously required if this component-based signal is to be carried on the digital communication network.

Several different techniques have been and are being developed for the bit-rate reduction of the video signal. Most techniques involve some reduction in picture quality resulting from the removal of information and the addition of noise and distortion in regions of the picture where the eye is least sensitive to the impairments. In general, the final picture quality depends on the transmission bit rate and on the complexity of the coding and decoding algorithms.

For broadcast quality applications there are two separate quality objectives, which are (a) distribution-link and (b) contribution-link quality. A distribution link carries the final network or studio output to the transmitter or the home after which no further processing of the signal will occur (except in the home receiver). A contribution link carries signals between studio centres or from outside broadcast units back to the studio and the quality requirements are higher than for a distribution link because it is likely that contribution signals will be processed further by special effects equipment which require a high quality of source signal.

At present, all signal distribution and contribution circuits use analogue composite signals and there is little distinction between these two types of link. The comparatively poor separation between the luminance and chrominance components in the composite signal limits the quality of any subsequent special effects processing. However, with the advent of a digital standard for component signals in the studio and with the increasing availability of digital links it is hoped that the quality of digital contribution links will be high enough to allow high-quality processing of the 'down-stream' signal.

An obvious bit-rate-reduction target is to compress the video signal into 34 Mbit/s for point-to-point transmission over the digital network. Whilst this compression can be achieved for a PAL signal without noticeable loss of quality it has not yet been shown that the full quality of the studio standard can be maintained at this low bit rate. However, at 140 Mbit/s the full quality can be maintained although the cost for the broadcaster of such a high-bit-rate link might be prohibitive. It should be possible to code the component signal at full quality in a bit rate for transmission of between 68 and 70 Mbit/s although, at the present time, no access at this bit rate to the European hierarchical network is provided.

Many good reviews have been published covering the field of digital coding and bit-rate reduction of television signals (for recent examples see [199–201]. It is a wide field and a review, of necessity, can only be a chart to guide the reader through the maze of a large bibliography. It is not the purpose of this chapter to give another review covering the whole subject. Rather, it will be an attempt to describe in a little more detail some of the techniques that have been studied at the BBC Research Department and that have proved or might be likely to prove successful and adequate for 'broadcast-quality' coding. The techniques covered in this chapter include sample rate reduction, differential coding (DPCM), transform coding, variable-length coding and vector quantisation.

Previous work at the BBC Research Department has produced several experimental systems, some of which have been used in field trials of digital television transmission. An early system coded one PAL video signal and six high-quality sound channels into a 60 Mbit/s package; two such packages could be multiplexed into 120 Mbit/s. This system was tested over both terrestrial and satellite links [7]. Next experimental equipment was constructed to reduce the bit rate of a digital PAL signal to within 34 Mbit/s [202]. After careful appraisal of the BBC's requirements it was decided to develop, however, PAL transmission equipment operating at 68 Mbit/s to convey one PAL signal plus many other services [203]. Two of these could be combined to form the standard 140 Mbit/s rate for connection to the national network. This system was tested on a nine-month field trial over a link between BBC premises in London and Birmingham [204]. Experimental equipment has also been constructed to form up a component signal into 140 Mbit/s and this equipment has also been tested on the London to Birmingham circuits. These equipments continue to be used to develop and improve the coding quality at lower bit rates.

SAMPLE RATE REDUCTION

Introduction

One method of reducing the bit rate of digitally coded signals is to reduce the sampling frequency in the encoder. According to the Nyquist sampling theorem the minimum sampling frequency that can be used, without introducing unwanted alias components into the decoded analogue signal, is equal to twice the highest frequency of the original analogue signal; this minimum frequency is often referred to as the Nyquist sampling frequency. In its simplest form this is illustrated in Figure 13.1a which shows the spectrum of a video signal after sampling (assuming that the samples are produced as narrow impulses). Provided that the bandwidth of the original signal is limited such that $f_{max} < f_s/2$ then the signal can be recovered after sampling without distortion by low-pass filtering (postfiltering).

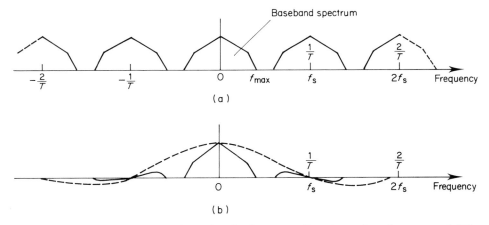

Figure 13.1 (a) Spectrum of a signal after sampling (sampling frequency $1/T$). (b) Spectrum of NRZ signal (sampling frequency $1/T$)

The process of sampling is equivalent to multiplying the analogue signal, $g(t)$, by a series of impulse or delta functions. The sampled signal, $s(t)$, is then given by

$$s(t) = g(t) \sum_{n=-\infty}^{n=\infty} \delta(t - nT) = \sum_{n=-\infty}^{n=\infty} g(nT) \, \delta(t - nT) \tag{13.1}$$

where T is the sampling interval. The spectrum of a series of delta functions is itself a series of delta functions in the frequency domain separated in frequency by an amount $1/T$ (see, for example, [206]). The spectrum, $S(\omega)$, of $s(t)$ is then obtained by convolving $G(\omega)$ with the delta function spectrum. $S(\omega)$ is then as shown in Figure 13.1a and is given by

$$S(\omega) = \frac{1}{T} \sum_{n=-\infty}^{n=\infty} G(\omega - n\omega_0) \tag{13.2}$$

where $\omega_0 = 2\pi/T$. The process of sampling adds energy to the original signal as each sampling delta function itself contains infinite energy. Normally, however, sampled signals are of a NRZ (non-return to zero) form. This NRZ signal can be obtained by convolving the impulse sampled signal $s(t)$ with a function, $c(t)$, where

$$c(t) = 1 \quad -T/2 < t < T/2$$
$$= 0 \quad |t| > T/2$$

The Fourier transform, $C(\omega)$, of $c(t)$ is given by

$$C(\omega) = \frac{\sin(\omega T/2)}{(\omega T/2)} \tag{13.3}$$

The spectrum of the NRZ signal is then as shown in Figure 13.1b and is given by

$$C(\omega)S(\omega) = \frac{\sin(\omega T/2)}{(\omega T/2)} \frac{1}{T} \sum_{n=-\infty}^{n=\infty} G(\omega - n\omega_0)$$

Often it is profitable to consider the video spectrum as a three-dimensional function and sampling as a three-dimensional process. The scene to be scanned can be considered as a function, $g(x, y, t)$, of x and y coordinates and time. This function has a three-dimensional spectrum, $G(\omega_x, \omega_y, \omega_t)$. The scanning device samples the scene in the vertical, Y, direction and in time. The signal is sampled in the horizontal direction during the process of analogue-to-digital conversion. The sampling causes the spectrum to be repeated at harmonics of the three-dimensional sampling frequencies. For example, line locked sampling combined with interlaced scanning gives the sampling structure shown in Figures 13.2a and 13.2b. Figures 13.2c and 13.2d give the vertical/horizontal and the vertical/temporal projections of this sampling pattern respectively in the frequency domain.

It can be seen that, just as in the one-dimensional case, if the three-dimensional bandwidth of the original signal is limited appropriately then the original signal can be recovered after sampling without distortion by suitable low-pass filtering in three dimensions.

For component coded signals there are many possible ways of choosing 3D sampling patterns. An optimum sampling pattern minimises the overall sample rate while

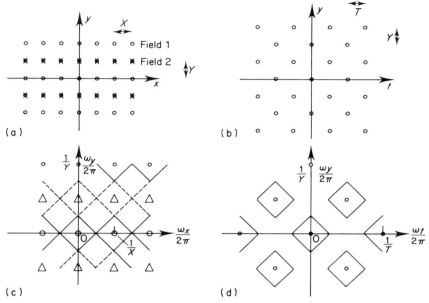

Figure 13.2 (a) Two-field sampling in horizontal/vertical plane. (b) Sampling structure in vertical/temporal plane. (c) Centres of repeat spectra in ω_y, ω_x domain (Δ represent spectra with $\omega_t/2\pi = 1/2T$) (d) Centres of repeat spectra in ω_y, ω_t domain

requiring that the signal bandwidth is restricted only in regions where the loss of bandwidth is least noticeable.

For any but the simplest one-dimensional sample rate reduction, the filtering is best performed digitally. The process of digital filtering combined with sample-rate reduction may be understood by reference to Figure 13.3. A signal, $g(t)$, is sampled at a frequency f_1 and filtered by a low-pass filter with an impulse response, $h(t)$. The output of this filter is resampled at f_2 and the final sample values will be a weighted sum of the f_1 impulse values. The coefficients in this summation depend on the impulse response, $h(t)$, and also

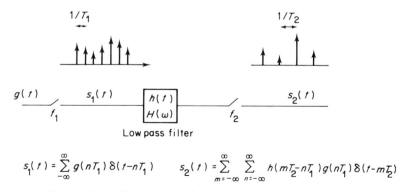

$$s_1(t) = \sum_{-\infty}^{\infty} g(nT_1)\delta(t-nT_1) \qquad s_2(t) = \sum_{m=-\infty}^{\infty} \sum_{n=-\infty}^{\infty} h(mT_2-nT_1)g(nT_1)\delta(t-mT_2)$$

Figure 13.3 Illustration of the process of sample-rate changing

on the position of an individual output sample in relation to the input samples. This process of downsampling may be seen therefore as one of interpolation [207]. Similarly, Figure 13.3 describes the process of postfiltering and sample-rate up-conversion if $f_2 > f_1$. The implementation of these interpolating filters is simplest when the ratio of f_1 to f_2 is 2:1 or 1:2.

When choosing a suitable reduced-rate sampling grid for a particular application, it is necessary to determine the centres of the repeat spectra in the frequency domain. For many repeating structures, the sampling grid can be described by the points of intersection of three sets of planes. If the distance separating the planes measured along the x, y and t axes is X_i, Y_i, and T_i respectively for the ith set of planes, then the centres of the repeat spectra in the frequency domain are described by vectors given by

$$\sum_{k_1=-\infty}^{\infty} \sum_{k_2=-\infty}^{\infty} \sum_{k_3=-\infty}^{\infty} (k_1\mathbf{V}_1 + k_2\mathbf{V}_2 + k_3\mathbf{V}_3)$$

where $\mathbf{V}_i$ is a vector with projections $1/X_i$, $1/Y_i$, $1/T_i$ on the f_x, f_y, f_t axes respectively and k_1, k_2 and k_3 are integers [98, 208, 209].

For more complicated sampling grids, the total sampling pattern can be considered as a superposition of identical rectangular grids which are shifted relative to one another. If the basic orthogonal grid of sampling impulses $d(x, y, t)$ is defined by

$$d(x, y, t) = \sum_j \sum_k \sum_l \delta(x - jX, y - kY, t - lT)$$

where j, k, l, are integers, the spectrum $D(f_x, f_y, f_t)$ of this grid is given by

$$D(f_x, f_y, f_t) = \frac{1}{XYT} \sum_p \sum_q \sum_r \delta(f_x - p/X, f_y - q/Y, f_t - r/T)$$

A second grid shifted by a_xX, a_yY, a_tT, where a_x, a_y, a_t, are less than unity, has a Fourier transform $D(f_x, f_y, f_t, \mathbf{a})$ given by

$$D(f_x, f_y, f_t, \mathbf{a}) = D(f_x, f_y, f_t) \exp\left[-j2\pi(a_xf_x + a_yf_y + a_tf_t)\right]$$

Then for a sampling pattern described by the superposition of N grids defined by $\mathbf{a}_i$ (where $\mathbf{a}_0 = 0$) the Fourier transform is

$$D(f_x, f_y, f_t)\left(1 + \sum_{i=1}^{N-1} \exp(-j\mathbf{f} \cdot \mathbf{a}_i)\right)$$

The magnitudes of the spectral impulses at $\mathbf{f} = (p/X, q/Y, r/T)$ are, therefore, modified by the function

$$1 + \sum_{i=1}^{N-1} \exp(-j\mathbf{f} \cdot \mathbf{a}_i)$$

which has zeros for certain values of p, q, r. The final sampled spectrum is then given by the convolution of this Fourier transform with the baseband signal spectrum.

Examples of sample-rate reduction

Chrominance components

In the studio standard for component coded signals [9], the chrominance bandwidth is determined by the possible requirement for special-effects processing of the signal within the studio. The chrominance bandwidth can be reduced further however by around a factor of two before any subjective impairment is perceptible on typical pictures. It is very common therefore in bit-rate-reduction systems for the chrominance sampling frequencies and bandwidths to be reduced from those of the source standard.

A common method for halving the sampling rate of the chrominance component is to reduce the vertical sampling frequency for both components by transmitting only U or V on each line. Provided that the components are vertically low-pass filtered before subsampling then the 2D spectrum will be as shown in Figure 13.4. The baseband spectrum can be recovered, without aliasing distortion, by intra-field interpolation to reconstruct the colour component on all field lines.

If the subsampling is as shown in Figure 13.5a and is not reset at the start of each frame then the spectral repeats in the vertical/temporal frequency plane are as shown in Figure 13.5b. Any residual alias components appearing in the baseband spectrum have a frequency of 12.5 Hz (for stationary detail). The most noticeable effect caused by this aliasing is that it gives rise to 12.5 Hz flicker on large amplitude horizontal chrominance transitions. However, in a 625-line system and with good pre- and post-vertical filtering, this method of chrominance sample-rate reduction gives rise to only a very slight loss in subjective picture quality on most pictures. The loss of vertical resolution is only visible at horizontal transitions where a boundary is defined by a chrominance change which is not accompanied by a luminance change. The resolution loss is also visible on some graphical material such as horizontal colour bars.

If 2:1 subsampling is achieved by a line-offset or 'line-quincunx' sampling pattern as shown in Figure 13.6a then the centres of the repeat spectra will be as shown in Figure 13.6b. This figure also shows a possible 2D boundary shape for a baseband spectrum that

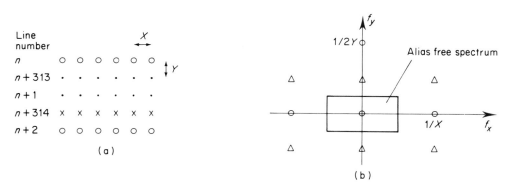

Figure 13.4 2:1 vertical sampling. (a) Sampling sites within one frame (○ = sampling sites field 1, x = sampling sites field 2, · = omitted sample). (b) Two-dimensional spectral repeats of vertically subsampled signal (△ represents centres of repeat spectra at 12.5 Hz)

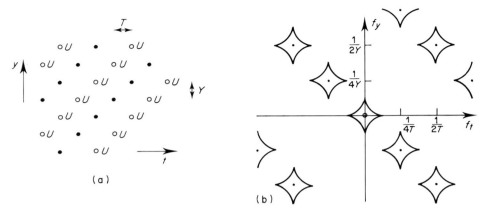

Figure 13.5 Example of line-subsampled chrominance. (a) 2:1 line subsampling structure for U chrominance component. (b) Spectrum of subsampled U chrominance

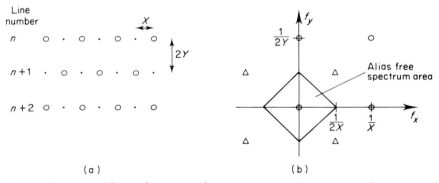

Figure 13.6 2:1 subsampling using a line-quincunx structure. (a) Sampling structure showing line-quincunx sampling within one field. (b) Two-dimensional pattern of repeat spectra and alias free area for baseband component

this line-quincunx sampling can support. Other shapes are possible and some different examples are given in following sections. The area of the diamond shape in Figure 13.6b is the same as the area of the rectangular shape shown in Figure 13.4b. However, it is probably the minimum value of the bandwidth in any direction which determines perceived image quality and not the area contained within the bandwidth boundary. For line-quincunx sampling as shown, the minimum value of the bandwidth is along the diagonal, and the maximum bandwidth, B, in this direction is given by

$$B = \tfrac{1}{2}[(1/2X)^2 + (1/4Y)^2]^{1/2}$$

For example, with 625-line interlaced scanning and vertical subsampling and a horizontal sampling frequency of 6.75 MHz, the maximum horizontal bandwidth is 3.375 MHz and the maximum vertical bandwidth is 78 cycles per picture height (c/ph) (which is equivalent to approximately 1.7 MHz in the horizontal direction). With line-quincunx sampling, the maximum value of the vertical bandwidth increases to 3.4 MHz (horizontal equivalent) and the maximum value of the bandwidth in the diagonal direction is 2.4 MHz. On critical pictures this increase in minimum bandwidth gives a perceptible improvement in quality. This improvement is obtained of course at the expense of an increase in complexity of the bandwidth-defining filters.

Another technique that has been used for reducing the sampling rate of the chrominance components is to halve the temporal sampling frequency, i.e. the chrominance information is transmitted for only one field per frame. This gives repeat spectra for each component as in Figure 13.7. In this case the maximum temporal bandwidth that can be supported is 12.5 Hz (for a 50 Hz system).

Reducing the temporal bandwidth has the effect of reducing the spatial bandwidth for moving images. For a given speed of motion, the spatial bandwidth in the direction of motion may be calculated as follows. Consider a vertical grating defined by $\cos(2\pi f_x x)$ moving with a velocity, u, in the horizontal direction. Then, at a given point in the picture, the magnitude of the temporal frequency, f_t, is

$$f_t = u f_x$$

For a maximum temporal bandwidth, B_t, the maximum spatial bandwidth, B_x, for a given speed of motion is then

$$B_x = B_t/u$$

If B_t is equal to 12.5 Hz, then the variation of B_x with u is given by the full curve in Figure 13.8. A convenient unit for measuring speed of motion is 'picture widths per second'. B_x is then in units of cycles per picture width (cpw). Considering displayed picture width (rather than total line length), 1 cpw corresponds to approximately 0.02 MHz.

Since the temporal subsampling and interpolating filters are constructed around field delays, practical filters will contain a minimum number of these large delays. Therefore, it is probable that the 6 dB temporal bandwidth will be close to 6.25 Hz rather than 12.5 Hz giving a spatial bandwidth as a function of speed as shown by the dotted curve in Figure 13.8. It can be seen that at movements as slow as 0.1 picture widths per second, the spatial bandwidth is reduced to 1 MHz. Temporal subsampling of the chrominance

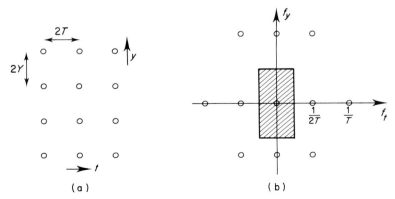

Figure 13.7 Spectrum for temporally subsampled chrominance. (a) Temporal subsampling pattern. (b) Vertical/temporal spectrum for temporal subsampling

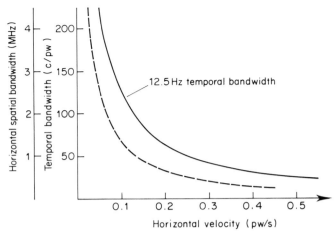

Figure 13.8 Horizontal bandwidth as a function of horizontal velocity for a temporal bandwidth of 12.5 Hz

gives therefore a perceptible loss of resolution on critical pictures containing movement. It should be noted that if this same form of straightforward temporal subsampling were to be applied to the luminance component then the resolution loss with motion would be very apparent.

Luminance component

The sample rate for the luminance component according to the studio standard for component coded signals is set at 13.5 MHz. Except for the highest quality requirements, the maximum bandwidth required for the luminance is around 5 MHz (in any direction). Then with line locked-sampling a sensible, reduced-rate (3/4) sampling frequency is 10.125 MHz. However, good quality pictures can be obtained using 9.0 MHz sampling in a line-quincunx pattern giving a supportable bandwidth shape of the form shown in Figure 13.9. This bandwidth shape is optimum because it maximises the minimum

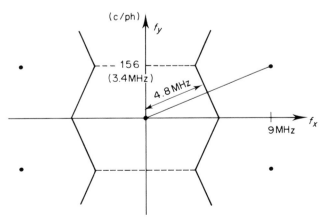

Figure 13.9 Optimum two-dimensional bandwidth for line-quincunx luminance sampling at 9 MHz

bandwidth. The minimum bandwidth is in the diagonal direction and this has a value which is equivalent to approximately 4.8 MHz. The maximum horizontal bandwidth is around 5.3 MHz. The effect on the zone plate test pattern filtered in this manner is shown in Figure 13.10.

An FIR filter having a two-dimensional response similar in shape to the optimum shown in Figure 13.9 can be constructed but this requires the addition of contributions from a large number of sample points taken over several lines. Alternatively, a simpler form of two-dimensional filter known as a 'comb filter' can be used. Comb filters are particularly suited to the sample-rate reduction of PAL signals and this is discussed in the next section.

Subsampling of PAL signals

In a PAL signal the chrominance information is added on a subcarrier at f_{sc} to the scanned luminance signal, where

$$f_{sc} = 4.433\ 618\ 75 \text{ MHz}$$
$$= (283 + 3/4) \cdot f_1 + f_p$$

where f_1 is the line frequency and f_p the picture frequency. Both the U and V components are derived from the same scanned source as the luminance and are placed on quadrature phases of subcarrier using double-sideband suppressed-carrier modulation. Before modulation the V component is multiplied by a square-wave of half line-frequency.

Although the subcarrier frequency is generated as a one-dimensional signal it can be interpreted as a scanned version of a moving spatial frequency [98]. The projections of the subcarrier positions in the vertical/horizontal and the vertical/temporal frequency are given in Figures 13.11a and 11b.

When PAL signals are digitised for transmission, the signal is usually sampled at a frequency which is locked to subcarrier frequency. For example, a common super-Nyquist sampling frequency is $3f_{sc}$ (13.3 MHz). However, it has been shown that the PAL signal may be satisfactorily reproduced after sampling at the 'sub-Nyquist' rate of $2f_{sc}$ (8.8 MHz) provided that the sampling frequency has a specified phase with respect to the colour subcarrier [51]. A $2f_{sc}$ sampling grid is almost exactly line-quincunxial. Therefore, for the luminance component, the baseband spectrum and the higher order spectra can be separated by using comb pre- and post-filters of the form shown in Figures 13.12a and 13.12b. These have a frequency response of the shape shown in Figure 13.13 when the delay element in the filter is equal to one line period. The shape shown in Figure 13.13 corresponds approximately to a 3 dB or a 6 dB contour for the response of a single filter or the pair of filters respectively. In the region where the spectra would overlap the vertical bandwidth is limited to 78 c/ph. No vertical filtering is required below about 3.3 MHz as the horizontal bandwidth of the baseband PAL signal is less than 5.5 MHz. The response of the pair of filters of Figure 13.12 has a uniform group delay equal to the length, T, of the delay element.

For the chrominance components, each comb-filter introduces an attenuation of 3 dB at the U and V colour-subcarrier frequencies. However, provided that the $2f_{sc}$ sampling instants are correctly timed, the alias component from each sideband adds in phase with the sideband symmetrically placed about f_{sc} thereby compensating for the loss in the

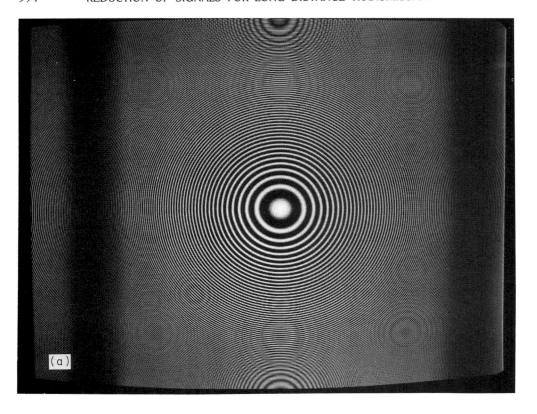

(a)

comb filter pair [51, 210]. The sampled signal is of the form

$$Y + C, Y - C, Y + C, Y - C, Y + C, Y - C, \ldots$$

where it is arranged, by selecting the timing of the sampling instants, that the chrominance component C is equal to $(U + V)/2$ on one line and $(U - V)/2$ on the next. This form of signal implies that both chrominance components are being carried on a single phase of subcarrier which is in phase with the sampling instants. On the whole the U and V components are still separate because their spectra are centred on different vertical frequencies. However, for chrominance detail containing high vertical frequencies the sub-Nyquist sampling causes a degree of U/V crosstalk. The comb postfilter conveniently reconstitutes this non-standard carrier into a standard PAL form with U and V in quadrature in plain picture areas. This results from the fact that the high-frequency group delay of the comb post-filter is equal to $T/2$ and also from the fact that the U and V spectra are separated in frequency by $1/2T$ (or multiples thereof). Therefore, the filter delays the phase of the V component by 90° relative to the U component (in plain picture areas) giving a chrominance signal in the standard PAL form.

The impairments introduced by sub-Nyquist sampling using line delay comb filters may be summarised as follows.

(a) The high frequency diagonal luminance resolution is reduced as shown in Figure 13.13; the minimum diagonal bandwidth is equivalent to approximately 3.0 MHz.

SAMPLE RATE REDUCTION 595

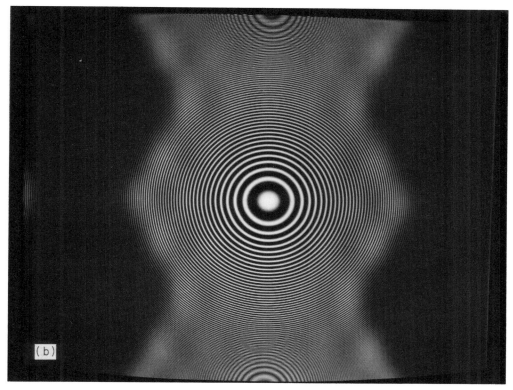

Figure 13.10 (a) Unfiltered zone plate test picture. (b) Two-dimensional filtered zone plate in the manner of Figure 13.9

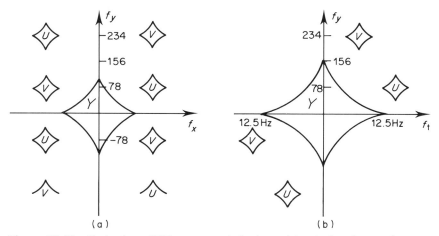

Figure 13.11 Projection of PAL spectrum in horizontal/vertical and vertical/temporal frequency planes. (a) Projection of PAL spectrum chrominance in two-dimensional spatial frequency plane. (b) Relative position of modulated U and V chrominance (and high-frequency luminance) in the vertical/temporal plane (shown for positive values of horizontal frequency)

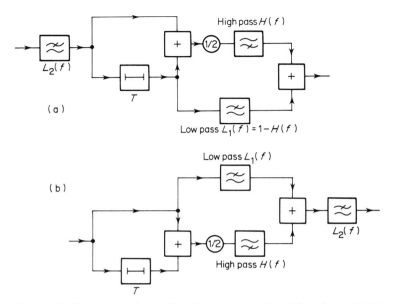

Figure 13.12 (a) Comb prefilter (low frequencies delayed by T, high frequencies delayed by $T/2$). (b) Comb postfilter (low frequencies undelayed, high frequencies delayed by $T/2$)

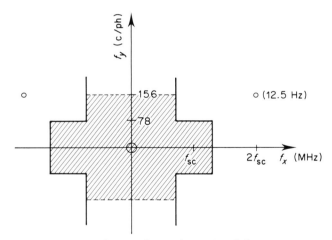

Figure 13.13 Shape of two-dimensional frequency response for comb filters

(b) The $2f_{sc}$ sampling can give rise to residual luminance aliasing which is not completely removed by the comb postfilter. This is because the comb filters which define the 2D baseband spectrum have a cosine-shaped boundary in the vertical frequency direction and not a sharp boundary separating the baseband spectrum from the higher order spectra. For stationary details, the alias components have a temporal frequency component of 12.5 Hz and can give rise to visible low-frequency patterning caused by beating between the baseband and alias components. However, the luminance frequencies which give rise to the most visible patterning also give rise to more objectionable cross-colour effects which mask any subsequent luminance aliasing.

(c) There is a degree of U/V crosstalk for high vertical chrominance frequencies. This cross-talk appears as 12.5 Hz flicker on some horizontal chrominance transitions.

In subjective tests, comparing the picture quality of sub-Nyquist PAL and analogue PAL systems, unprocessed analogue PAL pictures were graded at an average grade of 3.5 on the CCIR five-point quality scale. The digital PAL signal which had been filtered and sub-Nyquist sampled was graded on average at 3.4 on the same scale [210]. Thus the additional impairment of 0.1 grade introduced by the sub-Nyquist process is very small considering the imperfect quality of the PAL signal and also considering the significant bit-rate saving from $2f_{sc}$ sampling.

Other forms of comb filter

The comb filters just described average the high frequency signal information across a one-line-period delay. Alternatively, this delay may be of 'field' (313 line) or picture period duration [210]. The length of the delay affects the type of impairment introduced by the comb filters. For example, the picture-delay-based comb filters do not introduce any loss of resolution for stationary pictures. However, when averaging across a picture delay, there is inevitably a loss of resolution with picture motion. The impairments introduced by a field-delay filter fall between those of the line and picture-delay filters.

The loss and stop bands (at high horizontal frequencies) in the vertical/temporal frequency plane for 313-line and picture-delay based filters are shown in Figure 13.14a and 13.14b. Using these figures it can be shown that pre- and post-filtering (with filters of the same type) limits the presence of unwanted luminance alias components introduced by the subsampling. Also, as in the case of the line-delay-based comb filters, each filter has a loss of 3 dB at colour-subcarrier frequency.

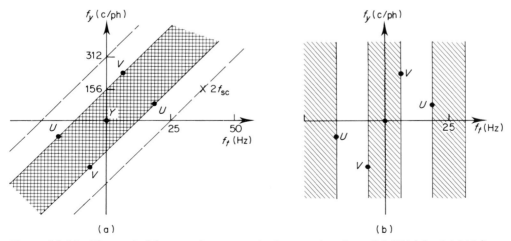

Figure 13.14 The vertical/temporal response (at frequencies above 3.3 MHz) for (a) 313-line-delay comb filter and (b) 625-line-delay comb filter. The passband is shown cross-hatched. The dashed lines show the null response of the filter

DPCM CODING

Introduction

A very successful technique for reducing the number of bits per sample is DPCM coding. In DPCM coding a prediction of the signal is formed from surrounding sample values and the difference between the input and the prediction is quantised and transmitted, rather than the signal itself. The difference quantiser is non-linear, coding small differences accurately and large differences, which occur less frequently, with less accuracy. Large differences or prediction errors tend to occur at edges or in detailed areas of the picture where the eye is less sensitive to quantisation distortion. The noise or distortion introduced by DPCM coding tends to be masked, therefore, because the noise is typically confined to a small fraction of the total picture area and also confined to those areas in which the distortion tends to be masked by the picture detail.

At the DPCM decoder, the quantised prediction error is added to the corresponding prediction value to form the coded output signal. In order that both coder and decoder are working with the same prediction value, it is necessary that the coder, in forming the prediction, uses only those same output sample values that are available to the decoder. A DPCM coder must, therefore, also contain within it a decoder, and block diagrams of the coder and the decoder are shown in Figure 13.15.

The prediction is formed by a sum of contributions from previous (output) sample values. The simplest prediction of the current sample is the previous (output) sample value. More complicated predictors include contributions from samples on the previous line, field or frame. When the number of coding levels is limited, the distortion in the reconstructed picture is minimised by 'optimising' the prediction, thereby minimising the power in the prediction-error signal. A technique for designing optimum predictors using picture statistics is described in this section. This method has been shown to give very good results for the coding of both composite PAL and component *YUV* signals [202, 203, 205] and also gives a transmission system which is robust in the presence of transmission errors. Prediction algorithms which adapt according to picture content might, in future, prove optimum at low bit rates. However, DPCM systems using non-adaptive predictors, consisting of contributions from previous elements from lines in the current field and lines in previous fields, have proved to be both practicable and give good

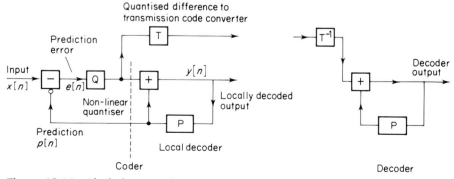

Figure 13.15 Block diagram of DPCM coder and decoder

performance for a wide range of picture material. The non-adaptive approach is discussed further below.

Linear prediction

The prediction is formed as a linear sum of previous sample values. If $x[n]$, $p[n]$, and $y[n]$ represent the input, prediction, and output sample sequences respectively (as shown on Figure 13.15) then

$$p[n] = \sum_{k=1}^{N} a_k y[n-k] \simeq \sum_{k=1}^{N} a_k x[n-k]$$

where N is the number of samples used in the prediction.

In order to design a predictor based on the input samples, $x[n]$, we have assumed $y[n] \simeq x[n]$ which is the case for high-quality coding. The prediction error, $e[n]$, given by

$$e[n] = x[n] - p[n]$$

has a mean square value, averaged over a picture, given by

$$\langle e^2[n] \rangle = \left\langle \left(x[n] - \sum_{k=1}^{N} a_k x[n-k] \right)^2 \right\rangle$$

where $\langle \; \rangle$ represents the average value. The mean square prediction error, as a function of the predictor coefficients, has a minimum when

$$\frac{\partial}{\partial a_j} \langle e^2[n] \rangle = 2 \left\langle e[n] \frac{\partial}{\partial a_j} e[n] \right\rangle$$

$$= 0 \quad \text{for } j = 1, 2, \ldots, N$$

$$\left\langle e[n] \frac{\partial}{\partial a_j} e[n] \right\rangle = \left\langle \left(x[n] - \sum_{k=1}^{N} a_k x[n-k] \right)(-x[n-j]) \right\rangle$$

$$= -\langle x[n]x[n-j] \rangle + \sum_{k=1}^{N} a_k \langle x[n-j]x[n-k] \rangle$$

$$= 0$$

Putting $\langle x[n]x[n-j] \rangle = R(j)$, which is the autocorrelation function of the input data sequence, this last equation gives

$$R(j) - \sum_{k=1}^{N} a_k R(j-k) = 0 \quad \text{for } j = 1, 2, \ldots, N$$

If the autocorrelation function $R(j)$ has been measured, this last set of N simultaneous equations can be solved for the N unknowns, a_k, giving the coefficients of the optimum predictor. A value for the mean square error, ξ, using the optimum predictor can then be calculated as follows

$$\xi = \langle e^2[n] \rangle_{\text{opt}} = \left\langle e[n] \left(x[n] - \sum_{k=1}^{N} a_k x[n-k] \right) \right\rangle$$

As

$$\langle e[n]x[n-k]\rangle = -\left\langle e[n]\frac{\partial}{\partial a_k}e[n]\right\rangle \quad k = 1, 2, \ldots, N$$

$$= 0 \quad \text{for the optimum predictor}$$

$$\xi = \langle e[n]x[n]\rangle$$

$$= R(0) - \sum_{k=1}^{N} a_k R(k)$$

The mean power, ξ, of the prediction error signal tends to be less therefore than the mean power, $R(0)$, of the original signal.

In the frequency domain, this process of optimisation ensures that the prediction will be accurate at frequencies with large component amplitudes, for example at frequencies around d.c. and around colour-subcarrier frequency in composite PAL signals. It can be shown that the statistically optimum predictor tends to make the power spectral density of the prediction error the same at all frequencies [211]. (In other words, the spectrum of the prediction-error signal tends to be 'white'.)

However, it is not possible to design a DPCM predictor which is accurate at all signal frequencies. To see this, consider the prediction error response, $E(\omega)$, to an input frequency, $\exp(j\omega t)$. Then for a sampling interval of T

$$E(\omega) = 1 - \sum_{s=1}^{N} a_s \exp(j\omega sT)$$

i.e.

$$|E(\omega)|^2 = \left(1 - \sum_{s=1}^{N} a_s \exp(j\omega sT)\right)\left(1 - \sum_{r=1}^{N} a_r \exp(-j\omega rT)\right)$$

which gives after some rearrangement

$$|E(\omega)|^2 = 1 - 2\sum_{s=1}^{N} a_s \cos(\omega sT) + \sum_{s=1}^{N} a_s^2 + 2\sum_{r=1}^{N-1}\sum_{s=1}^{N-r} a_s a_{s+r} \cos(\omega rT)$$

The average, prediction-error power spectral density up to half sampling frequency is then

$$\frac{T}{2\pi}\int_{-\pi/T}^{\pi/T} |E(\omega)|^2 \, d\omega = 1 + \sum_{s=1}^{N} a_s^2$$

This expression for the average value of $|E(\omega)|^2$ shows that if the prediction is good (i.e. $|E(\omega)|^2$ is small) for a wide range of frequencies then the prediction must be correspondingly poor (giving a large value for $|E(\omega)|^2$) in the remaining frequency range. This is illustrated in Figure 13.16. A good predictor, therefore, matches those spectral regions where the prediction is best with regions where there is, statistically, most energy. This matching, of the prediction to the power spectrum, is obtained via the autocorrelation function of an 'average' television signal as described above.

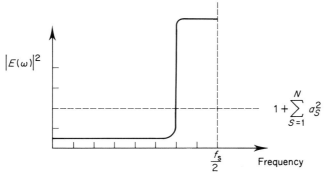

Figure 13.16 Example of a possible one-dimensional prediction error frequency response. Illustrating the fact that it is not possible to design a (non-adaptive) predictor which is good at all frequencies

Adaptive prediction

Many adaptive strategies have been proposed, for example by Graham [212], Zchunke [213] and Dukhovich and O'Neal [214]. These strategies involve adaption between various simple two-dimensional or one-dimensional predictors according to decisions, about the expected nature of the picture material, based on previously transmitted information. More recently, DPCM prediction strategies have been proposed which include motion compensated prediction [215]. Also DPCM systems which explicitly signal the predictor adaption information have been tested [216].

It is beyond the scope of this chapter to review here all these different systems. Preliminary comparative studies suggest that there is not a great deal of difference between the performance of the different systems [217]. For practical systems, therefore, other factors such as instrumental complexity and decoder stability in the presence of transmission errors assume more importance.

Non-linear quantisation

Introduction

A typical DPCM non-linear quantiser with fifteen output levels is shown in Figure 13.17. The impairments introduced by non-linear quantisation fall into three categories known as 'granular noise', 'edge busyness' and 'slope overload'. Granular noise appears typically in plain areas of the picture where the prediction error is small and is quantised with the more closely spaced, inner levels of the quantiser. Edge busyness results from larger quantising distortions added at less predictable edges. This quantising error can vary from line to line and field to field giving the appearance of edge-localised noise. The eye is less sensitive to noise localised on edges than in plain areas and the spacing of the mid-range quantiser levels can be increased accordingly. Slope overload occurs when the

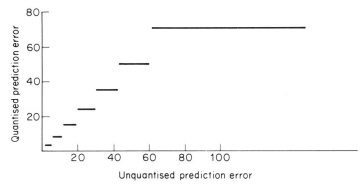

Figure 13.17 Example of a non-linear DPCM quantiser (positive half only)

prediction error at an edge is greater than the maximum magnitude of the quantiser output and gives rise to an apparent softening of such an edge.

The precise spacings of the output levels of a non-linear quantiser are not particularly critical. The arrangement of levels is a compromise, balancing the requirements to have a quantiser which spans a range which is sufficient to prevent slope overload and yet which does not introduce an excessive level of granular noise or edge busyness.

The noise and distortion introduced by the non-linear quantiser determine the subjective picture quality. However, it is the statistics of the prediction error signal which dictate the overall quantiser parameters. Ultimately, therefore, it is the predictor design which determines the picture quality at a given bit rate.

Two different approaches to the design of non-linear quantisers are described below.

Minimum mean square error (MMSE) design method

A common approach is to construct a quantiser which minimises, for a given predictor, the mean square quantisation error for a given number of output levels. Following this approach, the probability density function, $P(x)$, of the prediction error, x, is first measured for a given predictor and a range of picture material. The quantiser output levels, y_i, and decision levels, x_i, are then chosen such that the mean square error, σ_d^2, is minimised where

$$\sigma_d^2 = \sum_{i=1}^{N} \int_{x_i}^{x_{i+1}} (x - y_i)^2 P(x) \, dx$$

For a fixed number of levels, N, appropriate values of x_i and y_i can be found numerically [218]. Approximate values of these x_i and y_i can be determined as follows [219]. Assume $P(x)$ is approximately constant between x_i and x_{i+1}. Then

$$\int_{x_i}^{x_{i+1}} (x - y_i)^2 P(x) \, dx \simeq \frac{1}{12} P(y_i) \partial_i^3$$

where

$$\partial_i = x_{i+1} - x_i$$

i.e.

$$\sigma_d^2 = \frac{1}{12} \sum_1^N P(y_i) \partial_i^3$$

In addition,

$$\sum_1^N P^{1/3}(y_i)\partial_i \simeq \int P^{1/3}(y)\,\mathrm{d}y = K$$

Here, for a given distribution, K is a constant which is independent of the magnitudes of the individual ∂_i. Setting $\beta_i = P^{1/3}(y_i)\partial_i$, the problem of minimising σ_d^2 is equivalent to that of minimising

$$\sum_i \beta_i^3$$

subject to the condition that

$$\sum_i \beta_i = \text{constant}$$

Using the Lagrange method of undetermined multipliers, it follows that σ_d^2 is a minimum when

$$\beta_1 = \beta_2 = \beta_3 = \beta_4 = \cdots = \beta_N = \text{constant}$$

Therefore, the decision levels, x_i, can be determined by dividing the curve of the cube root of the probability density function, $P^{1/3}(x)$, into equal areas. In this approximate approach the output levels, y_i, can be chosen to lie midway between the decision levels.

Graphical design method

Figure 13.18 shows the magnitude of the quantisation error, $|q_e|$, plotted as a function of prediction error, p_e. The points where the quantisation error is zero correspond to the

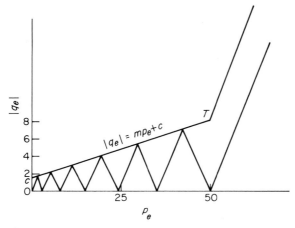

Figure 13.18 Example of simple graphical design of non-linear quantiser

output levels of the quantiser. A subjectively good quantiser can be constructed [220] by arranging that the maximum quantisation error is always less than the line given by

$$q_e = mp_e + c \quad \text{for } p_e \leqslant T$$

The values of m, c and T are easily adjusted to obtain a quantising law with the required number of output levels. The value of T must be chosen such that slope overload effects are not visible. The values, of c and m then affect the levels of granular noise and edge busyness respectively.

The graphical design method has been found to give subjectively slightly better quantisers than the MMSE method. This is because the MMSE method tends to give a wide spacing to the outer levels of the quantiser which can give rise to visible distortions on some picture material.

Much work has been done to include subjective factors and subjective measurements into the design of non-linear quantisers [221-224]. However, it has been found that given system constraints such as the number of bits available per sample, the two simple design techniques which have been described give a subjective performance near the optimum.

Adaptive quantisation

Quantisation noise tends to be masked in active picture areas. The picture 'activity' can be estimated and the quantiser adapted accordingly by altering the spacing of the inner levels of the quantiser such that the resulting level of granular noise always remains below the threshold of perception. Adaptation thereby allows the range of the quantiser to be increased in active areas.

Various activity measures are possible. For example, Pirsch [222] used as a measure the maximum magnitude of the differences between neighbouring picture elements. The value of the activity measure determines which quantiser, from a small selection, is chosen to code the following sample. In order to avoid the need to transmit quantiser selection information to the decoder, the activity measure is based only on information which would also be available at the decoder.

The prediction error itself is also a measure of local picture activity and an alternative activity measure could be formed by taking a weighted sum of the magnitudes of previously transmitted (quantised) prediction errors. Schafer [224] has performed subjective tests of DPCM coding using two-dimensional prediction and adaptive quantisation with up to four different quantisers. His results show that there is definitely some advantage, in terms of picture quality, of using adaptive non-linear quantisation. There are disadvantages, however, some of which are listed below.

(i) At the decoder, in the presence of transmission errors, there may be error extension problems caused by the incorrect selection of quantiser mode.

(ii) The quantiser can adapt too slowly to certain picture detail giving perceptible slope overload effects.

(iii) There are some regular, high-frequency patterns which give a high activity measure but which do not mask quantisation noise. For example if granular noise is added uniformly to test pictures such as BBC Test Card F or Test Card G which contain frequency gratings at several frequencies, the noise appears to be as visible in the highest frequency grating as it is in the plain areas of the picture. However, in the mid-range frequency gratings the noise is completely masked by the picture detail.

Within the scope of this chapter it is only possible to mention these difficulties but not to discuss them further.

Stability of (non-adaptive) DPCM decoders and coders

Both DPCM decoders and DPCM coders can exhibit instability. However, the meaning and manifestation of instability is not the same for both coder and decoder.

A stable *decoder* has an impulse response which decays away with time. For an unstable decoder the instability is excited by errors in the transmission path between coder and decoder. The decoder loop stability may be determined by examining the frequency response, $P(\omega)$, of the predictor. The overall frequency response, $A(\omega)$, of a recursive decoder loop is given by

$$A(\omega) = 1/(1 - P(\omega))$$

and if the plot of $P(\omega)$ (in the complex plane) encloses the point (1, 0) then the loop will be unstable and the effect of a transmission error can remain or grow. The design procedure described earlier for non-adaptive, linear predictors tends to give stable decoders.

With stable systems it is not necessary to transmit periodic resetting information to ensure that the coder and the decoder remain in step. Any difference between coder and decoder decays away except perhaps for some small differences resulting from the practical requirement to truncate the predictor output to a limited accuracy before addition to the quantised difference signal (Figure 13.15). This truncation adds random truncation noise to the prediction values and can be considered as equivalent noise added to the transmitted (quantised) difference signal. At low frequencies, where $P(\omega)$ is close to unity, this low level noise is amplified by the response, $A(\omega)$, of the loop which can give rise to visible low frequency noise patterning in the decoder output. The amplitude of this patterning can be reduced by increasing the accuracy of the arithmetic around the decoding loop (at both decoder and coder) or by incorporating 'truncation error feedback'. In this latter technique the prediction truncation error is added back into the prediction value, for example on the next clock period. This has the effect of putting a null at d.c. in the spectrum of the truncation noise thereby greatly reducing its visibility.

Coder instability results from an interaction between non-linear quantiser and predictor and manifests itself as a so-called 'limit cycle oscillation' which consists of a pattern of quantisation error which is self sustaining even in plain picture areas. Pirsch [225] has analysed in detail this process which may be outlined as follows. Consider, for example, a particular limit cycle which has an amplitude q_1. In a plain area the prediction error could then be as much as $q_1 \Sigma |a_i|$ where a_i are the predictor coefficients. If, at prediction errors up to this value, the maximum quantisation error is equal to (or greater than) q_1 then it is possible for the limit cycle to be self sustaining. In other words limit cycling can occur if at any region of the quantiser

$$fq_1 \sum_i |a_i| \geqslant q_1$$

or

$$f \geqslant 1 \Big/ \left(\sum_i |a_i| \right)$$

where f, a function of the prediction error p_e, is given by

$$f(e) = \frac{|\text{maximum quantisation error at } p_e|}{|p_e|}$$

Very large amplitude limit cycling can sometimes arise when the number of coefficients in the predictor is large and the output range of the quantiser is small. In this case the value of f increases for large prediction errors such that the above inequality is satisfied.

It should be noted that the inequality is only a necessary condition for instability to occur. In many cases limit cycles may not appear even though the above condition is satisfied.

Results of non-adaptive DPCM coding

At the BBC Research Department experimental, non-adaptive DPCM equipment was constructed to investigate the quality of coding sub-Nyquist sampled PAL television signals for transmission within 34 Mbit/s. Very good quality coding was obtained using a 14-element two-field predictor and a 22-level quantiser (4.5 bits per sample). The predictor is illustrated in Figure 13.19 which gives the weighted contributions taken from sample points from lines in two fields. This prediction was derived using average picture statistics from stationary pictures and assuming a temporal autocorrelation function with a decay factor of 0.85 between fields. This factor was derived experimentally and was chosen to limit the weight given to previous-field information in order to avoid introducing excessive prediction errors on moving material. On stationary pictures and for slow movement, the prediction error, an example of which is shown in Figure 13.20a, was small such that the loss introduced by the non-linear quantiser was virtually imperceptible. It was only by subtracting the system output from its input and displaying the difference on a television monitor, as shown in Figure 13.20b, that it was possible to know where to look for any loss. As movement speed increased, quantising errors did appear on the loss picture similar in appearance to the errors displayed in Figure 13.20a, but on the actual decoded pictures the movement was such that it was not possible to perceive the loss of resolution of the moving transitions. In contrast, the use of a previous picture element predictor gave rise to severe movement blurring (slope overload) and this predictor would have required additional movement compensating techniques for high-quality coding.

The system proved very tolerant to transmission errors. For example, informal measurements showed that a bit error rate of 1 in 10^6 (with any error correction turned off) gave only a 'perceptible' impairment (Grade 4 on the CCIR 5-grade impairment scale).

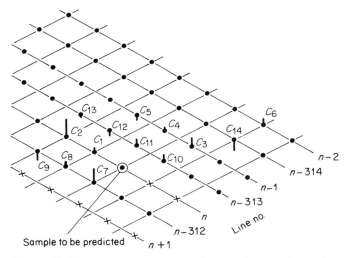

Figure 13.19 Isometric projection showing the sample weights of a 14-element predictor for composite PAL signals sampled at $2f_{sc}$. The weighting coefficients C_1 to C_{14} have the following values:

	Present field		Previous field
C_1	1/8	C_7	3/8
C_2	1/2	C_8	1/16
C_3	1/4	C_9	−1/4
C_4	1/16	C_{10}	3/16
C_5	−1/8	C_{11}	3/16
C_6	5/32	C_{12}	−5/32
		C_{13}	−3/32
		C_{14}	−5/16

These non-adaptive techniques have also proved effective in the coding of *YUV* signals, sampled with an orthogonal sampling structure according to CCIR Recommendation 601. Predictors which take contributions from samples within three fields and which are suitable for the luminance and chrominance components are described in Tables 13.1a and 13.1b respectively. In these tables a picture-based coordinate system is used as shown in Figure 13.21. Using such predictors, good quality coding can be obtained down to 4 bits per sample. At 3 bits per sample (or 8-level quantisation) granular noise and edge busyness become perceptible. If it is required to improve the quality further at this low number of bits per sample, further techniques, such as variable-length coding combined with adaptive prediction or quantisation, are required. Variable-length coding is discussed in a later section.

The stability of a DPCM system using multi-element prediction is demonstrated in Figure 13.22. This shows the impulse response of a luminance decoder using the prediction coefficients of Table 13.1a. The input to the decoder is zero except for a single impulse of 100 levels in the first field. It can be seen that the error decays away within a few frames.

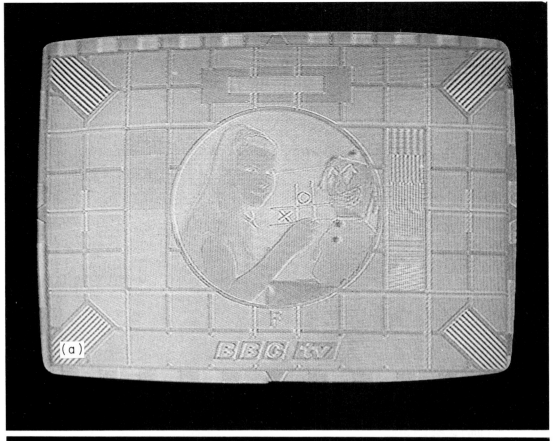

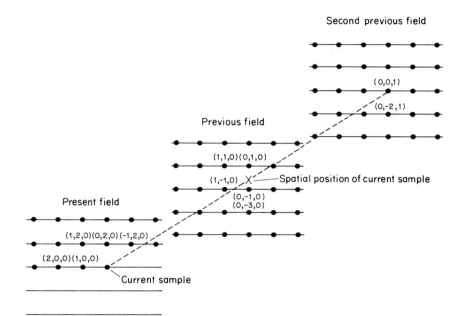

Figure 13.21 Predictor coefficient and sample numbering system

Table 13.1(a)
15-element luminance predictor

Predictor coefficients			
Coordinates			Value
i	j	k	
1	0	0	1.000
2	0	0	−0.437
3	0	0	0.235
0	2	0	0.327
1	2	0	−0.195
0	−1	0	0.201
1	−1	0	−0.158
0	1	0	0.172
1	1	0	−0.172
0	0	1	0.460
1	0	1	−0.497
2	0	1	0.277
3	0	1	−0.149
0	2	1	−0.291
1	2	1	0.198

Table 13.1(b)
7-element chrominance predictor

Predictor coefficients			
Coordinates			Value
i	j	k	
1	0	0	0.500
−1	2	0	0.194
0	2	0	0.272
0	0	1	0.655
1	0	1	−0.317
−1	2	1	−0.137
0	2	1	−0.191

(Note that the horizontal scale for the chrominance predictor coordinates is twice that of the luminance predictor coordinates.)

Figure 13.20 (a) Prediction error for 2-field, 14-element PAL predictor. (b) Coding loss for 14-element predictor and 4.5 bit quantiser obtained by subtracting the coded signal from the input. During moving signal sequences coding errors show up with an appearance similar to (a), but on the stationary test card they are hardly visible

Figure 13.22 Impulse (error) response of DPCM decoder using the multi-element predictor of Table 13.1(a). This illustration shows an error decaying away in a small portion of a picture. A sequence of nine successive frames is shown in the following order:

$$\begin{array}{ccc} 1, & 2, & 3 \\ 4, & 5, & 6 \\ 7, & 8, & 9 \end{array}$$

TRANSFORM CODING FOR BIT-RATE REDUCTION

Introduction

Transform coding is an increasingly popular technique for digital video bit-rate reduction. In this technique, a block of samples from the spatial domain is transformed into a second block of data consisting of a set of 'coefficients'. Each coefficient represents the amplitude of a particular pattern present within that block in the spatial domain. The original block of samples in the spatial domain can be recovered by performing the inverse transformation on the block of coefficients.

For typical blocks of picture data the amplitudes of many of the transform coefficients are small and can be approximated by zero or quantised using only a few numbers of bits per coefficient without introducing significant distortion when the coefficient block is transformed back to the spatial domain. Fewer bits overall are then required to transmit the picture block in the transformed domain, rather than in the spatial domain, for a given decoded picture quality.

Many different forms of transformation have been investigated for bit-rate reduction. The best transforms are those which tend to concentrate the energy of a picture block into a few coefficients. The discrete cosine transform (DCT) is one of the best transforms in this respect (see, for example, [226]. In addition, the DCT and its inverse can be carried out in a relatively computationally efficient manner [227]. Consequently at the present time, the DCT is the transform which is most widely used, or proposed for use, in transform-based bit-rate reduction systems.

In this section, therefore, the discussion will concentrate on the DCT in order to illustrate the sorts of processes that are used for bit-rate reduction. Firstly, a derivation of the cosine transform is given. This derivation is followed by a description of some of the different bit-rate-reduction methods that can be used on the transformed data.

The discrete cosine transform

Consider a signal, $f(x)$, with a Fourier transform, $F(\omega)$

$$F(\omega) = \int_{-\infty}^{\infty} f(x) \exp(-j\omega x) \, dx \qquad (13.4)$$

The inverse transformation recovers $f(x)$ from $F(\omega)$

$$f(x) = \frac{1}{2\pi} \int_{-\infty}^{\infty} F(\omega) \exp(j\omega x) \, d\omega \qquad (13.5)$$

Consider that a portion, $g(x)$ of length X, of the signal is selected as follows

$$g(x) = f(x)b(x)$$

where

$$\begin{aligned} b(x) &= 1 \quad 0 < x < X \\ &= 0 \quad x < 0 \text{ and } x > X \end{aligned} \qquad (13.6)$$

In order to obtain a transform containing only cosine terms a function $g_r(x)$ is first constructed by reflecting $g(x)$ about $x = 0$; i.e.

$$\begin{aligned} g_r(x) &= g(x) \quad x > 0 \\ &= g(-x) \quad x < 0 \end{aligned} \qquad (13.7)$$

Then, the Fourier transform $G_r(\omega)$ of $g_r(x)$ is given by

$$G_r(\omega) = \int_{-X}^{+X} g_r(x) \exp(-j\omega x) \, dx = 2 \int_0^X g(x) \cos(\omega x) \, dx \qquad (13.8)$$

If the original signal consisted of a sampled signal with N sampling intervals within the length X, then

$$g(x) = \sum_{k=0}^{N-1} s(k) \delta\left(x - \frac{kX}{N} - \tau\right) \qquad (13.9)$$

where τ represents the sampling phase. Most commonly for the cosine transform

$\tau = X/2N$. Then equations 13.8 and 13.9 give

$$G_r(\omega) = 2 \sum_{k=0}^{N-1} s(k) \cos\left(\frac{\omega X}{N}(k + \tfrac{1}{2})\right) \tag{13.10}$$

This last expression gives the value of $G_r(\omega)$ at all frequencies. In the discrete cosine transformation, ω takes only discrete values. The precise formulation of the transform can be considered to result from the following argument. Consider a periodic function, $r(x)$, generated by repeated shifting and adding of $g_r(x)$; i.e.

$$r(x) = \sum_{n=-\infty}^{\infty} g_r(x - n2X) \tag{13.11}$$

The Fourier transform, $R(\omega)$, of $r(x)$ can be obtained from $G_r(\omega)$ by using the Fourier 'shift' property that the transform of $g_r(x - n2X) = G_r(\omega)\exp(-j\omega n2X)$, i.e.

$$R(\omega) = G_r(\omega) \sum_{n=-\infty}^{\infty} \exp(-j\omega n2X) \tag{13.12}$$

The sum of exponentials in equation 13.12 has a non-zero value only at frequencies which are multiples of $1/2X$. At these multiples, the value of the sum is infinite. More precisely (see, for example, [206])

$$\sum_{n=-\infty}^{\infty} \exp(-j\omega n2X) = \frac{\pi}{X}\sum_{n=-\infty}^{\infty} \delta(\omega - n\omega_0) \quad \text{where } \omega_0 = \frac{2\pi}{2X}$$

i.e.

$$R(\omega) = \frac{\pi}{X} \sum_{n=-\infty}^{\infty} G_r(n\omega_0)\delta(\omega - n\omega_0) \tag{13.13}$$

Thus the spectrum of the repeating function, $r(x)$, consists of the spectrum, $G_r(\omega)$, sampled at intervals of the repeat frequency, ω_0. Alternatively, since $r(x)$ defined by equation 13.11 is a periodic function it can be represented by a traditional Fourier series expansion

$$r(x) = \sum_{n=-\infty}^{\infty} A(n) \exp(jn\omega_0 X) \quad \omega_0 = \frac{2\pi}{2X} \tag{13.14}$$

and

$$A(n) = \frac{1}{2X} \int_{-X}^{+X} g_r(x) \exp(-jn\omega_0 x)\,dx \tag{13.15}$$

Using equation 13.9 we obtain

$$A(n) = \frac{1}{X} \sum_{k=0}^{N-1} s(k) \cos\left(\frac{\pi}{N}n(k + \tfrac{1}{2})\right) \tag{13.16}$$

or

$$A'(n) = XA(n) = \sum_{k=0}^{N-1} s(k) \cos\left(\frac{\pi}{N}n(k + \tfrac{1}{2})\right)$$

Note that comparing equations 13.10 and 13.16 gives

$$A(n) = \frac{1}{2X} G_r(n\omega_0) \qquad \omega_0 = \frac{2\pi}{2X} \qquad (13.17)$$

The inverse of equation 13.16 is obtained by multiplying both sides by $\cos[(n/N)\pi(k + \frac{1}{2})]$ and summing over n, and by using the identity

$$\sum_{n=0}^{N-1} c(n) \cos\left(\frac{\pi}{N} n(k + \tfrac{1}{2})\right) \cos\left(\frac{\pi}{N} n(m + \tfrac{1}{2})\right) = 0 \quad k \neq m \qquad (13.18)$$

$$= \frac{N}{2} \quad k = m$$

where

$$c(n) = \tfrac{1}{2} \quad n = 0$$
$$= 1 \quad n \neq 0$$

This gives

$$s(k) = \frac{2}{N} \sum_{n=0}^{N-1} c(n) A'(n) \cos\left(\frac{\pi}{N} n(k + \tfrac{1}{2})\right) \qquad (13.19)$$

This last expression shows how the set of N sample values in the spatial domain can equally be represented by a sum of cosine terms of amplitude $(2/N)c(n)A'(n)$ where the $A'(n)$ are given by equation 13.16. Each cosine term is termed a DCT 'basis function'. The DCT coefficients $A'(n)$ are usually normalised such that the transform and its inverse appear more symmetrical in the following manner

$$S(n) = \left(\frac{2}{N}\right)^{1/2} C(n) A'(n)$$

where

$$C(n) = \frac{1}{\sqrt{2}} \quad n = 0$$
$$= 1 \quad n \neq 0$$

Then

$$S(n) = \left(\frac{2}{N}\right)^{1/2} C(n) \sum_{k=0}^{N-1} s(k) \cos\left(\frac{\pi}{N} n(k + \tfrac{1}{2})\right) \qquad (13.20)$$

and

$$s(k) = \left(\frac{2}{N}\right)^{1/2} \sum_{n=0}^{N-1} C(n) S(n) \cos\left(\frac{\pi}{N} n(k + \tfrac{1}{2})\right) \qquad (13.21)$$

With this normalisation, equal amplitude coefficients $S(n)$ correspond to equal power basis functions.

It is instructive to examine how the DCT, with only cosine basis functions, carries phase information in comparison with the discrete Fourier transform which has both cosine and sine basis functions. In the discrete Fourier transform one can consider that a length, X, of signal, $g(x)$, is repeated at intervals of length X. This periodic function can then be represented by a sum of cosine and sine terms with frequencies $\omega_n = n2\pi/X$ (compared with $n2\pi/2X$ for the DCT).

Considering then the example of a single Fourier component with phase ϕ

$$f(x) = \cos(\omega_n x + \phi)$$

giving

$$F(\omega) = \pi[\delta(\omega - \omega_n)e^{j\phi} + \delta(\omega + \omega_n)e^{-j\phi}]$$

Blocking the signal as in equation 13.6 gives for $g(x)$ and its Fourier transform $G(\omega)$

$$g(x) = f(x)b(x)$$

and

$$G(\omega) = F(\omega) * B(\omega)$$

$$= \pi[\delta(\omega - \omega_n)e^{j\phi} + \delta(\omega + \omega_n)e^{-j\phi}] * \left(\frac{X\sin(\omega X/2)}{\omega X/2}\exp(-j\omega X/2)\right)$$

$$= \pi X\left(e^{j\phi}\frac{\sin[(\omega - \omega_n)X/2]}{(\omega - \omega_n)X/2}\exp[-j(\omega - \omega_n)X/2]\right.$$

$$\left. + \frac{e^{-j\phi}\sin[(\omega + \omega_n)X/2]}{(\omega + \omega_n)X/2}\exp[-j(\omega + \omega_n)X/2]\right)$$

since

$$\delta(\omega - \omega_n) * F(\omega) = F(\omega - \omega_n)$$

As the Fourier transform of $g(-x)$ is $G^*(w)$, the spectrum $G_r(\omega)$ of the boxed and reflected waveform as required for the cosine transform (and as shown in Figure 13.23a) is given by

$$G_r(\omega) = 2\pi X\left(\frac{\sin[(\omega - \omega_n)X/2]}{(\omega - \omega_n)X/2}\cos[(\omega - \omega_n)X/2 - \phi]\right.$$

$$\left. + \frac{\sin[(\omega + \omega_n)X/2]}{(\omega + \omega_n)X/2}\cos[(\omega + \omega_n)X/2 + \phi]\right)$$

The components of $G_r(\omega)$ for $\omega > 0$ are illustrated in Figure 13.23b for the two cases where $\phi = 0$ and $\phi = -\pi/2$ respectively.

When the periodic function $r(x)$ is generated by repeated shifting and adding, as given by equation 13.11, the above spectrum is 'combed' or picked out only at frequencies given by $n\omega_0$ (with $\omega_0 = 2\pi/2X$ as in equation 13.17) to give values proportional to the DCT coefficients. With $\phi = 0$ the function $r(x)$ is continuous at $x = 0$ and $x = X$ and there is only one non-zero term in the combed spectrum (Figure 13.23c). With $\phi = -\pi/2$ the function $r(x)$ has a sharp discontinuity at $x = 0$ and $x = X$ giving several non-zero coefficients as shown in Figure 13.23d. These two cases correspond to cosine and sine basis functions of the discrete Fourier transform. Therefore, we can say that each Fourier *cosine* term is carried by a single even coefficient in the DCT and that each

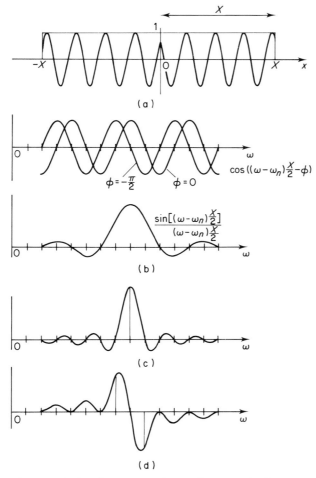

Figure 13.23 Illustration of DCT for a single frequency component. (a) Blocked and reflected waveform for $f_x = \cos(\omega_n x + \phi)$ (b) Component parts of $G_r(\omega)$ for $\omega > 0$, for $\phi = 0$ and $-\pi/2$ (c) DCT coefficients for $\phi = 0$ (Fourier cosine term) (d) DCT coefficient for $\phi = -\pi/2$ (Fourier sine term)

Fourier sine term is carried by a sequence of odd numbered coefficients as shown in Figure 13.23d.

Two-dimensional transform coding

The one-dimensional discrete cosine transform equations 13.20 and 13.21 can be extended straightforwardly to two dimensions. Consider a 2D block of sampled picture $s(j, k)$. Performing the discrete cosine transform separately for each row of the block gives a set of intermediate coefficients for each row. Each coefficient, corresponding to a given horizontal basis function, varies as a function of its vertical position or row number. A second discrete cosine transformation then be performed on each column of the block of intermediate coefficients. For a square block of N by N samples the resulting 2D transforms is given by

$$S(u, v) = \frac{2}{N} C(u)C(v) \sum_{j=0}^{N-1} \sum_{k=0}^{N-1} s(j, k) \cos\left(\frac{\pi}{N} u(j + \tfrac{1}{2})\right) \cos\left(\frac{\pi}{N} v(k + \tfrac{1}{2})\right) \quad (13.22)$$

and

$$s(j, k) = \frac{2}{N} \sum_{u=0}^{N-1} \sum_{v=0}^{N-1} C(u)C(v)S(u, v) \cos\left(\frac{\pi}{N} u(j + \tfrac{1}{2})\right) \cos\left(\frac{\pi}{N} v(k + \tfrac{1}{2})\right) \quad (13.23)$$

where

$$C(u) = \frac{1}{\sqrt{2}} \quad \text{for } u = 0$$
$$= 1 \quad \text{for } u \neq 0$$

This 2D transformation corresponds to the 2D Fourier transform of a block of 2N by 2N samples formed by reflecting the original sample block $s(j, k)$ about two axes as shown in Figure 13.24. The two-dimensional DCT basis functions are given by the product of a horizontally-varying cosine function and a vertically-varying cosine function sampled at points $j + \tfrac{1}{2}$ and $k + \tfrac{1}{2}$, i.e.

$$s_{uv}(j, k) = \frac{2}{N} C(u)C(v) \cos\left(\frac{\pi}{N} u(j + \tfrac{1}{2})\right) \cos\left(\frac{\pi}{N} v(k + \tfrac{1}{2})\right) \quad (13.24)$$

All such basis functions (for all combinations of u and v) defined by equation 13.24 have equal power.

The one- and two-dimensional DCTs are examples of 'orthogonal transformations' in which each individual basis function has zero component values (or zero coefficient

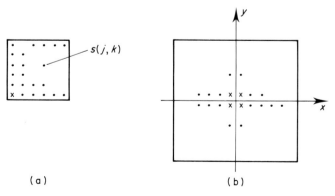

Figure 13.24 Illustration of two-dimensional blocking for DCT: (a) $N \times N$ block of picture samples with $s(0,0)$ marked by a cross. (b) $2N \times 2N$ sample block formed by reflections of original block

values) for all the other basis functions. This results from the fact that

$$\sum_{j=0}^{N-1}\sum_{k=0}^{N-1} s_{u,v}(j,k)s_{m,n}(j,k) = 0 \quad u \neq m \quad \text{or} \quad v \neq n$$

$$= 1 \quad u = m \text{ and } v = n \tag{13.25}$$

Bit-rate reduction in the transform domain

A block of N by N image samples in the spatial domain is transformed into a block N by N coefficients in the transform domain. Each coefficient represents the amplitude of a given basis function within the image data.

An example of how a picture is built up in terms of basis functions is given in Figure 13.25 for the DCT using 8 by 8 blocks. Each small picture has been coded by keeping, in the transform domain, only the largest amplitude coefficients in each block and putting

Figure 13.25 Pictures coded by retaining only the highest amplitude coefficients in an 8×8 DCT. The number of coefficients per block is as follows:

56,	48,	32
24,	16,	24
8,	4,	64 (original picture)

the remainder to zero. The number of coefficients retained per block is 56, 48, 32, 24, 16, 8 and 4 in the various subpictures. Figure 13.26 shows the respective differences between the original and the coded pictures. As the number of coefficients retained decreases, the pictures first become grainy as the small high frequency coefficients are discarded. When only a small proportion of the coefficients are retained in each block, then for some blocks comparatively large coefficients are discarded. For these blocks the individual basis functions become visible as can be seen in the blocks around some of the masts in the picture.

After transformation and before transmission, each coefficient is quantised and coded with as few bits as possible in order to minimise the average bit rate per block. Excess quantisation error in any coefficient can give rise to the visibility of the corresponding basis function within the decoded image block. In particular, comparatively small quantisation errors in the d.c. coefficient ($u = 0$, $v = 0$) can cause the block structure to become visible in the decoded picture.

For virtually all picture material the coefficients corresponding to the higher horizontal and vertical frequencies are consistently smaller and cover a smaller range than the lower frequency coefficients. Assuming that the coefficients require similar quantising accuracies then the number of bits required to code each coefficient will be proportional to the range covered by the given coefficient. Assuming further that the majority of

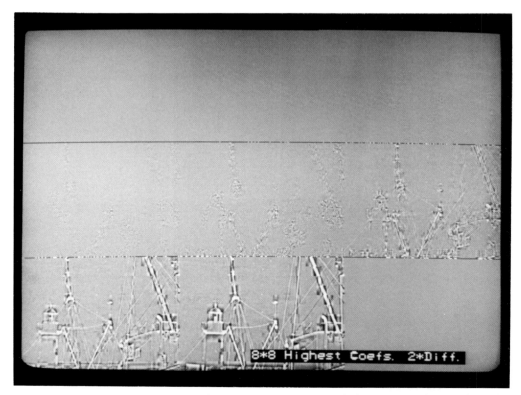

Figure 13.26 Picture showing the difference between the coded pictures and original for the sequence of pictures in Figure 13.25 (mid-grey equals zero difference)

coefficients have a similar form of amplitude probability distribution function then the coding range required for each coefficient will be proportional to the standard deviation of the amplitude probability distribution for that coefficient. Therefore, we can assign a number of bits, $n(u, v)$ to a coefficient (u, v) according to its standard deviation $\sigma(u, v)$, i.e.

$$n(u, v) = \log_2 [\sigma(u, v)] + \text{constant} \quad (13.26)$$

For example, the standard deviations of the distributions of DCT coefficient amplitudes were measured for the luminance components of several standard high-quality source pictures. The block size was 8 by 8 and the samples within each block were taken only from the lines of one field. The values that were obtained for the number of bits per coefficient according to equation 13.26 are given in Figure 13.27a. It can be seen that the number of bits per coefficient decreases with horizontal and vertical frequency. A practical bit assignment map that has been used in high-quality picture coding is given in Figure 13.27b.

Having assigned a given number of bits per coefficient, the average noise power in the decoded picture can be reduced by using a statistically optimum (minimum-mean-square) non-linear quantiser for each coefficient. However, care must be taken when using non-linear quantisers since they can produce occasional large quantising errors which are easily perceptible on the decoded picture.

Note that the bit assignment map of Figure 13.27b is for 8 × 8 field-based blocks. If, alternatively, the block had included samples from eight adjacent picture lines then any movement of the image between fields would give rise to larger amplitude high-frequency components, in particular components with high vertical frequency components.

	0	1	2	3	4	5	6	7
0	8.09	6.17	5.52	5.02	4.63	4.19	3.51	2.71
1	6.67	5.24	4.74	4.33	3.95	3.47	2.68	1.40
2	6.03	5.02	4.72	4.22	3.79	3.29	2.45	1.18
3	5.54	4.99	4.85	4.21	3.63	3.12	2.28	1.08
4	5.29	4.63	4.47	3.99	3.51	2.93	2.14	0.98
5	5.01	4.15	3.92	3.75	3.43	2.86	2.12	0.91
6	4.70	3.98	3.77	3.64	3.38	2.86	2.17	0.91
7	4.63	3.91	3.73	3.58	3.30	2.81	2.14	0.92

(a)

	1	2	3	4	5	6	7	8
1	9	6	5	5	4	4	3	2
2	6	5	4	4	3	3	2	1
3	6	5	4	4	3	3	2	1
4	5	4	4	4	3	3	2	1
5	5	4	4	3	3	2	2	1
6	4	4	3	3	3	2	2	0
7	4	3	3	3	3	2	2	0
8	4	3	3	3	3	2	2	0

(b)

Figure 13.27 Bit assignment maps for intra-field DCT. (a) $\log_2$ (standard deviation) for coefficient amplitudes. (b) Practical bit assignment map for an 8 × 8 intra-field DCT

Therefore, for picture-based blocks it is less straightforward to assign a bit map which allows significant bit-rate reduction and which takes into account a wide range of possible image movement.

The average bit rate per block for the bit map of Figure 13.27b is approximately 3.2 bits per coefficient. To reduce this average figure it is necessary to discard a significant proportion of the coefficients. Picture quality can only be maintained if this is done adaptively. For example, an average of 3/4 of the coefficients of an 8×8 block are zero after quantisation (for high-quality coding). These zero coefficients need not each be coded with the number of bits allocated by the bit assignment map. Many different methods have been proposed for signalling the positions of the zero coefficients. Some techniques signal 'zones' of zero coefficients and others run-length code strings of zero coefficients as the block is scanned for transmission. In order to maximise the length of the strings of zeros the block should be scanned in an order related to the standard deviations (or bit assignment) of the coefficient amplitudes. Typically, blocks are scanned diagonally.

This adaptivity gives a variable bit rate per block. Before transmission therefore, the variable-rate data is written into a buffer store and read from the store at a regular rate. Typically the buffer store would have a capacity of at least one field at the transmission data rate. In order to prevent the buffer store overflowing or emptying, a signal related to the buffer occupancy is fed-back periodically to control the level spacing of the coefficient quantisers. Increasing or decreasing the spacing of the quantiser levels increases or decreases the number of zero coefficients respectively which changes the average bit rate per block accordingly. Also, with variable bit-rate systems, extra signalling must be included to aid the recovery of the decoder after a transmission error. For example, some form of start-of-block code is necessary in order to prevent error extension from one block to the next.

An optimum system giving a variable bit rate would use variable-length or entropy coding to code the coefficients. Ideally, a separate code for each coefficient would be used, optimised for the amplitude distribution function of that coefficient. Also the quantising laws should be tailored such that the high-frequency coefficients are quantised more coarsely to take advantage of the eye's reduced sensitivity to high-frequency noise. Using such techniques, for example, the luminance component can be coded with almost imperceptible degradation at between 2.0 and 2.5 bits per sample.

The examples given in this section have concentrated on an 8×8 block size. Larger block sizes can give some improvement in performance but this is marginal considering the extra computational complexity involved.

Hybrid transform/DPCM coding

A technique which has been used for very low bit-rate coding and is now being applied at higher bit rates is interframe 'hybrid' transform/DPCM coding (also known as transform/predictive coding). Two different forms of the hybrid technique have been widely investigated. In the first approach, interframe DPCM coding is applied to the coefficients of a 2D intraframe transform (see, for example, [228]). In the second approach, 2D transform coding is applied to an interframe DPCM difference signal. Block diagrams for the two different sorts of coder are shown in Figures 13.28 and 13.29. One advantage of

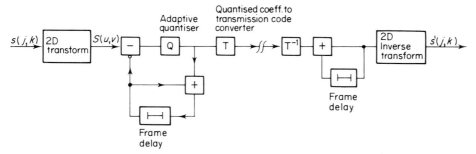

Figure 13.28 Block diagram of a basic hybrid transform-predictive coder–decoder

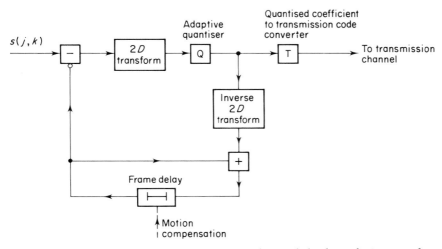

Figure 13.29 Block diagram showing principles at hybrid predictive–transform coding

the latter technique is that corrections for motion between frames can more easily be applied in the spatial domain. Very low bit-rate coding has been reported for these techniques for low-quality image applications such as teleconferencing and videophone.

VARIABLE-LENGTH OR ENTROPY CODING

Introduction

Variable-length coding is a bit-rate-reduction technique which can be used either alone or to augment the bit-rate reduction given by other techniques such as DPCM or transform coding. The technique exploits statistical redundancy in the signal or symbols to be transmitted where these symbols do not all occur with equal probability. The bit-rate saving that can be achieved depends on the symbol probability distribution. In many cases the technique can give a significant reduction in bit rate.

For example, suppose a DPCM coder produces a set of N levels, $[A_i]$, after non-linear quantisation. Some levels will occur more frequently than others. A variable-length coder assigns short code words, for transmission, to the most probable DPCM levels and longer codewords to the levels which occur less often. If L_i is the length of the codeword assigned to the level A_i, then the average bit rate, R, of the data signal for transmission is given by

$$R = \sum_{i=1}^{N} P_i L_i \qquad (13.27)$$

where P_i is the probability of occurrence of the level A_i.

The following section first describes how typical variable-length codes are constructed. A brief theoretical framework for variable-length coding is then given and some applications, in particular to DPCM coding, are described. Finally, some of the features and problems of a practical implementation of a variable-length coder and decoder are discussed.

Construction of typical variable-length codes

As a variable-length coder produces codewords which do not have a constant length, the codewords must be constructed such that the decoder, starting from the beginning of a new codeword, knows when it reaches the end of that word. This is ensured by arranging that each codeword is not equal to the first part, or prefix, of any longer codeword. Several procedures have been proposed for designing variable length codes, the most popular of which was proposed by Huffman [229]. This is described here, briefly, by reference to the example given in Figure 13.30. The eight symbols (or levels) to be transmitted occur with a set of probabilities, P_i, given in the first column of the figure. Step 1 involves adding together the two smallest probabilities to form a larger probability which is inserted into the list according to its size. The process is repeated until the set has only one member, whose value is 1 (since $\Sigma P_i = 1$).

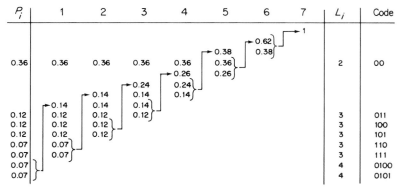

Figure 13.30 Table showing the derivation of a Hoffman code for an 8-symbol code

The length of the codeword for the symbol whose probability is x is given by the number of times that the probability is involved in a summation. By convention, if the probability is uppermost in the pair being added together a zero is assigned to that step. The code for the event with probability x is obtained by tracing out the path followed by x, working back from probability 1 to x, using the rules given above, giving the codewords in the final column of the figure. Note that the codes are uniquely decipherable. No codeword is a prefix of any other codeword.

The set of variable-length codes can be represented pictorially by a tree diagram, as shown in Figure 13.31. In this diagram the 'root' of the tree represents the start of every codeword and a terminal node represents the end of a codeword. Each codeword is represented by a path along the branches of the tree, an upward branch representing a 0 and a downward branch a 1. In the figure the probability of travelling along a particular branch is indicated. Intermediate nodes represent incomplete codewords. It can be clearly seen from this type of diagram that many variations in the arrangement of codewords are possible.

Entropy

The entropy of a signal can be used as a guide to the bit-rate savings that can be achieved through the use of variable-length coding. The entropy of a signal is equivalent to the average amount of 'information' in the signal. The 'information' content carried by a symbol or codeword, i, which occurs with probability P_i is *defined* by

$$\text{information content of } i\text{th codeword} = -\log_2 P_i$$

i.e. more information is carried by a codeword which is unexpected and occurs rarely than by a codeword which occurs more frequently. The average information content per

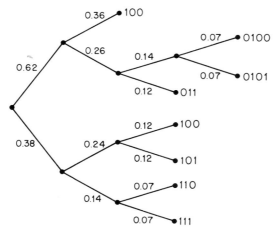

Figure 13.31 Tree diagram for the 8-symbol variable-length code example of Figure 13.29

codeword from a source is the entropy, H, of the signal and, in terms of bits per symbol, H is given by

$$H = -\sum_i P_i \log_2 (P_i) \qquad (13.28)$$

Shannon [230] showed that the above entropy figure equals the minimum number of bits per symbol into which the source can be coded assuming that the samples are independent. A variable-length Huffman code gives an average codeword length which approaches fairly closely this entropy figure. Therefore, the entropy of a signal gives a good, although slightly optimistic, guide to any gains that can be achieved through the use of variable-length coding.

Consider, as an example, the application of variable-length coding to the quantised difference signals produced by a DPCM coder as shown in Figure 13.32. If q_i is the probability of occurrence of the ith quantiser output, Q_i, then the entropy, H_O, of the quantised difference signal is given by

$$H_O = -\sum_i q_i \log_2 (q_i)$$

Each probability, q_i, can be given in terms of the probabilities, P_j, of the prediction-error signal at the input to the quantiser, i.e.

$$q_i = \sum_{\text{all } j \to i} P_j$$

where the summation is over all input levels that are quantised to the ith output level. Assuming that P_j varies relatively slowly and that ∂_i is the number of input levels within the ith partition, then $q_i = \partial_i P_i$ where P_i now represents an average P_j for the ith partition. Thus

$$H_O = -\sum_i P_i \partial_i \log_2 (P_i \partial_i) \qquad (13.29)$$

If the quantiser, Q, is linear then $\partial = \partial_i$ for all i, and equation 13.29 gives

$$H_O = -\partial \sum_i P_i \log_2 (P_i) - \log_2 (\partial)$$

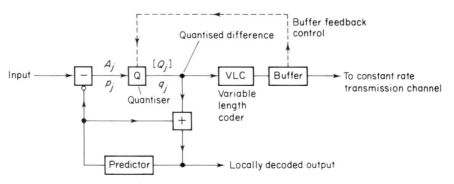

Figure 13.32 DPCM coder followed by variable-length coding

As

$$\partial \sum_{i\,\text{output levels}} P_i \log_2 (P_i) = \sum_{j\,\text{input levels}} P_j \log (P_j)$$

$$= \text{entropy of prediction error signal } H_s$$

$$H_O = H_s - \log_2 \partial \qquad (13.30)$$

Equation 13.30 shows how the entropy of the (unquantised) prediction error signal, H_s, is changed by linear quantisation. For example, increasing the spacing of the output levels by a factor of two reduces the entropy by 1 bit per sample.

For a non-linear quantiser with many levels it is possible to describe the quantiser approximately as consisting of different regions with uniform quantisation within each region. If the separation of output levels in region R_1 is ∂_1, and in region R_2 is ∂_2, etc, then it is fairly straightforward to show that the entropy of the quantised signal H_O is given by

$$H_O = H_s - I_1 \log_2 \partial_1 - I_2 \log_2 \partial_2 - \cdots - I_k \log_2 \partial_k \qquad (13.31)$$

where

$$I_k = \sum_{j\,\in\,\text{region }k} P_j$$

For typical predictors and prediction error statistics, the particular I_k corresponding to the region around zero prediction error has the largest value, which tends to mean that the entropy of the transmitted signal is dependent mainly on the source entropy and the spacing of the inner levels of the non-linear quantiser.

With a DPCM system which includes entropy coding as in Figure 13.32, we wish to minimise the noise in the decoded signal for a given average transmitted bit rate. If the mean square error is used to measure the noise, it has been found that the quantiser which is optimum and which minimises the noise is linear [231]. A non-linear quantiser formed, for example, by removing some of the levels of this linear quantiser would give a smaller entropy but at the expense of an increase in average noise level. Most commonly, the spacings of a DPCM quantiser are designed to be as coarse as possible whilst keeping the impairments of granular noise, edge busyness and slope overload below the threshold of perception. In these circumstances, if N levels are required in the non-linear quantiser, variable-length coding can give, typically, a saving of between 0.7 and 1.0 bits per sample on the $\log_2 N$ bits per sample required in the absence of variable-length coding.

In the above example, the application of variable-length coding to the quantised DPCM difference signal has been described. Variable-length coding could equally well be applied to the transmission of coefficients in a transform-based bit-rate-reduction system. In order to obtain optimum performance from a transform-based system, each quantised coefficient would have a variable-length code optimised for the amplitude distribution of that coefficient.

Practical considerations

In most transmission systems it is necessary to have a constant bit rate entering the transmission channel. Therefore, with variable-length coding, a buffer is required to smooth out the irregular data rate occurring per sample. Data is written into the buffer at a variable rate and read from the buffer at a constant rate for transmission. The average input and output rates must of course be equal.

In order to prevent the buffer overflowing or emptying on particular picture material, a signal describing the buffer occupancy is used to control the parameters of the source codes. Typically, the parameter that is varied, in DPCM and transform coders, is the spacing of the output levels in the prediction-error or coefficient quantisers. Increasing the spacing of the levels decreases the entropy of the signal and vice-versa. It is usual to have several quantiser modes in order to try to avoid sharp changes in picture quality when switching between modes.

In the limit, for very difficult pictures, there must be a 'fall-back' mode with a coarse quantiser which is guaranteed not to cause the buffer to overflow. Similarly, there must be a strategy for avoiding buffer underflow for plain picture material.

The buffer size which is required depends on several factors. These include (a) whether the coding scheme is intra- or inter-field, (b) how often the buffer occupancy signal is fed back to the source coder and (c) the maximum and minimum source bit rates from the coder. For example, intra-field coders only benefit from comparatively small buffers. In order to estimate the buffer size required, suppose that a proportion, x, of the field is very active and gives rise to codewords of the maximum length possible with a practical code, say s bits per sample. Also, suppose that the remainder of the field contains plain, inactive picture material and gives rise to codewords of the minimum length, r bits per sample. If the transmission channel removes samples at t bits per (input) sample and assuming that the picture is stable for a few fields, then there is no need for the buffer to be longer than the amount by which it can be emptied during the inactive picture area and filled during the active picture area. If N is the number of input samples per field, the maximum required buffer capacity, B_m, is given by

$$B_m = N(1 - x)(t - r) \text{ bits}$$

Since

$$(s - t)x = (t - r)(1 - x)$$

$$B_m = \frac{N(s - t)(t - r)}{(s - r)} \text{ bits}$$

$$= \frac{(s - t)(t - r)}{(s - r)t} \text{ fields at the transmitted bit rate.}$$

Taking a typical example with $r = 1$, $s = 16$, $t = 2.5$, then $B_m = 0.54$ fields *at the transmitted bit rate.*

A buffer store is also required at the decoder. In this buffer the data is written into the store at a constant rate from the transmission channel and read out from the store at a variable rate by the variable-length-codeword decoder. It is easy to show that, for a given point in a picture, the coder buffer occupancy and the decoder buffer occupancy are complementary. If $B_c(\mathbf{r}, n)$ is the coder buffer occupancy at the time when the data corresponding to the point $\mathbf{r}$ in frame number n is being written into the coder buffer and $B_d(\mathbf{r}, n)$ is the decoder buffer occupancy at some later time when the same data is being read from the decoder buffer then

$$\frac{d}{dt}B_c(\mathbf{r}, n) = -\frac{d}{dt}B_d(\mathbf{r}, n)$$

i.e.

$$B_c(\mathbf{r}, n) + B_d(\mathbf{r}, n) = \text{constant} \qquad (13.32)$$

Figure 13.33 gives an example of how the coder and decoder buffer states change with time. The coder read address and the decoder write address vary at a constant rate determined by the transmission channel. The coder write address and the decoder read address vary at a rate dependent on the picture content and state of the coder. The times T_c and T_d for which data corresponding to a given sample is in the coder and decoder buffers are given by

$$T_c = \frac{B_c(\mathbf{r}, n)}{t} \quad \text{and} \quad T_d = \frac{B_d(\mathbf{r}, n)}{t}$$

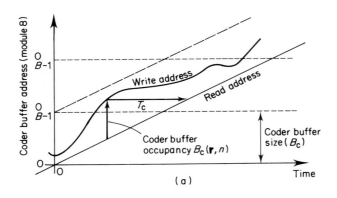

(a)

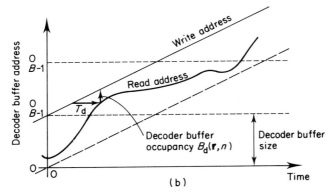

(b)

Figure 13.33 Diagrams illustrating buffer states in a variable-length-coding system. (a) Coder buffer, (b) decoder buffer

where t is the transmission rate in bits per sample. The total delay, T, through the system is a constant given by

$$T = T_c + T_d = \frac{1}{t}[B_c(\mathbf{r}, n) + B_d(\mathbf{r}, n)] \qquad (13.33)$$

In order that the decoder and coder buffers can be synchronised it is necessary to send periodic information to the decoder about the expected decoder buffer occupancy. This is known at the coder from equation 13.32.

In the presence of transmission errors, variable-length coding schemes give rise to error extension. If a codeword is received in error it may be interpreted as a codeword of a different length and therefore the next codeword will also be misinterpreted and so on, and the decoder could completely lose synchronisation. In general the variable-length decoder will eventually resynchronise and correct interpretation of the transmitted bit stream will be resumed. It is desirable to limit the effect of synchronisation loss as much as possible. The average time for resynchronisation depends on the construction of the particular variable-length code but it is possible to design codes with comparatively rapid error recovery properties [232].

In practical systems the decoder buffer is usually made larger than the coder buffer. This is because transmission errors cause the decoder buffer occupancy to deviate from what is expected according to equation 13.32. A larger decoder buffer allows the decoder buffer occupancy to vary over a wider tolerance before data is lost. Loss of data would complicate and inhibit any buffer resetting strategy. In addition to buffer resetting, strategies must be included to limit the impairments introduced by the variable-length-code extension. Typically, this might involve additional signalling to help locate, after transmission errors, the start of new lines or blocks. Error concealment techniques can then be used.

VECTOR QUANTISATION

The DPCM and transform coding systems described in the previous sections use what is often referred to as 'scalar quantisation'. This means that the amplitude of the prediction error at each pixel (or the amplitude of each coefficient) is quantised independently of the amplitudes of the prediction errors at neighbouring pixels (or the amplitudes of neighbouring coefficients).

In 'vector quantisation' a block of samples is quantised collectively. For example, consider the relatively simple case of a block of two luminance samples. If the two-sample combination is described as shown in Figure 13.34, where the amplitudes of the two samples are drawn along two perpendicular axes, then any pattern of two samples can be represented by a vector in the manner shown. Also, suppose that only one bit per sample was available for transmission, giving two bits per block-of-two-samples. Four different states or patterns can then be assigned to the block. These four representative states would be chosen according to the expected distribution of input vectors for blocks in the image to be coded. The process of coding an image at one-bit-per-sample then consists of comparing each two-sample block with each of the four representative states and choosing the state which gives the minimum error of distortion. Figure 13.35 illustrates

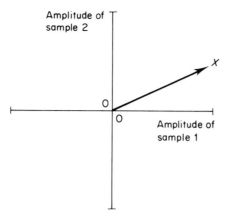

Figure 13.34 Two-dimensional representation of a block of two samples

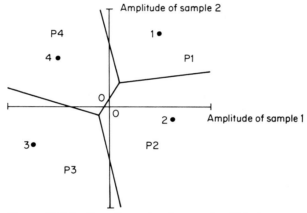

Figure 13.35 Two-dimensional example of block quantisation (all input blocks falling within region P1 are quantised as vector 1, etc)

this 'block quantisation' on the vector diagram. All input vectors within the partitions P1 to P4 are represented by the states 1 to 4 respectively. At the decoder one of only four representative states is reproduced corresponding to the codeword received from the coder. The set of representative vectors is referred to as a 'vector quantiser', 'block quantiser' or 'codebook'.

A block of two samples leads to a two-dimensional representation. Similarly, a k-sample block leads to a representation in which a k-dimensional vector corresponds to a particular pattern in the k-sample block. Following the example of Figure 13.34, the mth axis in k-dimensions describes the set of patterns in which all the k samples are zero (midgrey) except for the mth sample. The axes are said to be orthogonal because the mth coordinate value (or the mth sample amplitude) can be varied without altering the value of the other $k - 1$ coordinates (or sample values). Other axes could easily be chosen to describe the coordinates of the vector in k-dimensions. For example, any 'orthogonal' transformation such as the DCT represents a reorientation (e.g. rotation) of the k orthogonal axes.

The task of vector quantisation involves lengthy computations. Firstly, a set of representative vectors must be chosen to fit the block size, the average number of bits per sample and the expected distribution of image source vectors. Secondly, the coder must determine which of the representative vectors approximates most closely any particular input vector according to some distortion measure. Finally, the codeword of the appropriate codebook vector is sent to the decoder. Normally the codebook of vectors is predetermined. The task of the decoder is comparatively simple and is to reconstruct a pattern from the codebook corresponding to the codeword received.

The positioning of the codebook vectors depends on the distribution of input vectors in the images to be coded and on the choice of distortion measure. The images or parts of an image used to develop the vector quantiser are referred to as the 'training sequence'. The following section describes briefly a technique described by Linde, Buzo and Gray [233] for designing an optimal vector quantiser.

Codebook generation

For a given size of codebook, an optimum quantiser minimises the average distortion in the quantised signal measured over the training sequence. Many distortion measures have been proposed [233] but the most commonly used and most tractable is the mean-squared-error measure. Then, for a distribution of vectors, $\mathbf{x}_i$, within a specified partition, P_i, the codebook vector, y_i, which minimises the quantisation distortion for that partition is given by the 'centre of gravity' or 'centroid' of the distribution within P_i, i.e.

$$\mathbf{y}_i = \frac{1}{N_p} \sum_{\mathbf{x}_j \in P_i} \mathbf{x}_j \qquad (13.34)$$

where N_p is the number of unquantised vectors within P_i.

The technique for codebook generation described by Linde, Buzo and Gray is an iterative process which is illustrated here assuming a mean-square-error distortion measure. The process, for an N-vector quantiser, has the following stages:

(1) Firstly, an initial codebook or set of N vectors is assumed. This might for example be N points spread uniformly throughout a k-dimensional volume.

(2) The k-space is then partitioned (as illustrated in Figure 13.35) such that any input vector will be quantised to its nearest-neighbour codebook vector.

(3) Using the training sequence, the centroid of the vectors falling within each partition is calculated. The centroid minimises the distortion for the given partition and is taken as the new codebook vector for that partition. After this stage we have a modified codebook.

(4) Steps two and three are then repeated and so on.

On each iteration of the above process, the partitioning will change and each new partition will contain a slightly different selection of input vectors. The process converges to a quantiser giving a minimum average distortion for the training sequence. At each stage of the iteration the average distortion D_m is calculated and the process is stopped when the fractional change in distortion is small, i.e.

$$(D_{m-1} - D_m)/D_m \leqslant \epsilon$$

A variation of the above technique which is often used to generate M-level quantisers where $M = 2^R, R = 0, 1, 2, \ldots$, is as follows:

(1) The centroid, y, of the training sequence is taken and split into two close vectors $\mathbf{y} + \boldsymbol{\delta}$ and $\mathbf{y} - \boldsymbol{\delta}$.

(2) The training sequence is then partitioned between the two vectors and the previously described optimising procedure is followed to derive a two-level optimum quantiser.

(3) Each of the two vectors is split again and the training set partitioned into four and optimised to give an optimum four-level quantiser.

(4) The procedure is repeated until the distortion has been reduced to the required level or the number of codebook vectors has reached the required number.

Generation of a codebook is obviously a lengthy process particularly if a long training sequence is used. It is also a lengthy task for the coder to assign an input vector to its nearest-neighbour codebook vector. The following processes are required for each vector in the codebook:

(1) Subtraction of the codebook vector from the input block vector on a sample-by-sample basis.

(2) Summation of the squares of the sample-differences over the block.

(3) Finding the codebook vector that gives the minimum distortion.

In order to reduce the number of operations involved, tree-search routines have been proposed. In these routines an input vector is first compared with two vectors each representing half the codebook. The appropriate half of the codebook is then halved again and the comparison repeated, etc. Use of tree-search routines reduces the search time and ends giving a nearby codebook vector although it does not necessarily end giving the nearest-neighbour vector. The performance of the tree-search technique is not as good therefore as a full-search routine.

Vector quantiser performance

Because of the computational difficulty in codebook generation, the block sizes, in studies described in the literature, have been small containing typically less than 16 samples. Also, because of the heavy computational load involved in quantisation, the number of vectors in typical codebooks has been limited to give average bit rates of between 0.5 and 2 bits per sample. At these bit rates vector quantisation appears to perform only slightly better than other techniques such as the DCT [234]. However, the study of vector quantisation (VQ) is comparatively recent and many variations and new results are being reported in the literature. Developments include adaptive VQ, differential VQ, and colour VQ, in which the three *YUV* components are treated together as one vector [235].

At the time of writing VQ is not a practicable technique for real-time high-quality coding because of the high computational requirements. However, the development of sophisticated image-processing VLSI is progressing rapidly and VQ may prove to be a powerful bit-rate-reduction technique for future high-quality systems.

References

[1] Reeves, A. H. (1939) *British Patent* 535 860.
[2] Raimey, P. M. (1926) *US Patent* 1608 527.
[3] Cattermole, K. W. (1969) *Principles of Pulse Code Modulation* Iliffe, London.
[4] EBU Bulletin 15th January 1951.
[5] Davies, R. E., Edwardson, S. M. and Harvey, R. V. (1971) Electronic field-store standards converter. *Proc. IEE*, **118**, 460–468.
[6] Sandbank, C. P. (1980) *Optical Fibre Communication Systems* Wiley, New York.
[7] Reid, D. F., Stott, J. H. and Osborne, D. W. (1977) *Experimental digital transmission of multiplexed video and audio signals at 60 Mbit/s through a satellite* BBC Research Department Report RD 1977/32.
[8] Storey, R. (1980) Teletrack *BKSTS J.*, **62**, No 4.
[9] *Encoding parameters of digital television for studios* CCIR Recommendation 601-1 XVIth Plenary Assembly Dubrovnik 1986, Vol XI, Part 1, pp 319–328.
[10] *Interfaces for digital component video signals in 525-line and 625-line television systems* Recommendation 656, XVIth Plenary Assembly Dubrovnik 1986, Vol XI, Part 1, pp 346–358.
[11] *Digital television tape recording* CCIR Recommendation 657 XVIth Plenary Assembly Dubrovnik 1986, Vol X/XI, Part 3, pp 32–93.
[12] *Source encoding for digital sound signals in broadcast studios* CCIR Recommendation 646 XVIth Plenary Assembly Dubrovnik 1986, Vol X, Part 1, p 309.
[13] *A digital audio interface for broadcast studios* CCIR Recommendation 647 XVIth Plenary Assembly Dubrovnik 1986, Vol X, Part 1, pp 311–320.
[14] Fujio, T. (1980) High definition wide screen television systems for the future *IEEE Trans. Broadcasting*, **BC26**, 113–124.
[15] Sandbank, C. P. and Moffat, M. E. B. (1983) High-definition television and compatibility with existing standards *SMPTE J.*, **92**, 552–561.
[16] Sandbank, C. P. and Childs, I. (1985) The evolution towards high definition television *Proc. IEE*, **73**, 638–645.
[17] EBU Studies on High Definition Television *EBU Review (Technical)* no 219, October 1986.
[18] Windram, M. D., Tonge, G. and Morcom, R. (1982) *MAC - A television System for High Quality Satellite Broadcasting* IBA Report 118/82.
[19] EBU Specifications of the systems of the MAC/packet family Tech 3258-E October 1986.
[20] Sandbank, C. P., Childs, I., Storey, R. (1988) *British Patent* 2 197 561.
[21] Storey, R. (1986) *HDTV Motion adaptive Bandwidth Reduction using DATV* BBC Research Department Report RD 1986/5.
[22] Thomas, G. A. (1987) HDTV Bandwidth Reduction by Adaptive Subsampling and Motion-Compensated DATV Techniques *SMPTE J.*, **96**, 460–465.

REFERENCES

[23] Sandbank, C. P. and Stone, M. A. (1987) *The role of DATV in Future Television Emission and Reception* Proceedings Third Colloquium on Advanced Television Systems: Ottawa 1987, Vol. 1.

[24] *IEEE standard for performance measurements of A/D and D/A converters for PCM television video circuits* IEEE Std. 746–1984.

[25] *Dynamic performance testing of A to D converters* Hewlett Packard Product Note 5180A–2.

[26] Gardner, K. and Story, M. (1982) A test technique for high speed sampling systems *Electron Engng*, **54**, 44–51.

[27] Wilson, N. (1987) Testing flash A/D Converters *Electron, Engng*, **59**, 131–139.

[28] Goodall, W. M. (1951) Television by pulse code modulation *Bell Syst. Tech. J.*, **30**, 33–49.

[29] Carbrey, R. L. (1960) Video transmission over telephone cable pairs by pulse code modulation *Proc. IRE*, **48**, 1546–1561.

[30] Edson, J. O. and Henning, H. H. (1965) Broadband codecs for an experimental 224 Mb/s terminal *Bell Syst. Tech. J.*, **44**, 1887–1940.

[31] Devereux, V. G. (1968) *Experimental equipment for converting television signals into digital signals and vice-versa* BBC Research Department Report RD 1968/6.

[32] Devereux, V. G. (1970) *Pulse code modulation of video signals: 8-bit coder and decoder* BBC Research Department Report RD 1970/25.

[33] Fletcher, R. E. (1974) *A video analogue to digital converter* International Broadcasting Convention, IEE Conference 119, pp 47–57.

[34] Teesdale, R. R. and Weston, J. R. (1967) The encoding of broadband signals *Systems Commun.*, **3**, 18–22, 28–30.

[35] Johannesen, F. G. (1970) Successive approximation and feedback PCM encoder for television *Proc. IEE*, **117**, 671–680.

[36] Hanke, G. (1969). A PCM system using integrated circuits for television transmissions *NTZ*, **22**, 621–627.

[37] Goodman, D. J. (1969) The application of delta modulation to analog-to-PCM encoding *Bell Syst. Tech. J.*, **48**, 321–343.

[38] Kovanic, E. F. (1964) A high accuracy 9-bit digital-to-analog converter operating at 12 Mc *IEEE Trans. Commun.*, **71**, 185–191.

[39] Takemoto, T., Inove, M., Sadamatsu, H., Matsuzawa, A. and Tsuji, K. (1982) A fully parallel 10-bit A/D converter with video speed *IEEE J. Solid State Circuits*, **SC17**, 1133–1138.

[40] Yoshji, Y., Asano, K., Nakamura, M. and Yamada, C. (1984) An 8-bit, 100 Ms/s flash ADC *IEEE J. Solid State Circuits*, **SC19**, 275–279.

[41] Muto, A. S., Peetz, B. E. and Rehner, R. C. (1982) *Designing a ten-bit, twenty mega-sample/sec analog-to-digital converter system* Hewlett-Packard Journal, November, pp 9–20.

[42] Fleming, T. (1986) Analog/digital and digital/analog data converters *EDN*, **31**, 102–124.

[43] Blaesner, W. (1989) Analogue-to-digital and digital-to-analogue converters for video signal processing *Electronik*, **38**, 94–98 (in German).

[44] van de Grift, R., Rutten, I. W. J. M. and van de Veen, M. (1987) An 8-bit video ADC incorporating folding and interpolation techniques *IEEE J. Solid State Circuits*, **SC22**, 944–953.

[45] Roberts, L. G. (1962) Picture coding using pseudo-random noise *IRE Trans. Inf. Theory*, **IT-8**, 145–154.

[46] Limb, J. O. (1969) Design of dither waveforms of quantised visual signals *Bell Syst. Tech. J.*, **48**, 2555–2582.

[47] Devereux, V. G. (1974) Application of PCM to broadcast quality video signals *Radio Electron Eng.*, **44**, 373–381, 463–472.

[48] Allnat, J. W. and Prosser, R. D. (1966) Subjective quality of colour-television pictures impaired by random noise *Proc. IEE*, **113**, 551–557.

[49] Devereux, V. G. (1983) *Performance of cascaded video PCM codecs* EBU Review-Technical, No. 199, pp 114–131.
[50] Moore, T. A. (1974) *Digital video: number of bits per sample required for reference coding of the luminance and colour-difference signals* BBC Research Department Report RD 1974/42.
[51] Devereux, V. G. and Stott, J. H. (1978) Digital video: sub-Nyquist sampling of PAL colour signals *Proc. IEE*, **125**, 779–786.
[52] Rossi, J. P. (1976) Sub-Nyquist encoded PCM NTSC colour television *SMPTE J.*, **85**, 1–6.
[53] Salter, M. T. and Felix, M. O. (1976) Improved techniques for measuring phase and gain in digitised signals *Television J. Television Soc.*, **16**, (6) 7–10.
[54] Watson, D. (1985) At video bandwidths, flash A-D converters dictate stringent design *Electron Des. (USA)*, **33**, 143–148.
[55] Brockman, D. and Williams, A. (1985). Ground rules for high-speed circuits *Monitor (Australian)* May 1985, 9–10.
[56] Ahlquist *et al*, (1976) A 16384-bit Dynamic RAM *IEEE J. Solid State Circuits*, **SC11**, 570–574.
[57] May, T. C. and Woods, M. H. (1979) Alpha-particle-induced soft errors in Dynamic Memories *IEEE Trans. Electron Devices*, **ED26**, 2–9.
[58] Chan, J. Y. *et al*, (1980) A 100 ns 5 V only 64K × 1 MOS Dynamic Ram *IEEE J. Solid State Circuits*, **SC15**, 839-46.
[59] Shimonigashi, K. *et al*, (1982) An N-well CMOS dynamic RAM *IEEE Trans. Electron Devices*, **ED29**, 714–718.
[60] Yoshimoto, M. *et al.* (1983) *A 64Kb full CMOS RAM with divided word-line structure.* ISSCC Digest of Technical Papers, Feb. 1983, pp 58–59.
[61] Chwong, R. *et al*, (1983) *A 70 ns high-density CMOS DRAM* ISSCC Digest of Technical Papers, Feb. 1983, pp 56–57.
[62] Baba, F. *et al*, (1983) *A 35 ns 64K static column DRAM* ISSCC Digest of Technical Papers, Feb. 1983, pp 64–65.
[63] Walker, R. and McNally, G. W. W. (1977) *An experimental digital picture store* BBC Research Department Report RD 1977/9.
[64] Astle, J. M. and Shelton, W. T. (1980) *Digital Television Standards Converter with Random Access Memory storing Four Fields* IBC 1980, IEE Conf. Publication No. 191, pp 159–163.
[65] Riley, J. L. (1983) *Enhanced UK Teletext: experimental equipment for high-quality picture coding and other enhancements* BBC Research Department Report RD 1983/7.
[66] Croll, M. G. (1980) *A digital storage system for an electronic rostrum colour camera* IBC 1980, IEE Conf. Publication 191.
[67] Wolfe, M. J. and Kirby, D. G. (1982) *A Video Rostrum Camera and an Advanced Stills Store for Television Graphics* IBC 1982, IEE Conf. Publication 222.
[68] Storey, R. (1985) Electronic detection and concealment of film dirt *SMPTE J.*, **94**, 642–48.
[69] Devereux, V. G. (1971) *Pulse code modulation of video signals: subjective study of coding parameters* BBC Research Department Report RD 1971/40.
[70] Walker, R. (1971) *Digital line-store standards conversion: a feasibility study* BBC Research Department Report RD 1971/44.
[71] Devereux, V. G. (1973) *Digital video: differential coding of PAL signals based on differences between samples one subcarrier period apart.* BBC Research Department Report RD 1973/7.
[72] Chambers, J. P. (1972) *British Patent* 1415 519.
[73] Chambers, J. P. (1974) *The Use of Digital Techniques in Television Waveform Generation* IBC74, IEE Conf. Publication 119, pp 40–46.
[74] Devereux, V. G. (1975) *Digital video: differential coding of PAL colour signals using same line and two-dimensional prediction* BBC Research Department Report RD 1975/20.
[75] Devereux, V. G. (1975) *Digital video: sub-Nyquist sampling of PAL colour signals* BBC Research Department Report RD 1975/4.

REFERENCES

[76] Weston, M. (1976) *A PAL/YUV digital system for 625-line international connections* BBC Research Department Report RD 1976/24.

[77] Oliphant, A. (1980) *Weston Clean PAL* BBC Research Department Report RD 1980/1.

[78] Baldwin, J. L. E. and Lever, I. R. (1978) *Proposed digital television standards for 625 line PAL signals* IBC78, IEE Conf. Publication 166, pp 214–217.

[79] Drewery, J. O. (1984) *The present status of Clean PAL systems* BBC Research Department Report RD 1984/6.

[80] Baldwin, J. L. E. (1976) *Sampling frequencies for digital coding of television signals* IBA Technical Review, Vol. 9, pp 32–35.

[81] Clarke, C. K. P. (1982) *Digital standards conversion: comparison of colour decoding methods* BBC Research Department Report RD 1982/6.

[82] Clarke, C. K. P. (1979) *British Patent 2059 711.*

[83] Clarke, C. K. P. (1980) *British Patent 2061 053.*

[84] Jones, A. H. (1980) *Digital Television Standards* IBC 80 IEE Conf. Publication 191, pp 79–82.

[85] *Characteristics of television signals* CCIR report 624–3, XVIth Plenary Assembly Dubrovnik 1986, Vol XI, Part 1, pp 1–33.

[86] Chambers, J. P. (1972) *British Patent 1455 821.*

[87] Devereux, V. G. (1984) *Filtering of the colour-difference signals in 4:2:2 YUV digital video coding systems* BBC Research Department Report RD 1984/4.

[88] *Specification of television standards for 625-line System I transmissions in the United Kingdom* Radio Regulatory Division, Department of Trade and Industry, London, 2nd Impression, March 1985.

[89] Drewery, J.O. (1980) *British Patent 2073 535.*

[90] Weston, M. (1980) *A set of time varying television test patterns* BBC Research Department Report RD 1980/9.

[91] Macdiarmid, I. F. (1959) Waveform distortion in television links *The Post Office Electrical Engineer's J.* **52**, 108–114.

[92] Clarke, C. K. P. (1977) *Digital television: the use of waveform estimates in error correction* BBC Research Department Report RD 1977/27.

[93] Dorward, R. M. (1970) Aspect of the quantisation noise associated with the digital coding of colour television signals *Electron. Lett*, **6**, 5–7.

[94] Devereux, V. G. (1982) *Tests on eight video p.c.m. codecs in tandem handling composite PAL and monochrome video signals* BBC Research Department Report RD 1982/19.

[95] Gardner, F. M. (1967) *Phase Lock Techniques*, 2nd edn, Wiley, New York.

[96] Auty, S. J., Read, D. C. and Roe, G. D. (1977) Colour picture improvement using simple analogue comb filters *BBC Engineering*, **108**, 28–33.

[97] Clarke, C. K. P. (1978) *British Patent 2018 087.*

[98] Drewery, J. O. (1975) *The filtering of luminance and chrominance signals to avoid cross-colour in a PAL colour system* BBC Research Department Report RD 1975/36. (Also in *BBC Engineering*, 1976, **104**, 8–39).

[99] Lent, S. J. and Reed, C. R. G. (1974) *Reduction of cross-colour effects in PAL (System I) colour television by the use of low-pass and notch filters: subjective tests* BBC Research Department Report RD 1974/25.

[100] Clarke, C. K. P. (1982) *High Quality Decoding for PAL Inputs to Digital YUV Studios* IBC82, IEE Conf. Publication 220, pp 363–366.

[101] Rabiner, L. R. and Gold, B. (1975) *Theory and Application of Digital Signal Processing* Prentice Hall, New York.

[102] Dudgeon, D. E. and Mersereau, R. M. (1984) *Multidimensional Digital Signal Processing* Prentice Hall, New York.

[103] Mertz, P. and Gray, F. (1934) A theory of scanning and its relation to the characteristics of the transmitted signal in telephotography and television. *Bell Syst. Tech. J.*, **13**, 464–515.

[104] Oppenheim, A. V. and Lim, J. S. (1981) The importance of phase in signals *Proc. IEE*, **69**, 529–541.

[105] Gold, B. and Rader, C. M. (1969) *Digital Processing of Signals*, McGraw Hill, New York.

[106] Kaiser, J. F. (1974) *Nonrecursive digital filter design using the I_0-sinh window function* Proc. 1974 IEEE Int. symp. on circuits and systems Apr. 22–25, pp 20–23.

[107] McClellan, J. H., Parks, T. W. and Rabiner, L. R. (1973) A computer program for designing optimum FIR linear phase digital filters *IEEE Trans. Audio and Electroacoustics*, **AU-21**, 506–526.

[108] Speake, T. C. and Mersereau, R. M. (1981) A note on the use of windows for two-dimensional FIR filter design *IEEE Trans. Acoustics, Speech and Signal Processing*, **ASSP 29**, 125.

[109] Mersereau, R. M. and Mecklenbrauker, W. F. G. (1976) McClellan transformations for two-dimensional digital filtering: I-design *IEEE Trans. Circuits and Systems*, **CAS 23**, 405–414.

[110] Mecklenbrauker, W. F. G. and Mersereau, R. M. (1976) McClellan transformations for two-dimensional digital filtering: II-implementation *IEEE Trans. Circuits and Systems*, **CAS 23**, 414–422.

[111] McClellan, J. H. and Chan, D. S. K. (1977) A two-dimensional FIR filter structure derived from the Chebyshev recursion *IEEE Trans. Circuits and Systems*, **CAS 24**, 372–378.

[112] Weston, M. (1978) *Digital aperture correction* BBC Research Department Report RD 1978/10.

[113] Drewery, J. O. (1973) *A study of high order vertical aperture correction* BBC Research Department Report RD 1973/36.

[114] Oliphant, A. (1978) *A digital telecine processing channel* BBC Research Department Report RD 1978/9.

[115] Drewery, J. O., Storey, R. and Tanton, N. E. (1984) *Video noise reduction* BBC Research Department Report RD 1984/7.

[116] Lucas, K. and Windram, M. (1981) *Direct television broadcasting by satellite. Desirability of a new transmission standard* IBA Experimental and Development Report 116/81.

[117] Golding, L. S. and Garlow, R. K. (1971) Frequency interleaved sampling of a color television signal *IEEE Trans. Commun. Technol.*, **COM 19**, 972–979.

[118] Devereux, V. G. and Phillips, G. J. (1974) *Bit-rate reduction of digital video signals using differential p.c.m techniques* International Broadcasting Convention IEE Conf. Report 119, pp 83–89.

[119] Drewery, J. O. (1986) *A compatible improved PAL system* EBU Review Technical No 215, Feb. 1986.

[120] Commandini, P. 1977 Signal processing in the Image Transform System *SMPTE J.* **86**, 547–549. (See also *British Patent* 1402 609).

[121] Drewery, J. O., Chew, J. R. and Le Couteur, G. M. (1972) *Digital line-store standards conversion: preliminary interpolation study* BBC Research Department Report RD 1972/28.

[122] Le Couteur, G. M. (1973) *Digital line store standards conversion: determination of the optimum interpolation aperture function* BBC Research Department Report RD 1973/23.

[123] Panter, P. F. (1965) *Modulation, Noise and Spectral Analysis* McGraw-Hill, New York, p 513.

[124] Clarke, C. K. P. (1980) *British Patent* 2073 536.

[125] Clarke, C. K. P. (1982) *British Patent* 2126 450.

[126] Dalton, C. J. and Roe, G. D. (1979) *British Patent* 2013 067.

[127] Clarke, C. K. P. and Roe, G. D. (1978) *Developments in Standards Conversion* IBC 1978, IEE Conf. Publication 166, pp 202–205.

[128] Edwardson, S. M. et al, (1970) *Field store standards conversion: description of the CO6/506 converter* BBC Research Department Report RD 1970/37.

[129] Baldwin, J. L. E. et al, (1974) DICE: The first intercontinental digital standards converter *R. Television Soc. J.*, **15**, No 5.

[130] Baldwin, J. L. E. *et al*, (1976) *Digital video processing—DICE* IBA Technical Review 8, Sept. 1976.

[131] Roberts, A. (1983) *The improved display of 625-line television pictures* BBC Research Department Report RD 1983/8.

[132] Evens, R. C., Pexton, G. L. and Smith, T. B. (1980) *Computer Graphics in BBC Television. International Broadcasting Convention, Sept. 1980* IEE Conf. Publication 191, p 67.

[133] Lewill, J. (1981) Computer paint systems *BKSTS J.*, **63**, 746–8.

[134] Wood, J. H. (1983) Ten years of character generator development. Where Now? *BKSTS J.*, **65**, 84.

[135] Croll, M. G. (1985) *UK Patent Application* GB 2151431A.

[136] Bailey, M. E. (1977) *New block codes for digital tape recording* BBC Research Department Report RD 1977/33.

[137] Baldwin, J. L. E. (1979) *Codes for digital video tape recording at 10 Mbit/sq inch. International Conference on Video and Data Recording, Southampton, 1979* IERE Conf. Proceedings 43, pp 147–61.

[138] Mallinson, J. C. and Miller, J. W. (1977) Optimal codes for digital magnetic recordings *The Radio and Electronic Engineer*, **47**, 172–6.

[139] Ratliff, P. A. (1974) *Digital video: subjective assessment of an experimental Wyner-Ash error corrector* BBC Research Department Report RD 1974/41.

[140] Peterson, W. W. and Weldon, E. J. (1972) *Error Correcting Codes* MIT Press, Cambridge, MA.

[141] Lin, S. (1970) *An Introduction to Error Correcting Codes* MIT Press, Cambridge, MA.

[142] Runge, P. K. (1976) Phase locked loops with Signal Injection for Increased Pull-In Range and Reduced Output Phase Jitter *IEEE Trans. Commun.*, **COM 24**, 636–644.

[143] Velzel, C. H. F. (1978) Laser beam reading of video records *Appl. Optics*, **17**, 2029–2036.

[144] Bouwhuis, G. and Bratt, J. M. (1978) Video disk player optics, *Appl. Optics*, **17**, 1993–2000.

[145] Mason, B. R. (1984) *British Patent* 8432 365.

[146] Bellis, F. A. (1976) *An experimental digital television recorder* BBC Research Department Report RD 1976/7.

[147] Baldwin, J. L. E. (1981) A format for digital television tape recording. Television Technology of the 80's *SMPTE J.*, **90**, 24–33.

[148] Todorović, A. (1983) Digital video tape recorder *Digital Television Techniques* ASBU, Tunis.

[149] Todorović, A. (1985) *Background to the EBU standard for magnetic recording of digital component video signals* EBU Review-Technical No 213.

[150] (1985) *Standard for recording digital television signals on magnetic tape in cassettes* EBU Publication Technical No 3252.

[151] Chan, C. and Eguchi, T. (1987) Product Implementation of the 4:2:2 Digital Format *SMPTE J.*, **96**, 949.

[152] Dolby, D., Lemoine, M. and Felix, M. O. (1981) Formats for digital videotape recording *Ampex Horizons* No 5, pp 2–7.

[153] Habermann, W. (1983) *Progress in the development of the future digital video recording format* EBU Review-Technical No 198, pp 62–71.

[154] Heitmann, J. K. R. (1982) An analytical approach to the standardisation of videotape recorders *SMPTE J.*, **91**, 229–232.

[155] Enigberg, E. *et al*, (1987) The Composite Digital Format and its Applications *SMPTE J.*, **96**, 934.

[156] Zworykin, V. K. and Morton, G. A. (1954) *Television* Wiley, New York, Chap 10.

[157] Rainger, P. (1967) A camera tube head amplifier, *R.T.S.J.*, **11**, 230–231.

[158] Jones, A. H. (1965) *Use of a linear matrix to modify the colour analysis characteristics of a colour camera* BBC Research Department Technological Report RD T-157.

[159] Barbe, D. F. (1975) Imaging devices using the charge-coupled concept *Proc. IEE*, **63**, 38–66.
[160] Le Couteur, G. M. (1976) *Solid state image sensors: improvements in signal processing techniques* BBC Research Department Report RD 1976/4.
[161] Marsden, R. P. (1975) *Electronic masking for telecine: a review of masking for positive and negative film* BBC Research Department Report RD 1975/16.
[162] Nuttall, T. C. (1952) The development of a high quality 35 mm film scanner *Proc. IEE*, **99**, 136.
[163] (1976) The Rank Cintel Mark III flying spot dual gauge multistandard telecine *Television J. R. Television Soc*, **16**, 24–27.
[164] Millward, J. D. (1978) *The application of digital video techniques to telecine* Proceedings of the 1978 International Broadcasting Convention, IEE Conf. Publication 166, pp 227–230.
[165] Weston, M. (1975) *Pulse code modulation of video signals: visibility of level quantising effects in processing channels* BBC Research Department Report RD 1975/31.
[166] Wood, C. B. B. (1977) The evolution of the telecine. Verbal presentation of a paper delivered at the 1977 International Television Symposium at Montreux.
[167] Le Couteur, G. M. and Childs, I. (1982) *UK Patent* 2007 935 B.
[168] Childs, I. and Griffiths, M. J. (1982) *Novel uses of digital processing in a modern telecine* Proceedings of the 1982 International Broadcasting Convention, IEE Conf. Publication 220, pp 46–50.
[169] Wood, C. B. B., Sanders, J. R. and Griffiths, F. A. (1965) Electronic compensation for color film processing errors *SMPTE J.*, **74**, 755–759.
[170] Kitson, D. J. M., Sanders, J. R., Spencer, R. H. and Wright, D. T. (1974) Preprogrammed and automatic color correction for telecine *SMPTE J.*, **83**, 633–639.
[171] Marsden, R. P. (1978) An improved automatic color corrector for telecine *SMPTE J.*, **87**, 73–76.
[172] Kitson, D. J. *et al*, (1972) The pre-programming of film scanner controls *EBU Review* no 134.
[173] Ryan, J. O. (1980) *Recent advances in broadcast camera design* Proceedings of the 1980 International Broadcasting Convention, pp 27–30.
[174] Murakami, K., Wakui, K., Oonishi, K. and Ukigaya, F. (l980) *Automated set-up system for high sensitive handy camera* Proceedings of the 1980 International Broadcasting Convention, pp 31–33.
[175] Le Couteur, G. M. (1977) *The electronic concealment of blemishes on the output of solid state image sensors* BBC Research Department Report RD 1977/28.
[176] Kosonocky, W. F. *et al*, (1974) Control of blooming in charge-coupled imagers. *RCA Rev.*, **35**, 3–24.
[177] (1985) *Frame-transfer, X/Y and interline image sensors: how do they compare?* Philips Technical Publication No 170, (available from Philips International Business Relations, PO Box 218, 5600MD, Eindhoven, Netherlands).
[178] Horii, K. *et al*, (1987) 502(V) * 600(H) FIT-CCD image sensor *J. Inst. Television Eng Japan*, **41**, 1039–1046.
[179] Le Couteur, G. M. (1978) *Sampling frequencies and structures for solid state image sensing in broadcasting* BBC Research Department Report RD 1978/2.
[180] (1969) *UK Patent* 1308 962.
[181] Oliphant, A. (1982) *Sampling structures for solid-state area sensors* Proceedings of the 1982 International Broadcasting Convention, pp 29–32.
[182] Takemura, Y. and Ooi, K. (1982) New frequency interleaving ccd color television camera *IEEE Trans. Consumer Electron.*, **CE-28**, 618–623.
[183] Onishi, M., Ide, T., Kitamura, and Fujita, Y. (1982) New IC family for a single-tube color video camera *IEEE Trans. Consumer Electron.*, **CE-28**, 519–525.

[184] Mitchell, A. J. (1980) A personal history of video effects in the BBC *Int. Broadcast Eng.*, **173**, 6–16.
[185] Roe, G. D. (1980) Chroma-key development 1965–1980 *Int. Broadcast Eng.*, **173**, 19–20.
[186] Rawlings, R. (1980) Chroma-key in a digital system. *Int. Broadcast Eng.*, **173**, 30–33.
[187] Hughes, D. (1980) Ultimate Video Travelling Matte. *Int. Broadcast Eng.*, **173**, 22–25.
[188] Lucas, K. and Nasse, D. (1981) *Case studies in the implementation of chroma-key* EBU Review-Technical, 187, pp 135–141.
[189] Kirby, D. G. (1988) *Television animation store* BBC Research Department Report RD 1988/4.
[190] Oliphant, A. (1982) *Colour separation overlay and its relation to digital video sampling* BBC Research Department Report RD 1982/7.
[191] Croll, M. G., Devereux, V. G. and Weston, M. (1987) *Accommodating the residue of processed or computed digital video signals within the 8-bit CCIR Recommendation 601* BBC Research Department Report RD 1987/12.
[192] Croll, M. G., Devereux, V. G. and Weston, M. (1988). *Using the 8 bit CCIR Recommendation 601 Digital Interface International Broadcasting Convention, 1988*, IEE Conference Publication 293, pp 268–271.
[193] Stickler, M. J. (1984) *The EBU parallel interface for 625-line digital video signals* EBU Review 205, 102–110.
[194] Bradshaw, D., Nasse, D. and Stickler, M. J. (1985) *The EBU bit-serial interface for 625-line digital video signals.* EBU Review 212, August 1985, 181–187.
[195] Bajes, R., Grunaldi, J., Oyaux, J. and Vallais, J. (1984) Serial Interface within the digital studio *SMPTE. J.*, **93**, 1044.
[196] Bradshaw, D. J. and Mason, B. R. (1987) *Digital Electronic Caption Preparation Area* IEE Colloquium 'TV Studios-from A/D', Conference Digest 1987/11.
[197] Auty, S. J. and Bradshaw, D. J.; *Routing and Distribution of Digital Video Signals to the 4:2:2 Standard; IBC84*, IEE Conference Publication 240, pp 53–57.
[198] Oudin, M. (1987) The World's first All-digital Television Production *SMPTE J.*, **96**, 11.
[199] Netravali, A. N. and Limb, J. O. (1980). Picture coding: A review *Proc. IEE*, **68**, No 3, March 1980.
[200] Dubois, E., Prasada, B., and Sabri, M. S. (1981) in *Image Sequence Analysis* Vol 5, edited by T. S. Huang, Springer, Berlin, Chap 3.
[201] Musmann, H. G., Pirsch, P. and Grallert, H. J. (1985) Advances in picture coding *Proc. IEE*, **73**, 532–548.
[202] Stott, J. H. and Ratliff, P. A. (1983) *Digital television transmission: 34 Mbit/s PAL investigation* BBC Research Department Report RD 1983/9.
[203] Stott, J. H. (1984) *Digital transmission: 68 Mbit/s PAL field trial system* BBC Research Department Report RD 1984/5.
[204] Lewis, A. R., Osborne, D. W., Gooch, P. V. and Kearsey, B. N. (1984) Transmission of broadcast services over a 140 Mbit/s digital bearer: A joint BBC/BT pilot scheme *The Radio and Electron. Eng.*, **54**, 445–456.
[205] Wells, N. D. (1984) *Digital Video: YUV bit rate reduction for broadcasting applications* Proceedings of International Broadcasting Convention, Brighton, UK, IEE, London.
[206] Papoulis, A. (1977) *Signal analysis* McGraw-Hill, New York.
[207] Clarke, C. K. P. and Tanton, N. E. (1984) *Digital standards conversion: Interpolation theory and aperture synthesis* BBC Research Department Report RD 1984/20.
[208] Pearson, D. E. (1975) *Transmission and display of pictorial information* Pentech Press, London.
[209] Tonge, G. J. (1981) *The sampling of television images* IBA Experiment and Development Report 112/81.
[210] Wells, N. D. (1985) *Digital video: Assessment of comb-filter design for sub-Nyquist PAL codecs* BBC Research Department Report RD 1985/18.

[211] Makhoul, J. (1975) Linear prediction: A tutorial review *Proc. IEE*, **63**, 561–580.
[212] Graham, R. E. (1958) Predictive quantising of television signals *IRE Wescon Conv. Rec.*, **2**, 147–157.
[213] Zschunke, W. (1977) DPCM picture coding with adaptive prediction *IEEE Trans. Commun.* **COM25**, 1295–1302.
[214] Dukhovich, I. and O'Neal, J. B. (1978) A three-dimensional spatial non-linear predictor for television *IEEE Trans. Commun.* **COM26**, 578–583.
[215] Sabri, S. (1983) *Image sequence processing and dynamic scene analysis* edited by T.S. Huang, Springer, Berlin.
[216] Buley, H. and Stenger, L. (1985) Inter/intraframe coding of colour TV signals for transmissions at the third level of the digital hierarchy *Proc. IEE*, **73**, 765–72
[217] Knee, M. J. and Wells, N. D. (1987) *Comparison of DPCM prediction strategies for high-quality digital television bit-rate reduction* BBC Research Department Report RD 1989/8.
[218] Max, J. (1960) Quantising for minimum distortion *Trans. IRE*, **IT6**, 7–12.
[219] Panter, P. F. and Dite, W. (1951) Quantisation distortion in pulse-count modulation with nonuniform spacing of levels *Proc. IRE*, **39**, 44–48.
[220] Brainard, R. C., Netravali, A. N. and Pearson, D. E. (1980) Predictive coding of composite NTSC color television signals *SMPTE J.*, **91**, 245–252.
[221] Netravali, A. N. and Rubenstein, C. B. (1977) Quantisation of color signals *Proc. IEE*, **65**, 1177–1187.
[222] Pirsch, P. (1981) Design of DPCM quantisers for video signals using subjective tests *IEEE Trans. Commun.* **COM29**, 990–1000.
[223] Lukas, F. and Kretz, F. (1983) DPCM quantisation of color television signals *IEEE Trans. Commun.*, **COM31**, 927–932.
[224] Schafer, R. (1983) Design of adaptive and non-adaptive quantisers using subjective criteria *Signal Processing*, **5**, 333–343.
[225] Pirsch, P. (1982) Stability conditions for DPCM codecs *IEEE Trans. Commun.*, **COM30**, 1174–1184.
[226] Pratt, W. K. (1978) *Digital image processing* Wiley, Chichester.
[227] Kamangar, F. A. and Rao, K. R. (1982) Fast algorithms for the 2-D Discrete Cosine Transform *IEEE Trans. Comput.* **C31**, 889–906.
[228] Roese, J. A. (1979) *Image Transmission Techniques* edited by W. K. Pratt, Academic, New York.
[229] Huffman, D. A. (1951) A method for the construction of minimum redundancy codes *Proc. IRE*, **40**, 1098–1101.
[230] Shannon, C. E. (1948) A mathematical theory of communication *BSTJ*, **XXVII**, 379–423.
[231] O'Neal, Jr, J. B. (1976) Differential pulse code modulation (PCM) with entropy coding *IEEE Trans. Inf. Theory*, **IT22**, 169–174.
[232] Knee, M. J. (1987) *Designing variable-length codes with good error-recovery properties* BBC Research Department Report RD 1987/19.
[233] Linde, Y., Buzo, A. and Gray, R. M. (1980) An algorithm for vector quantiser design *IEEE Trans. Commun.* **COM28**, 84–95.
[234] Baker, R. L. and Gray, R. M. (1983) *Image compression using non-adaptive spatial vector quantisation* Conference Record of the 16th Asilomar Conference on circuit systems and computers, IEE, London.
[235] Gray, R. M. (1984) Vector quantisation *IEE ASSP Magazine*, April 1984.

Index

acceptance angle 542
access ports 94
access time 76, 79
ACE converter 362, 363
Active Play 412
adaptive decoding 208–9
adaptive interpolation 395
address down time 113
address strobe
  column 68
  independent 79
  row 68
addressing, multiplexed 79–81
afterglow correction 508–9
aliasing 41, 142, 146, 174, 175, 192, 193, 205, 206, 294, 302, 315, 344, 374
  alias components 22, 25, 43, 44, 292, 293, 335, 339, 346
  animated logo implementation 419
  anti-aliased graphics 419
  anti-aliasing 473
  as knotting 333
  beat patterns 43
  chroma-key and 552–3
  electronic graphics 465, 467, 471, 473
    causes of 474–80
    horizontal aliasing 476
    interlace twitter 474, 475
    movement judder 475, 476–9
    movement portrayal 476–9
    picture frequency flicker 474, 475
    stepping or jaggies 474, 475–6, 477
    stroke width modulation 475, 476–9
    temporal aliasing 478, 491
    vertical aliasing 476
  horizontal 476
  in cameras 536–7

judder and 356, 475, 476–9
  temporal 320, 372, 478, 491
  vertical 333, 334, 336–8, 350, 357, 476
alternate line switching 141
amplitude modulation 153
analogue-to-digital conversion 21–64
  analogue pre-filtering 22, 25, 43–5, 61
  circuit layout techniques 61–3
  filter specification 61
  fundamental principles 22–3
  horizontal interpolation in 134
  low-pass filters 22, 24, 43–5
  performance measurement
    linearity measurements 57–9
    signal-to-quantising-noise ratios 52–7
  range changing ADC 509, 516
  required performance 50–2, 59–61, 63–4
  subjective tests on codecs 48–50
  subjective tests on single codec 40–2
  techniques 30–6
  telecines and 509–18
  tests on codecs in tandem 42–50
animation of logos 375, 412–21
  Active Play and Long Play 412
  data recovery 414–18
  electron photomicrographs 414, 415, 417
  LaserVision 412–18
  pitshapes 414, 415
  pseudo-random data 414
  solid state implementation of logo 418–21

    aliasing 419
    end of line 420
    pixel bytes 420
    pixel mode 420
    run-length 418, 420
    spatial frequency response 417
  see also electronic graphics
animation store 375, 396–411
  buffers 106–9
  CSO key generator 399
  operation of store 400–2
    'Load' 400
    'Run Level 1' 400, 401
    'Run Level 2' 400
  recording on parallel transfer disc drive 402–11
    bit rate reduction 406–7
    block length 408
    data format 409–10
    error feedback 406–7
    error protection 407–9
    performance 410–11
    record channel code 402, 406
    redundancy 406–7, 408
  system description 398–400
    colour-difference signals 398
    CSO 398
    cylinder of tracks 399
    fades and cross-fades 399
    field blanking 399
    frame accurate synchronisation 399
    interpolator unit 399
    mixing 398
    television rostrum camera 397
  see also chroma-key
anti-aliasing 419, 473
antisymmetry, centre of 217
aperture 296
  fixed point aperture 298–300, 344, 345

aperture (cont.)
  interpolation aperture function 296, 297, 298–300, 307, 308, 311–13, 323, 332, 335
  loss, scanning and 468, 469–71
  optimisation 350–62
  quantisation 301–4, 346–9
  synthesis 298–300, 341–9
aperture correction 245–51
  cameras 501, 530
  display aperture 245
  error 29
  field pairing 251
  film motion 251
  horizontal 247
  jitter 29
  judder 251
  out-of-band 246
  source aperture distortion 245
  telecine 508, 513–15
  temporal 245, 250
  vertical 247, 250, 515
area averaging 483, 484–5
artwork see electronic graphics
aspect ratio 14
  conversion 372
Aston IV 497
audio signals
  audio interface specification 14
  digital audio 13–14
  digital standards 13–14
  DVTR see digital video tape recording
  sampling frequency 14
  telecine information transmission 507

bandwidth reduction 1, 4, 16, 18, 584, 591
'Battleground' 496
BBC analogue converters 335, 338–9
BBC experimental DVTR 425–7
beam coding tube 31
beat patterns 41, 395, 510, 513
  alias 43
  coarse, in transition region 257
  in coloured areas 38
binary-metric-modular racking system see BMM racking system
bit cells 402
bit rate reduction 16, 406–7
  DPCM see DPCM coding

sample rate reduction see sample rate reduction
transform coding see transform coding
see also long distance transmission
bit-parallel interface see interfaces
bits per sample 35, 40–2, 50–1
black-level clamping 41, 45–7, 160
blanking and picture movement block 109
block quantisation 628
BMM racking system 96, 104, 110
brightness see luminance
broadband NTSC 156–7, 180
bubble memory 66
buffers 625–8
  animation store buffers 106–9
  asynchronous buffer 99
  bidirectional buffers 392
  buffer occupancy 626, 628
  multipicture storage 381

cables 566–7
  equalisation 565–6
  RF interference 567
cameras 499
  aperture correction 501, 530
  charge-coupled device 500, 501, 534
  control systems 530–1
    differential width control 531
    zoom angle and 531
  digital processing channel 529–30
  digital techniques in 529–37
  display gamma 501
  frame transfer structure 534
  frame-interline transfer device 535
  gamma correction 501, 530
  hand-held 499
  integration 320, 356, 367, 468
  interline transfer structure 534
  optical pre-filtering 535
  painting 530
  panning 356, 358
  registration errors 530–1
  shading errors 530–1
  sidebands of sensor sampling frequency 535–6
  solid state area arrays 531–7
    dark current 533

integration time 533
optical overload 533–4
studio 499
television rostrum camera 397
types of 500–1
vacuum tubes 500
CAS (column address strobe) 68
CCD see charge-coupled device (CCD)
CCIR Recommendation 500 49
CCIR Recommendation 601 42, 51, 52, 61, 433, 442, 559, 560, 607
CCIR Recommendation 646 449
CCIR Recommendation 647 449
CCIR Recommendation 656 560
CCIR Recommendation 657 433
central limit theorem 265
centre of antisymmetry 217
centre of symmetry 217
channel coding, DTVR 455
charge-coupled device (CCD) 66
  64K 96–8
  cameras 500, 501, 534
  storage 75
Chebyshev polynomials 238
chroma-key 106, 539–57
  acceptance angle 542
  aliasing distortion 552–3
  control adjustment 556–7
  digital processing problems 552–6
  downstream operations 540
  error feedback 553–6
  filtering of key signals 546–8
  foreground suppressor 542–4
  foreground–background cross-fade 551–2
  fringe suppression 544
  garbage matte 550–1
  hue angle 540
  key colour suppression 542–4
  key gain 544
  key generator 540–2
  key lift 544
  key processor 544–5
  key signal 542
  keying from 12:4:4 signals 549–50
  linear key processing 539
  luminance keying 550
  luminance suppression 543–4
  mixer 546
  non-additive mixer 548–9, 551

other key facilities 548–52
principles of operation 540–8
quantising distortion 553–6
sample rate changing and filtering 546–8
shadows 544–5
stored key signals 551
suppression angle 544
chrominance 9, 108, 109
  demultiplex 18
  PAL chrominance modulation 142, 143
  resolution 9
  sample rate reduction 589–92
  vertical filtering 251–4
  see also colour coding
chrominance demodulation 157–80
  cross-colour 165
  decoder configuration 157
  line-locked sampling composite signals 158–62
  matrixing 173
  NTSC demodulation 178–80
  PAL 162–70
    coding range 174
    digital demodulation 165
    modulated chrominance processing 174–8
    PAL modifiers 174–7
    phase-shift filters 177–8, 179
    reference phase 164, 165
    signal relationships in decoders 170–4
    subcarrier-to-burst synchronisation 165–70
  phase comparator 159, 160, 161, 162
  quantising distortion 158–9
  ratio counters 169, 170
  reference phase 174, 175, 176, 179–80, 195
  sampling frequency stabilisation 159–62
  synchronous demodulation 157, 159, 163
chrominance–luminance separation 181–209
  3D spectrum of composite signal 183–7
    scanning effects 183–5
    vertical/temporal spectrum 183–7
    chrominance modulation in 186–7
    subcarrier in 185–6
  conventional decoding 181–3

multi-dimensional filters 187–208
  performance 188–91
  see also comb filters
  PAL decoding comparison chart 206
  phase shift 182–3
Chyron 'Scribe' character generator 497
circuit block, noise reduction 281–3
clamping
  black-level clamping 41, 45–7, 160
  clamp-streaking effects 46
clean coding 135, 141, 210–13, 251–2, 255
clipping 513
clock
  interface clock 566
  pulse jitter 55
  random clock jitter 61
CMOS technology 74
code size error 28
codebook codes 402
codec (PCM) 21
  errors caused by ideal 26–7
  input/output transfer characteristics 26
  subjective tests on 40–2
codecs
  quantising accuracy of 48
  subjective tests on 48–50
  tests on codecs in tandem 42–50
coding
  accuracies 59–61
  beam coding tube 31
  colour see colour coding
  composite coders 131
  DPCM see DPCM coding
  entropy see variable-length entropy coding
  parallel–serial type coder 32–5
  see also digital encoding and decoding: decoding
colour burst 133
  burst blanking 169
  NTSC 154
colour coding
  cross-colour see cross-colour
  digital PAL encoding 144–53
    filter implementation 152–3
    interpolation circuitry 146
    modulation 146–7
    signal relationships in coders 148–9
    synchronising pulses 146, 149–52

V-axis 150
NTSC encoding 153–7
  broadband NTSC 156–7
  crosstalk 157
  differential phase distortion 153
PAL signal features 141–4
see also chrominance demodulation
colour correction in telecines 526–9
colour difference 6
colour encoding 141–7
colour flicker 382
colour matrixing, telecine 508
colour operation, noise reduction see noise reduction
colour separation overlay (CSO) 6, 398
  see also chroma-key
column address strobe (CAS) 68
comb filters 142, 157, 159, 167, 176, 183, 187–208, 212, 213, 358, 593, 594, 596, 597
  adaptive techniques 208–9
  complex design 197–204
  field and picture delay based 159, 162, 169, 193–5, 382, 597
  improved 195–7
  interlaced and sequential scan 198, 201, 203
  line-based 169, 188–93
  performance comparison 204–8
  windowing 201
complementary filters 210
complementary signals 195
complementary metal–oxide–silicon technology see CMOS technology
component signals 6, 131, 132
composite signals 6, 133
  3D spectrum of 183–7
  clean 135
  non-mathematical 133, 134, 135
  PAL 132, 148–9
  see also coding: digital encoding and decoding: decoding
compression 314–16, 584
computer data 376
computer graphic devices see electronic graphics
connectors, interface connectors 567–8
consumer product 16
contouring 37, 38, 41, 556

646 INDEX

contribution-link quality 584
'Crankshaft' effect 336
cross-channel microwave links 3
cross-colour 144, 157, 165, 174, 179, 181, 182, 186, 191, 192, 193, 195, 204, 205, 206, 208, 209, 212, 359–60, 395
cross-fade 399, 546, 551–2
cross-luminance 144, 157, 169, 181, 182, 191, 192, 193, 195, 197, 206, 207, 208, 209–10, 212, 382, 395
crosstalk 61, 62, 141, 179, 182, 209
  NTSC colour encoding 157
  U–V 144, 147, 165, 191, 192, 193, 195, 203–4, 206, 252–3
CSO see colour separation overlay (CSO)

data output control, storage 86–7
data validity 376
DATV 18–20, 374
  motion vectors in 20
  techniques 209
DBS 16, 17–18
DDT (direct data transfer) 391
decoding
  adaptive decoding 208–9
  component signals 133, 134
  composite signals 131, 133
  MAC/packet decoder chips 18
  PAL
    with line-locked sampling 134–5
    with sub-carrier locked sampling 133–4
  see also digital encoding and decoding
definition
  enhanced definition 18
  HDTV 18, 20
deflection processor unit 17
deglitching 23, 37, 55
delay 4
  field and picture based 159, 162, 169, 193–5, 382, 597
  quartz delay lines 335
  storage 65, 109
    short video delay 122
    TV line delays 122–3
    TV picture delays 123–6
  timing correction 4
  ultrasonic delay line 4
demodulation 135

synchronous 157, 159, 163, 165, 212
  see also chrominance demodulation
demultiplex chrominance 18
demultiplex luminance 18
DICE converter 335, 339–41, 350
differential gains 28, 57–9, 61, 64
differential phase distortion 141, 163, 176, 177, 195, 197, 204, 205, 206
digital components, standards 8–11
digital encoding and decoding
  changeover to digital studios 213
  clean coding system 135, 141, 210–13, 251–2, 255
  colour encoding see colour encoding
  line-locked sampling 135–41, 149
  multiple decoding and recoding 209–10
  non-mathematical signal 169
  orthogonal subcarriers 141, 212
  PAL encoding 146
  subcarrier generation 135–41
    frequency relationships 135–7
    quadrature subcarrier generation 139–41
    ratio counters 137–9
    subcarrier reference phase 137
  use in studios 131–3
    sampling 131–3
    subcarrier-locked 132, 133–4
  see also chrominance demodulation: chrominance–luminance separation
digital special effects 6
digital studios 7, 131–3, 577–82
  audio studio standard 13
  changeover to 213
  clean coding systems 210–13
  decoders in 209–13
  digital interfaces in 577–82
  digital studio equipment 13
  multiple decoding and recoding 209–10
  sampling 131–3
  subcarrier-locked 132, 133–4

see also studio stills store
digital-to-analogue conversion
  analogue post-filtering 23, 25, 43–5
  filter specification 61
  fundamental principles 23
  low-pass filter requirements 43–5
  performance measurement
    linearity measurements 57–9
    signal-to-quantising-noise ratios 52–7
  required performance 61, 63–4
  techniques 36–7
  weighting and summing networks 36
digital video interfaces see interfaces
digital video tape recording 7, 12–13, 423–63
  audio cue 441, 455–6
  audio edits 432–3
  audio recording 449–53
    error protection 450–1
    position of audio sectors 439, 440–1
    shuffling 452
    synchronous and non-synchronous operation 450
  bandwidths 430
  BBC experimental DVTR 425–7
  bi-standard recorder 433
  bit rates, gross and net 430
  channel coding 455
  commercial products 457–63
    D-1 format 457, 458, 463
    D-2 format 462, 463
    recorder scanning hood 460
    rotary transformers 461
  control code 441
  control signals recording 456
  drop-out activity 435
  EBU requirements 427, 428–9, 433–5
  error concealment 427, 447–8
  error correction 427
  error protection
    audio signals 450–1
    Reed–Solomon code 446–7, 451
    video signals 445–7
  IBA DVTR 427, 428
  inching 429
  jogging 429
  lock-up time 429

minimum recorded
wavelength 430–1
packing density 430, 432, 434
picture-by-picture editing 429
portable recorders 436
potentialities for 425–9
recording format 429, 433–5
Reed–Solomon code 446–7, 451
scanning
  recorder scanning hood 460
  standard 433
shuffling
  audio signals 452
  video signals 444, 447–8
shuttle 429, 442
specific problems of 429–33
sub-Nyquist sampling 426, 427
synchronisation signals
  recording 453–5
    preamble and postamble 454
tapes
  cassette configuration 435–7
  cassette playing time 435–6
  guidability 436
  material selection 432, 434
  mechanical forces on 436
  programmable holes 437
  search time 436
  transport for portable recorder 436
  width 435, 436–7
time-code 429, 441, 456–7
track pattern 437–41
  audio sectors position 439, 440–1
  balanced 440
  control track 440, 440–1
  gap azimuth 439–40, 463
  video sectors 440
transparency 428, 442
variable broadcastable motion 429
video signals
  error concealment 447–8
  error protection coding 445–7
  head failure 442–3
  intersector shuffling 444
  intra-sector shuffle 447
  normal play 442
  recording of 442–9
  saturation recording 443
  source precoding or mapping 444

speeds other than normal 442
  video data words 444, 445
  video sectors 440
world digital studio interface 433
digitally assisted television *see* DATV
'Digivision' 17–18
direct data transfer (DDT) 391
disc drives 376
  Winchester 126, 127, 129, 378, 379, 381, 388–9
discrete cosine transform 611–15
display improvement 373–4
distortion
  phase *see* phase distortion
  quantising distortion 158–9, 174
distribution-link quality 584
dither 38–40, 45, 159, 516–17
  pseudo-random 39
domestic receivers 1, 14
  consumer product 16
  digital techniques in 16–18
  VLSI 18
DPCM coding 598–610
  adaptive prediction 601
  graphical design method 603–4
  hybrid transform/DPCM coding 620–1
  linear prediction 599–601
    autocorrelation function 599
  minimum mean square error (MMSE) design methods 602–3
  non-adaptive
    limit cycle oscillation 605
    results of 606–10
    stability of 605–10
  non-linear quantisation 601–5
  optimisation process 600
  prediction 132, 598
  prediction error response 600
DRAM 64, 68, 71
DVTR *see* digital video tape recording
dynamic memory 67–74, 76, 78–89

EAV (end of video) signal 564
edge busyness 601
edge rise time 253
EFP (Electronic Film Production) 375

electron photomicrographs 414, 415, 417
Electronic Film Production (EFP) 375
electronic graphics 465–97
  aliasing 465, 467, 471, 473
  anti-aliasing 473
  causes of 474–80
  horizontal 476
  interlace twitter 474, 475
  judder 475, 476–9
  movement portrayal 476–9
  picture frequency flicker 474, 475
  stepping or jaggies 474, 475–6, 477
  stroke width modulation 475, 476–9
  temporal 478, 491
  vertical 476
aperture loss 468, 469–71
applications of 496–7
  Aston IV 497
  'Battleground' 496
  Chyron 'Scribe' character generator 497
  Inmos Transputer 496
  Quantel 'Cypher' picture 497
  Quantel 'Paintboxes' 496, 497
  Rank Cintel 'Slide-File' 496
  'swingometer' 496
camera integration 468
character generator 465, 466, 493, 497
computer graphic devices 465
electronic captions 465
electronic painting system 465
electronically generated images 465–6
filter characteristics 465
filtered text images 480–2
  area averaging 483, 484–5
  combined filtering 490–4
  filter funtion choice 481–2
  'Flair' system 482
  for teletext and VDUs 494–5
  fringeing 482, 488
  gamma-correction 482
  Gaussian filters 483, 486
  linear interpolation 483, 486
  master fount data 480–1, 482

electronic graphics (*cont.*)
  monotonic step response filters 483–6
  movement portrayal 491
  multilevel text 481
  non-monotonic response 486–90
  practical investigation 482–94
  pulse response 482
  raised cosine transition 483
  source of data 480–1
  spatio-temporal filter 480
  step response 482
  windowing 489
flare 469
flicker 466
generator 103
image prefiltering 465–6
inter-character spacing 472
inter-line spacing 472
judder 469
movement portrayal 466
photoartwork 466
pixellated image 466
real artwork 465, 466
rolling captions 491
scanning artwork 466, 469–71
scanning effects 467–73
  aperture loss 468, 469–71
  keying 471–2
  multilevel digital images 472–3
  sampling theory 467
  scanning artwork 466, 469–71
  scanning as sampling 467–9
  soft keying 472, 482
  text with pictures 471–2
  two-level electronic images 472
slide scanners 465
spectral contributions 473–4
  moving detail 473–4
  stationary detail 473
stepping 466
television cameras 465, 467
twitter 469
Electronic News Gathering (ENG) 375
encoding *see* coding: digital encoding and decoding
end of video signal 564
ENG (Electronic News Gathering) 375
enhanced definition 18
enhancement *see* aperture correction

entropy 623–5
variable-length entropy coding 621–9
equalisation 402
$\sin(x)/x$ equaliser 23, 25
equi-ripple 237
errors
  aperture error 29
  burst error 407
  concealment 408
  correction 562
  feedback 406–7
    chroma-key 553–6
  protection 407–9
  soft errors 68, 71
  see also quantising errors/noise
European Broadcasting Union 8, 427–8, 433–5, 449–50
evaporated metal tapes 434
expansion 311–14

facsimile transmission 2
fade
  cross-fade 399, 546, 551–2
  fader 546
Fechner fraction 509
feedback
  error feedback 406–7, 553–6
  recursive filter feedback 270
field delay 159, 162, 169, 193–5, 382, 597
field frequency 15–16
field pairing 251
film dirt concealment 112–13
film grain 261
film motion 251
film replay speed, telecines 522
film scanning 321
  see also telecine
film shrinkage 524–5
filters
  2D luminance filtering 254–7
  beat patterns in transition region 257
  chroma-key and 546–8
  comb filter *see* comb filter
  diagonal frequencies removal 255
  digital filtering 4, 17, 65, 215–85
    and sample rate reduction 587–8
  filtered text images *see* electronic graphics, filtered text images
  Fourier transform *see* Fourier transform

Gaussian 142, 143, 146, 147, 152, 154, 155, 210, 264–5, 483, 486
half-line frequency offset 255
ideal filter 290–1, 292, 323, 333, 488
in digital PAL colour encoding 152–3
linear phase filter 233
loop filter 159, 164
low-pass filters *see* low-pass filters
monotonic step response filters 483–6
noise reduction *see* noise reduction
non-monotonic response filters 486–90
notch filter 181, 205, 250
phase-shift filters 177–8, 182–3
poles and zeros 232
recursive filter 231, 261, 262, 270
spatial filter 225, 265–70, 281–3
specification for A–D and D–A conversion 61
temporal filter 224, 225, 261
time series 230
transform transfer function 231–2
transformation filter 239
transversal *see* transversal filters
U–V filters 154
unit circle 232
vertical chrominance filtering 251–4
video applications 244–57
  real-time aperture correction *see* aperture correction
windowing 489–90
zone plate pattern 255–7
finite impulse responses 467
fixed-point aperture 298–300, 344, 345
'Flair' graphics system 482
flare 469
flicker 335, 357, 382, 395, 466
flying-spot cathode ray tubes 504–5
folding amplifier quantiser 31–2
founts 7
Fourier transform 202, 296, 303, 611–12
  2D

amplitude and phase
spectrum 217–21
centre of antisymmetry
217
centre of symmetry 217
ringing 221
spatial frequencies 216
3D 221–5
discrete, sampling and 225–
30
interpolation 303, 307, 339,
341, 349
inverse 296, 298
inverse discrete 229, 230
frame grabbing 7
frame transfer structure
cameras 534
frame-interline transfer device
cameras 535
fringeing 482, 488

gains, differential gains 28, 57–
9, 61, 64
gamma correction 482
cameras 501, 530
moiré patterns 513
telecine 503, 505, 508–9,
510, 513
garbage matte 550–1
Gaussian filters 142, 143, 146,
147, 152, 154, 155, 210,
264–5, 483, 486
glitches 23, 37
grabbing 120, 122, 395
granular noise 601
graphics *see* electronic graphics
gray code 30, 32
gridding 86

Hamming code 408, 410
Hanover bars 142, 163, 177,
192, 193, 195, 197, 208,
359
HDTV 20
2 Mbyte storage module 113,
115–17
conversions 362, 365–71
definition 18
high sampling frequencies 35
head crash 408
headroom 9, 14
high definition television *see*
HDTV
holding 22
sample-and-hold circuit 22,
32, 35, 54
horizontal resolution 16
hue angle 540
hysteresis 271–2

ideal filter 290–1, 292, 323,
333, 488
image spectrum 183
inching 429
injection locked loop 411
Inmos Transputer 496
integration times 3
inter-character spacing 472
inter-line spacing 472
interfaces
8 bit per sample 559
ancillary channel data 577
bit-parallel interface 560–8
electrical format 564–6
experience with 568
interface multiplex 561
mechanical implementation
566–8
signal format 560–1
synchronisation 561–4
bit-serial interface 568–70
8–bit–9–bit code 569–70,
572–3
8–bit–9–bit map 570
10–bit code 574
bit mapping 569
coding strategy 568–9
control of d.c. component
570
descrambling 575
electrical format 576
experience with 576–7
link synchronisation 570–
7
mechanical implementation
576
scrambling 574–5
cables 566–7
equalisation 565–6
RF interference 567
connectors 567–8
digital video interfaces 559–
82
interface clock 566
optical interface 577
parallel interface *see* bit-
parallel interface
requirements 559–60
serial interface *see* bit-serial
interface
storage 88, 106
studio applications 577–82
timing reference signals 561–
4
EAV signal 564
errors correction 562
format 562
preamble 561, 574
SAV signal 564
video standards 11–12
interference

RF 567
see also noise reduction
interlace flicker 335
interlaced scan 184, 197, 201,
203, 317, 318, 320, 321,
322, 323, 325–8, 330–1,
342, 366
interline transfer structure
cameras 534
international programme
exchange 3
interpolation 287–374
adaptive process of 395
aperture 296
aperture optimisation 350–
62
aperture quantisation 301–4,
346–9
aperture synthesis 298–300,
341–9
camera integration 356
fixed point aperture 298–
300, 344, 345
fixed ratio interpolation 304–
10
Fourier transform 296, 298,
303, 307, 339, 341, 349
ideal filter 290–1, 292, 323,
333
interpolation aperture
function 296, 297, 298–
300, 307, 308, 311–13,
323, 332, 335
interpolation coefficients 296,
304, 314, 347
linear 285–6, 287, 301,
302, 303, 306, 314, 316,
483, 486
movement 522–4
non-variables-separable 332–
3, 339, 350, 352, 353
recursive 237–8
sample rate changing 293–
304
resampling 293–5, 297,
302, 306
signal reconstitution 289–93
signal sampling 288–9
television scan conversions
see television scan
conversions
temporal 326, 327, 328,
329, 331, 340
time compression 314–16
time expansion 311–14
undersampling 356
up-conversion 307, 372, 374
variables-separable 324, 331,
333, 339, 374
vertical 329, 330, 331, 340
ITT 'Digivision' 17–18

650   INDEX

jaggies 474, 475–6, 477
jitter 55, 61, 402
   aperture 29
jogging 429
judder 251, 355–6, 475, 476–9
   electronic graphics 469
   telecines 522–4

Kaiser window 236
keying 471–2
   key 471
   soft keying 472, 482
   see also chroma-key
knotting 333

LAN (local area network) 129
large-scale integration 3, 34
LaserVision 412–18
late write 82
limiter, in spatial filter 283
line label 98
line trigger 98
line-based comb filter 169, 188–93
line-locked sampling 131–3, 149, 318
   chrominance demodulation 158–62
   PAL decoding with 134–5
   subcarrier generation from 135–41
line-quincunx sampling 589–91
linear interpolation 285–6, 287, 301, 302, 303, 306, 314, 483, 486
linear phase filter 233
linearity errors 27, 28, 57–9
local area network (LAN) 129
lock-up times 411
logos *see* animation of logos
long distance transmission
   bandwidth reduction 584
   bit rate reduction 583–632
   broadcast-quality coding 584
   contribution-link quality 584
   distribution-link quality 584
   DPCM coding *see* DPCM coding
   hybrid transform/DPCM coding 620–1
   sample rate reduction 585–97
      3D sampling frequencies 586
      chrominance components 589–92
      digital filtering and 587–8
      luminance component 592–3

      PAL signals 593–7
   transform coding *see* transform coding
   transmission errors 606–7
   variable-length entropy coding *see* variable-length entropy coding
   vector quantisation 628–32
      codebook generation 630–1
      vector quantiser performance 631–2
Long Play 412
loop filter 159, 164
low-pass filters 261
   analogue-to-digital conversion 22, 24
   digital-to-analogue conversion 23, 25
LSI (large-scale integrated devices) 3, 34
luminance 6, 9, 109
   2D luminance filtering 254–7
   component, sample rate reduction 592–3
   cross-luminance 144, 157, 169, 181, 182, 191, 192, 193, 195, 197, 206, 207, 208, 209–10, 212, 382, 395
   demultiplex 18
   keying 550
   modulation 212
   resolution 9
   suppression 543–4
   see also chrominance–luminance separation

MAC 251, 252
   packet decoder chips 18
magnetic data storage 377
mapping, one- to two-dimensional 238–9
matching and tracking 37
matrixing 154, 156, 173, 179, 213, 508
matte, garbage matte 550–1
memory 3
   3D solid state memory 3
   synchronisers 4–5
   see also storage
metal particle tapes 434
metal–oxide–silicon semiconductor memory devices *see* MOS technology
microwave links, cross-channel 3
mixer units 546
   cross-fade 546, 551–2

   non-additive mixer 548–9, 551
   output from fader 546
   see also chroma-key
modulation 135
   amplitude modulation 153
   colour coding 146–7
   pulse code modulation *see* PCM
moiré patterns 513
monochrome displays 144
monochrome receivers 142, 185
monochrome signals 133, 169
   colour burst 133
MOS technology 66–7
   CMOS 74
   N-MOS 68, 74
      scaled 71–3
   P-MOS 67, 74
movement compensation 20, 372
movement detection 208–9
   judder 355–6, 522–4
moving static detail 20
multipicture storage 375–421
   animation of logos *see* animation of logos
   animation store *see* animation store
   as picture library 376
   back-up devices 376–7
   buffer stores 381
   computer data 376
   data validity 376
   DC 300 tape cartridge 378, 379
   digital magnetic tapes 377
   disc drives 376
   disc rotation 381
   multi-head drives 381
   news central stills library 383–5
      digital video interfaces 384
      juke box arrangement of discs 383
      photographs 383
      studio stores 384
   open reel tape 377
   optical discs 378, 379, 383, 384
   PAL input to 381–2
   read–write circuitry 381
   removable disc drives 378
   removable media 376
   review of media and technologies 376–81
   Serpentine recording format 378, 390
   single reel cartridge 377
   slide scanners 375, 376, 385

still pictures 375
still store 381
studio stills store *see* studio stills store
Winchester disc drives 378, 379, 381, 388–9
see also storage
multiple decoding and recoding 209–10
multiplexed addressing 79–81
multiplexing arrangements 402
multiplier, floating point form 283

N-MOS 68, 74
  scaled 71–3
networks
  resistor ladder 36
  switched ladder 37
  weighting and summing networks 36
news central stills library *see* multipicture storage
NHK converter 371–2
nibble mode 84
noise reduction 2, 6, 257–85
  averaging the signal 261
  circuit block 281–3
  clock generation 281
  colour operation 276–80
    NTSC 279–80
    offset in colour subcarrier 276–7
    PAL 276–9
    predictor 277–9
  digital filtering methods 4
  film grain 261
  movement protection 263–76
    action of rectifier 263–5
    central limit theorem 265
    hysteresis 271–2
    non-linearity 270–6
    spatial filter 265–70
  photon noise 257, 261
  picture store 280–1
  prototype equipment 284–5
  recursive filter 261, 262
  storage 65
  thermal noise 261
  transmission path noise 261
  unweighted noise reduction factor 261–3
  see also filters: quantising errors/noise
non-variables-separable interpolation 332–3, 339, 350, 352, 353
notch filter 181, 205, 250
NTSC 6, 16

broadband techniques 156–7, 180
chrominance demodulation 178–80
colour coding *see* colour coding
colour standards 135, 136
noise reduction 279–80
ratio counter 137–9
subcarrier 358
system M 135
television standards 134
Nyquist sampling 22–3, 25, 42–50, 146, 585

optical conversion 322–3
optical discs 383, 384
  LaserVision 412–18
optical fibres transmission 4
orthogonal sampling 9, 11, 132, 133, 226
orthogonal subcarriers 141, 212
oversampling 476
oxide tapes, improved 434

P-MOS 67, 74
packing density, DVTR 430, 432, 434
page mode 82–5, 111
paintbox graphics 496, 497
painting 530
PAL 6, 16, 17
  chrominance demodulation *see* chrominance demodulation
  chrominance modulation *see* colour coding
  colour standards 135
  composite signals 132
  decoding 133–5
  input to multipicture storage 381–2
  modifiers 135, 183, 188, 193
  noise reduction 276–9
  non-mathematical signal 133, 134, 135, 169
  ratio counter 137–9
  reversible coding 132
  square wave signal 139
  switch 133, 168, 179, 185, 188, 193, 195, 209
  system I 135, 146
  V-axis switch 132
panning 356, 358
  detection signal 356
parallel interfaces *see* interfaces
PCM 2–3
Peltier devices 533
phase comparator 159, 160, 161, 162, 167, 168

phase distortion, differential 141, 153, 163, 176, 177, 195, 197, 204, 205, 206
phase lock loops 411
phase-locked loop 162, 167, 168, 169, 179
phase-shift filters 177–8, 179, 182–3
photoconductive telecine 504–5
photographs 383
photomicrographs 414, 415, 417
photon noise 257, 261
picture delay 159, 162, 169, 193–5, 597
picture frequency 15–16
picture impairment 516–17
  caused by quantisation 37–42
  dither signals and 38–40
  grading scale 40
  subjective tests on single codec 40–2
picture store, noise reduction 280–1
pixel bytes 420
pixel mode 420
pixellated image 466
POLY 68
POLY II 68, 71, 75
polysilicon material *see* POLY
postfiltering 23, 25, 43–5, 585, 588
prediction
  adaptive 601
  error 604
  linear 599–601
  optimisation process 600
  see also DPCM coding
predictor 277–9, 607
  optimum 600
  see also DPCM coding
programmable read-only memory devices *see* PROMS
programme exchange, international 3
PROMS 75, 283
propagation delay 109
pulse code modulation (PCM) 2–3

Quantel 6
  'Cypher' picture 497
  'Paintboxes' 496, 497
quantisation 139, 152, 159, 516
  accuracy of codecs 48
  adaptive 604–5
  aperture 301–4, 346–9
  block 629

652 INDEX

quantisation (*cont.*)
  DPCM non-linear quantiser 601–5
  folding amplifier 31–2
  process 2–3
  quantising distortion 158–9, 174
  scalar 628
  solid state quantiser 31, 34
  vector *see* vector quantisation
quantising errors/noise 38
  accuracy of codecs 48
  adaptive quantisation 604–5
  chroma-key 553–6
  dither signals and 38–40
  edge busyness 601
  filter implementation 152
  granular noise 601
  graphical design method 603–4
  ideal PCM codec 26–7
  linearity error
    differential 28
    independent integral 28
    terminal-based integral 27, 28
  measurement of 52–5, 61
  methods of specifying 26–9
  minimum mean square error (MMSE) design methods 602–3
  picture impairment caused by 37–42
  signal-to-noise measurements 29
  slope overload 601
  spectrum of 55–7
  static or low-frequency 27
  subjective tests 48–50
  tolerances to coding inaccuracies 59–61
quartz delay lines 335

Radio Regulations 12
raised cosine transition 483
random access storage 7
random clock jitter 61
Rank Cintel 'Slide-File' 496
Rank Cintel telecine 527
RAS (row address strobe) 68, 80
rates of cut 467
ratio counters 137–9, 169, 170, 311
read address down time 113
read cycle 79–81
read–modify–write cycle 81–2
reconstitution 289–93
recorders
  standards 12–13
  see also digital video tape recorder
recursive filter 231, 261, 262, 270
  clock generation 281
  feedback 270
recursive interpolation 237–8
redundancy 406–7, 408
Reed–Solomon code
  audio signals 451
  video signals 446–7
reference phase 164, 165, 174, 175, 176
  chrominance demodulation 179–80, 195
refresh 68, 87–8
reliability, storage 89
Remez exchange 237–8
resampling 293–5, 297, 302, 306
resistor ladder network 36
reversible PAL coding 132
RF interference 567
ringing 221, 253
rolling captions 491
rotary transformers 461
row address hold time, storage 81, 85
row address strobe (RAS) 68, 80
run length 418, 420

sample-and-hold circuit 22, 32, 35, 54
  see also sampling
sample rate changing 293–304
  interpolation process 295–8
  resampling 293–5, 297
sample rate reduction
  3D sampling frequencies 586
  chrominance components 589–92
  digital filtering and 587–8
  luminance component 592–3
  up-conversion 588
sampling 22, 288–9
  above Nyquist rate 165–60
  bits per sample 35, 40–2
  clock oscillator 162
  close to Nyquist limit 42–50
  discrete Fourier transform and 225–30
  downsampling 588
  frequency 2, 3, 9, 22, 23–5, 42, 51–2, 234–6
    audio 14
    line-locked 131–3, 149, 318
    chrominance demodulation 158–62
    PAL decoding with 134–5
    subcarrier generation from 135–41
  line-quincunx sampling 589–91
  loop 133, 134
  Nyquist limit 22–3, 25, 42–50, 146, 585
  orthogonal 9, 11, 132, 133, 226
  oversampling 476
  rate conversion 287
  resampling 302, 306
  sample rate reduction *see* sample rate reduction
  sub-carrier locked 132, 133–4
  sub-Nyquist sampling 23, 132, 426, 427, 593, 594
  subcarrier-locked 159
  subsampling 20, 589–90
    PAL signals 593–7
    temporal 591–2
  temporal sampling frequency 591
  undersampling 292, 356
  vertical subsampling 252
  see also sample-and-hold circuit
satellite broadcasting 4, 16
  direct broadcast satellite (DBS) 16, 17–18
SAV (start of video) signal 564
scalar quantisation 628
scaling 71–3
scan conversion 3
scanning
  DTVR recorder scanning hood 460
  effects 183–5
  film scanning 321
  see also telecine
  graphics *see* electronic graphics, scanning effects
  interlaced 184, 197, 201, 203, 317, 318, 320, 321, 322, 323, 325–8, 330–1, 342, 366
  sequential 15, 198, 201, 203, 317, 322, 324–5, 326, 372, 374
  slide scanners 375, 385, 465, 499, 508
  standard, DVTR 433
  see also television scan conversions
scrambling 574–5
  descrambling 575
SECAM 6, 16, 17
self-effects 204
semiconductor digital storage 3
sequential scanning 15, 198, 201, 203, 317, 322, 324–5, 326, 372, 374

sequential-to-interlace
    conversion 518–24
Serpentine recording format
    378, 390
set-top adaptors 18
shading errors 530–1
shadows, chroma-key and 544–5
shrinking 71, 395, 524–5
shuffling
    audio signals 452
    video signals 447–8
shuttle 429, 442
signal-to-quantising-noise ratio
    27, 52–3
silicon chips, VLSI 16
$\sin(x)/x$ equaliser 23, 25
size reduction *see* scaling:
    shrinking
slide scanners 375, 376, 385,
    465, 499, 508
'Slide-File' 496
    see also studio stills store
slope overload 601
SMD 389
Society of Motion Picture and
    Television Engineers 8
soft errors 68, 71
soft keying 472, 482
software control 17
solid state memory 3
solid state quantiser 31, 34
sound *see* audio signals
spacial resolution 20
spatial filter 225, 265–70,
    281–3
    limiter 283
special effects
    digital 6
    video 5–6
spectrum
    3D spectrum of composite
        signal 183–7
    amplitude and phase
        spectrum 217–21
    image spectrum 183
    of quantising noise 55–7
    of sampled signal 24–5
standards 7–14
    conversion 287
    DBS transmission 17–18
    digital audio 13–14
    digital components 8–11
    recorders 12–13
    video interface 11–12
start of video signal 564
stepping 466, 474, 475–6, 477
still pictures 375
    see also studio stills store
storage
    3D solid state memory 3

16K generation 98–109
    animation store buffers
        106–9
    microcomputer-controlled
        stores 103–5
    using 18 MHz sampling
        frequency 98–103
64K generation 109–22
    2 Mbyte storage module
        113–15
    compact random access
        picture store 117–22
    film dirt concealment
        application 112–13
    general purpose storage
        card 109–15
    HDTV application 113,
        115–17
    sequence store application
        116–17
    store board organisation
        119–20
    store control 120, 122
256K generation 126–30
    choice of memory devices
        127–8
    image processing tool 129
    local area network (LAN)
        129
    rack of storage 128–30
    store board organisation
        128
    Winchester disks 126, 127,
        129
access ports 94
access time 76, 79
address down time 113
address strobes 79
animation of logos *see*
    animation of logos
animation store *see* animation
    store
asynchronous buffer 99
bipolar technology 96
BMM 96, 104, 110
bubble memory 66
charge-coupled device (CCD)
    66, 75, 96–8
chroma-key 106
CMOS technology 74
column address strobe (CAS)
    68
data output control 86–7
delay applications
    short video delay 122
    TV line delays 122–3
    TV picture delays 123–6
delay elements 65
design philosophy 89–94
    multiplex factor 91–2
    store configuration 92–4

store size 90
developing trends 74–8
digital field delays 96–8
digital filters 65
DRAM 68
dynamic memory 67–74, 76,
    78–89
early read–write memories
    96
electronic graphics generator
    103
field store bays 95
gridding 86
interfaces 88, 106
late write 82
line label 98
line trigger 98
magnetic disc storage 7
memory cells
    dynamic 67–74, 76, 78–89
    static 66, 67
memory device choices 78–9
minimum line width 76
MOS technology 66–7
multipicture *see* multipicture
    storage
multiplexed addressing 79–81
N-MOS 68, 71–3, 74
nibble mode 84
noise reducers 65
P-MOS 67, 74
page mode 82–5, 111
PAL decoding 106–7
performance trend 75
picture store noise reduction
    280–1
POLY 68, 71
POLY II 75
power consumption 85
power distribution 85–6
PROMS 75
propagation delay 109
random access storage 7
random read–write cycle
    length 85
read cycle 79–81, 81, 111
read–modify–write cycle 81–2
refresh 68, 87–8
reliability 89
row address hold time 81, 85
row address strobe (RAS) 68,
    80
scaling 71–3
semiconductor 3, 65–130,
    65–74
sense amplifiers 81
shrinking operation 71
soft errors 68, 71

654   INDEX

storage (cont.)
  static memory 78, 89, 96
  television synchronisers 65, 98, 105
  TV picture stores 92–4
  video grab 120, 122
  video rostrum camera system interface 106
  Wardrobe stores 94–5
  write cycle 81, 111
  see also studio stills store
storage module drive-device interface 389
streaming tape drive 390–1
stroke width modulation 475, 476–9
studio stills store 385–96
  data locations 388
  data ports 388
  disc storage 388–90
    error check codes 390
    FIFO buffer 390
    SMD 389
  picture processing 395–6
    frame average 396
  polyphoto display 386
  semiconductor picture stores 387–8
  Serpentine recording format 378, 390
  'Slide File' 385–90
  SMD 389
  streaming tape drive 390–1
    QIC-02 and 24 390
  system bus 387
  system computer hardware 391–2
    bidirectional buffers 392
    direct data transfer 391
  system computer software 392–5
    cross-fade 392
    file status 395
    global variables 393
    picture transfer operations 393, 395
    repair 395
    shrink 395
studios, digital see digital studios
sub-Nyquist sampling 23, 132, 426, 427, 593, 594
subcarriers
  frequency relationships 135–7
  generation from line-locked sampling 135–41
  orthogonal subcarriers 141, 212
  quadrature subcarrier generation 139–41, 212
  ratio counters 137–9

subcarrier reference phase 137
subcarrier-locked sampling 159
subsampling 589–90
  PAL signals 593–7
  temporal 591–2
  vertical 252
summing networks 36
'swingometer' 496
switched ladder networks 37
switching, alternate line 141
symmetry, centre of 217
synchronisation 6
  bit-parallel interface 561–4
  subcarrier-to-burst synchronisation 165–70
  synchronising pulse, colour coding 149–52
  synchronous demodulation 157, 159, 163, 165, 212
synchronisers 4–5
  TV 65, 98, 105

tachometer 524
tapes for DVTR
  cassette configuration 435–7
  cassette playing time 435–6
  evaporated metal tapes 434
  guidability 436
  improved oxide tapes 434
  material choice 432, 434
  mechanical forces on 436
  metal particle tapes 434
  programmable holes 437
  search time 436
  tape width 435, 436–7
  transport for portable recorder 436
telecine 499
  A–D converters and 509–18
  afterglow correction 508–9
  aperture correction 508, 513–15
  audio information transmission 507
  CCD line-array sensor 502–3, 504
  colour correction 526–9
  colour matrixing 508
  colour operation 515–18
  control systems 524–9
    automatic colour correction 526–9
    film shrinkage 524–5
    scanning control 524–6
    tachometer 524
    variable speed operation 524–6
  digital procesing applications 505–8

  flying-spot cathode ray tubes 504–5
  gamma correction 503, 505, 508–9
  harmonic generation 510, 513
  logarithmic and exponential amplifiers 503
  photoconductive 504–5
  photomultiplier 467
  picture impairment 516–17
  processing channel 508–18
  Rank Cintel telecine 527
  sequential-to-interlace conversion 518–24
    film replay speed 522
    judder 522–4
    movement interpolation 522–4
    storage 520–1
    synchronisation 520
  types of 501–5
telephotography 2
teletext 494–5
teletrack device 5
television
  cameras see cameras
  digitally assisted television see DATV
  electronic graphics for see electronic graphics
  HDTV 14–16
  synchronisers 65, 98, 105
  see also television scan conversions
television scan conversions 317–74
  ACE converter 362, 363
  aperture optimisation 350–62
  aperture synthesis 341–9
  camera integration 320, 356, 367
  conversion methods 322–62
    optical conversion 322–3
    principles of conversion 322–35
  cross-colour reduction 359–60
  DICE converter 350
  display improvement 373–4
  HDTV conversions 362, 365–71
  image spectrum 318
  interlace-sequential conversion 373, 374
  interlaced scanning 317, 318, 320, 321, 322, 323, 325–8, 330–1, 342, 366
  flicker 335
  line-locked sampling 318

INDEX

line-rate conversions 363
movement compensation 372
panning 356, 358
previous converters 335–41
   BBC analogue converters 335, 338–9
   DICE converters 335, 339–41
   NHK converter 371–2
scanning film 321
sequential scanning 317, 320, 322, 324–5, 326, 372, 374
temporal frequency 318–20
test pattern generator 360–2
TV scanning 317–22
undersampling 356
up-conversion 372, 374
see also interpolation: scanning
temporal aliasing 320, 372, 478, 491
temporal filter 224, 225, 261
temporal frequency 184, 318–20
temporal resolution 20
terrestrial broadcasting 16
test pattern generator 360–2
time expansion 311–14
time-code recovery 429
time-code signal recording, DTVR 456–7
timing correction 4
timing jitter 55
timing reference signals 561–4
   EAV signal 564
   errors correction 562
   format 562
   preamble 561, 574
   SAV signal 564
track pattern *see* digital video tape recording, track pattern
transform coding 610–21
   bit reduction in transform domain 617–20
   discrete cosine transform 611–15
   hybrid transform/DPCM coding 620–1
   two-dimensional 615–17
   zones of zero coefficients 620
transmission
   625-line compatable 18
   errors 606–7
   long distance *see* long distance transmission
   optical fibres 4
   satellites 4, 16, 17–18
transparency 428, 442

transversal filters 146, 152, 160, 165, 178, 195, 231, 232–9
   coefficient pattern 233
   design of 234–9
   equi-ripple 237
   four quadrant symmetry 234
   frequency sampling 234–6
   hardware considerations 240–4
   accumulation ladder inputs 243
   parallel tree 240
   rounding errors 242
   serial accumulation ladder 240–1
   two's complement number representation 240
   interpolation 304, 310
   linear phase filter 233
   one- to two-dimensional mapping 238–9
   recursive interpolation 237–8
   Remez exchange 237–8
   sinc function 236
   symmetrical 233
   variables-separable frequency 234
   video applications 244–57
   windowing 236–7
TRS *see* timing reference signals
twitter 469, 474, 475
two's complement number representation 240

U–V crosstalk 144, 147, 165, 191, 192, 193, 195, 203–4, 206, 252–3
UARTS 391
ultrasonic delay line 4
undersampling 292, 356
universal asynchronous receiver transmitters 391
up-conversion 307, 372, 374

V-switch 132, 164, 174, 179, 182, 186, 277
variable-length entropy coding 621–9
   buffers 625–8
   construction of codes 622–3
   entropy 623–5
   practical considerations 625–8
variables-separable interpolation 324, 331, 333, 339, 374
VDUs 494–5
vector quantisation 628–32
   codebook generation 630–1

vector quantiser performance 631–2
vertical aliasing 333, 334, 336–8, 350, 357, 476
vertical chrominance filtering 251–4
vertical resolution 16
vertical subsampling 252
vertical/temporal spectrum chrominance modulation in 186–7
   subcarrier in 185–6
very large scale integration (VLSI) silicon chips 16
video interface
   parallel 12
   serial 12
   standards 11–12
   see also interfaces
video processor unit 17
video rostrum camera system 106
video special effects 5–6
video tape recording
   audio signals 424
   digital *see* digital video tape recording
   editing resolution 424–5
   limitations of 423–5
   multigeneration capability 424
   postproduction techniques and 424
   tape transport system 423–4
VLSI chips 16
voice transmission 2

Wardrobe stores 94–5
weighting and summing networks 36
wide screen 14
Winchester disc drive 126, 127, 129, 378, 379, 381, 388–9
window funtion 489
windowing 201, 489
   impulse response filter 489–90
   transversal filters 236–7
   two-dimensional 238
write address down time 113
write cycle
   late write 82
   read–modify–write cycle 81–2
   storage 81

zone plate pattern 255–7
zoom angle 531